HVDC GRIDS

HVDC GRIDS

For Offshore and Supergrid of the Future

Edited by

DIRK VAN HERTEM
ORIOL GOMIS-BELLMUNT
JUN LIANG

IEEE PRESS SERIES ON POWER ENGINEERING

IEEE
IEEE PRESS

WILEY

Published by John Wiley & Sons, Inc., Hoboken, New Jersey.
Published simultaneously in Canada.

Library of Congress Cataloging-in-Publication Data is available.

ISBN: 978-1-118-85915-5

10 9 8 7 6 5 4 3 2 1

To Els, Elena, Sílvia, Clara,
Rita, Chunmei, Qianyun,
our wives and daughters,
for keeping us onshore
and connected to ground

CONTENTS

PART III *PLANNING AND OPERATION OF HVDC GRIDS*

LIST OF FIGURES

LIST OF TABLES

CONTRIBUTORS

Oluwole Daniel Adeuyi Institute of Energy, School of Engineering, Cardiff University, Cardiff, Wales

Monica Aragues Penalba, Research Group CITCEA, Electrical Engineering Department, Technical University of Catalonia (UPC), Barcelona, Spain

Jef Beerten Research Group Electa, Department of Electrical Engineering, University of Leuven, Leuven, Belgium

Fernando D. Bianchi Research Group Wind Energy, Electrical Engineering Department, Catalonia Institute for Energy Research (IREC), Barcelona, Spain

Marc Cheah, Institute of Energy, School of Engineering, Cardiff University, Cardiff, Wales

Cedric De Jonghe Research Group Electa, Department of Electrical Engineering, University of Leuven, Leuven, Belgium

Mikel De Prada-Gil Research Group Wind Energy, Electrical Engineering Department, Catalonia Institute for Energy Research (IREC), Barcelona, Spain

Kristof De Vos Research Group Electa, Department of Electrical Engineering, University of Leuven, Leuven, Belgium

Erik Delarue Research Group TME, Department of Mechanical Engineering, University of Leuven, Leuven, Belgium

Francisco Diaz Gonzalez Electrical Engineering Department, Technical University of Catalonia (UPC), Barcelona, Spain

Jose Luis Dominguez Garcia Research Group Wind Energy, Electrical Engineering Department, Catalonia Institute for Energy Research (IREC), Barcelona, Spain

Agustí Egea Research Group CITCEA, Electrical Engineering Department, Technical University of Catalonia (UPC), Barcelona, Spain

Hakan Ergun Research Group Electa, Department of Electrical Engineering, University of Leuven, Leuven, Belgium

Chunmei Feng Engineering Department, Power Conversion, GE Energy Management, Stoke-on-Trent, United Kingdom

Oriol Gomis-Bellmunt Research Group CITCEA, Electrical Engineering Department, Technical University of Catalonia (UPC), Barcelona, Spain; and

Research Group Wind Energy, Electrical Engineering Department, Catalonia Institute for Energy Research (IREC), Barcelona, Spain

XIAOBO HU Power System Department, China Electric Power Research Institute (CEPRI), Bejing, China

DIYUN HUANG Research Group Electa, Department of Electrical Engineering, University of Leuven, Leuven, Belgium

WILLEM LETERME Research Group Electa, Department of Electrical Engineering, University of Leuven, Leuven, Belgium

CHUANYUE LI Institute of Energy, School of Engineering, Cardiff University, Cardiff, Wales

GEN LI Institute of Energy, School of Engineering, Cardiff University, Cardiff, Wales

JUN LIANG Institute of Energy, School of Engineering, Cardiff University, Cardiff, Wales

QING MU Power System Department, China Electric Power Research Institute (CEPRI), Beijing, China

EDUARDO PRIETO-ARAUJO Research Group CITCEA, Electrical Engineering Department, Technical University of Catalonia (UPC), Barcelona, Spain

ROBERT H. RENNER Research Group Electa, Department of Electrical Engineering, University of Leuven, Leuven, Belgium

JOHAN RIMEZ Division TGX, Elia Engineering, Brussel, Belgium

ANDREAS SUMPER Research Group Wind Energy, Electrical Engineering Department, Catalonia Institute for Energy Research (IREC), Barcelona, Spain; and Research Group CITCEA, Electrical Engineering Department, Technical University of Catalonia (UPC), Barcelona, Spain

MUHAJIR TADESSE MEKONNEN Research Group Electa, Department of Electrical Engineering, University of Leuven, Leuven, Belgium

CARLOS E. UGALDE LOO Institute of Energy, School of Engineering, Cardiff University, Cardiff, Wales

DIRK VAN HERTEM Research Group Electa, Department of Electrical Engineering, University of Leuven, Leuven, Belgium

TIL KRISTIAN VRANA Norwegian University of Science and Technology, Trondheim, Norway

SHENG WANG Institute of Energy, School of Engineering, Cardiff University, Cardiff, Wales

FOREWORD

Dear Reader,

The present book is one of the first and most significant studies on the concept of a European electricity Supergrid. The editors' work fills the scientific vacuum in this area. It is very much welcome!

The concept of an electricity Supergrid is now a well-routed expression in the sector, especially in Europe. However, clear definitions are still lacking, and one of the merits of this book is to tackle this. The notion of "Supergrid" has embraced various concepts. I would believe that it is correct to say that initially the Supergrid was limited to that part of the offshore grid which could—potentially—interconnect offshore wind farms located in the North and Baltic Seas. Gradually, the concept was enlarged in order to encompass all of Europe and now represents the very high voltage grid which would be needed to enable the transmission of significant volumes of renewables from where they would be produced to their consumption points. What is meant by "very high voltage" is at this very moment still not precisely defined as are the technical details of the needed technical equipment HVDC and/or others. Again, as mentioned, this work addresses many of these matters.

Today, we can affirm that (almost) nobody really opposes the need to build as soon as possible additional transmission infrastructure in Europe and that the "Supergrid" must become an essential part of that new transmission infrastructure.

The technical challenges are still numerous, but they are not the only ones. Financing the needed infrastructure is definitely another important challenge. The amounts needed to fund the infrastructure are not counted in the millions but in the hundreds of billions. Of course, transmission infrastructure is highly regarded by long-term investors, especially when the investment occurs in a regulated environment. Unfortunately, when one looks closer to the different regulatory regimes currently applicable in the 28 European countries, one will observe that, for the same kind of assets, the rate of returns approved by the respective authorities do greatly vary and could lead to forum shopping by investors, which will make the development of an Energy Union in Europe unfortunately more remote or at least more difficult. We believe it is urgent that Europe looks efficiently into the implementation of, at least, a harmonized regulatory system which will not only provide an attractive investment proposal but will also accelerate the development and construction of the infrastructure as well as foster Europe's security of supply.

Infrastructure, including high-voltage grids, can only be developed and built inasmuch as the developers have obtained the numerous environmental and other permits prior to construction. When it comes to supra-national infrastructure as is the Supergrid, the developers have to abide by the permitting laws and regulations in several countries. The rights protected by these permits as their respective procedures are more or less similar everywhere in Europe. However, the developer still needs to go country by country. Europe is certainly not (yet...) "lawyer-land" as is the United States of America, but everybody will agree that legal proceedings related to permits can and do take a lot of time (7 to 10 years is definitively not an exception, unfortunately!). This needs to be solved if one wants Europe to become less dependent on foreign energy sources.

The 2013 Electricity Infrastructure Regulation wants the Member States of the European Union to offer a "one-stop shop" body handling the various national permits. This is a strong indicator of Europe's will to try to clean up this situation without, however, limiting or reducing the individual rights of the European citizens. This initiative is mostly welcome and we hope that the Member States which today are trailing behind regarding for the implementation of this "one-stop shop" body will soon take the necessary steps to comply with this important European Regulation.

Even though permitting remains principally a national competence, we wonder if it would not make sense for Europe to think about an efficient and pragmatic way to have a "European One-Stop Shop" for essential infrastructure having a cross-border impact. A breakthrough herein would certainly significantly reduce the time but also the cost for new infrastructure in Europe.

It is now time to let you, dear Reader, dive into the important work of professors Van Hertem, Gomis-Bellmunt, and Liang and to hope that you too, once you will have been through this book, will definitely be a fan of the European electricity Supergrid!

Enjoy the read.

Pierre Bernard
Chairman of the Board of Friends of the Supergrid

PREFACE

THE FIRST electricity transmission line in the world was a DC line which was built in 1882. Since Westinghouse and Tesla won the "War of Currents" against Edison in the 1890s, AC technology has dominated the electric power transmission. However, DC transmission has enjoyed a revival since the middle of the twentieth century and, so far, more than 200 high-voltage DC (HVDC) projects have been built around the world and offer efficient transmission of large amounts of power over long distances. In the past couple of years, development of DC transmission has gained significant momentum as a result of two key drivers: firstly, the development of modern conversion technology (voltage source converter) and secondly, the need for transmission of offshore wind power. Plans for the interconnection of multiple renewable power sources, loads, and AC grids through DC technologies are leading to an exciting transmission concept: HVDC grids. HVDC grids and supergrids have sparked so much interest that researchers and engineers across the globe are talking about them, studying them, supporting them, or questioning them.

The editors of this book and their research teams have been carrying out extensive and collaborative research on HVDC grids for offshore wind power transmission and have achieved a series of outcomes. It is the editors' motivation to share their understanding and vision of HVDC grids. It is also the editors' privilege and responsibility to present a complete picture of HVDC grid technologies by collecting and summarizing recent technological advances, academic research development, and engineering applications.

The objectives of this book are to answer why HVDC is favorable over AC technologies for power transmission; what the key technologies and challenges are for developing an HVDC grid; how an HVDC grid will be designed and operated; and how future HVDC grids will evolve. The book also devotes significant attention to nontechnical aspects such as the influence of energy policy and regulatory frameworks. As such, this book is the first to discuss all relevant topics within the scope of HVDC for offshore grids and the supergrid. It provides valuable information for researchers, industry, and policy makers.

There are 5 parts in this book consisting of 22 chapters.

Part I, "HVDC Grids in the Energy Vision of the Future," discusses the reasons for and the necessity of building HVDC grids from the viewpoint of energy development and supply. There are two chapters in this part.

Chapter 1: Describes the motivations for developing new transmission infrastructure, including HVDC grids. Clean, cheap, and secure energy supply is a policy goal that drives grid development, requiring new transmission infrastructure. The development of offshore grids and of a supergrid will produce such a closely interconnected energy system.

Chapter 2: Elaborates on the basis of the transmission of electrical energy: generation and demand, and their evolution. The energy balance between supply and demand is emphasized, and its implication toward the DC grid development are investigated. Some future scenarios for the European energy mix are summarized.

Part II, "HVDC Technology and Technology for Offshore Grids," discusses two of the main technologies for future HVDC grids: HVDC and offshore wind power technologies. These are the foundations which will steer the management, operation, and control of HVDC grids. There are four chapters in this part.

Chapter 3: Gives an overview of both conventional LCC HVDC and newly emerging VSC HVDC technologies. The components of these HVDC technologies are analyzed. The benefits of VSC HVDC over LCC technology for grids are emphasized.

Chapter 4: Gives comprehensive technical and economic comparison between HVAC and HVDC transmission, indicating the advantages of HVDC over traditional HVAC.

Chapter 5: Briefly summarizes the wind turbine technology, with fixed-speed and variable-speed wind turbines.

Chapter 6: Discusses offshore wind power plants using AC or DC collection and transmission technologies.

From **Part III, "Planning and Operation of HVDC Grids,"** readers can expect to learn how an HVDC grid will be planned, designed, regulated, operated, and controlled. There are five chapters in this part.

Chapter 7: Focuses on the planning of HVDC grids. It provides answers for the four "W" questions on HVDC grid planning—that is, Where, What, When, and Who to invest HVDC grids. Design methods for HVDC grid expansion are also given in this chapter.

Chapter 8: Presents possible layouts of HVDC grids which can be designed in future. This chapter starts from HVDC substation configurations, to VSC pole and grounding configurations and extension, and to the network topologies of HVDC grids. Analyses of topologies particular suitable for offshore HVDC grids are also given.

Chapter 9: Discusses governance issues with national and transnational systems. Ownership, financing, and tariffs of such HVDC grids are described.

Chapter 10: Gives the operation and control principles for future HVDC grids, specifically looking at the interface with AC systems.

Chapter 11: Gives more specific operation and control methods for offshore wind power plants which are connected to HVDC links and HVDC grids.

Part IV, "Modeling HVDC Grids," presents models and modeling requirements of HVDC grids under various time frames for the purposes of power flow calculation, control design, protection, and system validation. There are seven chapters in this part.

Chapter 12: Defines different time frames for the modeling of HVDC grids for different applications. The objective is to highlight the need for different models within power system analysis and to indicate the specifics of HVDC grids with respect to modeling. The models themselves are introduced in greater detail in Chapters 13 to 18.

Chapter 13: Power flow calculations form the basis of all further analysis of the power system, including economics and stability. This chapter provides models and calculation methods of HVDC grids, particularly when HVDC grids are operating alongside AC grids.

Chapter 14: Deals with the theoretical and practical implementation of optimal power flow (OPF) for DC grids. Both the regular and security constrained OPF are presented as an extension to the deterministic AC optimal power flow case.

Chapter 15: Discusses the control principles of HVDC grids and covers requirements, modes, and strategies for the control of converters, DC networks, and hybrid AC/DC systems. The main focus lies on the power–voltage regulation in DC grids.

Chapter 16: Discusses state-space representation, which is a key tool for steady-state or dynamic analysis of power systems. The chapter describes a methodology to obtain a state-space model of HVDC grids. The procedure is based on combining the equations representing different parts of the transmission grid.

Chapter 17: Deals with DC grid protection, which is a major challenge for the development of an HVDC grid. The behavior of DC short-circuit faults is discussed in this chapter, as well as the special protection requirements of HVDC grids. The chapter summarizes the DC protection methods and algorithms and describes the DC relaying protection equipment.

Chapter 18: Describes the use of real-time simulation and physical experimental tests to validate the operation of HVDC grids and control systems which provide more reliable verification than the use of offline simulation. Physical experimental platforms are the most feasible representation of a real system and can be combined with real-time simulation.

Part V, "Applications," presents advanced ideas of operation and control HVDC grids. These include optimal control of VSC within HVDC grids to improve HVDC grid dynamics, flexible DC power flow control using advanced control devices, design of an offshore AC hub to remove the need for DC circuit breakers, and AC/DC grid interactions. There are four chapters in this part.

Chapter 19: Discusses the AC/DC grid interactions, focusing on the damping of AC system oscillations, low-frequency oscillation, and subsynchronous resonance (SSR) by using converter control of HVDC grids.

Chapter 20: Presents a design method to achieve optimal droop control of HVDC grid converters in order to improve the dynamic response of grid control.

Chapter 21: Summarizes advanced control devices which can be used in HVDC grids to increase control freedom and flexibility of power regulation. Device configurations, control systems, and control effectiveness are discussed and compared.

Chapter 22: Starts from the fact that, so far, HVDC circuit breakers are not yet commercially available. The chapter discusses a solution without DC circuit breakers for HVDC design, an AC hub, also referred to as the Super-node concept. This idea is particularly suitable for offshore applications. Design and control system of AC hubs is presented. Use of a variable nonstandard frequency idea is described in this chapter.

ACKNOWLEDGMENTS

THE WRITING of this book has only been possible thanks to the support of many. First and foremost, the editors would like to thank all authors of the individual chapters for their contribution. Writing a book in a collaborative manner is not a simple task, but we believe that together we achieved a very nice result. This was not possible without their work. We would also like to thank Toni Pérez-Villegas Morey for his help with the figures and our publisher for dealing with our (minor) delays. The editors would also like to gratefully acknowledge the support from ABB, Alstom Grid, Alstom Renewable Energy, National Grid UK, Nordex, and Siemens (in alphabetical order) for their kind consent of using their pictures and diagrams. Lastly, we would like to thank Pierre Bernard of the Friends of the Supergrid for contributing the Foreword to this book.

We hope that the book proves to be useful for researchers and engineers. The comments and discussions from readers with diverse backgrounds are highly appreciated as the field of HVDC grids is an emerging one and significant research and development is still required to reach the ultimate objective of a secure, sustainable, and cost-effective energy supply with the aid of HVDC grids.

ACRONYMS

AAC	Alternate arm converter
AC	Alternating current
ACE	Area control error
ACER	Agency for the Cooperation of Energy Regulators
AVR	Automatic voltage regulator
BAU	Business as usual
BLDFIG	Brushless DFIG
CACM	Capacity allocation and congestion management
CAES	Compressed air energy storage
CCC	Capacitor commutated converter
CCGT	Combined cycle gas turbine power plant
CCS	Carbon capture and storage
CIGRE	International Council on Large Electric Systems
CPI	Current policy initiative
CPU	Central processing unit
CSC	Current source converter
CSCC	Controlled series capacitor converter
CSP	Concentrated solar power
CTL	Cascaded two level
DACF	Day ahead congestion forecast
DAEs	Differential algebraic equations
DC	Direct current
DCC	Demand connection code
DCCS	Delayed CCS
DCPFC	DC power flow control devices
DFIG	Doubly fed induction generator
DP	Dynamic programming
DSO	Distribution system operator
DSP	Digital signal processor
DST	Diversified supply technologies
EB	Balancing network code
EENS	Expected energy not supplied
EHV	Extra-high voltage
EIB	European investment bank
EMT	Electromagnetic transients
ENTSO-E	European Network for Transmission System Operators for Electricity
ENTSO-G	European Network for Transmission System Operators for Gas
EPR	Ethylene propylene rubber

ER	Energy restoration
ESCR	Effective SCR
EU ETS	European Union Emission Trading Scheme
FACTS	Flexible AC transmission systems
FCA	Forward capacity allocation
FCG	Fully rated converter generators
FCR	Frequency containment reserves
FPGA	Field programmable gate array
FRR	Frequency restoration reserves
GHG	Greenhouse gas
GIL	Gas-insulated lines
GIS	Gas-insulated switchgear
GRT	General ring topology
GSC	Grid side converter
GTO	Gate turn-off thyristor
GUI	Graphical user interface
HEE	High-energy efficiency
HF	High frequency
HIL	Hardware in the loop
HRES	High RES
HTS	High temperature superconducting
HVAC	High-voltage alternating current
HVDC	High-voltage direct current
IEM	Internal electricity market
IGBT	Insulated gate bipolar transistor
IGCT	Integrated gate commutated thyristors
IGE	Induction generator effect
IRC	Inter-rack communication
ITC	Inter-TSO compensation
LAN	Local area network
LCC	Line commutated converter
LCOE	Levelized cost of electricity
LFCR	Low-frequency control and reserves
LFO	Low-frequency oscillation
LNUC	Low nuclear
MID	Mixed integer programming
MIND	Mass-impregnated nondraining
MMC	Modular multilevel converter
MOSFET	Metal oxide semiconductor field effect transistor
MPIC	Maximum power injection capabilities
MPPT	Maximum power point tracking
MTTR	Mean time to repair
NERC	North American Electric Reliability Corporation
NLP	Nonlinear programming
NPC	Neutral point clamped
NRA	National Regulation Authorities

OHL	Overhead AC transmission lines
OPS	Operational planning and scheduling
OS	Operational security
OTC	Over the counter
OWPP	Offshore wind power plant
PCC	Point of common coupling
PCI	Project of common interest
PFC	Power flow controlling devices
PLL	Phase-locked loop
PMSG	Permanent magnet synchronous generator
POD	Power oscillation damping
PPLP	Paper polypropylene laminate
PPT	Point-to-point topology
PSS	Power system stabilizer
PST	Phase-shifting transformer
PWM	Pulse-width modulation
RAB	Regulated asset based
RCP	Rapid control prototyping
RES	Renewable energy sources
RfG	Requirements for generators
RISC	Reduced instruction set computer
RMS	Root mean square
ROCOF	Rate-of-change of frequency relays
RoW	Right of way
RPC	RISC processor card
RR	Replacement reserves
RTDS	Real time digital simulator
RTS	Real–time simulator
SB	Slack bus
SCADA	Supervisory control and data acquisition
SCIG	Squirrel cage induction generator
SCR	Short-circuit ratio
SGRT	Star with a general ring topology
SIL	Software in the loop
SLPC	Single large power converter
SoS	Security of supply
SPWM	Sinusoidal PWM
SRF	Synchronous reference frame
SSCI	Subsynchronous control interactions
SSR	Subsynchronous resonance
SSRT	Substations ring topology
ST	Star topology
STATCOM	Static compensator
SVC	Static VAr compensator
SVS	Series voltage sources
TCSC	Thyristor-controlled series capacitor

TI	Torsional interaction
TSO	Transmission system operator
TYNDP	Ten-year network development plan
UGC	Underground cables
UHV	Ultra-high voltage
UPCTE	Union for the Coordination of Production and Transmission of Electricity
VF	Variable frequency
VSC	Voltage source converter
WFC	Wind farm converter
WFRT	Wind farm ring topology
WIF	Work station interface
WPP	Wind power plants
WRIG	Wound rotor induction generator
WRSG	Wound rotor synchronous generator
XLPE	Crossed-linked polyethylene

PART I

HVDC GRIDS IN THE ENERGY VISION OF THE FUTURE

THE RADICAL CHANGES experienced by the transmission system in recent decades, along with the fast increase of renewable (including offshore) power generation in the energy mix, have triggered the interest for offshore wind connections and for overlay HVDC grids.

Part I analyzes the main drivers that have contributed to the development of the HVDC grid concept and discusses the main economic and technical challenges that the development of such technologies will face. The worldwide energy scenarios driving investments in transmission infrastructure are presented, with special attention to the Europe, where the first initiatives are ongoing.

Part I sets the scene, where the technology, planning and operation, modeling, and some applications of HVDC grids for offshore grids and the supergrid will be developed in further parts.

HVDC Grids: For Offshore and Supergrid of the Future, First Edition. By Dirk Van Hertem, Oriol Gomis-Bellmunt, and Jun Liang.

CHAPTER 1

DRIVERS FOR THE DEVELOPMENT OF HVDC GRIDS

Dirk Van Hertem
Research Group Electa, Department of Electrical Engineering, University of Leuven, Leuven, Belgium

1.1 INTRODUCTION

For a long time, the electric power industry was seen as a stable and fully matured sector, which was operating well without much innovation. Over recent decades, the situation has changed drastically, with a rapid development and innovation process. There is a strong interest from the general public for a cleaner and cheaper supply of electric energy. At the same time, the security of energy supply must be kept at the existing level, or should even be increased. The consequences are seen throughout the industry and throughout the world, with as a consequence fundamental changes in the generation of electrical energy. From a transmission point of view, this results in the need to fundamentally invest in the transmission system and to move towards a smarter and more flexible use of the grid. Different geographic, social, and historic influences have led to a different evolution over the continents. In this chapter, the different drivers of the energy revolution are discussed and the effect that these drivers have on the development of the electrical energy system is sketched. The focus is on the developments on the bulk power transmission level.

This chapter describes the evolution and the drivers in energy policy, resulting in a strong move towards the development of the offshore grid and supergrid. As the perceived drive is strongest in Europe, the main focus is on the evolutions there.

1.2 FROM THE VERTICALLY INTEGRATED INDUSTRY TO FAST MOVING LIBERALIZED MARKET

1.2.1 Brief History of the Transmission System Before Liberalization

After the introduction of electricity, the power industry has been in continuous evolution throughout the twentieth century. The earliest systems consisted of a single

HVDC Grids: For Offshore and Supergrid of the Future, First Edition. By Dirk Van Hertem, Oriol Gomis-Bellmunt, and Jun Liang.

generator connected to one or more nearby loads. A generator company owned the generation unit and the grid connecting the loads. This generator company sold electricity directly to its customers. In order to increase reliability, facilitate a growing demand, and provide a sufficiently flexible grid operation with a minimum of assets, these local grids were connected to form interconnected grids. The outage of a single generator no longer led to supply interruption. As loads increased, more systems were interconnected, the total generated power grew and higher voltage levels had to be used. Longer distances were covered. Step by step, entire countries were electrified and smaller independently operated power systems were connected.

In continental Europe, the establishment of international interconnections really started after World War II. This led in 1951 to the establishment of the "Union for the Co-ordination of Production and Transmission of Electricity" (UCPTE) [1]. Since then, an international transmission system was developed with strong interconnections at the 380-kV level. UCPTE originally oversaw the development of economic activity through the improved exploitation of primary energy resources associated with the interconnection of electricity systems. Gradually, UCPTE also organized the international cooperation between the electricity system operators and set common operational rules, amongst which is a strict frequency control. NERC[1] fulfilled a somehow similar role in the United States, where different synchronous zones cooperated from the early 1960s.

On a local level, a consolidation of power companies led to the situation in the 1990s in which each country in Europe had one or more vertically integrated companies, each dealing with their own zone. They were responsible for generation and transmission of electricity, and in some countries also for distribution. This included the planning, operation, maintenance, and exploitation of the transmission system as well as the power plants. Often these companies were state-owned or state-controlled. The power system is still operated as such in a significant part of the world.

In the vertically integrated system, operation is done from a best engineering practice point of view. To ensure a high reliability, investments are done in a coordinated manner: Generation and grid development are planned by the same group of engineers or at least within one firm. The focus is to provide an adequate energy supply with a grid which is sufficiently secure. Also, the management of the entire power system is within a single entity. As the entire supply chain is within a single company, there is a tendency to favor generators which are "grid friendly." This generally means large, controllable generators with a high availability and predictable energy supply. In such a utility, "economies of scale" resulted in large centrally planned power plants, mostly using fossil fuels or nuclear energy, or hydro power where available.

As stated before, connections between zones also existed in the vertically integrated system. On the one hand, they allowed for support during events that threatened the secure operation of the power system, enabling each zone to operate more economically, with fewer reserves needed. On the other hand, long-term contracts between countries existed, allowing international trade. This made the cheap hydro power from the Alps and the cheap and abundant nuclear power from France

[1] North American Electric Reliability Corporation (NERC), http://www.nerc.com

available to other European countries. Mainly AC links were used, connecting the different zones into synchronous zones, but also asynchronous links existed in the form of HVDC connections between different synchronous zones.

As a whole, the vertically integrated power sector in industrialized countries was a slowly changing and well-established business, with limited, well-defined problems. This was especially so after the economic development in the industrial world became mature and the increase in energy consumption dropped to a few percent per year. However, as there was no competition, the power system was not necessarily operated at the highest techno-economic optimum.

1.3 DRIVERS FOR CHANGE

Although that the implementation and the priorities concerning energy policy differs between countries and regions, they are generally built around the same main pillars. Energy policy has the objective to ensure an energy provision that is reliable, cost-effective (cheap), and sustainable (Figure 1.1).

Given the importance of energy to the overal economy, a secure energy supply is crucial to the modern industry and society as a whole. The security of supply (SoS) requires the system to have sufficient resources available (adequacy), the infrastructure available to transmit the energy from the source to the consumer with sufficient redundancy, and the ability to deliver the energy at any given moment. Failure to do so leads to short-term economic and national security concerns in the form of (local or wide-scale) blackouts. The relative economic position of a country or region can be compromised if the security of supply is threatened over a longer period of time.

The second aspect is a cost-effective energy supply, which also has a direct impact on the economy of a country or region. A cost-effective energy supply requires access to the cheapest energy sources.

The last pillar of energy policy is a sustainable energy supply. Sustainable can be understood as an energy provision that can be maintained on a longer time horizon (e.g., using only sources that are non-exhaustible) or more narrowly as an energy provision that can be maintained over a longer time horizon taking into account the environment and society effects (e.g., using only renewable energy sources).

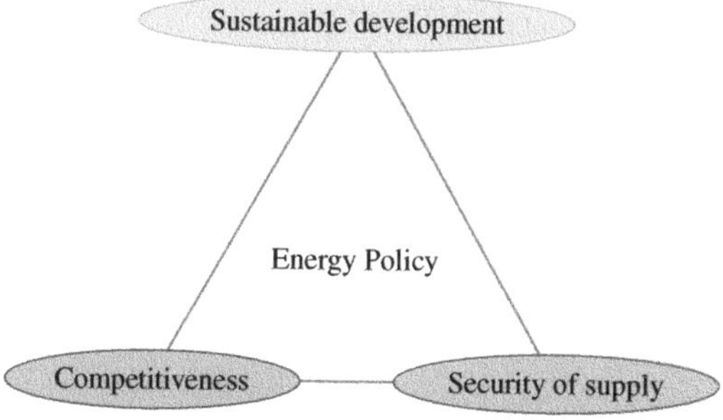

Figure 1.1 Policy drivers for the energy supply of the future: competitiveness, sustainable development, and security of supply [2].

The main concern dealt within this book is related to the reliable transmission of electrical energy from source to consumer using the appropriate transmission infrastructure for electricity. However, to understand the drivers for transmission developments, the entire energy system context needs to be considered. Over the past few decades, fundamental changes in energy policy and technology development have caused a significant change in the energy system, which in turn result in a energy system where HVDC connections are gaining importance and where HVDC grids can be developed. The specific drivers for HVDC grids are the liberalization of the energy sector and the increase of renewable generation in the energy mix.

1.3.1 Liberalized Energy Market

Economic theory suggests that competition leads to price reductions and higher economic efficiency [3–6]. With this theory in mind, different countries decided to introduce the concept of competition in the electricity sector. This process was started in the late 1980s in the United Kingdom, and it was soon followed in other countries.

1.3.1.1 Liberalized Energy Market in Europe Also the European Union decided to liberalize the electricity market in the entire EU. The basis was laid in the treaties of Rome (1957) and Maastricht (1993) as the foundations for the creation of an Internal European Market (IEM), where people, goods, and capital could freely move. In 1996, the Directive 96/92/EC [7] was published, which initialized the liberalization process for the electricity sector by setting the general conditions for the creation of an IEM. The directive included the opening of the market by slowly increasing the numbers of consumers that had the freedom to negotiate the sale and purchase of electricity. Customers were free to choose their suppliers in the European system. Furthermore, grid access was defined and regulated. Thirdly, the tasks and structure of the system operator were set. From then on, system operators had to be administratively unbundled from the generating companies [8]. The transmission and distribution of electric power was split from generation and brought into separate companies. The first directive was later replaced by a second and third energy package and is still under development. The current energy policy foresees among other things:

- Full unbundling of generation and transmission of both electricity and gas
- A (pan-)European Network for Transmission System Operators for electricity and for gas (ENTSO-E and ENTSO-G)
- A European regulatory body ACER,[2] with binding decision powers, to complement national regulators
- Increased power and independence of the national regulators
- Further promotion of cross-border cooperation and investments
- Greater transparency
- Increased solidarity
- Full internal European Market for energy

[2] ACER: Agency for the Cooperation of Energy Regulators http://acer.europa.eu/

While the vertically integrated system predominantly depended on long-term planning from within a single firm, the liberalized market encourages a more dynamic market mechanism, with more competing players. It becomes possible to trade electricity throughout the European Union, gradually leading to an increase in international flows. The interconnected transmission system has links between zones, originally intended for security and limited transfers, but not to enable large international transfers. As a result of these increased flows, congestion occurs in the transmission system, often at country borders.

1.3.1.2 Consequences of Liberalization The consequences of the liberalization process are far-reaching and have fundamentally altered the manner in which the power system is planned and operated. A first important change is the number of stakeholders involved. A complete new set of players was created, with different responsibilities, possibilities, and limitations compared to the earlier vertically integrated companies. On a local level, companies are split or new companies arise. On the international level, new organizations were created that aim at ensuring a system that is transparent and tries to steer the system towards an operation with maximal social welfare.

The following (new) entities are important to mention with respect to the development of the energy system after liberalization in Europe.

Generator company: Generator companies compete in the liberalized market. The companies are free to offer their generation portfolio (both in terms of quantity and duration) in the marketplace. There, generation and load bids are matched and a price is found against which the energy is traded. This transaction happens both on long and short term. Not all energy is traded on a market as also direct "over-the-counter" (OTC) trade is done. As the European market is open, every generator company is free to sell its energy to any market participant. This results in a competitive marketplace, which in principle leads to increased efficiency.

Investments in generation are done when perceived economically worthwhile. As such, only the most economical investments are done. This should benefit the end user, but also could lead to underinvestment in generation on the long run. On a national level, the self-sufficiency of energy supply in countries is no longer guaranteed during peak load conditions, possibly resulting in an increased dependence on import.

TSO: Transmission System Operator must ensure a safe operation of the transmission system (keeping the lights on), provide adequate asset management (investments and maintenance) and facilitate the market in a nondiscriminating manner. The TSO is responsible for balancing generation and load and within its control zone to match the scheduled import and export. The transmission system is operated as a monopoly under regulated supervision.

TSO boundaries often coincide with country (state) borders, although there are several exceptions where there are multiple TSOs per country. The TSO can be split into a system owner and a system operator.

In Europe, ENTSO-E[3] is the organization joining the different TSOs in Europe (within EU and beyond).

Merchant investors: These are investors that have received exemption from normal regulations are not owned by the system operator and remuneration of their assets goes through special (usually market-based) schemes.

DSO: Distribution System Operator is the entity responsible for the distribution network.

Consumers/retailers: These individuals have the right to choose their provider of electrical energy. Smaller consumers are supplied by retailers, while larger ones can buy directly in the commodity market, not bounded by country borders. The time frame in which contracts take place can be long or short. As a result of both the variable source and sink location, power flows throughout the power system become less predictable.

Regulator: Next to the companies involved in the trading process, a regulator is appointed to supervise the market and to advise the government on the organization and functioning of the electricity market. As such, the regulator sets the tariffs for transmission and distribution. They also approve the investment plans proposed by the regulated and nonregulated investors.

Throughout Europe, as a rule, one regulator is assigned per country for the transmission system. These regulators do not all have the same powers, competences, and legislative framework. The proposed third energy package of the EU intends to improve this aspect with a better cooperation between regulators, or even an overseeing regulatory agency (ACER).

It is clear that the interaction between several stakeholders, which were formerly operating under a single organization, is more difficult and can lead to inefficiencies.

1.3.2 More Renewables in the Energy Mix

The energy mix influences the future grid in a number of ways that are addressed here. A more in-depth discussion of the foreseen changes in the energy mix can be found in Chapter 2.

1.3.2.1 Increasing the Share of Renewable Energy Sources Throughout the world, a growing awareness concerning the environmental footprint of the human species is noticed. Especially greenhouse gases are considered problematic for contributing to global warming. As the generation of electricity from fossil fuels is an important contributor to CO_2 emissions, these sources are increasingly replaced by renewables. Europe has been a strong advocate of the development of renewables for electricity generation by publishing a directive to increase the renewable energy production from 14% to 22% between 1997 and 2010 [9]. In January 2008, a new directive was proposed and adopted by the Commission, and binding targets for individual countries were set in the so-called 20–20–20 targets [10]:

[3]ENTSO-E: European network of transmission system operators for electricity, http://www.entsoe.eu

- A 20% reduction in EU greenhouse gas emissions from 1990 levels
- Raising the share of EU energy consumption produced from renewable resources to 20%
- A 20% improvement in the EU's energy efficiency

Also post 2020, a continuation of the effort for more renewables in the energy supply is foreseen, with binding targets of a share of 27% of energy consumption generated from renewable energy sources by 2030 [11].

Although electric power from renewable energy is not directly related to the liberalization of the electricity market, the influence of the latter may not be underestimated. During the pre-liberalization period, generation was usually based on well-known technologies with a high-capacity factor. Typical examples of such power generators are nuclear, coal, combined heat and power, and large-scale hydro. Variable renewable energy like wind and solar was usually seen as low power, inefficient, difficult to operate in the system, and above all costly. The liberalization ensured that any amount of energy, no matter the scale, could be sold with potential profits. In the wind sector, small-scale generation became popular and increased rapidly, partly driven by support mechanisms. This led to the continued development of increasingly larger and more economic installations. Today, wind farms in the same range as medium-sized fossil fuel-based power plants are no oddity.

The different possible renewable sources for electricity generation are solar, geothermal, hydro, tidal, wave, biomass, and wind energy. The use of geothermal, tidal, and wave energy strongly depends on the location, and the majority of large-scale European hydro power reserves are already in use today. This leaves biomass, solar, and wind energy as the main possibilities to increase the use of renewable energy sources in Europe on the short term. Although biofuels can be and are used for electricity generation, the main aim is currently to use biofuels as a renewable source of energy for the transport sector. As a result, the main increase in electricity generation from renewables within the EU is coming from wind and solar energy. It is important to note that the development of these technologies has been strongly driven by policy and, most importantly, by the support schemes and subsidies for these energy sources. The same increase in renewable energy sources, albeit a bit delayed, is also seen in other parts of the world.

1.3.2.2 Foreseen Changes in Energy Mix and the Effect on Grids In 2013, wind energy passed the 120-GW installed capacity threshhold in Europe [12]. In Germany alone over 34 GW of wind and 35 GW of solar power was installed by end 2013 [13], which equals to over 17% of the installed capacity in Germany each. Especially for wind and solar power generation, a strong further increase of generation capacity is expected until the middle of this century [14–16]. A strong increase in generation from renewables is also expected in the rest of Europe, with an anticipated increase in installed capacity from offshore wind power to exceed 150 GW [17]. At that moment, the installed onshore wind power generation will amount to about 250 GW. As such, a big part of the electricity generation mix will be provided by wind power [17]. The best sites for wind power plants are situated in the North Sea region and the western part of Europe (Figure 1.2a). On the other hand, the best sites for large

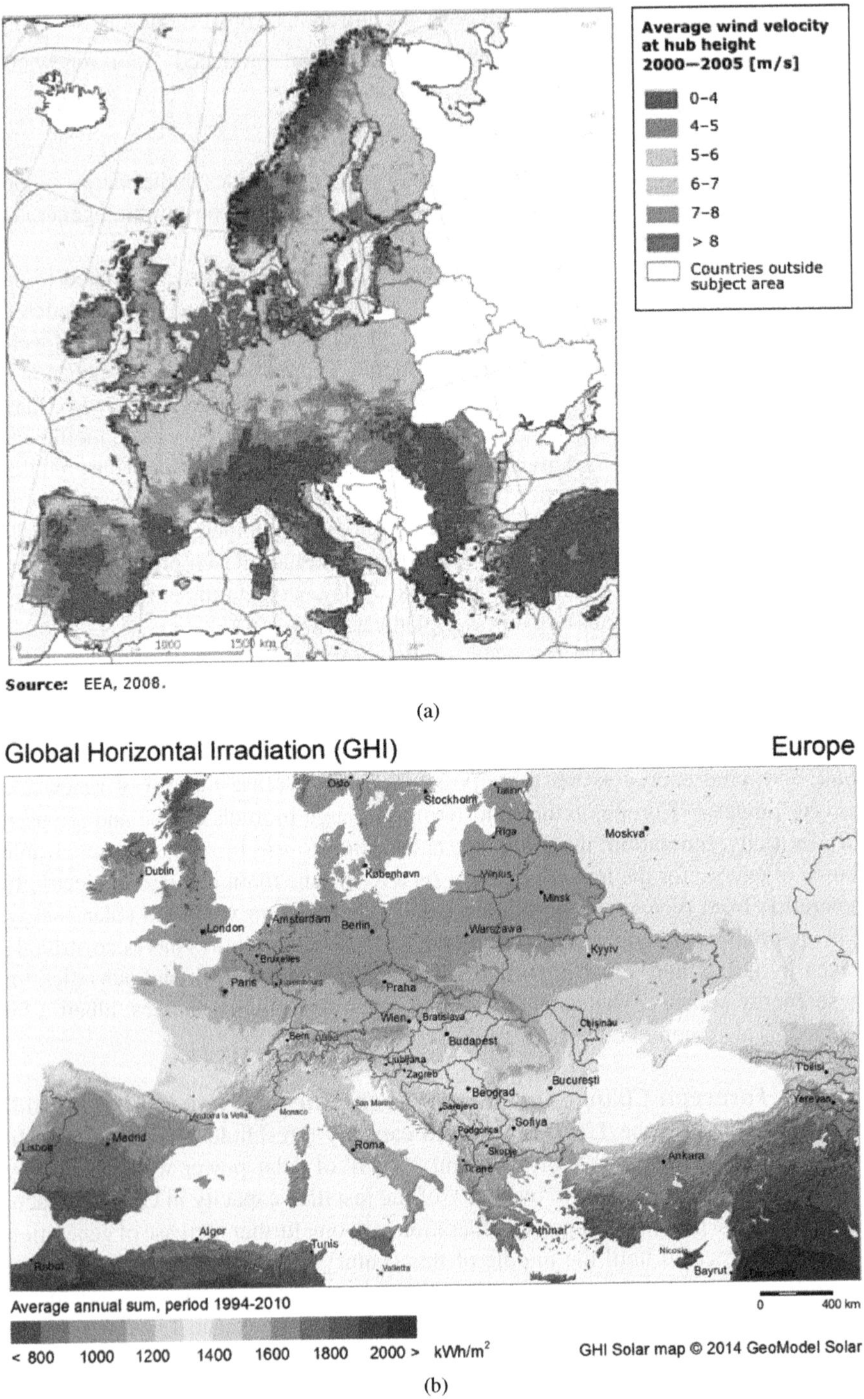

Figure 1.2 Location of wind and solar resources. (a) Average wind profile in Europe. (*Source*: EEA) [18]. (b) Average solar radiation in Europe. (*Source*: SolarGIS. Copyright ©GeoModel Solar.)

scale solar power plants are situated around the Mediterranean Sea: in Spain, Italy, and Northern Africa (Figure 1.2b). Classical generation (fossil fuels and nuclear) will not disappear and completely be replaced by RES on a short time horizon, but the system will evolve towards a new generation mix, with the traditional generation not necessarily located at the same location as it is now. Where the generation originally was placed near the load center or near the source of primary energy, a shift is noticed towards the latter. For the European energy supply, this means a shift of classical generation towards the borders of the system, specifically towards harbors. Overall, it can be seen that electrical energy is transmitted over longer distances, a tendency which is expected to continue.

A second consequence of an energy mix with more renewable energy sources is the more pronounced need for balancing services (not necessarily local) and the limited capacity factor of these energy sources. While the transmission investments need to accommodate the generation capacity (power rating), the transmission investments connecting towards that generator will be less efficient as the link will have a lower amount of energy to transport. Not only this effect is valid for direct connections to renewable energy sources, but throughout the grid a higher reserve transmission capacity is needed to accommodate for an equal amount of energy transferred. When hydro energy from Scandinavia or the Alps is used to balance the fluctuating generation from renewable energy sources, they also need sufficient transmission capacity to perform this balancing.

Much of the renewable energy is connected to lower-voltage networks at the distribution grid. These generators are generally not controllable by the grid operators, and their operation is even not transparant to the grid operator. As a consequence, each connection to the distribution grid, which was always acting as a slowly varying and predictable load, is now a much more variable and uncertain. Furthermore, the generation and consumption of energy is not simultaneous. Solar integration is a good example of this change. Although the load profile of solar largely coincides with that of the general system load, it does not coincide with that of the residential customers who have these devices mounted on their rooftops and generally consume less during the day. As a result, the load profile of that residential area shows even larger differences in the load profiles over the course of a day. The power system acts as a buffer. An option which receives much attention at the moment is the inclusion of integrated local energy storage to compensate for peaks in generation (and load). Such systems can limit the need for investments in grids, specifically at the distribution level.

Because the load centers are situated in central Europe [the region roughly comprising the Benelux, the north of France (including Paris), (Western) Germany, and north of Italy] and England (Figure 1.3), it will be inevitable to build new north–south and west–east connections within Europe to secure electricity supply to the demand centers. Furthermore, the load centers are getting more concentrated as there is a noticeable shift of the population and industry towards the city.

Although the energy consumption is getting increasingly efficient, the consumption of electrical energy is expected to continue to rise slowly over the next decades. Not only is there a general tendency towards a higher level of comfort, there is also a fundamental shift of energy consumption towards electricity—for example, through the anticipated introduction of electric vehicles.

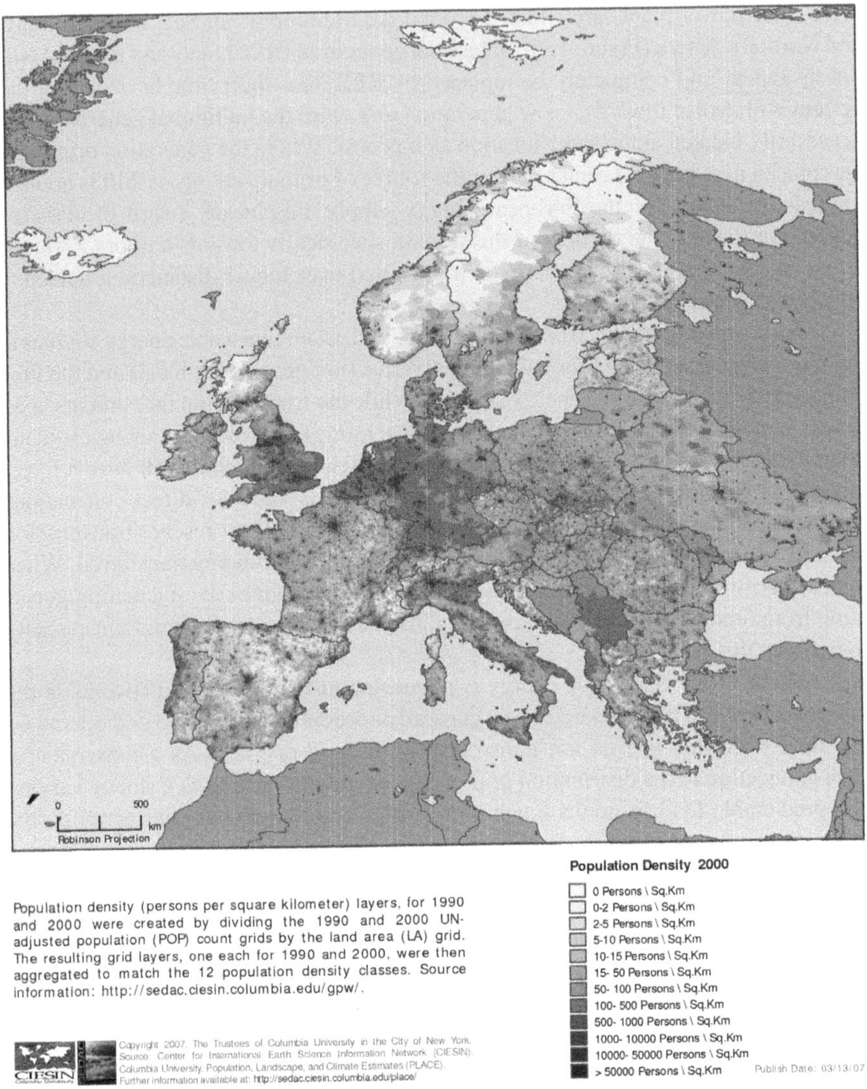

Figure 1.3 Population density in Europe. (*Source*: Columbia University.)

1.4 INVESTMENTS IN THE GRID

1.4.1 Why Investments Are Needed in the Transmission System

After the introduction of more renewables and liberalization, system operation has become more challenging due to more variable energy flows through the power system, which is furthermore less controllable due to a reduced number of control options available to the system operator. At the same time, renewables integration and the new market environment not only increase the flows on the international transmission system, but also make those flows more variable in time. Add to these facts that the demand for electric power keeps increasing, although at a reduced pace in industrial countries, and that there is increased transmission distance between generation and load; it is obvious that the system is used closer to its operational and security limits while uncertainty increases. Investments in new transmission lines form a clear and necessary solution to overcome the aforementioned problems. The correct investments can alleviate the flows in the system and remove congestion. The safety margin increases, and higher transmission capacity becomes available to the market. Additional transmission capacity between zones is needed to remove bottlenecks in the transmission system. Doing so, prices converge, when all bottlenecks are removed, a so-called *copper plate* system is reached, without congestion and a single price zone. In such a system, each user has access to the cheapest energy source. However, it is important to note that the formation of a copper plate transmission system is in fact an overinvestment, and an economic balance between investments, reliability, and congestion costs results in the most cost-effective energy system.

Not only new grid requirements drive the need for new investments. The existing transmission system is aging and slowly reaching its design life. The current system was largely constructed in the 1960s, 1970s and 1980s. As such, a considerable part of the current system is approaching the end of life and requires a fundamental refurbishment in the decades to come.

Grid interconnection capacity is also a fundamental contributor to security of supply, allowing exchange of energy over a larger system—for example, by reducing local effects of renewables through spatial balancing. With this in mind, the European Commission proposed to set a target for the interconnection capacity between member states to 15% by 2030 (where it currently stands at an average of 8%) [19].

The advantages to increased interconnection capacity include [20]:

- Improved reliability and increased capacity reserve
- Improved load factor and increased load diversity, leading to reduced investments in generation capacity
- Economies of scale in new generation units
- Diversified generation mix and supply security
- Increasing market potential
- Environmental dispatch and new plant siting
- Coordination of maintenance schedules

1.4.2 Difficulties with New Transmission Lines

Even when the reasons for additional transmission capacity are clear, in practice these investments are not straightforward. Several obstructions can cause severe delays or even cancellation of projects. These obstructions can be divided into social, environmental, political, legal, and economic [20].

Although new interconnections often improve overall social welfare in the corresponding regions, public resistance is common in industrialized regions.

Transmission lines are considered non-aesthetic and suffer from the *not in my backyard* (NIMBY) concepts and related syndromes.

Environmental concerns exist on the effects of electromagnetic fields on human health (although scientific consensus supporting these concerns is not available), or bird population. Acoustic noise is also a concern. Natural reserves are considered off limit for a transmission corridor. Oil-filled cables exhibit potentially dangerous or hazardous leakages, as do Gas Insulated Lines (utilizing the SF_6 gas). This makes it difficult to obtain permits for crossing natural reserves or residential areas, even when undergrounding parts of the corridor.

Legally and politically seen, several entities are involved in the process of permitting a new interconnection: international, national, and local authorities, ministries, and other public bodies, each with their own requirements, rules, and agendas [21]. The companies involved include not only transmission system owners, but also generator companies. The latter experience a different market situation after the construction of new capacity. Furthermore, several institutions or pressure groups at different hierarchical levels get involved. The result is a difficult and a complex permitting process with several facets that intertwine:

- Policy makers are greatly influenced by public concerns and tend to postpone any decision making as they could have a negative effect on forthcoming elections.
- Transmission investments are subject to a plethora of regulations and requirements, with possibly different and even conflicting requirements and time frames of handling. The fact that requirements are often linked does not help either (e.g., when the trajectory has to be altered because of problems obtaining an environmental permit, reapplication for a new building permit and afterwards an environmental permit can be required).

As such, obtaining a permit for an overhead transmission line takes normally about 7–10 years, but can take significantly longer [22]. In the power industry, there is a broad consensus that complex processes and unpredictable timetables for authorization and building of new lines are major issues in terms of building necessary transmission infrastructure.

Economically, not all players necessarily benefit from extra interconnections. Zones that expect a higher degree of export after the additional capacity may expect prices in their region to rise. In the importing region, the existing generator firms may see increased competition. For those players, a perfectly competitive market can potentially lead to lower profits. For system developers, the development of additional transmission infrastructure is governed by the regulatory framework and the remuneration of transmission investments. Financing and risk of such investments plays an important role (see also Chapter 9).

For international connections, finding new transmission paths requires the involvement of even more stakeholders, generally operating in a different framework. In references 21 and 22, different requirements for the application for new lines in different member states are given.

It becomes even more problematic when the most optimal investment in the transmission system for a certain TSO is not located within its own control area.

There is no incentive for a foreign party to invest in transmission capacity within its own zone that would mainly benefit an other party. This is a so-called *regulatory gap*, where there is currently no incentive for member state regulators to consider issues outside the borders of their country.

1.4.3 Available Investments Technologies

Several technologies exist to expand the available transmission capacity. A short overview is given here; a more in-depth coverage of this topic can be found in reference 23.

1.4.3.1 New Overhead Lines The traditional approach in transmission system enforcement is three-phase overhead AC transmission lines (OHL). This solution is very cost effective and robust. It uses technology that is known and used for decades. There are currently no technologies that can compete with OHL on a purely economic basis, especially not in rural areas [23]. Although OHL often experience short outages—for instance, due to lightning strikes—they can be restored with a simple line reclosure, making this a non-issue in meshed transmission systems. However, OHL have a high visual impact and are considered not appealing and possibly hazardous by public opinion. New tower types are available which are more visually appealing than lattice towers, but the towers can still be seen from a distance. This results in tough siting opposition and long permitting processes. The construction of a new transmission system demands a new transmission path, which requires a significant right-of-way (a corridor of up to hundreds of meters wide).

1.4.3.2 Uprating or Upgrading Existing Circuits A first option to add capacity to existing transmission paths is by making optimal use of the existing capacity through the use of dynamic line rating. A second option is to add capacity by adding an additional circuit to existing pylons or by increasing the voltage when possible. This is only possible when this upgrade was already conceived at the planning stage. Alternatively, the existing conductors may be replaced by new conductor types able to carry higher currents. New materials are constantly developed, which allow higher temperatures and, consequently, higher permissible currents for equal cross sections. These approaches can lead to significantly shorter permitting procedures as the right-of-way remains the same, and the visual impact of the transmission line is not greatly affected. Changing a double circuit three-phase AC system (6-wires) to an HVDC system can increase the capacity of a transmission corridor by a factor of 2 up to 3.5 [24,25]. The German system operators consider using such conversion from AC to DC for the upgrade of the German transmission system [26].

1.4.3.3 Underground Cable Connections In order to avoid most of the visual impact, undergrounding transmission assets are considered the ideal solution. However, technically and economically, this is not necessarily the case. Next to being several times more expensive than OHL, high-voltage cables for transmission systems act as capacitors, requiring large compensation units, and hinder system operations. Cable systems for transmission system voltages are limited in length due to the charging current. A special type of cables are gas insulated lines (GILs) which allow longer lengths, higher transmission voltages, and higher power ratings. At this moment there are no long-distance connections using GIL installed. GIL is more expensive than underground cables. Serious advances are being made in the field of high-temperature superconducting (HTS) cables, which promise high ratings under low losses. At this moment though, the high-temperature supercondicting technology is not yet ready for use in the transmission system.

Underground connections are several times more expensive than overhead lines, while less troublesome concerning permitting and social requirements. Therefore, often a compromise solution is found, where part of the connection is overhead and part underground. The combined solution requires a substation at each transition. Each transition also comes with a change in the characteristic impedance, requiring additional equipment to protect it from voltage surges in the transmission system.

1.4.3.4 Grid Flexibility Through Power Flow Control The solutions proposed above all add capacity by adding new paths or strengthening existing paths. In the case of international connections, permits at both sides of the border are needed to achieve this incremental capacity. Power-flow-controlling devices (PFCs) such as phase-shifting transformers and HVDC connections offer a completely different approach as they consist of devices placed at one specific point. As such, they can be placed in existing interconnectors, influencing the flow through that path. The additional space required by devices such as the PST is normally limited to the device itself, which can be placed in an existing substation. When used as a reinforcement of an interconnector, only the investing zone/system owner is involved. This makes the permitting process much easier. The devices themselves allow the control of the power flowing through the devices. As such, they can redistribute the line flows in the system, alleviating stress on the heavily loaded lines and thus increasing the overall transfer capabilities of the grid. As this technology presents a single investment in a single device, the solution is in relatively cheap compared to other ones. PFCs receive lots of attention of TSOs that need increased transmission capacity for a short time. A more in depth coverage of the principles and the technologies is presented in reference 27. With respect to the applicability for grid investments, the power-flow-controlling devices are being categorized into two types: flexible AC transmission systems (FACTSs) [28] including phase-shifting transformers (PST) and high-voltage DC (HVDC) solutions. The HVDC solution combines the power flow control aspect and the additional transmission capacity.

1.4.4 HVDC Technology

Next to providing additional flexibility to the system through flow control, HVDC provides additional transmission capacity in itself. HVDC transmission systems transmit electric power at zero frequency and use power electronic converters to interface between the AC grid and the DC grid. Traditional applications of HVDC are:

- Bulk transmission of energy over long distances
- Interconnection of asynchronous systems (possibly back-to-back)
- Undersea connections

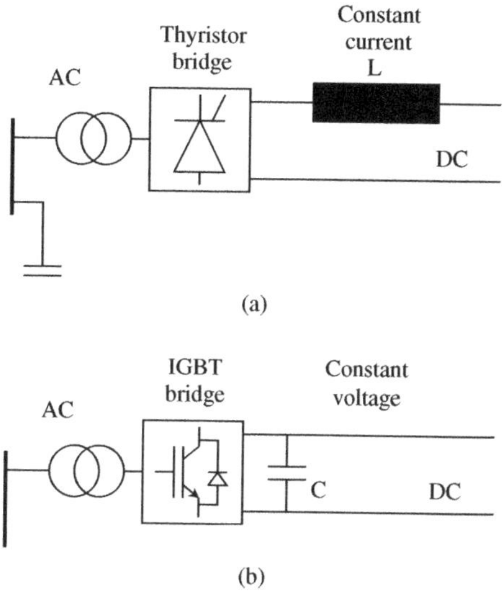

Figure 1.4 LCC- and VSC-HVDC one-line diagrams. (a) Current source converter. (b) Voltage source converter.

The most common and oldest technology is called line-commutated converter (LCC) HVDC, sometimes also referred to as current source converter (CSC) HVDC. It uses thyristors as switches and operates using a constant DC current characteristic. The DC voltage is controlled to alter the power flow.

Voltage source converter (VSC) HVDC was developed in the mid-1990s and uses IGBTs. The voltage source converters have a number of advantages over LCC technology. The VSC converter delivers a constant DC voltage, and the current is controlled to alter the power flow. This allows the use of the easier and cheaper XLPE cables instead of mass-impregnated oil-filled cables. Furthermore, the VSC converter can build an independent rotating field at the AC side. This allows the connection to weak islanded grids and offers black-start capability. An example is the connection to offshore wind power plants, which is virtually impossible when using LCC HVDC. Figure 1.4 shows the one-line schematics of the two technologies.

HVDC technology is briefly introduced in the following section and is discussed in far greater detail in the remainder of the book, more specifically in Chapters 3 and 4.

1.5 TOWARDS HVDC GRIDS

For many, VSC-HVDC technology is seen as an enabler for the future power system, and more specifically a technology that allows the massive integration of renewable energy sources in the system [29]. This is especially so for Europe, where large amounts of renewable energy are available on remote locations, often offshore or near the sea. The long-distance transmission of energy from source to load, but also for balancing, puts extra pressure on the already heavily loaded transmission system. Because of the variability of the renewable energy sources, more transmission lines are needed for the same amount of energy delivered when compared to classic energy sources [30, 31]. The lack of support for new transmission lines, especially overhead ones, require solutions other than the traditional AC overhead line [32]. An additional problem is the location of offshore resources, which are increasingly difficult to realize with AC technology.

HVDC lines, and by extension an HVDC grid, has the potential to address the aforementioned problems.

1.5.1 Transmission Technology

New grid investments are required for, on the one hand, fundamental upgrade of the existing transmission system, possibly evolving into a backbone or overlay grid. Such a system is also referred to as a supergrid. A second type of new investments is specific for connection of offshore generation and load, eventually leading to an offshore grid.

A new backbone transmission system can be built using traditional alternating current (AC) or direct current (DC) technology. AC systems are a well-known technology, offering a cheap and reliable solution for high power transmission using classic overhead lines. DC systems, on the other hand, experience much fewer problems with long-distance power transmission, especially when cables are used. As such, a DC overlay grid seems to be the most appropriate solution when cables are needed.

VSC HVDC is the most appropriate technology for multi-terminal applications as it uses a common DC voltage, making parallel connections easy to build and control. LCC HVDC is much more troublesome to control in a parallel multi-terminal configuration, and especially changing the power direction in a single converter without interruption is problematic. Operating a multi-terminal LCC-HVDC system with many terminals with constantly changing power flows might result in an unreliable system. The possibility to connect islanded grids to a VSC-HVDC converter is another important advantage.

Two LCC-HVDC systems are in operation which are used in a multi-terminal HVDC configuration, namely the system between Hydro-Quebec to New England and the connection between Italy, Corsica, and Sardinia. Both have three terminals (although the first system actually has five terminals, only three are in operation). At the time of their construction, VSC-HVDC technology was not yet developed.

New multi-terminal systems using VSC-HVDC are built in a pilot phase (e.g., the system connecting Nanao island to the mainland Guangdong power grid and the Zhoushan 5-terminal system in China) and several projects are under study (e.g., the Moray Firth Offshore HVDC hub in Scotland, the Cobra cable between the Netherlands and Denmark with an offshore wind farm connected and the Atlanic Wind Interconnector in the USA).

1.5.2 Why Not AC?

The use of AC is disregarded as a potential option for a future supergrid in Europe. However, AC systems are capable of carrying large quantities of electric power over large distances when ultrahigh voltage (UHV, 1000 kV AC or higher) is used. AC systems using UHV voltages were developed in the 1970s, and they are currently planned and installed in China. The technical reasons why UHV AC is not seen as a potential technology for the supergrid are as follows:

- DC line losses are lower (no skin effect, no proximity effect).
- AC cable solutions for the needed high voltages are not yet available.
- AC cables experience a high charging current which limits their length. Long AC cables at very high voltages are difficult to construct and expensive.
- Offshore resources, as well as connections outside the main continent, are virtually inaccessible when using AC.
- HVDC offers an inherent active power control, making it more flexible in use and easier to limit overloads in the system.

There are also nontechnical reasons which are in favor of DC over AC technology. By using cables which cause no visual pollution and emit no varying electromagnetic fields, much less opposition and problems with licencing and construction are expected. Overhead lines are very difficult to construct because of nontechnical issues. Furthermore, using sea cables allows a fast and relatively cheap cabling because less joints are needed [32].

Figure 1.5 shows a graphic representation of the relative costs of HVDC and HVAC systems. The difference in cost price for an installation in systems without constraints (easier permitting process and the lower amount of "special" considerations to be taken into account such as deviations from the ideal path, sound protection, etc.) and in systems that experience significant constraints cannot be underestimated.

In short, AC overhead lines can be an option from a technical point of view, but constructing them is often an issue because of political and environmental concerns and AC cables are not suited for long-distance bulk power transfer. In case these disadvantages are deemed less important, AC overhead lines remain a valid option. This is also the case for regions with relatively seen more open space to place the transmission lines and where fewer offshore connections are needed. Both northern America and China might benefit more from additional AC reinforcements, combined

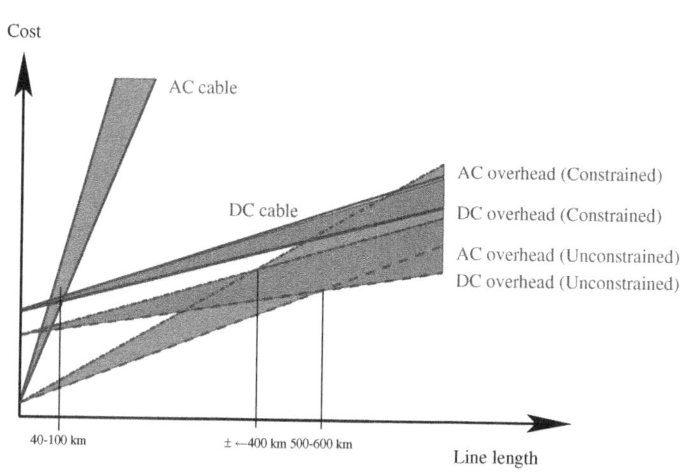

Figure 1.5 Cost breakdown for AC and DC systems [33].

with specific HVDC installations, than from a DC supergrid. Even in continental Europe, this would be possible if the opposition against new lines would cease. Offshore and other remote resources would still require cable connections, which makes AC no longer possible. A future "supergrid" which uses a mix of AC and DC technology is a realistic option.

HVDC is discussed in further detail in the remainder of the book, and more specifically in Chapters 3 and 4 on the comparison between AC and DC.

1.5.3 HVDC Grids as a Supergrid

A new overlay grid should allow an increased integration of all resources, including renewables. The variability of the renewable energy sources can be reduced because of the limited correlation of different weather system through connection of different energy sources such as wind, hydro, and solar, at different locations. Wind energy from the north of Europe can be partially balanced with wind from Spain or solar energy from the Sahara dessert. The remainder can be balanced by hydro energy, possibly from Scandinavia or the Alps. It could even allow the connection to the vast geothermal energy resources of Iceland.

Definition. A supergrid can be defined as a transmission solution which allows the massive integration of renewable energy sources in the European power system. It connects the different remote energy sources to the existing grid while offering additional control, it offers balancing through geographic spread, and allows a more diversified energy portfolio. In the meanwhile it increases the security of supply [33].

Many supergrid topologies have been proposed by different organizations [34, 35] and also by environmental organizations [36]; this is considered to be one of the most important options to reach a cleaner energy provision. These projects and ideas have received widespread attention from both politics and the press. The developments in Europe [33] have received most attention because of their leading role in renewable integration and their specific concerns with respect to transmission expansion. However, also in other regions we see the necessity to expand the power system. In China, long-distance energy transmission is vital to bring the energy resources from the west to the main load centers at the coast, most importantly the regions of Beijing and Shanghai. The transmission system in China is therefore playing a key role in the rapid growth in the country [37] and can be considered to be new overlay grid. India is in the process of connecting its different subsystems and reinforcing the interconnection in between with the focus on achieving a energy system which is more reliable. The Japan Renewable Energy Foundation [38] has proposed to interconnect the Japanese islands with mainland Asia (with China, India, South Korea, and Thailand) as well as Indonesia and Taiwan (Figure 1.6). In the United States, the energy system is up to a fundamental upgrade with a considerable need for new invesments at the highest transmission level [39]. In the eastern parts of the United States, the Atlantic Wind Connection consortium was formed to develop an offshore system which should strengthen the aging infrastructure, alleviate congestion, and lower energy prices and improve the reliability of the region [40] (Figure 1.7).

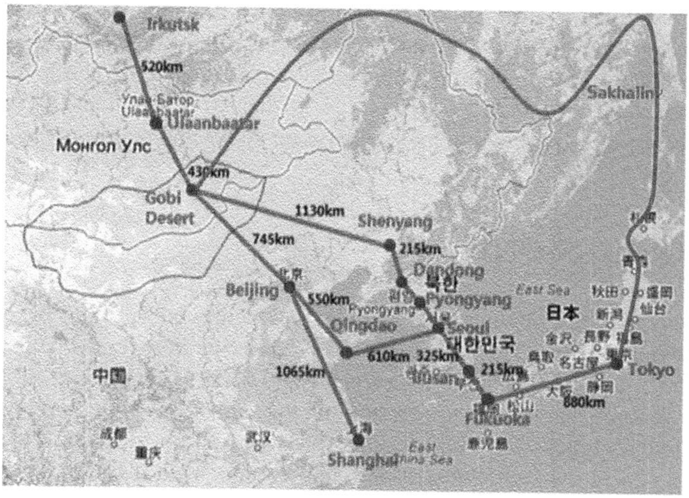

Figure 1.6 Plans for an Asian super grid [41].

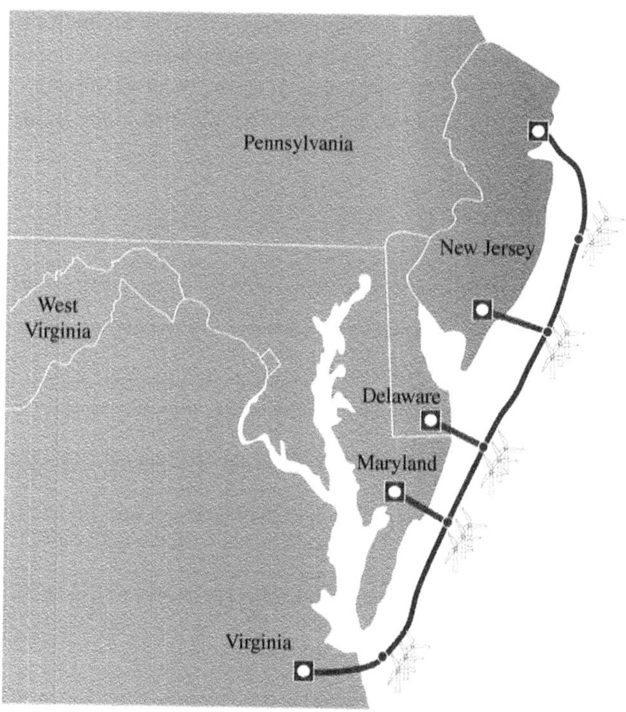

Figure 1.7 Schematic overview of the announced Atlantic Wind Interconnection offshore system connecting the east of the United States (*Source*: www.atlanticwindconnection.com.)

However, within the technical community, a lot of skepticism exist with respect to the development of HVDC grids. These visions do not contain sufficient detail in either technological, economic, or power systems engineering to allow full-scale development on a very short time frame. Nevertheless, research and technological innovations are quickly bridging the gap between vision and reality.

1.6 CONCLUSIONS

The efforts towards a sustainable, competitive, and secure energy supply have changed the requirements for the grid of the future. This has led to more flows on the transmission system which are furthermore increasingly variable. This has caused the power system to become increasingly congested. At the same time, structural investments in transmission capacity have been lacking over the past few decades. The foreseen changes in generation and load will only increase the need for transmission, especially if the massively available offshore energy resources are to be connected in the next decades.

Different technologies already exist to reinforce the transmission system. Most common technologies are AC overhead lines and cables, system uprating, power flow controlling devices, and HVDC connections. Of these technologies, HVDC is seen as very promising as it allows structural upgrades using cable technology. More specifically, the evolution of a DC grid is seen as the most promising option which can cause a paradigm shift and act as an enabler for the grid of the future with a high share of renewables.

REFERENCES

1. UCTE, "Articles of Association," Union for the Co-ordination of Transmission of Electricity (UCTE), Technical Report, June 12, 2005.
2. Single market for gas & electricity: Third package for electricity & gas markets. European Union [online]. Available: http://ec.europa.eu/energy/gas_electricity/legislation/third_legislative_package_en.htm
3. D. S. Kirschen and S. Goran, *Fundamentals of Power System Economics*. Chichester, UK: Wiley, 2004.
4. T. Jamasb and M. Pollitt, "Electricity market reform in the European Union: Review of progress toward liberalization & integration," Center for Energy and Environmental Policy Research, Technical Report, 2005, doc: 05-003 WP.
5. IEA, *Lessons from Liberalised Electricity Markets*, series Energy Market Experience. Paris: International Energy Agency, 2005.
6. P. Joskow, "Lessons learned from electricity market liberalization," *The Energy Journal*, vol. 29, no. 2, pp. 9–42, 2008.
7. "Directive 96/92/ec of the European Parliament and the counsel of 19 December 1996, concerning common rules of the internal market in electricity," European Union, 1996 [online]. Available: http://eur-lex.europa.eu/LexUriServ/LexUriServ.do?uri=CELEX:31996L0092:EN:HTML
8. L. Meeus, K. Purchala, and R. Belmans, "Development of the internal electricity market in Europe," *Electricity Journal*, vol. 18, no. 6, pp. 25–35, July 2005 [Online]. Available: http://www.esat.kuleuven.be/electa/publications/fulltexts/pub_1467.pdf

9. European Commission, "Directive 2001/77/EC of the European Parliament and of the Council of 27 September 2001 on the promotion of electricity produced from renewable energy sources in the internal electricity market," *Official Journal of the European Union*, pp. 33–40, Oct. 2001 [Online]. Available: http://eur-lex.europa.eu/pri/en/oj/dat/2001/l_283/l_28320011027en00330040.pdf
10. The 2020 climate and energy package. European Union [online]. Available: http://ec.europa.eu/clima/policies/package/index_en.htm
11. European Commission. (April 2014) 2030 framework for climate and energy policies. Brussels, Belgium [Online]. Available: http://ec.europa.eu/clima/policies/2030/index_en.htm
12. European Wind Energy Agency (EWEA), "Wind in power—2013 European statistics," EWEA, Technical Report, 2014 [Online]. Available: http://www.ewea.org/fileadmin/files/library/publications/statistics/EWEA_Annual_Statistics_2013.pdf
13. Renewable Energy Policy Network for the 21st century (REN21), "Renewables 2014—global status report," REN21, Technical Report, 2014 [Online]. Available: http://www.ren21.net/Portals/0/documents/Resources/GSR/2014/GSR2014_full%20report_low%20res.pdf
14. German Energy Agency, "Dena grid study II—Integration of renewable energy sources in the German power supply system from 2015–2020 with an outlook to 2025," Technical Report, Nov. 2010.
15. J. De Decker and P. Kreutzkamp, "Offshore electricity grid infrastructure in Europe—Final Report," Technical Report, Oct. 2011.
16. European Comission, DG energy in collaboration DC climate action DG and DG mobility and transport, "EU Energy Trends to 2030," Technical Report, 2009.
17. European Wind Energy Association (2010), Factsheet: Statistics & targets [Online]. Available: http://www.ewea.org/fileadmin/ewea_documents/documents/publications/factsheets/Factsheets.pdf
18. European Energy Agency (EEA), "Europe's onshore and offshore wind energy potential: *An assessment of environmental and economic constraints*," EEA, Technical Report ISSN 1725–2237, 2009.
19. European Commission, "Communication from the commission to the European parliament and the council—*European Energy Security Strategy*," European Commission, Technical Report, 2014. [Online]. Available: http://eur-lex.europa.eu/legal-content/EN/TXT/PDF/?uri=CELEX:52014DC0330&from=EN
20. United Nations: Economic & Social Affairs, "Multi dimensional issues in international electric power grid interconnections," United Nations, Technical Report, 2006.
21. ETSO, "Overview of the administrative procedures for constructing 110 kV to 400 kV overhead lines," European Transmission System Operators, Technical Report, Dec. 5, 2006.
22. Nascimento Baptista (Ed.), "Public acceptance for new transmission overhead lines and substations," Eurelectric, Technical Report, 2003.
23. S. Cole, D. Van Hertem, L. Meeus, and R. Belmans, "Technical developments for the future transmission grid," in *International Conference on Future Power Systems, ISBN 90-78205-01-6*, Amsterdam, the Netherlands, 16–18 Nov. 2005, p. 6 [Online]. Available: www.esat.kuleuven.be/electa/publications/fulltexts/pub_1503.pdf
24. J. Lundquist, L. Barthold, A. Beutel, A. Britten, D. Douglass, J. Iglesias, V. Jankov, J. Jardini, D. Muftic, and S. Steevens, "Guide to the conversion of existing ac lines to dc operation," Cigré, Technical Brochure 583, May 2014, wG B2.41.
25. J. Lundkvist, I. Gutman, and L. Weimers, "Feasibility study for converting 380 kV lines to hybrid AC/DC lines," in *High-Voltage Direct Current & Flexible AC Transmission Systems*. EPRI, Nov. 5–6 2009, p. 25.
26. P. Fairley, "Germany jump-starts the supergrid," *Spectrum, IEEE*, vol. 50, no. 5, pp. 36–41, May 2013.
27. D. Van Hertem, "The use of power flow controlling devices in the liberalized market," Ph.D. dissertation, KU Leuven, Jan. 2009.
28. N. G. Hingorani and L. Gyugyi, *Understanding FACTS*. New York: IEEE Press, 2000, iSBN 0-7803-3455-8.
29. S. Gordon, "Supergrid to the rescue," *Power Engineer*, vol. 20, no. 5, pp. 30–33, Oct.–Nov. 2006.
30. ETSO, "European wind integration study (EWIS)—*Towards a Successful integration of Wind Power into European Electricity Grids*," Project report, Jan. 2007 [Online]. Available: http://www.wind-integration.eu/downloads/library/EWIS-phase-I-final-report.pdf

31. P. Buijs, D. Bekaert, D. Van Hertem, and R. Belmans, "Needed investments in the power system to bring wind energy to shore in Belgium," in *IEEE PowerTech 2009*, Bucharest, Romania, June 28–July 2, 2009, p. 6 [Online]. Available: www.esat.kuleuven.be/electa/publications/fulltexts/pub_1884.pdf
32. P. Buijs, D. Bekaert, S. Cole, D. Van Hertem, and R. Belmans, "Transmission investment problems in Europe: Going beyond standard solutions," *Energy Policy*, vol. 39, no. 3, pp. 1794–1801, 2011 [Online]. Available: http://www.sciencedirect.com/science/article/pii/S030142151100022X
33. D. Van Hertem and M. Ghandhari, "Multi-terminal vsc hvdc for the european supergrid: Obstacles," *Renewable and Sustainable Energy Reviews*, vol. 14, no. 9, pp. 3156–3163, 2010 [Online]. Available: http://www.sciencedirect.com/science/article/B6VMY-50PJ72B-8/2/83156f1dd30bf5615ce11bf4ab2e99dc
34. Desertec Foundation, "Clean power from deserts—*The DESERTEC concept for energy, water and climate security*," Trans-Mediterranean Renewable Energy Cooperation (TREC), Whitebook, 2007 [Online]. Available: http://www.desertec.org/fileadmin/downloads/DESERTEC-WhiteBook_en_small.pdf
35. J. De Decker, A. Woyte, B. Schödwell, J. Völker, and C. Srikandam, "Directory of offshore grid initiatives, studies & organisations," OffshoreGrid, Technical Report, Sep. 2009.
36. A. Woyte, J. De Decker, and T. Vu Van, *A North Sea electricity grid [R]evolution—Electricity output of interconnected offshore wind power: A vision of offshore wind power integration*. Greenpeace—3E, 2008 [Online]. Available: http://www.greenpeace.org/raw/content/eu-unit/press-centre/reports/A-North-Sea-electricity-grid-(r)evolution.pdf
37. Stage Grid Corporation China, "Sgcc framework and roadmap for strong & smart grid standards," SGCC, Technical Report, 2009.
38. T. Kåberger, "Asian super grid—vision for Mongolia and Asia," in *Mongolian Renewable Energy—International Conference*, Ulaanbaatar, Mongolia, May 2012.
39. J. Lawhorn, D. Osborn, J. Caspary, B. Nickell, D. Larson, W. Lasher, and M. Rahman, "The view from the top," *Power and Energy Magazine, IEEE*, vol. 7, no. 6, pp. 76–88, 2009.
40. The Chertoff Group, "Grid resiliency, its economic and security impacts, and the implications of AWC," The Chertoff Group, Technical Report, Oct. 2012.
41. S. Mano, B. Ovgor, Z. Samadov, M. Pudlik, V. Jülch, D. Sokolov, and J. Yong Yoon, "Gobitec and Asian super grid for renewable energies in northeast Asia," Energy Charter, Technical Report ISBN 978-905948-143-5, 2014.

CHAPTER 2

ENERGY SCENARIOS: PROJECTIONS ON EUROPE'S FUTURE GENERATION AND LOAD

Erik Delarue[1] *and Cedric De Jonghe*[2]
[1]Research Group TME, Department of Mechanical Engineering, University of Leuven, Leuven, Belgium
[2]Research Group Electa, Department of Electrical Engineering, University of Leuven, Leuven, Belgium

2.1 INTRODUCTION

Electricity is a commodity with specific characteristics. Supply and demand of electricity have to be continually in perfect balance. Maintaining this balance requires careful system operation, as both demand fluctuates (daily, weekly, seasonal patterns) and supply can fluctuate (variable renewables, unforeseen outages of power plants), and it is still difficult to store electricity at a large enough scale in an economic viable way. Furthermore, a grid is required to transport electricity from where it is generated to where it is consumed. The technical limitations of such a grid together with the required continuous demand–supply balance create specific needs towards optimization and operation of electric power systems.

This chapter will focus on the context of supply and demand evolutions, along with the role of the grid. In essence, the first "task" of the grid is to transport electricity from where it is generated to where it is consumed. In addition, connecting different areas also improves security of supply and system reliability. In a market context the grid infrastructure also contributes to improve competition in the power industry, where networks enable electricity generation with the lowest cost facilities and transport it to where it is needed.

First, the general system setting is described, with supply, demand, the grid, and relevant policies in this regard. Second, the focus is on scenario analysis, a tool making projections on certain evolutions of the electric power system. Focus mainly is on existing scenarios from a European perspective, as well as on the role of the

HVDC Grids: For Offshore and Supergrid of the Future, First Edition. By Dirk Van Hertem, Oriol Gomis-Bellmunt, and Jun Liang.

grid. Finally, some concluding reflections on a wider international (global) context are formulated.

2.2 SYSTEM SETTING

This section describes the relevant technologies and principles for supply and demand of electricity, together with the different flexibility options in ensuring the continuous instantaneous balance. The current setting and trends in Europe are briefly discussed.

2.2.1 Supply

Electricity can be generated through different sources. A classification can be made between sources that are dispatchable and those that have an intermittent character (i.e., being not or only partially dispatchable). Dispatchable means that one can fully control the output of the power plant. Intermittent means that the power plant output cannot or only limitedly be controlled and that it is not fully predictable. Another classification can be based on centralized sources versus decentralized generation. The classification that is followed here is based on the primary energy source. Basically, a distinction can be made between electricity generation from nuclear origin, fossil generation, or renewables. Over time, technological evolutions have been made, but these can be considered as rather gradual improvements of existing principles or technologies (e.g., photovoltaic effect has been known for far over a century). The discovery of nuclear fission in the 1930s can be considered the last fundamental discovery, triggering a step-change innovation in energy-related science.

2.2.1.1 Nuclear Nuclear power plants are based on a Rankine steam cycle. Heat is provided by nuclear fission. The size of a nuclear power plant can vary between a few hundreds of MWs, up to typically 1000 MW or even 1600 MW for new plants. Characterized by high investment and low operational costs, nuclear power plants are typically developed and operated in base load—that is, running at continuous maximum output. However, in technical terms these plants can be controlled in a quite flexible manner, comparable to other classic power plants [1]. In countries with a high share of nuclear power (such as France), some of these plants are typically operated in load-following mode. Nuclear power plants are CO_2-free, making them one of the options towards decarbonization of the power system. Clearly, nuclear power plants face risks related to the use and production of radioactive material (waste, accidents, and proliferation).

The future development will be mainly determined by policy and public acceptance. After a long period of nuclear submissiveness, 68 new reactors were under construction at the moment of the Fukushima incident, mainly in China, Russia, and India [2, 3]. Whether a so-called nuclear renaissance actually will occur in North America and the EU remains to be seen, especially after the Fukushima incident. It might already be difficult to keep the current amount of nuclear stable, as many units are approaching the end of their lifetime [4]. Finland and France are currently building a new reactor, and the United Kingdom is still planning for new capacity.

TABLE 2.1 CO_2 Content of Fossil Fuels and Corresponding CO_2 Emissions[a]

Fossil Fuel	CO_2 Content [kg CO_2/MWhp]	Conversion Efficiency [MWhe/MWhp]	CO_2 Emission [kg CO_2/MWhe]
Lignite	364	35%	1039
Coal	342	45%	760
		35%	977
Oil	272	35%	777
Natural gas	207	60%	344
		35%	590

[a]MWhp stands for the primary energy content (heating value), whereas MWhe stands for electric output.

Other countries like Germany, Switzerland, and Belgium, on the other hand, have decided on a nuclear phase out.

The cost of new nuclear power is very uncertain, and wide ranges and regional differences are reported (see, e.g., [3] and [5]).

2.2.1.2 Fossil Fuels Fossil fuels can be burned to generate heat, which is used in a thermodynamic cycle to generate electricity. Typical fossil fuels are hard coal, lignite, peat, natural gas, and oil. When fossil fuels are burned, CO_2 is formed. Coal, lignite, and peat have the highest carbon content, followed by oil and then natural gas. The more efficient the power plant, the lower the CO_2 emission per unit of generated electricity will be. The CO_2 content of various fuels is listed in Table 2.1; together with a possible conversion efficiency, the CO_2 emission per unit of electrical output is determined.

Coal and lignite-fired power stations are typically relatively large Rankine-cycle-based units. Coal, and speficically lignite power plants are not rarely situated in the near vicinity of where the resource is mined, as large volumetric quantities of these fuels are required, and transport costs can become significant. Alternatively, these plants will be located at harbors or near large waterways, or at a railway connection. Gas-fired power plants are typically deployed in different setups. A combined cycle power plant combines one or more gas turbines (Brayton cycle) with a steam cycle (CCGT). This yields high efficiencies of up to 60%. CCGTs are flexible and can be easily used in load following mode. A basic (open cycle) gas turbine is more simple in implementation and has a lower efficiency, but is very flexible in operation and has low investment costs. Oil can also be used in gas turbines (e.g., turbojets) or in diesel engines, typically used as peak-power plant (high operational cost, low investment).

All plants fired on fossil fuels emit CO_2. To make such plants usable in a low carbon framework, carbon capture and storage (CCS) can be a future option. The theory of CCS is clear: Keep using fossil fuels and capture and store the released CO_2 underground. Given the worldwide massive reserves of easy available coal, CCS seems indispensable for a low-carbon future. However, the technology is currently still in a developing stage and has to be demonstrated to be feasible on a large scale at acceptable cost. Apart from significant investment, a high CO_2 price will be required to trigger actual use of the CCS part, as a significant share of the power is needed to drive the CCS part, thereby lowering the plant's net output and efficiency.

2.2.1.3 ***Renewables*** Renewable energy streams are streams that are not "exhausted" when being harvested; that is, these sources are continually replenished. The crucial parameter related to RES is the potential, both technical and economic (what is available and at what cost). The current economically most attractive technologies are hydro, wind, solar PV, and some sorts of biomass. Concentrated solar power (CSP), wave/tidal power, and geothermal power are very location-specific, and in general still currently too expensive to make them competitive with classic generation. Enhanced geothermal power is attributed a very high potential, but many uncertainties (also related to cost) remain.

Biomass-fired power plants and hydro power (with a dam) are dispatchable plants; that is, the output can be actively controlled (like conventional power plants). Other sources such as wind and solar PV have an intermittent character, being variable in output (only generating when the wind blows or when the sun shines) and not perfectly predictable. Load factors for these technologies (20–40% for wind, 10–20% for solar PV) are low compared to conventional technologies, meaning that installed capacities need to be high to achieve a certain level of electricity generation over time. Diversification of different RES might help to mitigate the overall intermittency (i.e., possibly reducing the aggregated variability and unpredictability).

In the current (global) RES generation portfolio, hydro still has the largest share. The potential for further large-scale hydro deployment is, however, limited in the developed regions where only some potential for small-scale installations remains. The deployment of wind is or has been rapidly increasing worldwide. Compared to onshore wind, the share of offshore developments is still minor. The deployment has, however, been increasing over the past few years [6]. For example, in 2013 the share of offshore wind deployment (i.e., installed MW's in 2013) in Europe was about 85% onshore and 15% offshore. The European countries with highest offshore wind installed capacities are currently the United Kingdom, Denmark, Belgium, and Germany [7]. Solar PV has faced a major rise in the past years, especially in Europe. CSP requires direct sunlight, and hence the potential is limited to certain countries/locations. Biomass currently is widely used worldwide, but mostly for heating, with only a very limited share for electricity generation. The further potential is uncertain, given the possibly limited availability, concerns related to sustainable production, and security of supply and conflicts with food production. Wave/tidal and enhanced geothermal are still in an immature state, and the actual potential and costs remain uncertain.

Figure 2.1 presents an overview of the installed generation capacity by fuel type in Europe, for the years 2000 and 2013. Figure 2.2 presents indicative ranges for the levelized cost of electricity (LCOE) per technology. For the dispatchable technologies, a base-load operational mode is considered (85%). When the load factor drops, LCOE rises significantly above the levels presented.

2.2.2 Demand

Both the aggregated and the instantaneous demand for electricity are important for the optimal design of the supply side and the network.

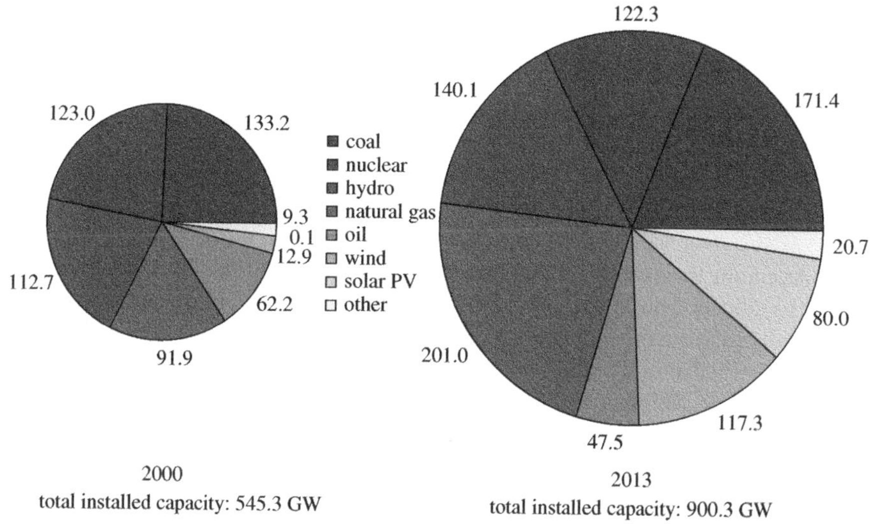

Figure 2.1 Installed electricity generation capacity in [GW] in the EU, for the year 2000 and the year 2013. (*Data source:* [6].)

The aggregated demand for electricity is often described in long-term scenarios. It is determined by several factors, being population, wealth, the energy intensity, and the degree of electrification. It can be written as follows:

$$\text{Electricity demand} = \text{population} \cdot \text{welfare} \cdot \text{energy intensity} \cdot \text{degree of electrification} \tag{2.1}$$

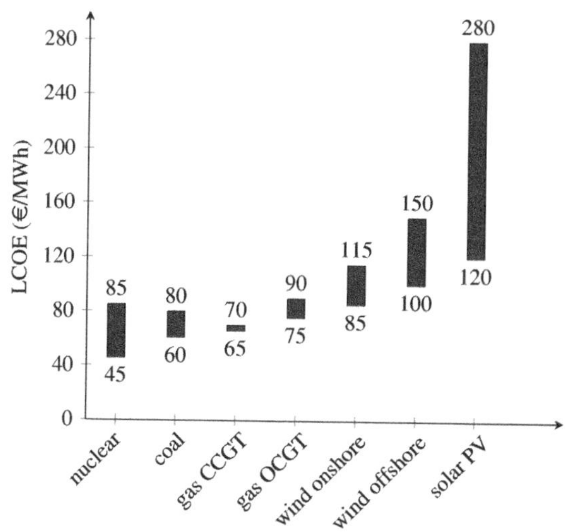

Figure 2.2 Indicative range for levelized cost of electricity (LCOE) for different technologies. Values are mainly based on [3] for European countries, and they are complemented with data from [5].

These parameters are expressed in the following units:

- Electricity demand [MWhe]
- Population [capita]
- Welfare [GDP/ capita]
- Energy intensity [MWhp/GDP]
- Electrification [MWhe /MWhp]

Assuming the trends in population and welfare as given, only the last two factors are left to influence the electricity demand by policy makers. The energy intensity is the inverse of the energy efficiency on the demand side. Increasing energy efficiency is often identified as one of the priorities in energy or climate action. Current electric devices (lighting, fridges, motors, etc.) leave room for improvement. However, some barriers are often identified as preventing to tap the full potential (e.g., high upfront investment costs).

The degree of electrification plays an important role because the electricity sector is a sector where decarbonization is feasible, both technically and institutionally. The options in the other energy sectors (i.e., transport and heating and cooling) are much more limited. Hence, within a decarbonization framework, an increasing share of electrification might be expected, due to a shift to electric vehicles in transport and to heat pumps in heating.

From a technical perspective, meeting the instantaneous demand for electricity is crucial when designing the electric power system. Both demand fluctuations and peak demand levels must be instantaneously balanced. This matching of supply and demand is discussed in the following section.

2.2.3 Matching Supply and Demand

The increasing share of intermittent renewables requires a power system with sufficient flexibility—that is, to deal both with forecasted variations in output and with sudden unexpected fluctuations. The participation of the demand side in the energy market and the development of smarter grids on the other hand is causing the load to become more controllable. The energy system must manage the fluctuations ranging from quickly varying phenomena (number of cycles) up to slowly varying phenomena (cold spell of some weeks or even seasonal effects). This required flexibility can be provided by different means. These are not to be seen as distinct options, as future systems will have to rely on a combination of them. Clearly, these options are also not all equivalent:

- A first option is concentrated on the supply side and consists of flexible power plants (so-called back-up capacity), which would be typically combined cycle or open cycle gas turbines, although existing coal-fired plants and even nuclear-power plants may have to participate in the load balancing. These conventional power plants can be actively dispatched, to deal with the more variable and unpredictable residual demand profile. Power plants can ramp their output while staying online, or could start/stop more frequently. This higher "cycling" behavior typically comes at a cost: higher aggregated start costs, ramp costs, higher maintenance costs, possibly higher failure rates and outages, potential decrease of rated efficiency, shorter lifetime of power plant components due to for instance higher thermal stresses, and so on.

- A second option, also on the supply side, is curtailment of nondispatchable renewables (effectively reducing the output of RES, thus regulating in a downward direction). Through curtailment, zero marginal cost energy would be "wasted." This might be required to maintain system security, or could lead to a lower cost on overall system's level as this might prevent large power plants from shutting down. Currently, priority access for RES or specific subsidy mechanisms might hamper RES curtailment from taking place. Furthermore, RES need to be "controllable" to allow for actual curtailment. This is possible by turning wind turbines (partly) out of the wind or by blade pitch control. For solar PV deployed at household level, additional control devices would typically need being implemented to actually make curtailment possible.
- Recent evolutions (both technical and in terms of communication) have also activated the demand side. This constitutes the third option for flexibility. Dynamic tariff schemes, especially with large industrial consumers, may lead to more possibilities for controlled load interruptions, and the rolling out of smart grids could stimulate further demand response and demand shifting also at the household level. Smart appliances could be developed, consuming electricity when most optimal. Washing machines and dryers could for instance provide some flexibility of several hours. In the future, the roll-out of (plug-in hybrid) electric vehicles might also provide flexibility. As a car typically needs to charge for several hours, flexibility becomes available when a certain time is specified by the owner when the car needs to be charged (rather than specifying when the car needs to charge). Similarly, when heat pumps are deployed, these would also provide some flexibility, as the thermal inertia is typically higher, making it relatively easy to shift electricity for heating by up to a few hours.
- Storage is the fourth option. Flexibility is provided by a temporal relaxation of the supply–demand balance (to some extent similar to demand shifting), that is, arbitraging in the time dimension. Depending on the storage-device characteristics, different kinds of fluctuations can be absorbed—for example, from milliseconds up to days (each with a certain efficiency). The storage option currently most widely deployed (though still limited compared to the overall electricity generation system size) is pumped hydro storage. Some Compressed Air Energy Storage (CAES) facilities also exist. These types of storage are able to store electricity at quite a high rate, and store it also for relatively long time periods. An opposite example is, for instance, a flywheel, storing energy in the form of kinetic energy and basically used to dampen, absorb and smooth out short fluctuations (very short timescales). Batteries (in a wide set of different forms) might play an increasingly important role in the future. Batteries of electric vehicles connected to the grid might possibly serve as local storage devices.

- The fifth option for flexibility is the expansion of the grid. By wide geographic aggregation, intermittency of renewables can be smoothened out to some extent, that is, arbitraging in the spatial dimension. This relates to both the variability and the unpredictability (forecast errors are smoothed out when larger areas are being considered). Towards this aim, networks need to be set up in the horizontal dimension. Also in the vertical dimension (from high to low voltage grids), networks need to be reinforced and expanded. This overall grid expansion should be seen as a requirement and/or facilitator for the other options.

As indicated above, future systems need to deploy a diversified mix of these flexibility options, as they often are dependent on one another. Each of these options comes at a certain cost, and all have a specific efficiency. Some need to be implemented at local (low voltage) or system level (high voltage) and, as such, require corresponding investments. Finally, all need to be connected adequately to the electric network, requiring both significant horizontal and vertical expansion of the network.

2.2.4 European Energy Policy

Energy policy strongly affects the above-mentioned supply side, demand side, and the matching of both. The decisions taken by policy makers have a direct and indirect impact on the developments of the electric power system. For most of the European countries, energy policy is a competence on the level of the European Commission. The commission has built up its policy on three pillars, namely, competitiveness, security of supply, and reduced climate and environmental impact.

With respect to renewables integration that the European Commission has set forward a target for 2020, and post-2020 targets are under discussion. The European objective is to achieve a 20% share of renewables in the final energy consumption by 2020. This overall European target is split up into different member-state-specific targets. In this regard, European member states have developed National Renewable Energy Action Plans (NREAPs), outlining the envisaged plan and policies to realize these national targets. This approach has led to a diversity of national policies rolling out various supporting schemes. In addition, member states are also free to further develop their own overall energy policy, within the European framework. A clear example is the ongoing German "Energiewende," where Germany is aiming for 60% renewable energy by 2050, which translates into at least 80% of RES for the electricity sector (with intermediate targets of 35% in 2020, and 50% for 2030, both regarding electricity).

Especially market pull instruments have been implemented providing additional revenue streams to renewable energy projects. The two most well-known market pull instruments to promote renewable energy sources are the Feed-in-tariffs and the Tradable Green Certificate mechanisms. The former oblige electric utilities, typically the regulated system operators, to buy all generated electricity at a fixed price. Ideally, this fixed price is able to support technologies at varying stages of maturity on different terms by closing the "financial revenue cap" relative to conventional sources of energy. The remuneration decreases over time to take the maturity of the technology into account. An alternative to tariffs are the Tradable Green Certificates for which a market is set up. Demand for these certificates is driven by a predetermined target or quota for renewables in the final energy delivery. The supply of these certificates, guaranteeing that renewable energy is produced, and the demand for these certificates finally determine the certificate price, providing the additional support to the renewable energy technologies. In addition, renewable energy is sold at the same price in the market as the energy produced through conventional technologies.

Clearly, when renewables are introduced into the system (in part through subsidy mechanisms), this also impacts the other power plants. Renewables generate electricity at zero marginal cost and as such lower the wholesale price for electricity. This puts pressure on the conventional plants. As a result, in several European Member States, power plants are mothballed (when no longer profitable). Discussions on capacity remuneration mechanisms are currently ongoing. Also regarding new investment, power plants need a sufficient number of operating hours and high enough electricity prices, to represent a cost-effective investment.

With respect to carbon mitigation, the EU has set up the European Union Emission Trading Scheme (EU ETS), initially as an instrument to meet its obligation under the Kyoto protocol. As of 2005, this cap and trade system entered into force. The EU ETS is currently Europe's main instrument in the fight against climate change. According to economic theory, a cap-and-trade system with a cap progressively decreasing would set an appropriate price on CO_2 and thereby would trigger the optimal amount of investment in and deployment of low-carbon technologies. This would yield the optimal (lowest cost) low-carbon generation mix. A consistent price for CO_2 is an important and crucial requirement in the path of decarbonization.

To provide some insight in the different drivers of CO_2 emissions from electricity generation, an analytical formula is presented. This is similar to the formula for electricity demand (2.1), now with one factor added, introducing CO_2 emissions:

$$\begin{aligned} CO_2 \text{ emissions} = {} & \text{population} \cdot \text{welfare} \cdot \text{energy intensity} \\ & \cdot \text{degree of electrification} \cdot CO_2 \text{ intensity} \end{aligned} \qquad (2.2)$$

These parameters are expressed in the following units:

- CO_2 emissions [kg CO_2]
- Population [capita]
- Welfare [GDP/ capita]
- Energy intensity [MWhp/GDP]
- Electrification [MWhe /MWhp]
- CO_2 emission intensity of electricity [kg CO_2/MWhe]

The presented formula allows making a clear distinction between the demand for electricity and the corresponding supply. The demand for electricity is determined by the product of the first four parameters and is expressed in [MWhe], being the demanded amount of electric energy (after conversion). The supply side is reflected by the last parameter. This factor determines the CO_2 intensity, or the amount of CO_2 emitted per [MWhe] of electricity generated.

When aiming for nearly full decarbonization, the left-hand side of this equation (i.e., the emissions of the power sector) should be zero. This implies that one of the factors on the right-hand side should be zero as well. The only possible option is to put the CO_2 emission intensity equal to zero, or to have an electricity system that exists solely of carbon-free technologies. The size of the electricity system is determined by the demand (product of first four factors) and can be affected by improving the energy efficiency. However, independent of this demand, the supply will have to be zero carbon and can therefore be "decoupled" to some extent, that is, looked at separately. Similar alternatives for this formula can be found in [8].

2.3 SCENARIOS FOR EUROPE'S ENERGY PROVISION

2.3.1 The Role of Defining Scenarios

Scenarios can generally be defined as *"An internally consistent picture of a sequence of events or situation, based on certain assumptions and factors chosen by its creator."* Scenarios are not forecasts. They rather provide a projected evolution of the system, under a specific set of predefined assumptions and boundary conditions. The first important element is the creator and his intentions. When scenarios are being defined, the creator of those scenarios always has an objective in mind. This also demonstrates why the scenarios of one report or project may not be applied literally in other reports or projects. The following examples of long-term energy analyses illustrate this perfectly. The Power Choices report by Eurelectric, the Union of the European electricity industry, wanted to define an "optimal portfolio of power generation" and the technological developments needed to reach the Greenhouse Gas (GHG) reduction target [9]. The EC's Energy Roadmap 2050 has as objective to identify "no regret" options in the European energy system through a scenario analysis of an "illustrative nature," focusing on the common elements of ways to modernize the energy system [10]. The aim is to assist policy making, analyze possible pathways, and show impacts (economic, social, and environmental). The Battle of the Grids report assesses the need for extra grid reinforcements and production capacity expansions to ensure equal security of energy supply [11].

Maybe even more important than the goal of the creator are the assumptions and factors put forward. This means that a set of parameters and variables must be defined. The parameters are exogenous, predefined by the creator, whereas the variables are the result of the long-term analysis.

When defining energy scenarios, the following elements are typically included (either as constraint or as explicit output of the model): GHG emission reduction level, CO_2 emission price, share of renewable energy sources, shares of conventional sources of electricity generation, and consideration of the electric power sector only versus the inclusion of the heating/cooling and transportation as well. In addition, some assumptions must be made with respect to (a) the relevant region and time horizon, and (b) commercial availability of new technologies like advanced nuclear or CCS.

Over the past 5 years, a wide set of scenario analysis for Europe's energy (or electricity) future, typically towards 2050, has been published [9–15]. Most scenarios consider strong decarbonization pathways, in line with Europe's ambition. The aim of this section is to illustrate the use of scenarios and to report on some general trends (rather than going into comparing different studies or providing a full overview). Focus in the end here is on the impact of different evolutions on the high-voltage grid. In this regard, the scenarios developed in the EC Energy Roadmap 2050 will be mainly used as an example [10]. This roadmap was released by the EC at the end of 2011. The aim of this roadmap is to explore the options and challenges towards the objective of decarbonization (based on current EU objectives of sustainability, security of supply, and competitiveness). Taking the current policies and 20–20–20 targets as given, the roadmap explores 7 scenarios (2 reference scenarios and 5 decarbonization pathways). The entire energy provision is considered, meaning electricity supply, heating and cooling, and transport. The partial equilibrium model Primes is used for the analysis. In the discussion provided below, focus is mainly on the electric power sector.

2.3.2 Supply Side

On the generation side, different technologies are typically considered: nuclear, fossil fuels (possibly with CCS), and renewables. Dependent on the scenario boundary conditions, scenarios might exclude nuclear, make assumption on CCS availability, consider certain fixed RES trajectories, and so on. Clearly, such assumptions greatly affect and determine the scenario simulation outcome.

In the EC Energy Roadmap 2050, seven scenarios are developed [10]. There is one original reference scenario, based on business as usual. Next to this reference, a scenario with current policy initiatives is considered (which can be seen as an updated reference scenario). Both of these scenarios reflect evolutions as they might take place when no additional specific climate or energy policies would be implemented on the EU level. Note, however, that for instance the EU ETS in its current setting would remain active (up to 2050 and longer).

Besides these two reference cases, five decarbonization scenarios are considered, all facing a domestic 85% CO_2 emission reduction (energy-related). The key assumptions of the different scenarios are summarized in Table 2.2.

The shares of the different low carbon technology options (i.e., nuclear, renewables, and CCS) in total electricity generation in the different scenarios, is presented for the years 2030 and 2050, in Figure 2.3. In 2010, the share of nuclear in electricity generation amounted to 27.5%, while that of RES was about 21%.

The share of RES has grown significantly in Europe over the past 5 years. For 2020, a binding target of a 20% share in the gross final energy consumption is put forward, which translates into a little more than a 35% share in the electric power sector [16]. The future development and deployment of nuclear face uncertainty due to risk perception and lack of implementations of solutions for waste. CCS, on the

TABLE 2.2 Overview of Scenarios Put Forward in the EC's Energy Roadmap 2050 [10]

Pathway	Scenario	Key assumptions
Reference	Business as usual (BAU)	Current trends and projections. No GHG or RES policies/targets after 2020
Updated Reference	Current policy initiatives (CPI)	Updated to policies and technology assumptions (e.g., nuclear after Fukushima incident and latest CCS)
Decarbonization	High energy efficiency (HEE)	Very high primary energy savings (stringent requirements on appliances and buildings, roll-out of smart grids, etc.)
Decarbonization	Diversified supply technologies (DST)	Technologies compete on market basis, carbon price is sole driver
Decarbonization	High RES (HRES)	High ambition for RES deployment by 2050 (60%)
Decarbonization	Delayed CCS (DCCS)	Technologies compete on a market basis, but deployment of CCS faces difficulties
Decarbonization	Low nuclear (LNUC)	Technologies compete on a market basis, but deployment of nuclear is low due to a lack of public acceptance

other hand, might be facing difficulties in financing and public acceptance and still requires a lot of R&D efforts.

Dependent on the scenario assumptions, the share of the different technologies evolves over time. Some general trends can, however, be observed. An increase of the share of RES can be seen in all scenarios. Even in the CPI scenario (i.e., without specific policy measures beyond 2020), the share of RES in electricity generation will reach about 50%. Given the limited load factors of two important RES contributors, namely wind and solar PV, the total installed system capacity reaches high values in the different scenarios (see further).

In the reference scenarios (BAU and CPI), CO_2 emissions from the power sector decrease by about 67% from 2010 to 2050 (overall energy related CO_2 emissions decrease by about 40% over this period). The EU ETS remains in place (as currently set up), and the CO_2 price reaches a value of about 50 €/ton in 2050. After 2020, no specific targets or support for RES are foreseen. Import dependency increases for gas. In these scenarios, Europe's efforts in recent years do not achieve the long-term targets of decarbonization.

All decarbonization scenarios feature a CO_2 emission reduction of 85%. Accordingly, these all face a higher CO_2 price in 2050 compared to the reference cases, ranging between 234 and 310 € /ton (€ 2008). The share of RES (in terms of electricity generation) is higher and is situated around 60%, except for the "high RES" scenario, where it amounts to about 85%. To achieve such relatively high shares in terms of electricity generation, high installed capacities are required for those RES that have intermittent profiles, namely wind and solar PV. Table 2.3 presents these

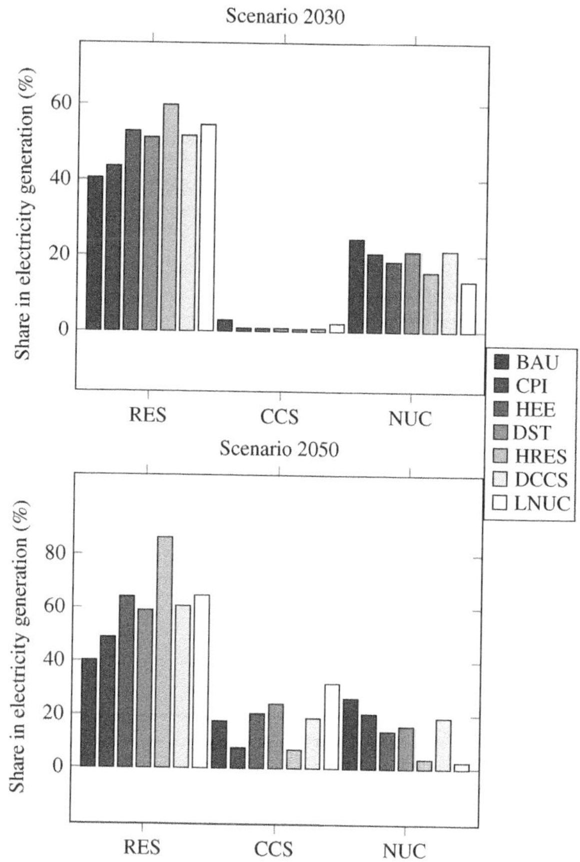

Figure 2.3 Share of different technologies (RES, fossil with CCS, and nuclear (NUC)) in the overall electricity generation for 2030 and 2050, for different scenarios (abbreviated as defined in Table 2.2).

values for the different scenarios, as well as for some of the most well-known 2050 studies.

2.3.3 Demand Side

In the two reference scenarios, final energy consumption peaks in 2020 and then stabilizes. An economic growth of 1.7% per annum is assumed for the period 2010–2050, which is counterbalanced by energy efficiency improvements. Despite this trend in final energy consumption, the demand for electricity is projected to increase quite substantially towards 2050, due to an increased level of electrification. Electricity demand (the measure of "electricity generation" is actually taken here—also considering, for example, losses) is projected to increase to 4932 TWh in 2050 for the reference scenario, and to 4621 TWh in 2050 for the CPI scenario, coming from a historical level of 3274 TWh in 2005.

TABLE 2.3 Installed Capacity [GW] of Wind, PV, and Overall System, in Different Scenarios of the EC Energy Roadmap and in Some Illustrative Scenarios of Other Studies[a]

			2050						
			EC Roadmap						
GW	2010	2013	BAU	CPI	HEE	DST	HRES	DCCS	LNUC
Wind	82	117	382	432	548	595	984	609	674
Solar PV	23	80	171	224	330	351	603	348	381
Total system	853	900	1454	1502	1473	1621	2219	1639	1721

(a) EC Energy Roadmap

	2050					
		ECF			Greenpeace & EREC	
GW	PC	40% RES	60% RES	80% RES	ER	AER
Wind	382	165	295	435	448	483
Solar PV	125	195	555	815	348	510
Total system	1318	1280	1700	2020	1252	1537

(b) Other Studies

[a]The scenarios from the EC Roadmap are indicated with the abbreviations as defined in Table 2.2. PC stands for Power Choices, ECF stands for European Climate Foundation, ER stands for the Energy Revolution scenario, while AER stands for the Advanced Energy Revolution [6, 9, 10, 12, 15, 16].

In the decarbonization scenarios, energy efficiency improvements are quite high, reducing the primary energy demand in 2050 between 32% and 41% (depending on the scenario) compared to the 2005 level, while GDP more than doubles over this same period. This increased energy efficiency is due to technological changes on both the supply and demand side. The final energy demand is lower accordingly (compared to the reference scenarios).

Considering the demand for electricity, this again increases. The electricity share in the final energy consumption increases significantly, as electricity is an energy carrier that can be made carbon-free in a relatively efficient way. It increased from about 20% in 2005 to 36–39% in 2050. This indicates the transition towards an electrified society, with a plug-in hybrid and electric vehicle roll out, and wide heat pump deployment. Electricity generation in these scenarios amounts to 4281–5141 TWh in 2050 (compared to 3274 TWh in 2005).

2.3.4 Implications Towards the Grid

The sections above have outlined projected implications of future energy scenarios on both the supply and demand side. The electric network will need to accommodate these scenarios, ensuring that electricity can be transported from where it is generated to where it is consumed. Increasing the ability to transport power from generation to load centers requires actions on different fronts.

2.3.4.1 *Optimal Use of the Existing Grid Capacity* The best short-term option is to make efficient use of the current capacity of the grid. By applying smart technologies, the existing interconnections could be used closer to the technical limitations. The installation of power electronic technologies such as HVDC (high-voltage direct current) transmission and flexible alternating current transmission system (FACTS) devices will help improving the degree of flexibility and self-healing nature of the grid. This means that those technologies support the increasing usable capacity without harming secure operation of the grid. Such flexible system operation will be required to balance short-term variation in the power generation and consumption profiles in different grid nodes. Instead of having longer-term import or export situations, grid flows may change rapidly. This will occur not only intraday, but even within periods of less than one hour. In order to support the efficient operation of the existing grid, the development of new monitoring and control systems will also be required.

Making optimal use of the existing grid capacity should not be considered from a transmission grid perspective only. The transmission grid will increasingly be interacting with the distribution grid. Whereas this low- and medium-voltage grid was initially a rather passive asset, it will increasingly be operated as an active component facilitating bidirectional flows affecting the high-voltage grid. In addition, not only within the power sector but also the coupling with gas and heating networks would create an opportunity for an overall more efficient grid use.

2.3.4.2 *Grid Capacity Investments* Even when taking the appropriate measures for the optimal use of the existing grid capacity, investments will be required in additional grid capacity. Although this is suggested as a necessary step in the long-term, actions should already be taken today considering the long lead time between the investment decision and the moment of taking into service of new grid infrastructures.

From a technology perspective, required grid updates give us the opportunity to evaluate both AC and DC grid solutions. This decision depends on the regional grid characteristics, but also on the long-term vision on grid development. Are we expecting continuous commissioning of additional interconnections, or is constructing a new overlay grid the better solution to deal with future energy scenarios? Also the development of an offshore grid connecting offshore wind farms should be discussed within this context.

When looking back at alternative scenarios, it is clear that the evolutions regarding the European electric power system will require substantial grid investments. After all, the amount of RES is projected to increase substantially in all scenarios, requiring grid integration and additional flexibility in the power system, which in turn will further require stronger and expanded grids.

In the EC Roadmap, costs are presented for grid investments between 2010 and 2050 (Table 2.4).

TABLE 2.4 Projected Grid Investment Costs for the Different Scenarios of the EC Energy Roadmap 2050 [10]

	10^9 € 2005
BAU	1269
CPI	1357
HEE	1518
DST	1712
HRES	2195
DCCS	1717
LNUC	1793

Entso-E, the European Network of Transmission System Operators for Electricity, provides an outlook regarding the European grid for the next decade in its 10-year network development plan (TYNDP). Regarding RES, "*the TYNDP 2012 identifies the need to invest €104 billion in the refurbishment or construction of roughly 52,300 km of extra-high-voltage power lines clustered into 100 investment projects across Europe. Eighty percent of the identified 100 bottlenecks are related to the direct or indirect integration of renewable energy sources (RES) such as wind and solar power. Such massive development of RES is the main driver behind larger, more volatile power flows, over longer distances across Europe*"[17].

2.3.5 International Cooperation and Market Perspective

Technical decisions only will not suffice if the grid would be considered as the facilitator of the integration of variable RES and decentralized energy systems. Innovative technical solutions also support an efficient market framework. But the reverse is also true. The market framework helps to valorize the optimal use of the existing grid. Smart congestion management and capacity allocation schemes, but also electricity market clearing algorithms considering the technical constraints, contribute to efficient grid operation.

Efficient market frameworks should be implemented on an international level in order to prevent from having a fragmented energy market. International cooperation will strongly contribute to the creation of an efficient electricity market providing a level playing field where regulatory harmonization will trigger the right investments.

2.4 CONCLUSIONS

Currently, a major transition is ongoing in the power industry in Europe. The integration of renewable power generation technologies, particularly wind and solar energy, increases the variability of power output at the system supply side. Simultaneously new types of electricity consumption are being introduced in the system. When using electrical heating through heat pumps and electrical mobility, energy consumption is shifted towards the electricity sector. This transition is referred to as the electrification of energy needs.

Today, the power grid is playing a major role in transporting power from electricity generation centers to load centers. This need to balance increasingly variable supply with new sources of demand will be the main challenge of the power grid of the future. This future grid requires smart and flexible technologies to make efficient use of the existing infrastructure. However, towards the future, a major investment wave is forthcoming in the development of supergrids connecting, for example, offshore wind farms. This does not happen only at the transmission level; also at the distribution level, new sources of electricity generation are being connected requiring active bidirectional distribution grids.

Although this chapter focuses on Europe, we are also witnessing significant changes in the power sector in other parts of the world. In the United States, the developments in shale gas exploration present a real game changer, affecting fossil fuel prices worldwide. Gas and coal prices affect power markets, as well as the corresponding generation investments. Alternatively, India and China are facing a massive growth in electricity demand. Even though renewables are present in the mix of new investments, coal-fired power plants still dominate the investment scene.

It is clear that the particularities of the power system developments driven at the supply and demand side are requiring a more active role of the power grid. Depending on those requirements, the power system needs to grow or adapt drastically.

REFERENCES

1. Nuclear Energy Agency (NEA), "Technical and economic aspects of load following with nuclear power plants," NEA-OECD, Paris, Technical Report, June 2011 [Online]. Available: http://www.oecd-nea.org/ndd/reports/2011/load-following-npp.pdf
2. World Nuclear Forum (March 2013). Plans for new reactors worldwide. [Online]. Available: http://www.world-nuclear.org/info/current-and-future-generation/plans-for-new-reactors-worldwide/
3. International Energy Agency/Nuclear Energy Agency, "Projected costs of generating electricity," IEA/NEA, Technical Report, 2010.
4. F. P. Sioshansi, "De-carbonizing electricity generation: It won't be easy, cheap, nor enough," *Utilities Policy*, vol. 17, no. 3, pp. 217–224, 2009.
5. A. Schröder, F. Kunz, J. Meiss, R. Mendelevitch, and C. von Hirschhausen, "Current and prospective costs of electricity generation until 2050," Data Documentation, DIW, Technical Report, 2013.
6. European Wind Energy Association, "Wind in power, European statistics," EWEA, Technical Report, Feb. 2014.
7. ———, "The European offshore wind industry—key trends and statistics 2013," EWEA, Technical Report, Jan. 2014.
8. C. Jones and J.-M. Glachant, "Toward a zero-carbon energy policy in Europe: Defining a viable solution," *The Electricity Journal*, vol. 23, no. 3, pp. 15–25, 2010.
9. Eurelectric, "Power choices 2050," Eurelectric, Technical Report, 2010.
10. European Commission, "COM(2011) 112 final, A roadmap for moving to a competitive low carbon economy in 2050," EC, Technical Report, 2011.
11. Greenpeace, "Battle of the grids—how europe can go 100% renewable and phase out dirty energy," Greenpeace, Technical Report, 2011.
12. European Climate Foundation, "Decarbonization roadmaps for the eu-27," ECF, Technical Report, 2010.

13. European Gas Advocacy Forum, "Making the green journey work—optimised pathways to reach 2050 abatement targets with lower costs and improved feasibility," EGAF, Technical Report, 2011.
14. European Renewable Energy Council, "RE-thinking 2050—A 100% renewable energy vision for the european union," EREC, Technical Report, 2010.
15. Greenpeace and EREC, "Energy [r]evolution—a sustainable world energy outlook," Greenpeace and EREC, Technical Report 3rd edition, 2010, ISBN: 978-90-73361-90-4.
16. European Commission, "EU energy, transport and GHG emissions trends to 2050," EC, Technical Report, 2013.
17. European Network of Transmission System Operators for Electricity, "10-year network development plan 2012," ENTSO-E, Technical Report, 2012.

PART **II**

HVDC TECHNOLOGY AND TECHNOLOGY FOR OFFSHORE GRIDS

PART II INTRODUCES the technology that can be used in the future transmission system. The technology overview starts with a short overview of HVDC technology with special emphasis on the technology most relevant for HVDC grids: voltage source converter (VSC) technology. The different relevant components are introduced, including converter stations, filters, transformers, cables, capacitors, choppers, switchgear, cables, and lines.

HVAC and HVDC technologies are compared based on technical and economic criteria. Relevant comparative graphs are used to illustrate that above a critical break-even distance, HVDC tends to be more economical than HVAC.

The technology of wind turbines and wind power plants is also discussed. Offshore wind power is one of the main drivers for offshore grids and therefore it is relevant to explore the main characteristics of offshore wind turbines, which are significantly larger than their onshore counterparts. Furthermore, offshore wind power plants have some distinct technical characteristics, especially when they are connected to HVDC systems (and operate nonsynchronous with the main AC grid).

HVDC Grids: For Offshore and Supergrid of the Future, First Edition. By Dirk Van Hertem, Oriol Gomis-Bellmunt, and Jun Liang.

CHAPTER 3

HVDC TECHNOLOGY OVERVIEW

Gen Li,[1] Chuanyue Li,[1] and Dirk Van Hertem[2]
[1]Institute of Energy, School of Engineering, Cardiff University, Cardiff, Wales
[2]Research Group Electa, Department of Electrical Engineering, University of Leuven, Leuven, Belgium

3.1 INTRODUCTION

High-voltage direct current (HVDC) has been used in power transmission systems for over 50 years; its main applications are the interconnection of nonsynchronous networks, long-distance transport of electrical energy, and submarine and underground cable transmission. The most commonly used systems are based upon current source converter (CSC) or line commutated converter (LCC) HVDC, using thyristor-type valves. Voltage source converter (VSC)-HVDC technology, which uses faster power electronic switches, is a recent development which is seen as a game changer and as the key enabling technology for future (DC) grids [1, 2]. This chapter discusses the overview of different HVDC technologies. The main emphasis is on VSC HVDC.

This chapter provides an introduction for the technology of both converter technologies and the system aspects, as well as the transmission line technology (especially cables). The discussion in this chapter is focused on the technology of point-to-point connections. For more elaborate discussions on the technological aspects of HVDC systems, the reader is referred to the references, specifically references [3]–[7].

3.2 LCC-HVDC SYSTEMS

Line commutated converter (LCC) HVDC or classic HVDC is built using thyristors in a current source converter (CSC) topology. Thyristors can be switched on, but need the current to pass through zero in order to switch off. The angle at which the thyristors are switched in is called the firing angle, α. Figure 3.1 depicts the basic circuit of a HVDC connection: an AC transformer connected to a six-pulse thyristor bridge. The typical waveform of such a device is shown in Figure 3.2.

HVDC Grids: For Offshore and Supergrid of the Future, First Edition. By Dirk Van Hertem, Oriol Gomis-Bellmunt, and Jun Liang.

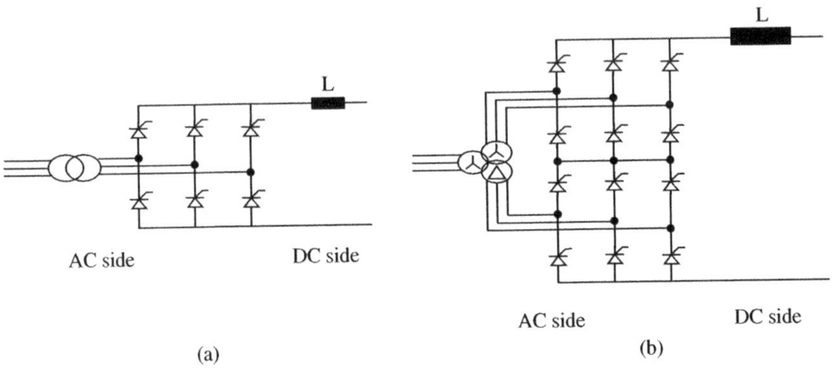

Figure 3.1 Converter topology for the line commutated converter. (a) Six-pulse LCC converter. (b) Twelve-pulse LCC converter.

Figure 3.2a shows the waveform for a system without commutation overlap, while in Figure 3.2b the effect of commutation is shown. The commutation process is caused by the system inductance and causes a short overlap period during the transfer of the current between two branches of the converter where both branches conduct. This commutation period is expressed as the commutation angle, μ. In order to reduce harmonics in the AC system, additional harmonic filters are needed. A reduction of the harmonics can also be obtained by utilizing converters with a higher order of pulses: 12-, 18-, or even 24-pulse converters. Most classic HVDC configurations use 12-pulse converters (Figure 3.1b), which cancels out the 5th, 7th, 17th, 19th,…, $(6 \pm 1 + k \cdot 12)$ harmonics.

The transformer of a 12-pulse bridge has a Y–Y–Δ three-winding configuration, or a combination of Y–Y and Y–Δ transformers. Both single-phase and three-phase units are used. The leakage reactance of the transformers is typically 10–18% in order to limit the current during a short-circuit fault of the bridge arm [7].

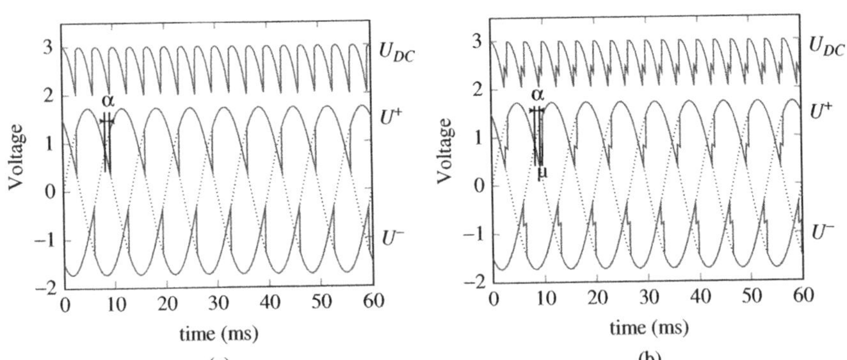

Figure 3.2 Basic waveforms for the six-pulse thyristor converter. (a) Six-pulse converter waveform, neglecting commutation ($\alpha = 27°$, $\mu = 0°$). (b) Six-pulse converter waveform, with commutation ($\alpha = 27°$, $\mu = 9°$).

In order to reach the required high voltages, several thyristors are placed in series. A single thyristor can block up to 8.5 kV and can carry currents up to 5 kA. There is normally redundancy built in so that a single thyristor failure does not lead to a failure of the entire HVDC link.

The current through the classic HVDC system is always in the same direction. Reversing the power flow is done by changing the polarity of the voltage.

3.2.1 Configurations

Several configurations are possible: monopole, bipole, both with or without ground return, and combinations of these (Figure 3.3).

A monopolar link has one conductor at a high DC voltage. The other conductor can use either the ground or sea as the return path or a metallic return wire. The latter can be used when harmonic interference and/or corrosion are an issue. Since the corona effect is more important with positive than with negative voltages, the monopolar link is normally operated with negative polarity in case of overhead lines.

Although systems with a single conductor and earth or sea return are a technically easy and economically beneficial solution, they cannot always be used because of the possible effects of corrosion on metallic pipes in the ground and the possible

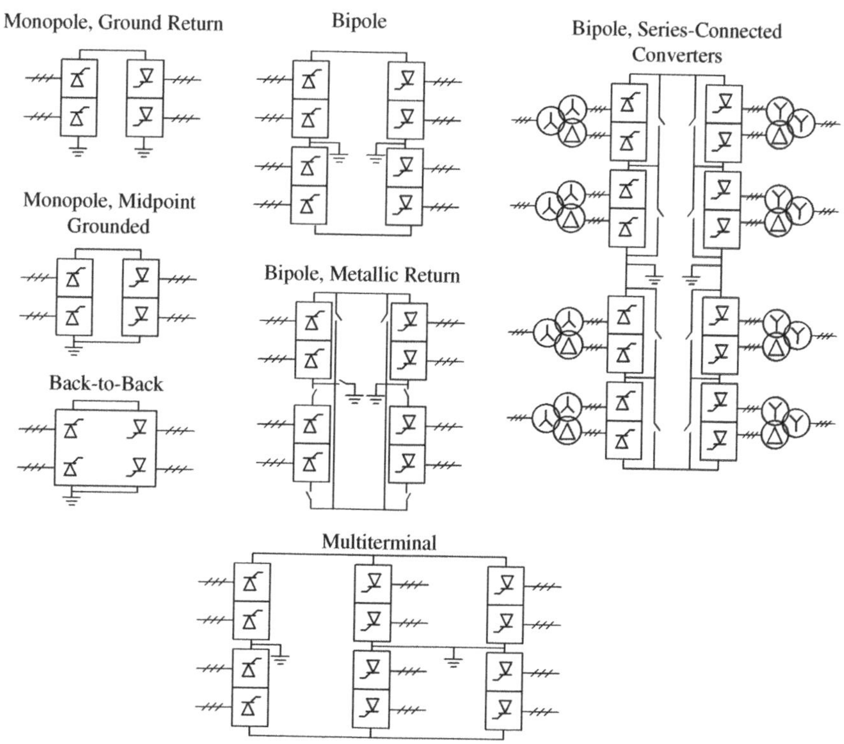

Figure 3.3 Different configurations for HVDC [8].

negative environmental effects on animals, especially fish [9] and boats [10]. However, several of these monopolar configurations are currently in operation, amongst others the recent Baltic cable (connecting the German and Swedish power systems) [11] and the GRITA connection between Italy and Greece. Systems with a metallic return wire have the advantage that the metallic return only has to be insulated for a very low voltage ($R \cdot I$). This significantly reduces the need for insulation material around that wire, and hence the cable cost. The Basslink, the connection between Australia and Tasmania which is in use since 2005, is also of the monopolar type with metallic return [12]. Also the connection between Sweden and Poland (SwePol) is of this type.

The bipolar link is the most commonly used topology for modern LCC HVDC connections of high power ratings. Each terminal has two converters of equal rating. Between the converters, the midpoint can be grounded on one or both sides. In case both sides are grounded, there is virtually no current flowing through the ground during normal operation, since the two converters are operated symmetrically. Bipolar systems require lower voltage ratings for equal power transfer compared to monopolar systems. The monopolar connection requires a pole of double voltage rating, while the bipolar system would require two cables of single voltage rating. A second advantage comes from the ability to use the 50% of the system during maintenance/fault of one module/cable of a bipolar link, by using the ground or a metallic return as return path. For this reason, bipolar systems can be equipped with a metallic return as well. An example of a bipolar connection, without metallic return, is the NorNed link between Norway and the Netherlands [13].

In a back-to-back connection the two converters are located in close proximity of each other. The typical application is the separation of two independent power systems. These systems are not synchronized or on a different nominal frequency. The DC voltage in the intermediate circuit can be selected freely at HVDC back-to-back stations because of the short conductor length. Because of the short distance, the line resistance is low, and the DC voltage can be kept relatively low, resulting in a smaller valve hall.

3.2.2 Reactive Power Properties of LCC HVDC

3.2.2.1 Basic LCC-HVDC Reactive Properties When neglecting the commutation, the required active and reactive power at the terminals of an LCC-HVDC station can be approximated as [14]

$$P = \frac{3\sqrt{2}}{\pi} \cdot U_{ac} \cdot I_{dc} \cdot \cos(\alpha) \tag{3.1}$$

$$Q = \frac{3\sqrt{2}}{\pi} \cdot U_{ac} \cdot I_{dc} \cdot \sin(\alpha) \tag{3.2}$$

where the firing angle, α, of the rectifier should be replaced with the extinction angle, γ, at the inverter side. When no reactive compensation is used, the reactive power absorbed at the terminals is 0.5 to 0.6 pu at rated active power. This has serious consequences for the control of the voltage at the end terminals, the higher currents (and losses) in the AC grid, and the voltage stability of the system. Therefore, it is often stated that LCC HVDC links can only be inserted when the short-circuit ratio (SCR) is high enough [15].

3.2.2.2 *Short-Circuit Ratio (SCR)* The SCR is defined as follows:

$$SCR = \frac{\text{Short-circuit MVA of AC system}}{\text{DC converter MW rating}} \tag{3.3}$$

in which the nominator of the fraction in (3.3) is given by (3.4):

$$\text{SC MVA} = \frac{U_{ac}^2}{Z_{th}} \tag{3.4}$$

where U_{ac} is the commutation bus voltage at rated DC power, and Z_{th} is the Thevenin equivalent impedance of the AC system. Kundur et al. [15] also refer to the effective SCR (ESCR), including the effects of the applications related to the DC link, such as transformers, shunt capacitors, filters, and so on.

$$SCR = \frac{\text{Short-circuit MVA of AC system} - \text{Reactive power compensation}}{\text{DC converter MW rating}} \tag{3.5}$$

The SCR and ESCR can be classified as follows:

High: $SCR \geq 5$ and $ESCR \geq 4.5$

Moderate: $3 < SCR < 5$ and $3 < ESCR < 5$

Low: $SCR \leq 3$ and $ESCR \leq 2.5$

Very low: $SCR \leq 2$ and $ESCR \leq 1.5$

The control of the HVDC link has an important impact on its interaction with the AC system. At low and very low (E)SCR, the connection of the LCC-HVDC link becomes difficult and requires specific tuning, or might require additional installations such as a STATCOM. However, specific studies have to be performed in order to assess the full impact.

The HVDC link itself does not contribute significantly to the short-circuit power of the AC power system.

3.2.2.3 *Reactive Power and AC System Strength* Since the AC system has to deliver a significant amount of reactive power, this will have consequences when the system itself is weak. The following problems can arise [15]:

- Dynamic overvoltages in the case of DC transfer interruption
- Voltage stability issues, especially at the inverter side
- Harmonic resonance, as the voltage distortion is higher
- Voltage flicker

3.2.2.4 *Reactive Power Compensation* The traditional solutions to cope with the high reactive power demands at the end terminals of an HVDC link are an adjusted control of the HVDC link, installation of switched capacitors, or dynamic

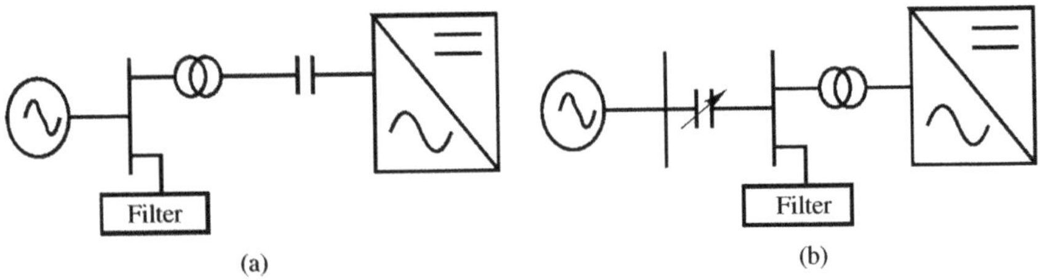

Figure 3.4 Principal CCC- and CSCC-HVDC topologies schematics. (a) CCC converter module. (b) CSCC converter.

reactive compensation such as SVCs (static VAr compensator), STATCOM (static compensator) or when possible the excitation control of a nearby generator.

3.2.2.5 Advanced Converters for LCC-HVDC Systems "New" thyristor-based HVDC systems use different converter topologies, available since the early 1990s. Depending on the manufacturer, several alternative configurations are possible.

The CCC, or capacitor commutated converter, is a configuration used by certain manufacturers, in which a series capacitor is inserted between the transformer and the converter (Figure 3.4a). Other manufacturers use a different configuration (Figure 3.4b), with similar results. This solution is called a controlled series capacitor converter (CSCC) and uses a thyristor-controlled series capacitor (TCSC) placed between the power system and the AC transformer.

Other topologies have been proposed [16]. These advanced configurations have the advantage that they can operate in weaker AC grids, and that they can also withstand voltage drops much better. Furthermore, the reactive power requirements are significantly lower.

LCC-HVDC converter reactive power requirements can be significantly limited using these alternative schemes (Figure 3.5). Also an example of switched capacitors

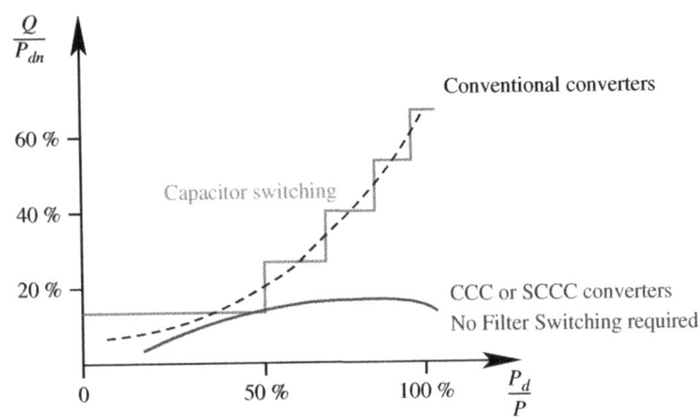

Figure 3.5 Reactive power demand with different converter topologies.

is shown, with the resulting reactive demand, the difference between the conventional converter characteristic, and the switched capacitor characteristic. With the advent of VSC converters, the application of these devices has been limited.

3.3 LCC-HVDC CONVERTER STATION TECHNOLOGY

The main components that comprise the HVDC system are the converter station, the transmission medium (cable) and the electrodes. A schematic overview of an LCC-HVDC station is given in Figure 3.6.

3.3.1 Converter Station

Both identical converter stations convert AC to DC and vice versa. The individual components of the converter station are described below.

The thyristor valves have to perform the following tasks:

- Connect AC and DC sides
- Conduct a high current (up to 5 kA)
- Block a high voltage (up to 8.5 kV)
- Control the DC voltage via the firing angle
- Operate fault tolerant and robust

Since the DC voltage is higher than the maximum blocking voltage of a single thyristor, several thyristors are placed in series. Similary, the current capabilities of a single thyristor can ask for the connection of several thyristors in parallel, which, however, is normally not needed. Normally the thyristors are installed in modules, which are placed in a stack, making maintenance straightforward. The thyristors can be electrically triggered or by using light.

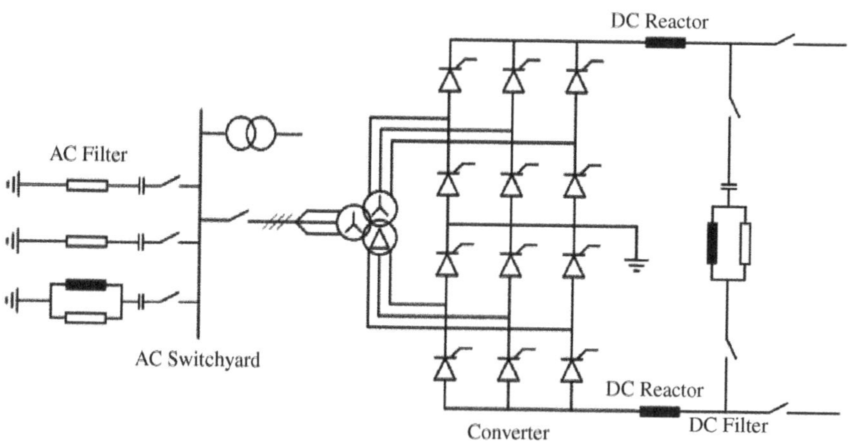

Figure 3.6 Schematic view of an LCC-HVDC converter station.

3.3.2 Transformers

Strictly speaking, transformers are not always needed in an HVDC installation. However, they are in practice always used for the following reasons:

- They allow optimizing the DC system voltage independent of the AC transmission system
- Tap-changers can be used to manage the voltage, which can reduce the losses
- Using rectifiers with a higher pulse number is possible
- They Limit the short-circuit current into the valve, working as a smoothing reactor

Depending on the application, different transformers and rectifiers can be used to obtain a techno-economically optimal solution. Because of its size, the choice for a transformer often depends on the available transport means, the location of the substation and its space requirements, or the construction site. The larger HVDC systems mostly make use of multiple single phase units. Although the principles for LCC-HVDC transformers are quite similar to those of traditional AC transformers, there are several practical aspects of HVDC which cause the transformers to be more complicated, such as oversaturation of the core through DC currents, higher test voltages, and specific arrangements to connect the windings [17].

3.3.3 Filters and Reactive Compensation

An HVDC link requires several filters and reactive compensation devices. As stated in Section 3.2.2, LCC HVDC requires a significant compensation for reactive power. If there are no nearby generators that can compensate for this reactive power, additional capacitors, a SVC, or STATCOM has to be installed. Furthermore, switching thyristors causes high harmonic distortions on the power system voltage, possibly influencing nearby installations. These harmonics have to be mitigated by additional filters. The DC filter is formed by the DC smoothing reactor which flattens the DC current.

3.3.4 Other Required Components

Next to the already discussed components of LCC-HVDC systems, the following parts are also important, but not treated into detail in this work [4, 6, 18]:

- Earthing electrodes
- Protection devices
- Surge arresters
- AC and DC switchyards
- Auxiliary equipment
- Cooling system
- Control system

3.4 VSC-HVDC SYSTEMS

Recent developments of power electronic devices have led to the rise of self-commutated voltage source converters (VSC) in power system applications. This is mainly attributed to the development of the insulated gate bipolar transistor (IGBT), a device which combines the controllability of the MOSFET with the reliability and power rating of the BJT. Together with the increased computational power of digital signal processors (DSPs), it was possible to use this technology in power transmission. The first VSC-HVDC application was built at the end of the 1990s [19, 20].

The full controllability and switching symmetry through both turn-on and turn-off operation allows the device to regulate power flow much more quickly than its predecessors. The utilization of solid-state devices virtually eradicates the risk of commutation failure. This technology also has the ability to absorb and generate both active and reactive power independently of one another. This eliminates the requirement for the expensive reactive power compensators used extensively within LCC-HVDC projects. Further advantages are that there is no requirement for regulation of the short-circuit level as commutation can operate without an AC system voltage source. Also, the generation of harmonics is greatly reduced, minimizing the volume of filters required to absorb them. An additional feature of the VSC is the capability to offer "blackstart" support. That is, restoring power without the aid of an external power source. This is advantageous in a wide area power outage as these power sources can be brought online first to begin the progressive resurrection of the network grid [21].

The VSC-HVDC transmission technique is experiencing great development. The first commercial voltage source converter (VSC)-HVDC was built 1997 on the island of Gotland. Since then, ratings and applications have progressed rapidly. The power and voltage for Gotland Light^{TM}was 50 MW and $\pm$ 80 kV. Fifteen years later, power and voltage ratings have increased by a factor of 20 and 6 times, respectively—for example, the 2 × 1000 MW INELFE project between France and Spain (Figure 3.7) and the 500-kV Skagerrak 4 project between Norway and Denmark [21]. The first VSC-HVDC offshore converter station was installed in 2005 on the Troll A oil platform, while the first connection to an offshore wind farm was commissioned in 2010 with the Borwin system (Figure 3.8).

Although there is obvious potential for VSC-HVDC schemes to supersede the LCC-HVDC technology in certain instances, the foreseeable future of HVDC transmission will be a landscape whereby static power converters are chosen on a project by project basis depending on system requirements and control philosophy.

3.5 VSC-HVDC CONVERTER STATION TECHNOLOGY

A complete VSC-HVDC system consists of (a) AC transformers, AC filters, phase reactors at the AC side and (b) DC capacitors, DC cables, DC chopper, and DC switchgear at the DC side. They are connected by the DC converter, which is built using power electronic switches. A typical topology of VSC-HVDC converter station is shown in Figure 3.9.

Figure 3.7 Converter station of the INELFE project. (Photo courtesy of Siemens.)

Figure 3.8 Borwin offshore VSC-HVDC converter station. (Photo courtesy of ABB.)

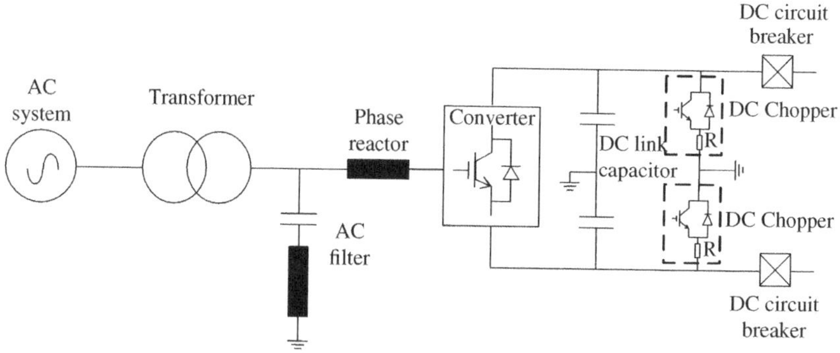

Figure 3.9 Main components of the VSC-HVDC system. The DC capacitors can be integrated in the converter itself, while the DC chopper and the DC circuit breaker are not standard components in a VSC-HVDC station.

Compared to the traditional LCC-HVDC system, the VSC-HVDC system is significantly smaller. This is due to the reactive power compensators and AC filters that are required in the LCC arrangement which are more voluminous compared to the VSC solution, resulting in a significantly smaller footprint.

The remainder of this section discusses different components in the VSC-HVDC system.

3.5.1 Converter Configurations

Generally, three types of converter configurations have been mainly utilized for VSC-based HVDC systems: two-level converter, three-level neutral point clamped (NPC) converter, and modular multilevel converter.

3.5.1.1 Two-Level Converter with Pulse-Width Modulation (PWM) The purpose of a power converter is to produce a sinusoidal AC voltage from a DC voltage which can be considered to be constant (inverter mode) and to reverse this process (rectifier mode). An effective technique to achieve the conversion can be the use of pulse-width modulation (PWM). Switching devices at high frequency and averaging the output can build a near-sinusoidal wave at a specified frequency, which is achieved by modulating the width of the voltage pulse. If the input signal is a sinusoidal wave, this is called sinusoidal pulse-width modulation (SPWM). Figure 3.10 gives the waveforms of instantaneous output voltage and the fundamental output sinusoidal generated from a two-level converter using PWM. The output pulses have the same amplitude and different widths.

A key feature of PWM is that the triangle (carrier) signal frequency f_{tri} must be higher than the frequency of the input signal f_{in}. For a single-phase PWM the frequency modulation ratio m_f is defined by

$$m_f = \frac{f_{tri}}{f_{in}} \tag{3.6}$$

However, increasing the switching frequency will result in an increase of losses. A compromise for the switching frequency must be found to ensure that the PWM is efficient with reasonable low losses.

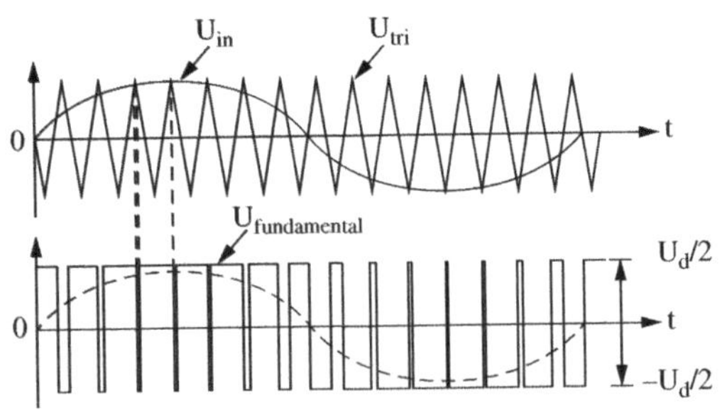

Figure 3.10 Typical waveforms of pulse-width modulation (PWM).

The amplitude of the sinusoidal signal is determined by a modulation ratio m_a. which is expressed using the maximum amplitude of the input signal voltage U_{in} and the triangle (carrier) signal amplitude U_{tri}:

$$m_a = \frac{U_{in}}{U_{tri}} \tag{3.7}$$

The first VSC-HVDC-based technology, trademarked HVDC Light, was developed by ABB in 1997. A SPWM control philosophy is used to control the IGBT's gate switching frequency producing a two-level AC waveform.

Switching frequencies between 1 and 2 kHz are used to balance the compromise of harmonics and switching losses. The switching instances can be varied in a style called optimum PWM. This method reduces the switching frequency as the current increases, which efficiently manages the switching losses across the converter valves while maintaining harmonic elimination. In order to efficiently eliminate harmonics, switching frequencies are concentrated into a narrow band where small filters can be used. The Estlink between Estonia and Finland uses a two-level converter (with optimum PWM).

In two-level configurations, IGBTs are connected in series to increase the voltage level of the converter. As the instantaneous voltage over the individual components cannot exceed the component limits, these series-connected IGBTs need to switch at exactly the same moment, so that each individual IGBT carries the same voltage.

A high switching frequency can lead to high loss: For two-level PWM converters the losses can be up to 1.4% per converter, which is significantly higher than 0.8% of LCC HVDC technology [3].

3.5.1.2 Three-Level Neutral Point Clamped (NPC) Converter PWM strategies used in a conventional inverter can be modified to use in multilevel converters. The switching frequency can be up to 2 kHz. This high frequency reduces the generated harmonics significantly, so that transformers with tertiary windings (such as with the 12-pulse configuration) are not needed. However, as each switching operation causes losses, the high switching frequency increases total power losses. Switching is the highest loss element in the converter station. In order to lower the harmonic distortion of the output voltages, higher level converter bridges are used. A three-level neutral point clamped is used for the Murray Link (Australia). In Figure 3.11a a three-level VSC bridge is depicted, and a typical output voltage waveform of a three-level VSC can be seen in Figure 3.11b.

Higher-level converter topologies (e.g., five-level, seven-level, etc.) are able to lower the harmonic distortion further and can lower the number of switching operations per switch. These higher-level converter topologies also promise lower switching losses. However, it becomes clear that the system complexity increases rapidly with the number of voltage levels. In practice, only two-and three-level converters are currently in use. When higher-level topologies are desired, the current industry practice is to resort to multilevel converters that stack different modules.

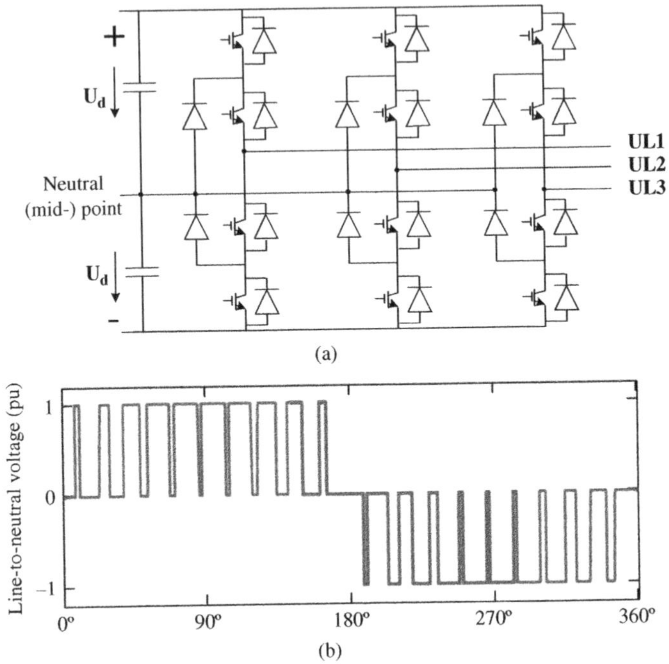

Figure 3.11 Three-level VSC-HVDC converter topology [5]. (a) Scheme of a three-level three-phase VSC. (b) Voltage waveform of a three-level three-phase VSC with single phase output voltage ($f_{switch} = 21 \cdot f_n$).

3.5.1.3 Modular Multilevel Converters A PWM scheme utilizing a two-voltage-level arrangement can suffer from high power losses. Modular multilevel converters (MMCs) are used in VSC HVDC to reduce losses by stepping through voltage levels to build output waveform. The MMC concept was proposed in Germany by Marquardt [22] and was firstly implemented by Siemens in the Trans Bay Cable project in the United States. The converters can be based on two different cell principles: half-bridge and full-bridge. The VSC-HVDC power converters from Siemens, Alstom grid, and ABB [4th generation: cascaded two-level (CTL)] are based on this technology.

Half-Bridge Configuration A multilevel converter which utilizes half-bridge modules in series is shown in Figure 3.12. There is a separate small capacitor of each module. Each bridge is supplied by a separate DC source.

The positive voltage of each module is achieved by turning on the upswitch—for example, S1 in Figure 3.12a, while a "zero" voltage can be achieved by turning on the low-switch, S2 in Figure 3.12a. The half-bridge module cannot generate a negative voltage. The converter output voltage can be increased by many modules connected in series.

By switching each module on/off once every period of the sinusoidal signal at given instances, many small voltage steps are formed, which then build the stepwise AC waveform, as shown in Figure 3.13.

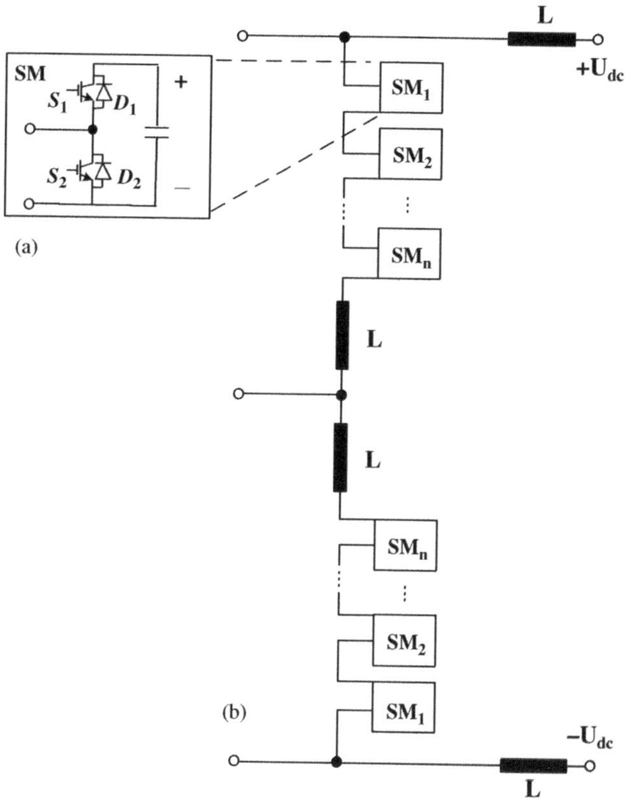

Figure 3.12 Schematic of a half-bridge module and a cascaded configuration.

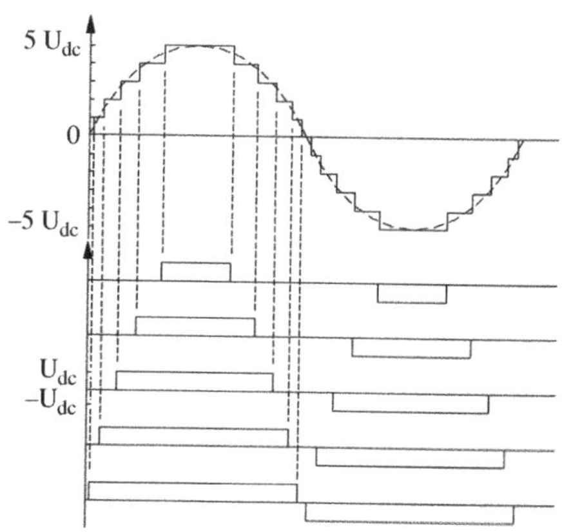

Figure 3.13 Output waveform generated from an 11-level cascade converter with five separate DC sources [23].

In comparison of conventional two-level converter technologies, this multi-modular configuration shows many advantages including low switching losses and low level of HF noise because of low switching frequency. The modular nature of the converter allows easy manufacture and maintenance; however, the quantity of the separate DC sources required can be problematic.

This approach of using modules maximizes the output voltage of the converter because many modules can be used in series; however, there are some limitations due to the conductor rating and speed of the logic hardware. A methodology to control each individual submodule is needed to calculate the required converter leg voltages, choose which submodules to switch, allow the control of active and reactive power, and balance the submodules' voltage. The Trans Bay Cable connection, which was commissioned in 2010, utilizes this technology [24].

Full-Bridge Configuration The module in a multilevel converter can also be achieved by employing full bridge (also called H-bridge) modules in series as shown in Figure 3.14. The positive and negative voltage levels are achieved by charging the mid-capacitor in either direction—for example, by turning on S1/S4 or S2/S3 in

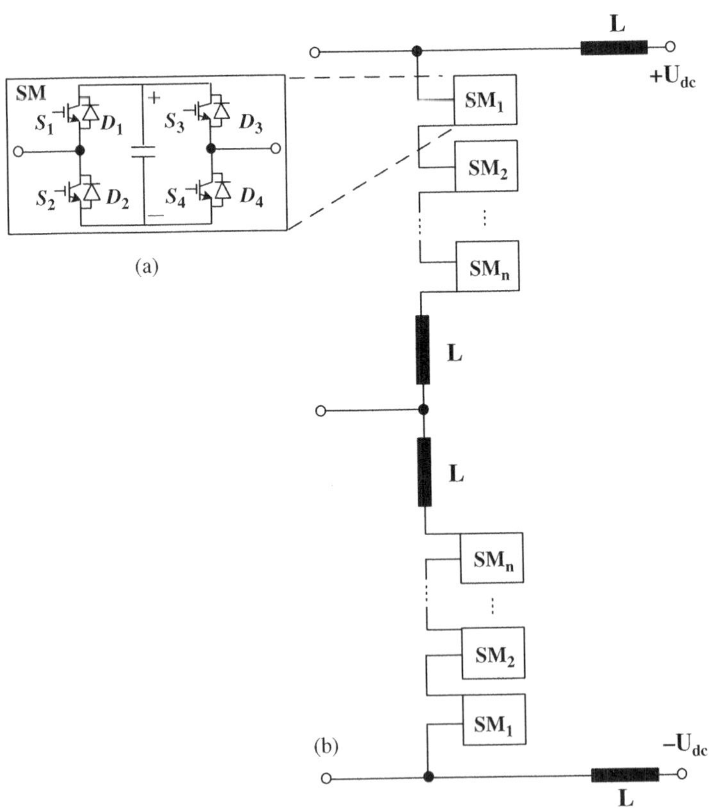

Figure 3.14 Schematic of full or H-bridge module and a cascaded configuration.

Figure 3.14a—while a "zero" voltage can be achieved by bypassing the capacitor altogether—for example, by turning on S1/S3 or S2/S4 in Figure 3.14a. The modular nature of the converter allows easy manufacture and maintenance; however, the quantity of the separate DC sources required can be problematic as well, which is the same with the cascaded half-bridge configuration.

By switching each module on/off at given instances, many small voltage steps are formed, which then build stepwise AC waveforms by the addition of small voltage steps generated by different full-bridge. Increasing the number of modules will lead to a larger number of smaller steps and closer synthesis of the desired sinusoidal wave.

Alternate-Arm Converter (AAC) Alstom Grid has developed an alternative concept called Alternate-Arm Converter (AAC) (commercial name: HVDC MaxSine™) [25]. Other objectives of the AAC are to push further the characteristics of the traditional VSC by cutting the number of H-bridge cells, reducing overall losses, and gaining the ability to control DC fault current.

As illustrated in Figure 3.15, each phase of the converter consists of two arms, each with a stack of full-bridge cells, a director switch, and a small arm inductor. The stack of cells is responsible for the multi-step voltage generation, as in a

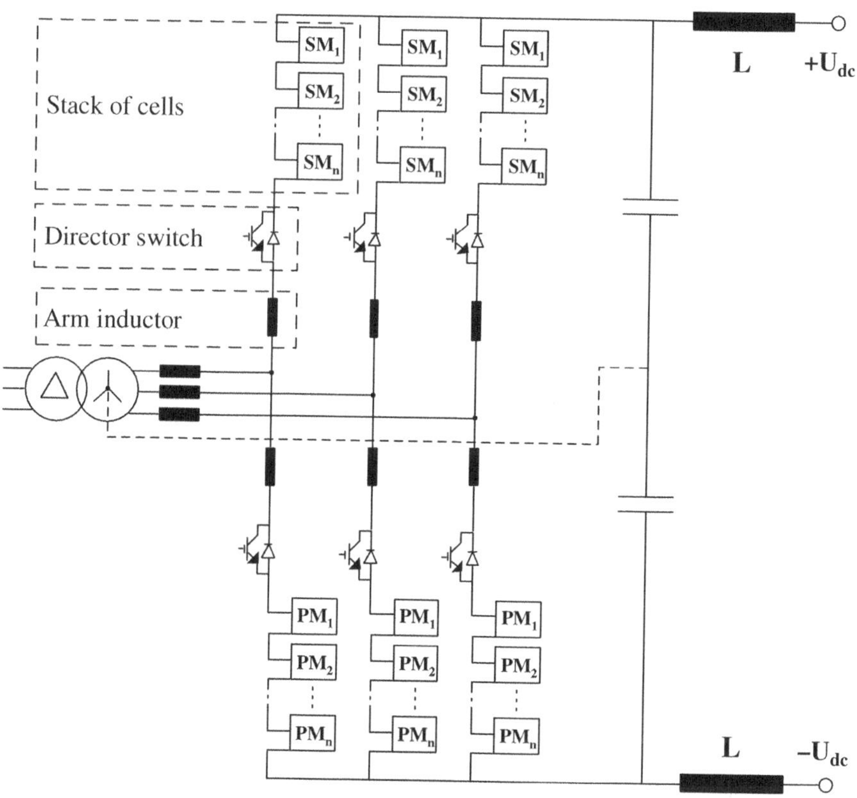

Figure 3.15 Schematic of the alternate arm converter [25].

multi-level converter. Since full-bridge cells are used, the voltage produced by the stack can be either positive or negative; thus, the converter is able to push its AC voltage higher than the DC terminal voltage if required. The director switch is composed of IGBTs connected in series in order to withstand the maximum voltage which could be applied across the director switch when it is in the open state. The main role of this director switch is to determine which arm is used to conduct the AC current. The key feature of this topology is to use essentially one arm per half-cycle to produce the AC voltage. By using the upper arm to construct the positive half-cycle of the AC sine wave and the lower arm for the negative part, the maximum voltage that each stack of cells has to produce is equal to half of the DC bus voltage, which is approximately half the rating of the arm of the MMC. The aim of the AAC is to reduce the number of cells, hence the volume and losses of the converter station [26].

Furthermore, the converter is still very responsive in case of a fault. Half-bridge cells are normally used in preference to full-bridge cells in order to reduce the number of devices in conduction at any time and, therefore, reduce the conduction power loss. Even if this choice is justified by the large cost associated with the power losses, it also means that the converter is vulnerable to a dc-side fault in a similar way to a two-level converter whereas an H-bridge version would not be. The inability of half-bridge cells to produce a negative voltage results in the conduction of the antiparallel diodes connected to the IGBTs, thus creating an uncontrollable current path in case of a collapse of the DC bus voltage. DC circuit breakers for high-power applications are still under development. Therefore, H-bridge MMC or AAC will be options for future MTDC grid.

3.5.1.4 Comparison A summary of comparison between VSC HVDC based on two-level converters and multilevel converters is given in Table 3.1. In general, multilevel converters have the disadvantage of involving a larger footprint than their two-level counterparts while maintaining an advantage in minimizing stresses. The commissioning projects using multilevel topologies in the coming years is a testament of the success of the technology's approach.

3.5.1.5 VSC-HVDC Practical Applications VSC-HVDC schemes are commonly referred to with their brand names: HVDC Light (ABB), HVDC Plus (Siemens), or HVDC MaxSine (Alstom Grid).

The first VSC HVDC installation started operation in 1997 and has a rating of 50 MW, with a ± 80 kV DC voltage, and was a prototype. Currently, there are over 10 active installations worldwide with different technologies (Table 3.2). The

TABLE 3.1 Comparison of VSC-HVDC Topologies

Two-Level Converter	Multilevel Converter
Series-connected IGBTs and PWM	Multi-modular in cascaded connection
Conceptually simple circuit	Easily "scalable"
Requires PWM	More complex controls
High switching losses	Low switching losses
Harmonic problems from PWM	Virtually no harmonics

TABLE 3.2 The Evolution of VSC-HVDC Technology [27]

Project	Year	Converter Type	Power Rating (MW)	Converter Losses (%)	DC Voltage (kV)	Supplier
Gotland	1997	Two-Level	50	3	±80	ABB
Eagle Pass	2000	Three-Level Diode NPC	36	2.2	±15.9	ABB
Estlink	2006	Two-Level with PWM	350	1.4	±150	ABB
Trans Bay Cable	2010	MMC	400	1	±200	Siemens
Nanao 3-terminal DC grid	2013	MMC	200/100/50	—	±160	Multiple
SW Link	2014IP	MMC	2×720	—	±300	Alstom Grid
Dolwin2	2015IP	MMC	900	—	±320	ABB

table shows the evolution in converter technology used in a selection of commercial projects: from two-level and three-level NPC towards multilevel converters.

3.5.2 Switching Components

Although theoretically other semiconductor valves, such as MOSFETs or GTOs, could be used, all existing VSC-HVDC systems use voltage-driven IGBT (insulated gate bipolar transistor) technology. An IGBT operates in two discrete states: a non-conducting off state and a saturated on state. By controlling the operation of these switches, an AC waveform can be constructed from a constant DC signal by "chopping" the incoming signal. To increase the voltage rating of the valve unit, IGBTs are organized into a series arrangement comprising hundreds of modules in some cases. As the valve "building blocks" are connected in series, it is critical that the transistors fail to a conducting mode, therefore, in the event of a device failure, the circuit remains complete given that sufficient redundancy is available. The number of redundant modules determines the actual rating of the converter and the maintenance requirements of the converter stations.

In order to make the converter work efficiently, an auxiliary circuit is required to assist each IGBT. A resistor is usually connected in parallel with the transistor to improve voltage sharing across all the series connected transistors. The gate of each switch needs a control circuit and protection snubber for over-current, over-voltage, and under-voltage faulting on the gate. To minimize over-temperature defects, each transistor block can be mounted on individual heat sinks dissipating heat energy efficiently.

To ensure that a VSC has bidirectional capabilities, the reverse direction diodes in the bridge compose a rectifier allowing the current path from the AC side to the DC side to charge the capacitor(s). A typical IGBT used for VSC HVDC and a schematic diagram of a valve unit is shown in Figure 3.16. A complete power module is shown in Figure 3.17.

3.5.3 AC Filters

Contrary to LCC, VSC HVDC does not generate harmonics of lower orders. For PWM-based converters only the high-frequency PWM signals have to be filtered.

Figure 3.16 A 4.5 kV IGBT (photo courtesy of ABB) and a schematic presentation of an IGBT.

Figure 3.17 Power module employed with MMC. (Photo courtesy of Siemens.)

For this, a single, relatively small AC filter can be used. In order to use standard transformers, AC filtering circuits are installed between the transformers and phase reactors. This is to prevent damaging DC voltage stresses and harmonics affecting the transformers operation [21].

With the use of PWM, passive high-pass damped filters are selected to filter the high-order harmonics. Normally, a passive second-order high-pass filter, of which the characteristic frequency is selected based on the switching frequency, is used in VSC-HVDC systems. In RMS-type simulations the filter solely injects reactive power at fundamental frequency and does not need to be represented in full detail. The design of AC filters should depend on harmonic requirements of the specific system under study. AC filters at Woodland HVDC Light station are shown in Figure 3.18.

As for MMC, the multilevel modulation can eliminate the need for output filters; therefore the filter yard for MMC is even smaller than for two-level converters. This reduction in size due to the filter is compensated by a considerably larger converter size.

3.5.4 Transformers

An AC transformer is usually installed between the AC system and the converter of a VSC HVDC system. Its functions are similar to the transformer in LCC-HVDC system.

Figure 3.18 AC filters at Woodland HVDC Light station. East West Interconnector, Ireland. (Photo courtesy of ABB.)

VSC systems are standardly operated as a symmetric monopole and do not need a special three-winding transformer. In case a different configuration is used (see also Section 8.2), the transformers are subjected to a DC offset voltage and need to be designed to withstand those. Otherwise, the VSC-HVDC transformer can be similar to normal AC transformers, which is a significant cost advantage over LCC HVDC.

The VSC-HVDC transformer is normally also equipped with a tap-changer to take part in the voltage control. Example VSC-HVDC converter transformers are shown in Figure 3.19.

3.5.5 AC Phase Reactor and Arm Inductor in a Multilevel Converter

Phase reactors within the converter serve multiple purposes:

- They allow the control of reactive and active power.
- They reduce harmonic current components at the AC side of the converter.
- They reduce any fault currents, separating the AC system (and its short-circuit current) from the converter that it is connected to.

The choice of the size of the phase reactor depends on the switching frequency, converter saturation, and control algorithm, with the converter saturation being the dominant determinant factor.

Figure 3.19 Spare tranformer in the Caprivi VSC-HVDC converter station. (Photo courtesy of ABB.)

Comparing with the two-level VSC topology, there are two arm inductors inserted separately between the upper and lower arms in a multilevel converter (Figures 3.12 and 3.14). The arm inductors have two functions: (i) suppress balancing currents caused by the inequality of the three generated phase voltages and (ii) reduce fault currents inside or outside of the converter.

3.5.6 DC Capacitors

The primary application of the capacitors on the DC side of the converter is to stabilize this DC voltage. Operation of converters produces harmonic current and DC voltage ripples on one terminal which can propagate further into the DC network to other terminal(s). Sizing these capacitors should be in accordance with the switching frequency for optimum performance and economy: the faster the switching frequency, the smaller the capacitor can be.

Due to PWM switching action in VSC HVDC, the current flowing to the DC side of a converter contains harmonics, which will result in a ripple on the DC side voltage. The magnitude of the ripple depends on the DC side capacitor size and on the switching frequency. The design of the DC capacitor should not only be based on the steady-state operation. During disturbances in the AC system (faults, switching actions), large power oscillations may occur between the AC and the DC side. This in turn will lead to oscillations in the DC voltage and DC overvoltages that may stress the valves. It is important to consider the transient voltage variation constraint when the size of the DC capacitors is selected. Here, a small DC capacitor C_{dc} can be used, which should theoretically result in faster converter response and to provide an energy storage to be able to control the power flow. The DC capacitor size is characterized as a time constant τ, defined as the ratio between the stored energy at the rated DC voltage and the nominal apparent power of the converter. The trade-off relation for the design of the DC-link capacitor in the back–back converter is described in reference 28 as follows:

$$C_{dc} = \frac{S_N}{u_{dc} \cdot \Delta u_{dc} \cdot 2 \cdot \omega_e} \tag{3.8}$$

$$\tau = \frac{C_{dc} \cdot u_{dcN}^2}{2 \cdot S_N} \tag{3.9}$$

where u_{dcN} denotes the nominal DC voltage and S_N stands for the nominal apparent power of the converter. The time constant is equal to the time needed to charge the capacitor from zero to rated voltage u_{dcN} if the converter is supplied with a constant active power equal to S_N. The time constant τ can be selected less than 5 ms to satisfy small ripple and small transient over voltage on the DC voltage. This relatively small time constant allows fast control of active and reactive power [28].

As for MMC, there is no need for this DC link capacitor, as one capacitor is placed in each submodule providing DC voltage for each module. These capacitors require both large voltage capacity as the fundamental current is passing through them and large capacitance to avoid energy ripple. The typical parameters of submodule capacitor are around 3 mF [28].

3.5.7 DC Chopper

VSC HVDC systems are prone to severe overcurrents and overvoltages during DC line or cable faults. Also for HVDC systems used to connect offshore wind energy to onshore AC grid, the offshore wind turbines may be vulnerable to overspeed damage in the event that the receiving (onshore) AC system suddenly becomes unable to accept the power infeed—for example, because of a fault. Although wind turbines normally include protections to deal with such an event, the HVDC system itself must be capable of dissipating the entire generated power of the wind power plant for a short period of time so that the offshore wind turbines can ride through an onshore AC system fault with minimal disturbance. Such systems usually take the form of a braking resistor connected to a power electronic switch, located on the DC side of the onshore converter station, and are frequently referred to as a DC chopper which is also called chopper-controlled resistor or dynamic breaker resistor [29]. A classic buck DC–DC converter can be used for such a dynamic braking system; but when scaled up to the levels required for HVDC, this requires an IGBT valve and a diode valve, each of which contains hundreds of semiconductor devices in series.

Figure 3.20 shows the schematic diagram of a DC chopper. The DC chopper consists of an IGBT switch T with anti-parallel diode D and a series resistor Rc. The DC choppers are used across the DC capacitors (in Figure 3.9). The aim here is to suppress the overvoltages during faults (because the one connecting the wind power

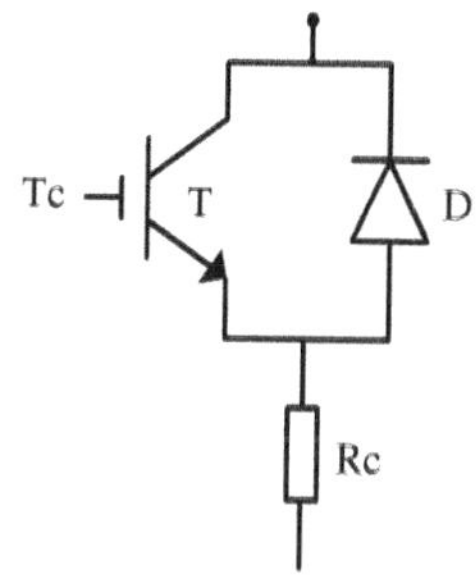

Figure 3.20 Basic diagram of a DC dynamic breaking resistor or DC chopper [30].

plant is to isolate the wind plant from faults on the mainland AC grid) and make the recovery of the system smooth.

The operation of DC chopper is simple. The DC line fault causes the voltage across the DC capacitors to increase to higher value, and the switch *T* is switched ON through a triggering pulse, T_c. The voltage across the DC capacitor is discharged through resistor R_c. When the DC voltage reaches to acceptable range, the switch *T* is turned OFF. Thus, by using DC chopper across each DC capacitor, the DC line voltages can be brought to normal values in a controlled manner. The value of resistor R_c decides the rate at which voltage is discharged and the discharging current [30].

DC choppers are most likely placed onshore to avoid additional costs at the offshore platform. In a large DC grid such as a supergrid, the rating and the distribution of the DC choppers is not trivial.

3.5.8 HVDC Switchgear

The absence of cyclic moments of current zero in a DC system inherently makes DC current switching more difficult than for AC systems, as arcs require a current zero in order to extinguish. There are three main types of HVDC circuit breaker: electromechanical, solid-state, and those that are a combination of both.

In a multi-terminal HVDC system without HVDC circuit breakers the whole system voltage would have to be brought to zero in order to clear a fault. In the case of multi-terminal LCC HVDC systems, the over-current can be limited by its DC current control function and the fault can be cleared by the action of thyristor valve control and protection. After a de-ionization period, the system can be restarted in 100–300 ms [7, 31]. In the case of multi-terminal VSC- HVDC systems, such a fast re-start would be impossible without the use of HVDC circuit breakers. If a fault occurred on a multi-terminal VSC HVDC system, the free-wheeling diodes used in the VSC would cause DC current to continue to flow into the fault even if the IGBTs were blocked. If HVDC circuit breakers are not used, it would be necessary to open the AC circuit breakers at all converter terminals. It would take a relatively long time to restore the system since it would be necessary to recharge the DC system to rated voltage before restarting.

In the case of a large multi-terminal VSC-HVDC system, it would not be feasible to allow the whole system voltage to go to zero in the event of a fault, as this would result in a significant loss of power. DC circuit breakers which can clear the fault within a few milliseconds are required (see also Chapter 17). This is to meet the requirements of preventing the fault current rising very fast towards very high values due to the low system impedance and preventing voltage collapse of the whole HVDC system.

3.5.8.1 Mechanical HVDC Breaker There have been many electromechanical HVDC circuit breakers developed over the past 5 decades. These can be grouped into three main interruption techniques: (1) inverse voltage generating method, (2) divergent current oscillating method, and (3) inverse current injecting method [30]. Of these, the inverse-current injecting method is suitable for higher voltage and current ratings. This creates a current zero by superimposing a high-frequency inverse current on the DC current by discharging a pre-charged capacitor through an inductor. A conventional AC circuit breaker can be selectively utilized in this method, and the cost of components required for such an electromechanical DC circuit breaker would not be significantly higher than that of an AC circuit breaker. Typical applications of electromechanical HVDC circuit breakers can be found in the neutral bus of bipolar LCC-HVDC schemes [32]. Traditional electromechanical HVDC circuit breakers are available up to 500 kV, 5 kA and have a fault-clearing time of the order of 30–100 ms [33, 34].

The possible development of VSC-HVDC grids requires faster fault clearing, which has led to the development of faster mechanical breakers—for instance, through the use of a resonance circuit. In [35], a DC breaker is proposed that interrupts a fault current in approximately 10 ms. The voltage level in the DC grid can be retained at 0.8 p.u. through the use of large series inductors (e.g. 300 mH). An alternative approach is proposed in [36], where fast electric actuators are used to break the DC current for a prototype with 80 kV DC, 10.5 kA and an interruption time below 5 ms.

3.5.8.2 Solid-State HVDC Breakers Solid-state circuit breakers are suitable and able to interrupt the current within a couple of milliseconds [37, 38]. They are generally based on integrated gate commutated thyristors (IGCTs) as they have lower on-state losses than IGBTs [39]. Current flows through IGCTs during on-state operation. In order to interrupt the current, the IGCTs are turned off and the voltage quickly increases until a varistor starts to conduct. The varistor is designed to block a voltage above the system voltage level. The main disadvantages of the solid-state circuit breakers are their high on-state losses, inherent cooling requirements, and capital costs. A technical difficulty is the use of series components for higher voltages. The switches would require to have identical parameters and switch as a single component to avoid voltage unbalances and over-voltages on the slowest component.

Typical ratings of solid-state circuit breakers in operation are 4 kV, 2 kA (single component) [38, 39], although ratings of up to 150 kV, 2 kA were considered [40].

3.5.8.3 Hybrid HVDC Breakers Considering the drawbacks of electromechanical breakers and power electronic breakers, the concept of hybrid DC breakers is proposed. Advances in hybrid DC breaker technology have recently been reported

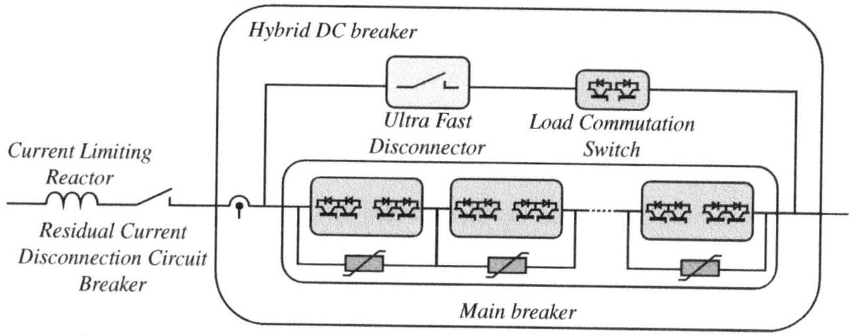

Figure 3.21 Schematic of a hybrid HVDC breaker proposed by ABB [41]. (Copyright ©Cigré.)

[41]. ABB and Alstom have each announced a prototypes of a hybrid DC circuit breaker, as presented in Figures 3.21 and 3.22, respectively. The main concept of the two hybrid HVDC breakers is comparable—that is, the combination of fast electromechanical breakers for low losses in normal operations and power electronic breakers to make the actual interruption.

The hybrid HVDC breaker in Figure 3.21 consists of a main branch and an additional branch, a bypass formed by a semiconductor-based load commutation switch in series with a fast mechanical disconnector. The main semiconductor-based HVDC breaker is separated into several sections with individual arrester banks dimensioned for full voltage and current breaking capability, whereas the load commutation switch matches lower voltage and energy capability. After fault clearance, a disconnecting circuit breaker interrupts the residual current and isolates the faulty line from the HVDC grid to protect the arrester banks of the hybrid HVDC breaker from thermal overload [41].

During normal operation the current will only flow through the bypass, and the current in the main breaker is zero. When an HVDC fault occurs, the load commutation switch immediately commutates the current to the main HVDC breaker and the fast disconnector opens. With the mechanical switch in open position, the main HVDC breaker breaks the current.

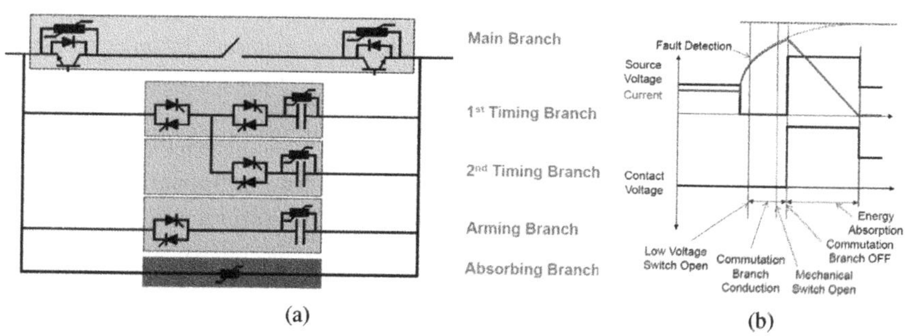

Figure 3.22 Schematic of a hybrid HVDC breaker proposed by Alstom Grid [42]. (a) Schematic. (b) Control strategy. (Photo courtesy of Alstom Grid).

Also, Alstom Grid has proposed a hybrid HVDC breaker, combining a mechanical HVDC switch with a semiconductor breaker.

The structure of this hybrid DC breaker is shown in Figure 3.22 [42]. Apart from the mechanical contacts, there is a low-voltage switch in the main branch. The other two branches are the commutation branch (or arming branch) and the energy absorption branch.

During normal conditions, the load current flows through the main branch. The contact voltage is zero and the power losses produced by the low-voltage switch are low. There are several components in series for voltage rating and in parallel for current rating. If a fault occurs, the large fault current will flow through the main branch. After a short period for fault detection, the low-voltage switch will open and the commutation branch will conduct. At this moment, the majority of the fault current will commutate to the commutation branch. As long as the fault current passing through the main branch is reduced to nearly zero, the mechanical switch can be opened with no arcing across its contact. When the commutation branch turns off, the fault current will be forced to flow in the energy absorption branch. The voltage that is generated across the energy absorption branch must be higher than the source voltage to be able to reduce the fault current and lower the breakdown voltage in the mechanical disconnector. In an inductive circuit the difference in voltage magnitude between the source and that generated across the energy absorption branch will determine the rate of fall of the fault current and thus the total time to current extinction. The generally held view is that the energy absorption branch will generate a voltage within the range of 1.5 to 1.8 times the rated source emf. Consequently it is this voltage which determines the voltage ratings of all of the breaker components. Figure 3.22b [42] shows the control mechanism of this hybrid DC breaker.

3.6 TRANSMISSION LINES

Both DC overhead lines and cables are used in practice. One of the key advantages of HVDC over AC is the reduced Right-of-Way (RoW). This aspect is covered in greater detail in Section 4.3.7.

The permissible load of an overhead line (OHL) is limited by the thermal expansion (sag) and the annealing temperature of the conductor. Cables on the other hand, are mainly limited by the aging of the insulator material. In order to assure the same life expectancy of cables and OHL, the cable must be operated at lower temperatures than bare overhead conductors [43].

Both for cables and overhead lines the conducting material can be copper or aluminum. Copper has a higher conductivity, resulting in thinner cables, while aluminum has a lower density and cost per kilogram. This lower density is especially important for OHL as it reduces the weight. Currently the relative pricing of both materials makes aluminum the preferred choice, especially for overhead lines where almost no copper is found anymore.

Copper remains the preferred choice at higher (cable) ratings when the diameter of the conductor becomes an issue.

3.6.1 HVDC Overhead Lines

HVDC can make use of overhead lines, similarly to AC systems, and the design of these towers is very similar to AC systems. However, the power-carrying capability of HVDC lines is substantially higher than for AC systems. Gains of up to 80% in transmission capacity can be reached when converting existing AC transmission lines (double circuit) to HVDC systems [44].

When comparing AC and DC overhead lines, DC lines are more susceptible to flashovers due to the dust on insulators and the constant polarization of the air surrounding the wires [3]. Of course, the overhead line is also more susceptible to lightning strikes.

3.6.2 HVDC Cables

The heat produced by cable systems has to be transferred through the insulator and, in the case of direct burial, the surrounding soil. Since the heat requirements of a cable system are more stringent than that of an OHL, a cable system must have a higher conductor section for an equal power. As a consequence, the resistance for cable systems is typically lower than that of overhead lines. Because of the narrower spacing possible with cable systems, the inductance is generally lower. On the other hand, cable systems have a much higher capacitance to ground as the distance between conductor and ground/shielding is much smaller. This shunt capacitance causes a charging current in cable systems which is significant, even up to a degree where 400 kV AC systems need reactive compensation every 20–40 km. Of course, inductances, capacitance, and charging current are of limited importance when considering the steady-state behavior of DC connections, and one of the main applications of HVDC is therefore long interconnections where no overhead is possible, such as long-distance submarine cables.

There are three main types of cable used for HVDC connections: mass-impregnated, self-contained fluid-filled, and extruded. Of these, the mass-impregnated cable and the extruded cable are the most commonly used types [40, 45, 46].

Currently, submarine and underground cables are being more widely used in Europe offshore HVDC grids. The advantages of DC cables are [47] as follows:

- DC cable lengths are not limited by charging currents, and no reactive compensation is required at intermediate points as in the case of AC transmission systems.
- DC cables are more economical when compared with the corresponding AC cables for the same power rating due to conductor and insulation savings.
- HVDC cables are especially practical for long-distance underwater bulk energy transmission.

The maximum length of a single section segment of a land cable is not limited by production limitation or the reactive properties of the cable, but by the maximum weight and diameter of the cable drum. This means that the thickest cables are

Figure 3.23 Cable prepared in the boat for being laid out. (Photo courtesy of ABB.)

limited to about 1 km. The different pieces are connected using joints. The jointing of the different pieces forms an important part of the costs and the duration of the laying process and the reliability of the cable. Sea cables can be much longer (up to 30–50 km) as they can be placed using special cable boats (Figure 3.23).

3.6.2.1 *Paper Lapped Cables: Mass-Impregnated Non-Draining (MIND)*

The mass-impregnated non-draining (MIND) type of cable is currently the most used for LCC HVDC. It has been in service for more than 40 years and has a proven track record as a highly reliable technology. Voltages up to 600 kV and current ratings of 1800 A DC are available, corresponding to a maximum pole rating of 1100 MW and bipole rating of 2200 MW. The insulation material is paper, impregnated with a high-viscosity compound. The conductor sizes are typically up to 2500 mm^2.

The mass-impregnated cable has some specific disadvantages: The low operating temperature tolerance reduces the power capacity, and the cable is expensive and heavy compared to polymeric extruded cables. The cable is also very difficult to install as the jointing process is difficult. Some new developments of this kind of cable include the using of paper polypropylene laminate (PPL). This cable can weigh up to 30–60 kg/m.

3.6.2.2 *Polymeric, Extruded Cables: Cross-Linked Polyethylene* As with AC cables, also polymeric cables are developed for HVDC. The insulating material in this case is crossed-linked polyethylene (XLPE).

Figure 3.24 XLPE HVDC cables for VSC-HVDC application. (Photo courtesy of ABB.)

This cable is not well suited for LCC HVDC as space charges in the insulation material might cause damage to the cable at current reversals. XLPE DC cables are mainly used for VSC HVDC as there is no voltage polarity reversal when operating a VSC-HVDC link. XLPE cables used for VSC HVDC are shown in Figure 3.24.

At the time of writing, this technology has been applied at voltages up to ±200 kV (in service with a power capacity of 500 MW), with several larger projects at an advanced construction stage which use voltages of ±320 kV and have power ratings between 800 MW and 1000 MW per converter.

The current rating of the cables is limited by the thermal losses in the cable, which are a function of the cable resistance and the thermal heat conduction from the conductor to the surroundings. It is assumed that there are little improvements to be expected with respect to the thermal ratings as the outer conductor diameter is limited by practical constraints. Furthermore, there is often a strong correlation between thermal and electrical conductivity in insulator material. The allowable cable voltage is determined by the capabilities of the insulator material. The pole-to-ground voltage stands over a few centimeters of insulating material. In order to increase the voltage, the insulation thickness can be increased or material with a higher dielectric strength can be used. Currently, higher XLPE cables (and accessories) at higher voltages (up to 500 kV) are being tested and expected for the near future [40, 48].

Compared to MIND cables, XLPE have some advantages: lighter weight and more flexible, smaller cable drums, smaller bending radius, lower manufacturing cost. The installations of this type of cable is also faster and less expensive. The joining of extruded cables is much simpler and requires less skill. Furthermore, a higher conductor temperature can be used, giving an overall more compact cable for the same power rating.

For onshore systems, both cables can also be buried closer to each other in one trench, there is no oil present, and the operation temperature is lower. For subsea systems, the latter advantages over MIND do not hold due to the extra lead and armoring requirements to prevent compression of the x-linked voids at high pressure.

3.7 CONCLUSIONS

Two types of HVDC technologies are described in this chapter: LCC and VSC-HVDC. For both technologies, the relevant components in the system are described: the converter, transformers, filters, overhead lines and cables, ... Also, HVDC breakers, which are considered to be the "missing link" for the development of HVDC grids, are discussed. The focus in this chapter is on the VSC-HVDC technology.

LCC-HVDC technology uses thyristor-based current source converters. The DC voltage can be either polarity to change power direction, but the current direction does not change. LCC HVDC also has lower station losses, lower cost, and higher converter ratings compared with VSC HVDC technology, which is suitable for long distance bulk power transmission. There are also some disadvantages of LCC HVDC: requires strong AC systems, requires additional equipment to realize "black" start capability, generates harmonic distortions, and requires large AC and DC harmonic filters.

VSC HVDC technology is based on converters that make use of IGBTs, components which can be both switched on and off. This results in systems that can manage both active and reactive power at the AC side and which operate with a fixed voltage at the DC side, allowing the use of XLPE cables and the possibility to add multiple converters in a straightforward manner.

REFERENCES

1. D. Van Hertem and M. Ghandhari, "Multi-terminal vsc HVDC for the European supergrid: Obstacles," *Renewable and Sustainable Energy Reviews*, vol. 14, no. 9, pp. 3156–3163, 2010. [Online]. Available: http://www.sciencedirect.com/science/article/B6VMY-50PJ72B-8/2/83156f1dd30bf5615ce11bf4ab2e99dc
2. D. Van Hertem and M. Delimar, *Electricity Transmission, Distribution and Storage Systems*, series Energy. Cambridge, UK: Woodhead Publishing, 2013, no. ISBN 978-1-84569-784-6, chapter High voltage direct current (HVDC) electric power transmission systems, pp. 143–173.
3. J. Arrillaga, Y. H. Liu, and N. R. Watson, *Flexible Power Transmission: The HVDC Options*. Wiley, Aug. 2007, no. ISBN: 978-0-470-05688-2.
4. E. Kimbark, *Direct Current Transmission*, series Direct Current Transmission. Hoboken, NJ: Wiley-Interscience, 1971, no. v. 1.
5. A. ABB, "Grid systems HVDC. it's time to connect—technical description of HVDC light® technology," ABB, Technical Report, 2013.
6. ALSTOM Grid, Ed., *HVDC*—Connecting to the future. ALSTOM Grid, 2010.

7. B. R. Andersen, "VSC transmission," Cigré, Workgroup B4: *HVDC and Power Electronics*, Technical Report 269, Apr. 2005, working group B4.37.
8. M. P. Bahrman and B. K. Johnson, "The abcs of HVDC transmission technologies," *Power and Energy Magazine, IEEE*, vol. 5, no. 2, pp. 32–44, 2007.
9. M. O'Brien, M. Zavahir, M. Clark, D. Boyle, and P. Milosevic, "New Zealand bipolar hvdc earth/sea return operation—environmental experiences," *Cigré sessions 2006*, no. B4-206, 2006.
10. P. M. Smith, I. W. Whitlock, and A. P. Balkwill, "The effects of HVDC submarine cables on shipping with magnetic autopilots," in *AC and DC Power Transmission, Sixth International Conference on (Conference Publication no. 423)*, Apr. 1996, pp. 76–80.
11. Baltic Cable. Baltic cable AB website [Online]. Available: http://www.balticcable.com/
12. T. Westerweller and J. Price, "Basslink HVDC interconnector-system design considerations," in *IEE Conference Publication*, vol. 513. London; Institution of Electrical Engineers, 1999, 2006, p. 121.
13. J.-E. Skog, K. Koreman, B. Pääjärvi, T. Worzyk, and T. Andersröd, "The norned HVDC cable link—a power transmission highway between Norway and the Netherlands," *Proceedings of Energex 2006*, 2006.
14. N. Mohan and T. M. Undeland, *Power Electronics: Converters, Applications, and Design*. Hoboken, NJ: John Wiley & Sons, 2007.
15. P. Kundur, N. J. Balu, and M. G. Lauby, *Power System Stability and Control*, vol. 7. New York: McGraw-hill, 1994, vol. 7.
16. V. K. Sood, *HVDC and FACTS controllers: Applications of Static convertors in Power Systems*, series Power Electronics and Power Systems, M. A. Pai, ed. Boston/Dordrecht/New York/London: Kluwer Academic Publishers, 2004, ISBN 1-4020-7890-0.
17. K. Eckholz and P. Heinzig, "HVDC transformers—a technical challenge," in *Power System Technology, 2002. Proceedings. PowerCon 2002. International Conference on*, vol. 1, Oct. 2002, pp. 547–551.
18. J. Arrillaga, *High Voltage Direct Current Transmission*, 2nd ed., series Power & energy. The Institution of Electrical Engineers, 1998, no. ISBN 0-85296-941-944.
19. K. Eriksson, "Operational experience of HVDC light," in *AC–DC Power Transmission, 2001. Seventh International Conference on* (Conference Publication no. 485), Nov. 2001, pp. 205–210.
20. U. Axelsson, A. Holm, C. Liljegren, M. Aberg, K. Eriksson, and O. Tollerz, "The Gotland HVDC light project—experiences from trial and commercial operation," in *Electricity Distribution, 2001. Part 1: Contributions. CIRED. 16th International Conference and Exhibition on* (IEE Conference Publication no. 482), vol. 1, 2001, p. 5.
21. L. Livermore, J. Liang, and J. Ekanayake, "MTDC VSC technology and its applications for wind power," in *Universities Power Engineering Conference (UPEC), 2010, 45th International*. New York: IEEE, 2010, pp. 1–6.
22. R. Marquardt, "Current rectification circuit for voltage source inverters with separate energy stores replaces phase blocks with energy storing capacitors," German Patent (DE10103031A1), Technical Report, 2002.
23. S. Khomfoi and L. M. Tolbert, *Power Electronics Handbook*, 2nd ed. Amsterdam: Elsevier, 2007, ISBN 978-0-12-088479-7, chapter Multilevel Power Converters, pp. 451–482.
24. J. Dorn, H. Gambach, J. Strauss, T. Westerweller, and J. Alligan, "Trans bay cable—a breakthrough of VSC multilevel converters in HVDC transmission," in *Cigré Colloquium*, San Francisco, CA, Mar. 2012, p. 7.
25. M. Merlin, T. Green, P. Mitcheson, D. Trainer, D. Critchley, and R. Crookes, "A new hybrid multi-level voltage-source converter with dc fault blocking capability," in *AC and DC Power Transmission, 2010. ACDC. 9th IET International Conference on*. IET, 2010, pp. 1–5.
26. M. Merlin, T. Green, P. Mitcheson, D. Trainer, R. Critchley, W. Crookes, and F. Hassan, "The alternate arm converter: A new hybrid multilevel converter with dc-fault blocking capability," *Power Delivery, IEEE Transactions on*, vol. 29, no. 1, pp. 310–317, Feb. 2014.
27. M. Barnes, "Multi-level converter topologies," EES-UETP course "HVDC and HVDC Grids for Future Transmission," Leuven, Technical Report, Dec. 2013.
28. T. W. Shire, "VSC-HVDC based network reinforcement," Ph.D. dissertation, M. Sc. thesis Electrical power Engineering department, Delft University of Technology, 2009.

29. C. C. Davidson and R. A. Mukhedkar, "A comparison of different types of dynamic breaking systems for HVDC systems for offshore wind power," in *Conference on Innovation for secure and efficient transmission grids*. Brussels: Cigré, March 2014, p. 8.
30. W. Ahmed and P. Manohar, "Dc line protection for VSC–HVDC system," in *Power Electronics, Drives and Energy Systems (PEDES), 2012 IEEE International Conference on*. New York: IEEE, 2012, pp. 1–6.
31. W. Long, "A study on some switching aspects of a double circuit hvdc transmission line," *Power Apparatus and Systems, IEEE Transactions on*, no. 2, pp. 734–741, 1973.
32. D. Andersson and A. Henriksson, "Passive and active dc breakers in the three gorgeschangzhou hvdc project," in *Proceedings of the International Conference on Power Systems. Wuhan, China: CIGRE*, 2001, pp. 391–395.
33. K. Nakanishi, *Switching Phenomena in High-Voltage Circuit Breakers*. New York: McGraw-Hill Mining Information Services, 1991.
34. B. Pauli, G. Mauthe, E. Ruoss, G. Ecklin, J. Porter, and J. Vithayathil, "Development of a high current HVDC circuit breaker with fast fault clearing capability," *Power Delivery, IEEE Transactions on*, vol. 3, no. 4, pp. 2072–2080, 1988.
35. K. Tahata, S. Ka, S. Tokoyoda, K. Kamei, K. Kikuchi, D. Yoshida, Y. Kono, R. Yamamoto, and H. Ito, "HVDC circuit breakers for HVDC grid applications," in *Cigré AORC Technical Meeting*, no. 1120. Tokyo: Cigré, May 2014.
36. T. Eriksson, M. Backman, and S. Halén, "A low loss mechanical HVDC breaker for HVDC grid applications," in *45th Cigré Sessions*, no. B4-303, Paris, Aug. 2014.
37. T. Podlesak, H. Singh, K. Fonda, J. Creedon, G. Schofield, and F. Johnson, "Megawatt high speed solid state circuit breaker for pulse power applications," in *Pulsed Power Conference, 1993. Digest of Technical Papers, Ninth IEEE International*, vol. 2. New York: IEEE, 1993, p. 984.
38. C. Meyer, M. Kowal, and R. W. De Doncker, "Circuit breaker concepts for future high-power dc-applications," in *Industry Applications Conference, 2005. Fourtieth IAS Annual Meeting. Conference Record of the 2005*, vol. 2. IEEE, 2005, pp. 860–866.
39. B. Oedegard, T. Stiasny, E. Carroll, and M. Rossinelli, "An application-specific asymmetric igct," *Proceedings PCIM, USA*, 2001.
40. Power cable systems. ABB. [Online]. Available: http://new.abb.com/systems/high-voltage-cables
41. J. Häfner and B. Jacobson, "Proactive Hybrid HVDC Breakers – A key innovation for reliable HVDC grids," Cigré symposium the electric power system of the future – integrating supergrids and microgrids", September 13–15, 2011, Bologna, Italy.
42. R. Whitehouse, "HVDC breakers & grid protection: Why, what & when?" Alstom Grid, Leuven, Technical Report, 2013, lecture notes EES-UETP: Course on HVDC and HVDC Grids for Future Transmission, KU Leuven.
43. R. Bartnikas and K. D. Srivastava, Eds., *Power and Communication Cables—Theory and applications*, series Power Engineering. New York: IEEE Press, 2000, ISBN 0-7803-1196-5.
44. P. Pramayon, S. Guerard, G. Aanhaanen, K. Bakic, P. Catchpole, M. Norton, R. Puffer, A. Sorensen, and M. Weibel, "Increasing capacity of overhead transmission lines—needs and solutions," Cigré, Paris, Technical Brochure 425, 2010, jWG B2/C1.19.
45. T. Worzyk, *Submarine Power Cables: Design, Installation, Repair, Environmental Aspects*. Berlin: Springer, 2009.
46. Europacable, "An introduction to high voltage direct current (HVDC) underground cables," Europacable, Technical Report, Oct. 2011 [Online]. Available: http://www.europacable.com/images/Document_Uploads/Introduction_to_HVDC_Underground_Cables_October_2011.pdf
47. M. S. Khalil, "International research and development trends and problems of HVDC cables with polymeric insulation," *Electrical Insulation Magazine, IEEE*, vol. 13, no. 6, pp. 35–47, 1997.
48. M. Jeroense, "HVDC, the next generation of transmission: Highlights with focus on extruded cable systems," *IEEJ Transactions on Electrical and Electronic Engineering*, vol. 5, no. 4, pp. 400–404, 2010.

CHAPTER 4

COMPARISON OF HVAC AND HVDC TECHNOLOGIES

Hakan Ergun and Dirk Van Hertem
Research Group Electa, Department of Electrical Engineering, University of Leuven, Leuven, Belgium

4.1 INTRODUCTION

Alternating and direct current (AC and DC) solutions are the two main options to upgrade the transmission system. However, in the context of grid reinforcements, one needs to consider more options than more than only AC and DC, but also properties such as overhead or underground, the technology used, and the additional features, both technical and nontechnical. In general, there are six technology options that can be compared:

- Classic three-phase AC transmission using overhead lines (OHL)
- Three-phase AC transmission using underground cables (UGC)
- AC system with power-flow-controlling devices (PFCs) using FACTS or phase shifters
- Classic thyristor-based line commutated converter (LCC)-HVDC transmission
- Voltage source converter (VSC) HVDC transmission
- Low-frequency AC offshore cable connections

As shown further in this chapter, the choice of specific equipment influences the system behavior and the economics significantly. The economics of low frequency AC offshore cable connections are not taken into account within this chapter, but more information on low-frequency transmission can be found in Section 22.3.

4.2 CURRENT TECHNOLOGY LIMITS

In terms of transmission capability, the power rating of equipment is limited by the voltage and current that are allowed on each specific component. There is a significant gap between the capabilities of overhead and underground transmission

HVDC Grids: For Offshore and Supergrid of the Future, First Edition. By Dirk Van Hertem, Oriol Gomis-Bellmunt, and Jun Liang.

TABLE 4.1 Limits of AC Onshore Transmission Systems

System	Voltage Rating	Power Rating
Underground cable	Up to 500 kV	Up to 1600 MVA per system
Overhead line	Up to 765 kV (1200 kV in experimental stage)	Up to 3550 MVA per system (4 conductors per phase)

systems, although cable technologies are evolving fast. The two main reasons are (a) the limited voltage stress in the insulation of the cable material and (b) the possibility of bundling overhead line conductors into a single circuit which increases the transmission capacity of overhead systems significantly.

4.2.1 Onshore Equipment

Table 4.1 illustrates current technology limits for high-voltage AC onshore transmission systems. With one system of 400-kV underground cables, ~1600 MVA can be transmitted. Using overhead lines, the same amount of power per system can be transmitted if two conductors per phase are used. The advantage of overhead lines is that more conductors per phase can be placed in a single circuit, and multiple circuits can placed on the same transmission tower to increase the transmission rating of one corridor.

Table 4.2 shows the limits of onshore HVDC transmission systems. Mass-impregnated cables are available until 500 kV and connections with ratings of 1000 MW are existing. The technical limit of the power rating is up to 2500 MW per system. For overhead line systems, connections up to 800 kV and 7200 MW (with up to 8 conductors per pole) per system exist. HVDC converters with LCC technology are available for voltages up to 800 kV and 7200 MW, with developments up to 1100 kV. VSC converters are only available up to ± 320 kV and power ratings of 1200 MW, but higher ratings are anounced also for VSC HVDC.

4.2.2 Offshore Equipment

The fact that overhead lines cannot be used, limits the application of offshore transmission to AC and DC cable systems. Consequently, the available ratings for offshore transmission systems are lower as well. As in case of onshore transmission, mass-impregnated cables up to 500 kV and 2500 MW per system can be used for submarine applications. As the installation of three single core cables in each others vicinity

TABLE 4.2 Limits of Currently Installed DC Onshore Transmission Systems

System	Voltage Rating	Power Rating
Underground cable	Up to ± 500 kV	Up to 2500 MW per system
Overhead line	Up to ± 800 kV	Up to 7200 MW per system

TABLE 4.3 Limits of Offshore Transmission Systems (Available)

System	Voltage Rating	Power Rating
DC submarine cable mass-impregnated	Up to ± 500 kV	Up to 2500 MW per system
DC submarine cable extruded	Up to ± 525 kV	Up to ±2650 MW per system
AC submarine cable	Up to 275 kV	Up to 400 MVA per three-phase cable
Offshore DC converters (VSC)	Up to ± 320 kV	Up to 1200 MW per converter

is difficult in submarine applications, three core cables are often used. Three core cables have a limited cross section, which currently does not allow higher voltages than 275 kV and 400 MVA per system for submarine AC cables.

In combination with VSC technology, voltages up to 320 kV are being installed at the time of writing, having a rating of over 1200 MW per system. XLPE cable systems with a voltage rating of 525 kV DC and a single circuit rating of over 2500 MW are announced [1].

The actual ratings of the various cable systems depend on the actual placement and use of the systems, which results in case-dependent capabilities.

4.2.3 Current Ratings for HVDC Technology

In Figure 4.1, the current ratings for HVDC systems are given, both for LCC and VSC systems and for overhead lines and for XLPE and mass-impregnated cable

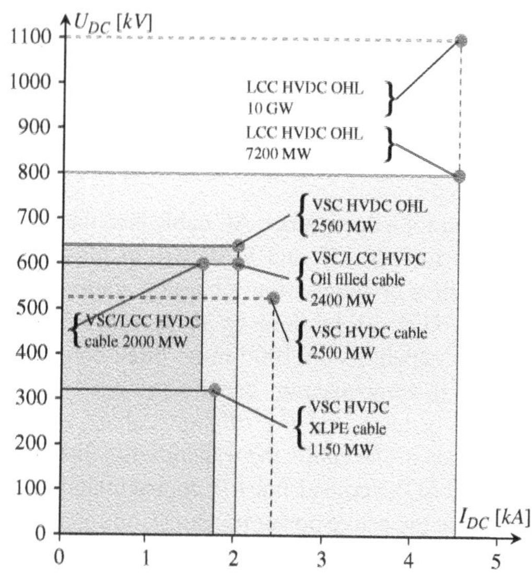

Figure 4.1 Current possible ratings for HVDC systems (U_{DC} refers here to the pole voltage, in a bipolar or symmetrical monopole setup, $P = 2 \cdot U_{DC} \cdot I_{DC}$) [2]. The dashed curves indicate announced maximum rates for LCC and VSC HVDC, respectively.

types. Overhead lines for LCC HVDC are commonly used, although mostly outside of Europe. Only one VSC-HVDC system with overhead lines exists, the Caprivi link between Zambia and Namibia. The figure shows that the voltage limit for cables is the limiting factor for the development of VSC HVDC of high power ratings. Higher voltage ratings (up to 500 kV) for DC XLPE cables are under development. Mass-impregnated cables are able to withstand higher voltages. However, they are more cumbersome in installation, more expensive, and more difficult to maintain.

The power ratings provided are indicative as they are subject to the installation and the environmental conditions. Specifically for cables, the actual system ratings depend on a number of factors such as the burial (direct burial, sea burial, tunnel,etc.), the material used (copper or aluminum), and so on. Oil-filled cables are not suited for long-distance transmission as they require regular oil refilling [2].

4.3 TECHNICAL COMPARISON

4.3.1 Charging Currents—Transmission Distance

A cable can be modeled as a long cylindrical capacitor. Per unit length, the capacity C' is about 160–600 nF/km [3]. This value depends on the dielectric, the conductor crosssection, and the thickness of insulation. The charging current of a cable can be expressed as follows:

$$\begin{aligned} I_c &= U_0 \cdot \omega \cdot C' \cdot l \\ &\sim U_0 \cdot \omega \cdot \epsilon_r \cdot l \end{aligned} \tag{4.1}$$

In (4.1), U_0 is the phase voltage, ω the angular frequency, ϵ_r the relative dielectric constant, and l the cable length.

This relation increases linear with the frequency. For DC, with a frequency of 0 Hz, the charging current is zero. For AC systems on the other hand, the charging current cannot be neglected.

The charging current for a three-phase AC cable is of the order of 1.5 MVAr/km for 150 kV, 3 MVAr/km for 220 kV, and 9 MVAr/km for 400 kV. It is therefore necessary to provide reactive compensation (charging stations) at suitable intervals (20–40 km for 400 kV). This can be done by using reactors or SVCs. Figure 4.2 shows the active transfer capability of 1000-mm^2 three-core submarine cables for different voltage levels. If compensation can be provided in short intervals, the transmission distance can be increased. For offshore cables, where it is only feasible to compensate on both ends of the cable, the transmission distance is below 150 km.

The capacitance of AC overhead lines is approximately 20 times lower than AC cables. This means that the active power transmission capability of long overhead lines is not compromised by charging currents for several hundreds of kilometers. Nevertheless, for long AC overhead lines, stability issues can occur as the voltage angle between sending and receiving end grows with the transmission length. In the case of DC overhead lines, there is virtually no length limitation.

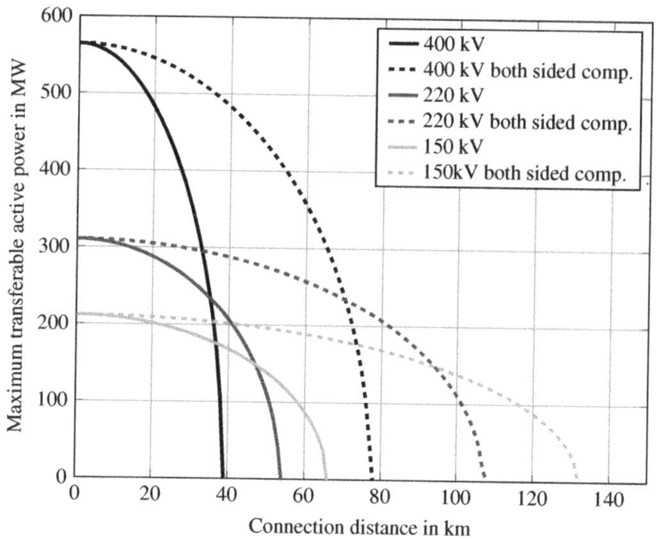

Figure 4.2 Maximum transferable active power as a function of transmission distance for three-core 1000-mm^2 XLPE submarine cables.

For long transmission lines, if not closed with the surge impedance,[1] reflection effects can occur due changes in the voltage. These effects can cause over-voltages at the end of the line and hence stability problems. Besides large transmission losses and voltage drop along the line, this is one of the reasons why for long transmission lines (above 500 km) HVDC technology is used.

4.3.2 Asynchronous Networks

In some cases, there are no alternatives for DC connections. For example, the connection of two asynchronous networks, such as between England and the European mainland or even two systems that are operating on a different fundamental frequency such as between Peru and Brazil at the Itaipu power plant.

4.3.3 Power Flow Control Capability

DC transmission is inherently suited to control the power flow through a line. At any time, the active power flowing through the connection can be set to a desired value within the DC system limits, as long as the AC grid allows this. When using VSC HVDC, the reactive power at both terminals can also be controlled dynamically.

In the case of AC transmission, power flow control is only possible with usage of FACTS devices and phase-shifting transformers. As phase-shifting transformers are mechanical devices, the change of tap settings can take up to several minutes, which constraints the speed in power flow control. Another issue is that the maintenance

[1] Which is generally the case as the impedance at the end of the line is determined by the connected grid.

interval of such transformers decreases with increasing number of switching actions. FACTS devices are mainly used for voltage control and stability. They can also can be used for power flow control in a limited range.

4.3.4 Voltage Support

With classic AC transmission, voltage support can only be provided by fixed elements. Using FACTS devices, the voltage stability of transmission grids can be improved in a fast way. LCC HVDC cannot provide voltage support to the grid without additional FACTS devices. VSC HVDC can provide voltage support independent from the transmitted active power, which makes it an attractive solution for transmission grids with a high need for flexibility.

4.3.5 Dynamic System Performance

The ability to control the power flowing through the DC connection dynamically allows the active damping of power system oscillations. The power electronic converter makes control possible in less than a second. This makes the HVDC system much more flexible than conventional power flow controlling technology such as phase-shifting transformers.

4.3.6 Stability Limits

The power transferred by a purely inductive AC line depends on the phase angle between the voltages at both line ends:

$$P_{12} = \frac{|U_1| \cdot |U_2|}{X_{line}} \cdot \sin(\delta_{12}) \tag{4.2}$$

For a given power level, this angle increases with rising distance as the line impedance becomes higher. The maximum power transfer is therefore limited because of steady state and transient stability. The power carrying capability of an AC line is inversely proportional to the transmission distance. DC lines, however, are not affected by this. Furthermore, by the decoupling through a DC link, oscillations are not (or to a lower degree) transmitted through the link. When properly controlled, the DC link can even contribute to damping of oscillations.

4.3.7 Right-of-Way

The right-of-way (RoW) needed for a DC circuit is generally smaller than that of an equivalent AC connection. Although the required space for the converter stations is significantly larger, the same power transfer requires significantly less space for DC transmission compared to AC transmission. For instance, the transport of approximately 2000 MW with ± 500 kV DC requires 30% less space than an equivalent AC solution at 800 kV. The same power transport using 500 kV AC would require 40% less space when using DC [4].

For all cable connections, the RoW is significantly smaller than for overhead connections. For DC connections specifically, there is no proximity effect, and only the thermal properties have to be brought into account. Also there are fewer cables to be placed. The RoW of AC cables connection is generally a few meters per circuit, while DC cables even require less space.

Figure 4.3a illustrates the possibilities to fulfill HVDC 5-GW power transmission requirements. (1) overhead line at 600 kV HVDC; (2) overhead line at 800 kV HVDC; (3) mass-impregnated cables and LCC technology: 3 bipoles at 500 kV HVDC; (4) extruded cables and VSC technology: 5 bipoles at 320 kV HVDC [5]. Figure 4.3b shows different overhead structures and their RoW for a 2-GW corridor.

4.3.8 Black Start Capability

When using VSC HVDC, the two electricity systems connected by the DC link are electrically separated. However, when one of these systems experiences a blackout, this system can be restarted through energy transfer by the DC link.

4.3.9 Electromagnetic Fields

In terms of magnetic fields, HVDC overhead lines are comparable to HVAC overhead lines [7, 8]. For a rated current of 1–1.5 kA, the highest magnetic flux density between 20 and 25 μT is under the overhead line. The flux density reaches values below 1 μT at 40 m distance for HVDC overhead lines and at ca. 60 m for HVAC overhead lines. The electrical field strength for a ±600-kV DC overhead line can reach peak values up to 20 kV/m [7], whereas for 380-kV overhead line systems the peak electrical field strength is below 10 kV/m [9].

Depending on the current flow, the magnetic flux density above HVDC cables can reach values up to 20 μT, whereas flux densities above HVAC cables can attain values of up to 60 μT [8].

4.3.10 Insulation Requirements

For DC overhead line systems the length of insulators have to be longer than in case of AC transmission. For a 400-kV AC transmission system, the highest phase to ground voltage in normal condition is 326.6 kV. In the case of DC transmission voltage the rated voltage will appear against earth. Additionally, the pollution of the insulators is higher in the case of DC transmission because of the electrostatic fields. This requires longer insulators for DC overhead line systems in comparison to AC.

In contrast, the amount of required insulation material for underground cables is less than that in the case of AC cables. As no skin effect and cable charging occurs in DC cables, the conductor sizes for the same ampacity can be chosen smaller, which also decreases the amount of insulation material for the same voltage rating [9].

4.3.11 Reliability

The total availability of an HVDC link is determined by two separate kinds of outages: scheduled and unscheduled outages. Scheduled maintenance has a much

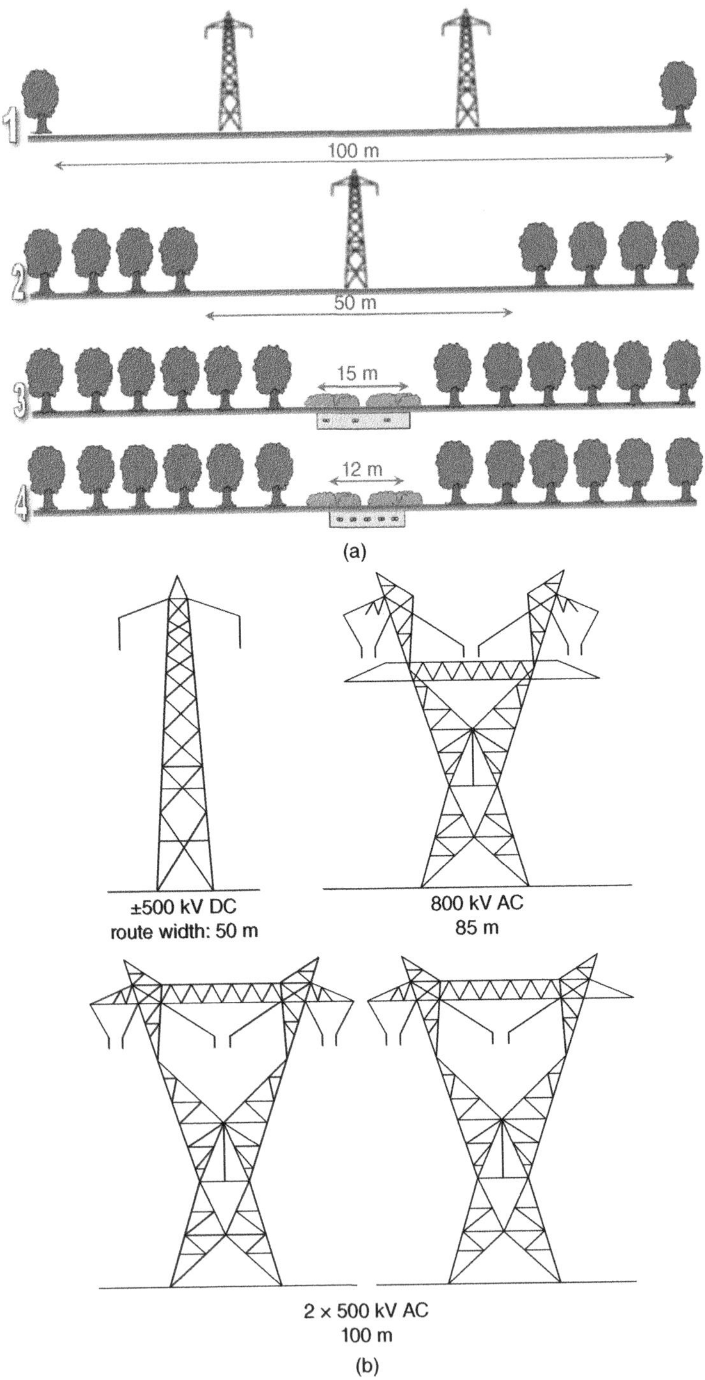

Figure 4.3 Right-of-way for different transmission options. (a) The possible land occupation for 5-GW HVDC transmission [5]. (b) Typical tower structures and RoW for different transmission options for 2-GW capacity. (Copyright © The Institution of Engineering and Technology [6].

lower impact. A thyristor-based system generally has an availability above 90%. For 2003 and 2004, Cigré reported [10] an energy availability of respectively 92.6% and 95.2%. The remainder consisted, on average, of respectively 2.2% and 4.1% of scheduled unavailability.

Although VSC HVDC is a new technology, experiences have shown that the reliability of this technology is quite high and comparable to other power electronic equipment.

In general the reliability of AC equipment is higher than that of DC equipment, due to long operational experience. For scheduled outages in the most cases, enough redundancy is foreseen, so that these outages do not cause contingencies in grid operation. For 380-kV AC OHL the unscheduled unavailability is in the range of 1.18×10^{-6}/(km·a). This corresponds to an availability of 99.98% for a transmission line of 200 km [11]. In the case of 380-kV underground cables the unscheduled unavailability is 1.12×10^{-4}/(km·a), which corresponds to an availability of 97.76% for a 200 km cable. The average unavailability of a transformer is 99.98% [11]. Similarly, in reference 12, an expected availability in the range of 97.1–97.8% is given for the NEMO link between Belgium and the United Kingdom (rated at 320 kV and 1000 MW).

4.4 ECONOMIC COMPARISON

In this section a cost comparison between HVAC and HVDC transmission solutions is given. The breakeven length between HVAC and HVDC transmission is depends on the transmission system length and the rating of the transmission system. There is also a significant difference between the breakeven lengths between offshore and onshore transmission system, as overhead lines are not suitable for offshore transmission.

4.4.1 Onshore Transmission

Figures 4.4 and 4.5 show the breakeven distances for an onshore transmission system with a rating of 2 GW (GVA) for both OHL and underground cabling. In the presented total costs, the investment costs and the costs for losses are taken into account.[2] 400 kV and ±500 kV are used for AC respectively DC transmission voltage. The highest available cross sections of overhead lines and cables has been used during the calculations. Figure 4.4 shows that the breakeven point between AC and DC options for this specific study case is 310 km. In the case of underground cabling, the breakeven point is between 250 km. Although the loss ratio between AC and DC underground cables is smaller compared to the loss ratio between AC and DC OHL, smaller investment and installation costs of DC cables shorten the breakeven distance. In case of underground cables, it should be noted that the cables are compensated every 40 km, in order to minimize charging currents.

[2] To calculate the cost of losses, an average link utilization factor of 0.7 is used. Line, cable, and converter parameters are obtained from literature, and a cost of 100 €/MWh is assumed.

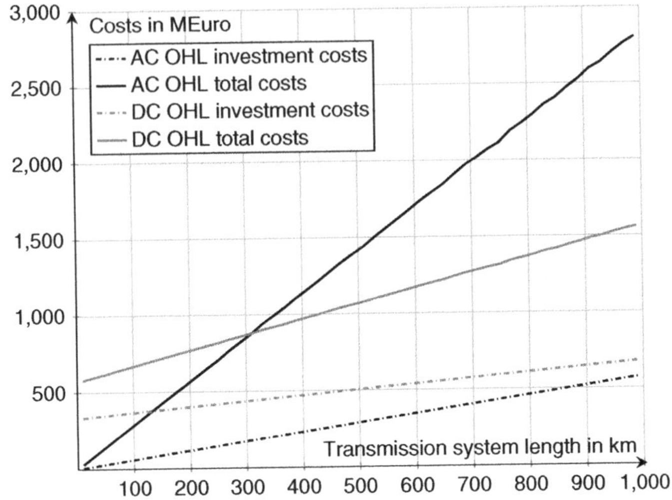

Figure 4.4 Cost comparison between onshore AC and DC OHL transmission.

4.4.2 Offshore Transmission

In offshore transmission, no overhead lines can be used. In the case of AC transmission, charging current compensation can only be provided from two both sides of the cable, which limits the transmission distance for AC systems. In the case of HVDC offshore transmission, a distinction should be made between transmission systems connecting two onshore substations and transmission systems connecting an offshore substation with an onshore substation.

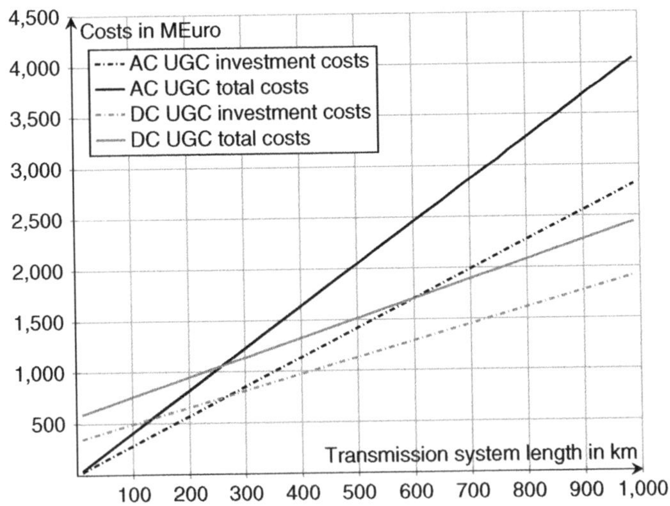

Figure 4.5 Cost comparison between onshore AC and DC underground cable transmission.

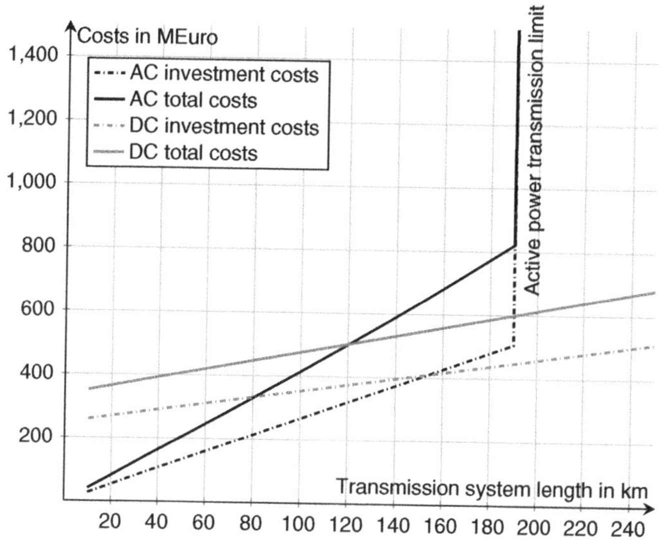

Figure 4.6 Cost comparison between AC and DC submarine transmission to connect two onshore substations.

In the first case, LCC converter stations can be used, which are available up to ±500 kV. For connection of offshore substations (such as wind farm clusters) only VSC converters are feasible, where the current voltage rating is at ± 320 kV. On the other hand, VSC converters allow the use of extruded HV cables, which are cheaper than mas-impregnated HV cables. Figure 4.6 shows the breakeven distance for the submarine connection of two onshore substations. In this case a 1500-MW (MVA) transmission system is considered. As AC transmission voltage, 245 kV is chosen. The picture shows that the breakeven point is reached before the active power transmission capability of the AC system is reached.

The study case depicted in Figure 4.7 analyzes the connection of an offshore substation to the shore. In this study case, the investments for the offshore substations are taken into account. As DC transmission voltage, ± 320 kV is used, which results in a higher number of conductors (two systems, in contrast to one) for the DC option. The higher number of conductors together with the more expensive converters shift the breakeven point to an active power transmission capability point of the AC system.

4.4.3 AC Transmission Losses

4.4.3.1 Power Transformers There are three main sources of losses in power transformers. First of all, due current flow in the windings resistive losses occur. These losses are proportional to the square of the load current (load losses). Secondly, losses are caused due to eddy currents in the magnetic conductive parts of transformers, such as the core and the housing. These losses are dependent on the frequency and therefore constant (no load losses). Additionally, dielectric losses occur at insulating parts of the transformer. These losses are also constant, as they depend on the voltage.

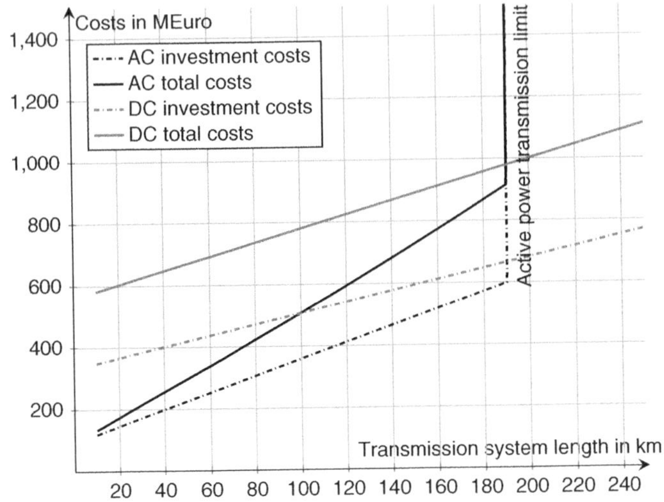

Figure 4.7 Cost comparison between AC and DC submarine transmission to connect an offshore substation to shore.

The ratio of the dielectric losses to the resistive and eddy current losses is small and can be neglected for approximate calculations.

Table 4.4 shows transformer losses for different ratings [13]. The resistive losses are given for a utilization factor of 1.

4.4.3.2 AC Overhead Lines The losses on AC overhead lines are dependent on the current flow of the lines. Due to the skin effect, the AC resistance of the conductors is significantly higher than the DC resistance. The AC resistance can be calculated with (4.3) based on the approximation given in reference 9. In (4.3), f is the rated frequency, and μ_r and μ_0 are respectively the relative the absoulte permeability of the conductors.

$$R_{AC} = R_{DC} \cdot \left(1 + \frac{x^4}{3 + 0.8 \cdot x^4}\right) \tag{4.3a}$$

$$x = \sqrt{\frac{f \cdot \mu_r \cdot \mu_0}{4 \cdot R_{DC}}} \tag{4.3b}$$

TABLE 4.4 Resistive (p_{cu}) and Eddy-Current (p_{fe}) Losses of Power Transformers [13]

S_N	p_{fe}	p_{cu}
100 kVA	0.25%	1.3%
1 MVA	0.12%	0.7%
10 MVA	0.08%	0.5%
100 MVA	0.05%	0.35%
1000 MVA	0.02%	0.15%

For an aluminum conductor with 680 mm^2 and a relative permeability of 1.000022, the resistance at 50 Hz is 66% higher than the DC resistance.

In general, aluminum is used as a conductor for overhead lines, which increases the resistance by 63% in comparison to a copper conductor of the same cross section.

4.4.3.3 AC Underground Cables Due to high capacitance of underground cables, the charging currents can cause significant losses. The charging losses occur independent of the load current of the cable and can be considered as no-load losses. In this context the way of providing charging currents determines the amount of losses. In general the charging current losses can be calculated with Eq. (4.4), where $I_C(l)$ is the charging current as a function of the line length.

$$P_{V,Ch} = 3 \cdot R \cdot \int_0^L I_C^2(l) \cdot dl \tag{4.4}$$

If the charging current is provided from one side and a linear decrease of the charging current along the length of the line L is assumed, the charging current can be written as $I_C(l) = \frac{I_{C0}}{L} \cdot l$ with I_{C0} as given in (4.1). In this case, the charging losses are $P_{V,Ch} = R \cdot I_{C0}^2 \cdot L$. The charging losses decrease with the number of compensation units along the cable. If the charging current is provided from both sides of the cable, the losses due to charging are $P_{V,Ch} = \frac{R \cdot I_{C0}^2 \cdot L}{4}$.

4.4.3.4 Losses in FACTS Devices The losses in FACTS devices depend heavily on the operating principle and the operating point of the device itself (Figure 4.8). In general the maximum losses in FACTS devices are below 1.5% of the MVAr rating of the devices [14, 15].

4.4.4 DC Transmission Losses

4.4.4.1 DC Overhead Lines and Cables In the case of DC overhead lines, the skin effect does not occur as the transmission frequency is 0 Hz. In the case of underground or submarine cables the cable capacitance is only charged once, when

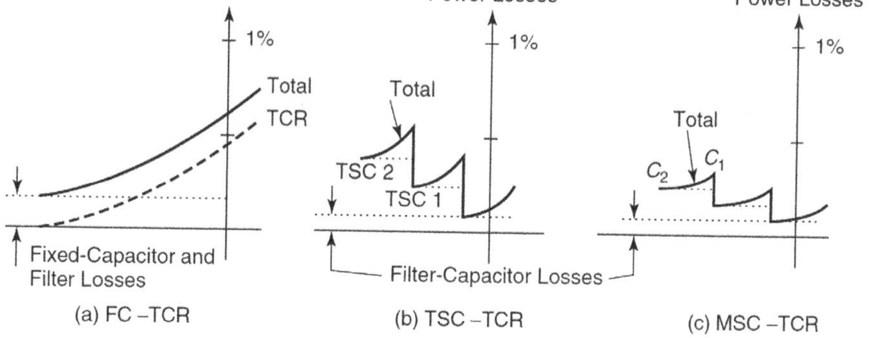

Figure 4.8 Losses of FACTS devices [15].

TABLE 4.5 Typical Breakdown of LCC HVDC Converter Station Losses [18]

Item	Total Losses in %
Converter transformers	
No-load losses	12–14
Load losses	27–39
Thyristor valves	32–35
DC smoothing reactors	4–6
AC filters	7–11
Auxiliary losses	4–9

the cable is put under voltage. This means that in contrast to AC cables the charging current has only to be provided for a very short time.

4.4.4.2 DC Converters

Losses of LCC-HVDC converters consist of converter transformer losses, thyristor valve losses, DC smoothing reactor losses, AC filter losses, and additional auxiliary losses. A breakdown of LCC converter losses in given in Table 4.5. The table shows that 23–34% of the total converter losses are independent of the load. The total losses of state-of-the-art LCC converters are in the range of 0.65–0.75% of the rated power at full load [16, 17].

To determine the losses under partial loading situations, the load losses can be scaled by the square of the utilization factor.

There are four main components causing losses in VSC-HVDC converters, namely the valves, the converter transformer, the phase reactors, and auxiliary equipment. The VSC converter station losses can be broken down as given in Table 4.6. In recent years the losses of VSC-HVDC converter stations have been dropped from 1.9% to 1% at full load due to introduction of the multilevel converter topologies [19–22]. For VSC-HVDC converter stations, the no load losses are in the range of 0.2% [20].

4.4.5 Comparison of AC and DC Equipment Losses

In this section, HVAC and HVDC transmission losses are compared in the example of a 1000-MW transmission system. First the transmission losses with overhead lines are

TABLE 4.6 Typical Breakdown of VSC-HVDC Converter Station Losses [23]

Item	Total Losses in %
Converter transformers	13
Semiconductors	70
Phase reactors	8
Auxiliary losses	9

TABLE 4.7 Assumptions for Loss Calculation for AC and DC Overhead Transmission

Item	AC	DC
Transmission voltage [kV]	400	±320
OHL cross section [mm²]	680	680
Conductors per phase	2	2

compared. Therefore the transmission distance is varied and the annual energy losses in dependence of the transmission distance are calculated. Table 4.7 shows the used assumptions for the results depicted in Figure 4.9. The power flow on the transmission line is assumed to be Gaussian distributed with an expected value of 700 MW and a standard deviation of 90 MW. The transmission losses for AC transmission consist of load losses and charging current losses on the lines, whereas in the DC case the losses consist of losses in two converters in addition to resistive losses on DC lines.

Figure 4.9 shows that the breakeven point between AC and DC transmission in terms of losses is situated at 300 km for LCC technology and at 370 km for VSC HVDC. As voltages up to ± 600 kV are possible with LCC technology, the transmission losses on the DC lines can be reduced by using the higher voltage.

In the case of underground cables, the breakeven distance for losses is increased. The reason therefore is that losses both in AC and DC cables are significantly lower than in overhead lines (due to higher cross sections). As the losses in cables are lower, the effect of converter losses on the total losses gets higher, which increases the breakeven point. For the depicted example in Figure 4.10, two systems of cables with a cross section of 2000 mm^2 are used for both AC respectively DC connections. For

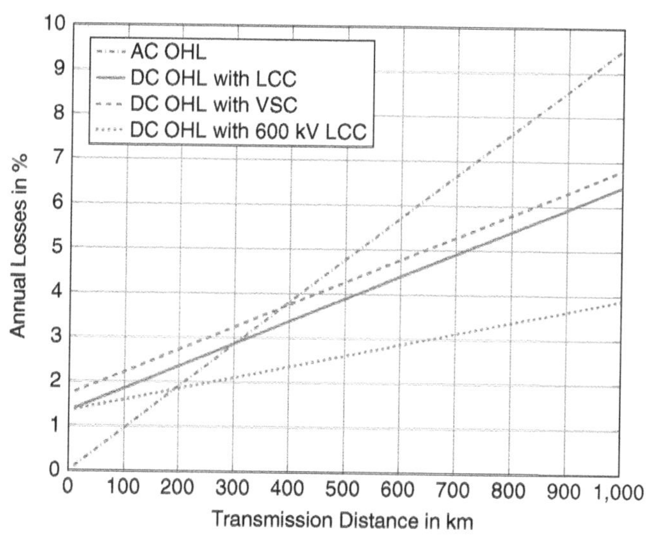

Figure 4.9 Comparison of annual losses for a 1000-MW OHL connection.

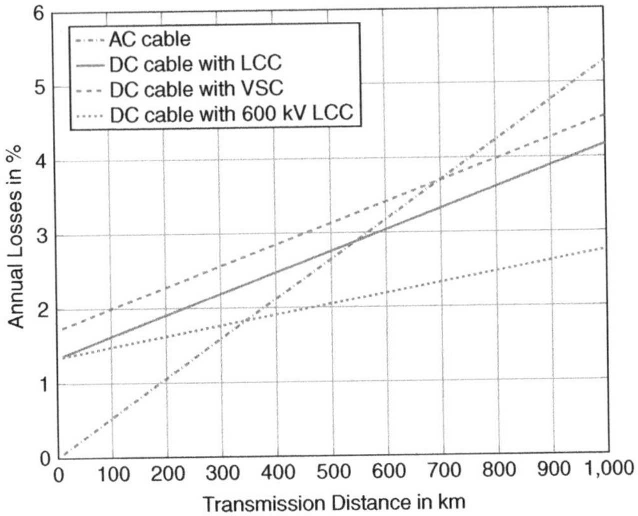

Figure 4.10 Comparison of annual losses for a 1000-MW cable connection.

the AC connection the charging current has been compensated every 40 km. Although the capacitance of underground cables are 15–20 times higher than overhead lines, due to periodic compensation, the charging losses are only 6 times higher than in the case of overhead lines.

4.5 CONCLUSIONS

Overhead AC transmission has been the default option for transmission systems and remains a viable candidate to reinforce the power system. Nevertheless, different technical solutions exist which are an alternative to investing in traditional AC extension. Both AC and DC solutions, underground and overhead, are discussed. The techno-economic elements that influence the investments decisions are discussed. Next to the investment costs, several other parameters influence the investment decision, both technical and nontechnical. The onshore and offshore cases are discussed separately.

Summarizing, Table 4.8 gives a comparison of different AC and DC transmission technologies on technical, social, and economical issues. This matter is dealt with in detail in reference 24.

The properties discussed before lead to the following applications for HVDC transmission:

1. Underground or submarine cables
2. Long-distance bulk power transmission
3. Asynchronous interconnection of AC systems
4. Stabilization of power flows in integrated power systems

TABLE 4.8 Comparison of Classic AC Transmission and, Power Transmission Using FACTS or PST for Power Flow Control (PFC) and HVDC, both VSC and Classic (LCC)

	AC OHL	AC Cable	AC PFC	LCC HVDC	VSC HVDC
Power control: active	No	No	Yes	Yes	Yes
Power control: reactive	No	No	Dependent	No	Yes
Grid interconnections	Synchr.	Synchr.	Synchr.	Any	Any
Losses	Low	Low	Low+	Medium	Medium+
Power oscillation damping	No	No	Possible	Limited	Yes
Power reversal	Fast	Fast	Fast	Slow	Fast
Social implications	High	Low	Low	Low[a]	Low[a]
Cost	Low	High	Medium	medium	High

[a] The impact of HVDC is considered low when considering cable connections [24].

With the increasing difficulties in getting permits for onshore transmission lines, HVDC is increasingly considered as an alternative option, specifically because of the additional advantages it has over traditional AC systems.

REFERENCES

1. A. Gustafsson, M. Saltzer, A. Farkas, H. Ghorbani, T. Quist, and M. Jeroense, "The new 525 kV extruded HVDC cable system—World's most powerful extruded cable system," ABB, Technical. Report., Aug. 2014, aBB White Paper.
2. D. Van Hertem and M. Ghandhari, "Multi-terminal vsc hvdc for the European supergrid: Obstacles," *Renewable and Sustainable Energy Reviews*, vol. 14, no. 9, pp. 3156–3163, 2010.
3. E. Peschke and R. von Olshausen, *Cable Systems for High and Extra-High Voltage: Development, Manufacture, Testing, Installation and Operation of Cables and Their Accessories*. Publicis MCD Verlag, 1999.
4. V. K. Sood, *HVDC and FACTS Controllers: Applications of Static Convertors in Power Systems*, M. A. Pai, Ed. Boston/Dordrecht/New York/London: Kluwer Academic Publishers, 2004.
5. Europacable, "An introduction to high voltage direct current (HVDC) underground cables," Europacable, Technical Report, Oct. 2011 [Online]. Available: http://www.europacable.com/images/Document_Uploads/Introduction_to_HVDC_Underground_Cables_October_2011.pdf
6. J. Arrillaga, *High Voltage Direct Current Transmission*, series IEE power and energy series. London: Institution of Electrical Engineers, 1998.
7. C. Kim, V. Sood, G. Jang, S. Lim, and S. Lee, *HVDC Transmission: Power Conversion Applications in Power Systems*. Hoboken, NJ: Wiley, 2009.
8. R. Conti, A. Giorgi, R. Rendina, L. Sartore, and E. Sena, "Technical solutions to reduce 50 Hz magnetic fields from power lines," in *Power Tech Conference Proceedings, 2003 IEEE Bologna*, vol. 2, June 2003, p. 6, pp. Vol. 2.
9. D. Oeding and B. Oswald, *Elektrische Kraftwerke und Netze*, 6th ed. Springer-Verlag, 2004, ISBN 3-540-00863-2.
10. I. Vancers, D. J. Christofersen, A. Leirbukt, and M. Bennett, "A survey of the reliability of HVDC systems throughout the world during 2003--2004," in *Cigré Sessions 2006. Paris*, 2006.
11. M. Obergünner, M. Schwan, C. Krane, K. Von Sengbusch, C. Bock, and D. Quadflieg, "Ermittlung von Eingangsdaten für Zuverlässigkeitsberechnungen aus der VDN–Störungsstatistik," July 2004.
12. M. Hook, "Calculating target availability figures for HVDC interconnectors," Sinclair Knight Merz, Technical Report, Dec. 2012.

13. V. Crastan, *Elektrische Energieversorgung 1*, 2nd ed. Heidelberg, Germany: Springer-Verlag, 2006, ISBN 978-3-540-69439-7.
14. N. G. Hingorani and L. Gyugyi, *Understanding FACTS*. New York: IEEE Press, 2000, ISBN 0-7803-3455-8.
15. R. Mathur and R. Varma, *Thyristor-Based FACTS Controllers for Electrical Transmission Systems*, 1st ed. Hoboken, NJ: Wiley-IEEE Press, 2002.
16. A. Giorgi, R. Rendina, G. Georgantzis, C. Marchiori, G. Pazienza, S. Corsi, C. Pincella, M. Pozzi, K. Danielsson, H. Jonasson, A. Orini, and R. Grampa, "The Italy–Greece HVDC link," *Cigre Sessions 2012*. Paris, 2012.
17. V. Prasher, D. Kumar, C. Bartzsch, V. Hartmann, and A. Mukherjee, "HVDC East–South interconnector II in India: 2000 MW, ±500 kV," in *AC–DC Power Transmission, 2001. Seventh International Conference on (Conference, Publication. no. 485)*, Nov. 2001, pp. 78–83.
18. "IEEE recommended practice for determination of power losses in high-voltage direct-current (HVDC) converter stations," *IEEE Std 1158-1991*, p. 1, 1992.
19. B. Railing, G. Moreau, L. Ronstrom, J. Miller, P. Bard, J. Lindberg, and P. Steckley, "Cross sound cable project—second generation VSC technology for HVDC," in *Cigre Sessions 2004. Paris*, 2004.
20. D. Van Hertem, J. Verboomen, S. Cole, W. Kling, and B. R., "Influence of phase shifting transformers and HVDC on power system losses," in *PES 2007, IEEE Power Engineering Society General Meeting, Tampa, Florida USA*, 24–28 June 2007.
21. O. A. Giddani, G. P. Adam, O. Anaya-Lara, and K. Lo, "Efficiency evaluation of a dc transmission system based on voltage source converters," in *Power Electronics, Machines and Drives (PEMD 2010), 5th IET International Conference on*, Apr. 2010, pp. 1–6.
22. C. Oates and C. Davidson, "A comparison of two methods of estimating losses in the modular multi-level converter," in *Power Electronics and Applications (EPE 2011), Proceedings of the 2011—14th European Conference on*, Aug. 30, 2011-Sept. 1, 2011, pp. 1–10.
23. H. Pang, G. Tang, and Z. He, "Evaluation of losses in VSC–HVDC transmission system," in *Power and Energy Society General Meeting—Conversion and Delivery of Electrical Energy in the 21st Century, 2008 IEEE*, July 2008, pp. 1–6.
24. D. Van Hertem, J. Verboomen, R. Belmans, and W. Kling, "Power flow control devices: An overview of their working principles and their application range," in *International Conference on Future Power Systems, Amsterdam, the Netherlands*, 16–18 Nov. 2005, p. 6.

CHAPTER 5

WIND TURBINE TECHNOLOGIES

Eduardo Prieto-Araujo[1] *and Oriol Gomis-Bellmunt*[1,2]

[1]Research Group CITCEA, Electrical Engineering Department, Technical University of Catalonia (UPC), Barcelona, Spain

[2]Research Group Wind Energy, Electrical Engineering Department, Catalonia Institute for Energy Research (IREC), Barcelona, Spain

5.1 INTRODUCTION

In the early stages of the electric power generation, resources as coal and oil fuels presented a more consistent production of electricity than the fluctuating generation of wind power. This consistency was the key factor that positioned fossil fuels as the basic resources for electricity production for many decades. Although the first developments of electric power generation through wind turbines started during the nineteenth century, it was not until the 1970s and specially during the 1990s that wind power became a more important energy source. Today, it is one of the most widely used sources of electric power [1].

Focusing on the wind turbine technology, in 1881 Poul LaCour built one of the firsts wind turbine generators to produce electric power [2]. Since then, many different topologies have been developed, ranging from a few to thousands of kilowatts, showing different types of rotor blades, including different generator topologies, among other variations. Despite the different proposals and manufacturers, wind turbine power has always been increasing. For instance, in 1989 a common wind turbine could produce up to 300 kW and it had 30-m rotor blades. A decade later, in 1998, the state of the art was a 1500-kW turbine with 70 m of rotor diameter [3], and today the largest wind turbines can produce more than 8 MW of electric power. Analogously, there is also an increasing trend of the wind power installed worldwide, growing each year, with a installed capacity of 240 GW in 2012 [4].

In recent years, wind turbines have been installed in the main land where wind speeds are more favorable. However, there is a current trend towards the installation of wind turbines offshore. Offshore wind shows several advantages in comparison with onshore locations, such as better wind climate, higher average wind speeds, reduced visual impact, and nonlimited turbine dimensions due to logistics [5]. On the

HVDC Grids: For Offshore and Supergrid of the Future, First Edition. By Dirk Van Hertem, Oriol Gomis-Bellmunt, and Jun Liang.

other hand, these installations present some drawbacks, such as complex wind turbine marine support structures and the logistics for installation, operation, and maintenance. These new conditions are leading wind power manufacturers to develop and design new turbine topologies to be installed in the future offshore wind power plants.

5.2 PARTS OF THE WIND TURBINE

Different wind turbines types have been proposed during the last decades. A first division can be made between machines including a vertical axis, such as Savonius, Darrieus or Giromill, or an horizontal axis turbines. Specifically, the most common structure installed worldwide is the well-known three-bladed rotor, horizontal axis turbine.

In this section, the main parts of this concept are described in detail, considering different possible layout variations. Figure 5.1 shows a conceptual scheme of a generic horizontal axis turbine. The rotor is mainly compounded by the hub, the blades, and the mechanical shaft. The blades, which capture energy from the wind, are attached to the hub by mechanical joints, and this energy is transferred to the generator by means of the shaft. Also, placed in the rotor, the pitch system allows us to limit the incoming power from the wind changing the blade angle, in order to maintain the machine under its power limits. Besides, mechanical brakes are installed to restrict also the speed of the machine during the operation. Usually, wind turbines include a gearbox in order to adapt the slow angular speed of the rotor blades to the high speed of the generator. This gearbox could also be removed in some topologies, if the generator is modified to adapt itself the rotor speed to the generator electrical speed—for instance, through multipolar structures.

The high-speed gearbox output shaft is directly connected to the generator, where the mechanical to electrical power conversion is performed. Different generator types are tipically included, mainly induction or synchronous machines. Depending on the generator installed and the turbine structure considered, power electronics converters could be connected at the machine rotor or stator terminals. By means of this converter, the generator can be controlled, achieving a better wind power extraction, together with a proper grid integration of the produced energy. Conventionally, the generator output is connected to a transformer in order to adapt the voltage level, and its corresponding switchgear is placed before the connection to the wind farm grid.

Regarding the distribution of the described elements inside the wind turbine, commonly the gearbox and the generator are embedded inside the nacelle, and other parts as the converter or the transformer can be located either inside the nacelle or the tower. Figure 5.2 shows a nacelle scheme of an actual wind turbine including different of elements presented above.

This previous component description applies to onshore and offshore wind turbines. The main difference between them is the support structure employed. Onshore wind turbines are installed over a conventional concrete basement, whereas offshore wind turbines show different possibilities. Two main offshore structures can be differentiated, the foundation and floating ones. Foundation structures are based on supporting the wind turbines through an structure attached to the sea bed. Whereas

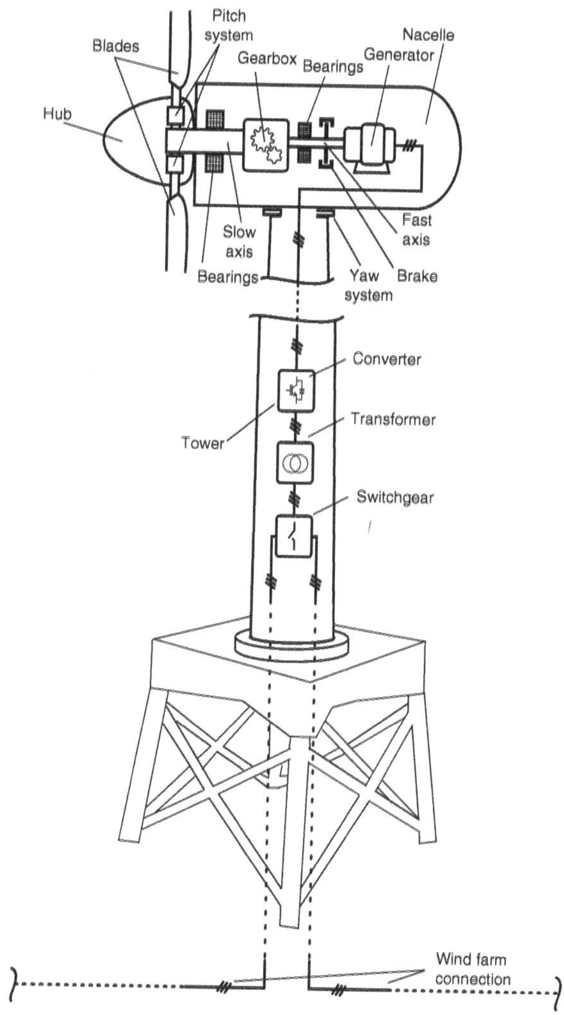

Figure 5.1 Horizontal axis wind turbine conceptual scheme.

floating structures, provide support systems without a physical joint with the sea bed, employing buoyancy or ballast techniques.

5.3 WIND TURBINE TYPES

This chapter deals with the most important types of wind turbines that have been proposed since 1970. They can be classified as [3]

- Fixed-speed wind turbine
- Limited speed wind turbine

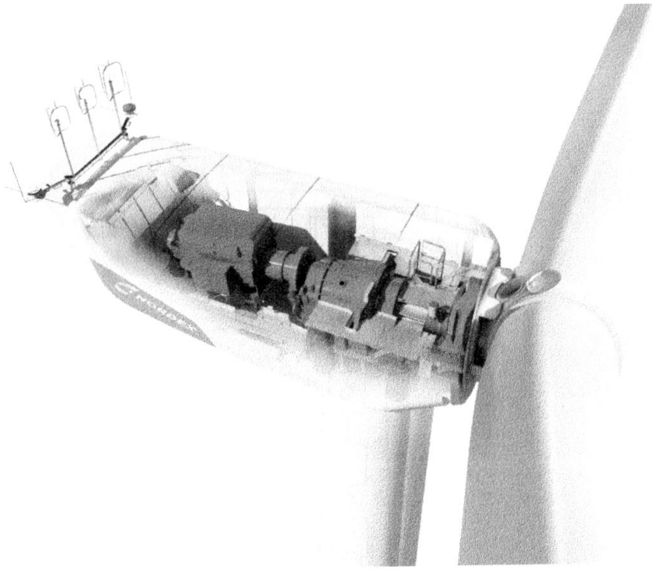

Figure 5.2 Horizontal axis wind turbine view. (Courtesy of Nordex.)

- Variable speed with partial-scale converter
- Variable speed with full-scale converter

Each of these types is described, showing its possibilities, advantages and drawbacks.

5.3.1 Fixed-Speed Wind Turbines

During the 1980s and 1990s, the most relevant machine topology was the fixed-speed wind turbine (Figure 5.3). This solution, also known as "the Danish concept," is usually compounded by a three-bladed rotor, a multiple-stage gearbox, and a squirrel cage induction generator (SCIG) connected directly to the grid by means of a transformer [6]. This machine topology is operated over and near the synchronous speed, where the induction machine works as a generator, fixed due to its direct connection to the grid.

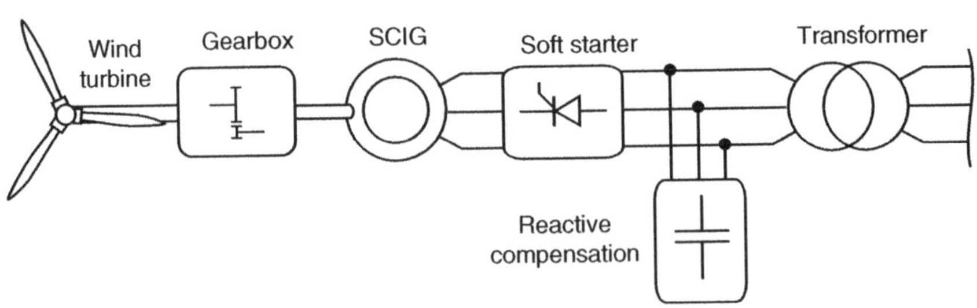

Figure 5.3 Fixed-speed wind turbine—SCIG—Multiple-stage gearbox.

During its operation, the SCIG needs to absorb reactive power to magnetize the machine, usually provided by a capacitor bank. Regarding the power control, these turbines usually include a thyristor-based soft-starter used to start up the machine when it is connected to the grid in order to avoid large currents. In order to increase the power extraction possibilities, some generators incorporate two winding sets with a different number of pole pairs, to be able to change the synchronous speed. These machines can be divided into three different types, depending on the power limiting system implemented, which can be passive or active stall and pitch [3].

The fixed-speed main advantages are the robustness of the system, the relatively low production costs, and its straightforward control [7]. These advantages were the main reasons that prompted the manufacturers to install this technology in the first wind farms. However, it also presented several disadvantages that initially caused the change to other topologies. The first drawback is related to the generator constant speed operation, which does not allow the machine to extract the maximum available power from the wind. Also, a fixed speed implies high mechanical stress considering that any wind speed fluctuations are transmitted in terms of torque, causing possible failures of the drive train. Other drawbacks were the requisites imposed by the grid operator as voltage/frequency support and fault ride-through requirements. The SCIG not only cannot provide voltage support injecting reactive power, but also needs to absorb it during its normal operation. Regarding the fault ride-through capability, SCIG wind turbines are connected directly to the grid and it is difficult to avoid a disconnection due to high current flowing through the generator during a voltage sag [7].

5.3.2 Limited Variable-Speed Wind Turbines

During the 1990s, the limited variable speed was proposed, changing the generator type from the conventional SCIG to a wound rotor induction generator (WRIG) [8]. This technology (Figure 5.4), also known under the commercial name *Optislip* (Vestas manufacturing), introduced a variable rotor resistance to increase the speed operation range (typically from 0% to 10%). By changing the rotor resistances value, through a power electronic converter, the slip of the machine (therefore the speed) can be modified. However, in order to achieve this level of control, some power has to be lost in the wound rotor resistances. In summary, these machines solved partially the demand of the need of a variable speed to increase the aerodynamic efficiency, but they still had mainly the same problems than the fixed-speed concept. This was

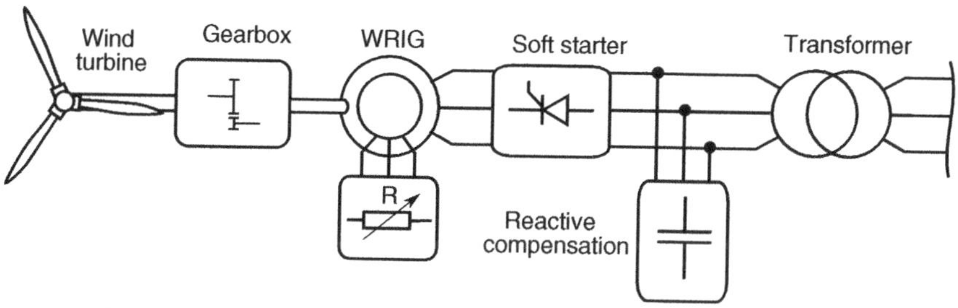

Figure 5.4 Limited variable speed—SCIG—Multiple-stage gearbox.

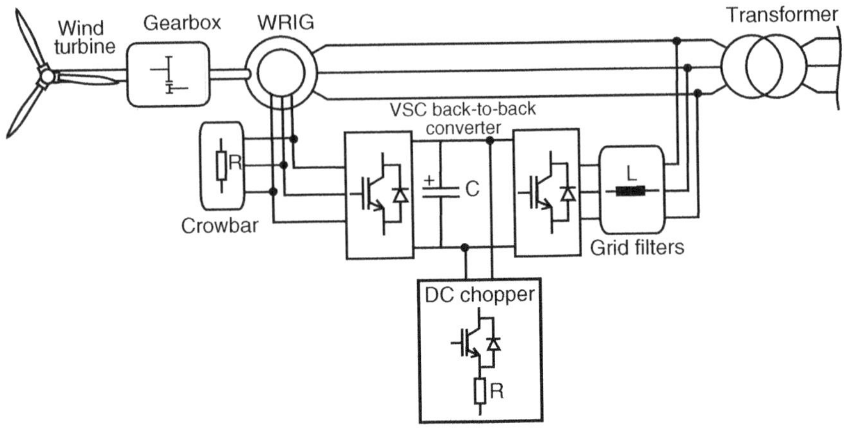

Figure 5.5 Variable speed—DFIG—Multiple-stage gearbox.

one of the first steps towards the variable-speed concept, which is leading the wind turbine market today.

5.3.3 Variable Speed with Partial-Scale Converter Wind Turbines

The variable-speed concept appeared in order to increase the operation performance of the wind turbines. Specifically, this concept (Figure 5.5) consists of a WRIG generator connected to the grid directly by the stator and through the rotor by a partial-scale power converter, resulting in the so-called doubly fed induction generator (DFIG) [9]. This power converter allows us to control the machine in a wider speed range (typically from −40% to +30%) than the limited speed topologies. Using the converter, a proper active and reactive power control on the machine can be performed in order to optimize the wind power extraction. Regarding the grid integration, the partial-scale converter increases the system possibilities. The grid side part, operated independently from the machine side, can be controlled to achieve a better ride-through capability of the wind turbine [10] and for voltage support.

However, this concept shows different drawbacks [3]. Slip rings are needed in order to extract the power from the rotor, one of the components that is causing machine operation failures. Moreover, the stator is still connected directly to the grid, so during grid faults it is complicated to manage a proper ride-through operation.

5.3.4 Variable Speed with Full-Scale Converter Wind Turbines

The variable-speed topology with full-scale converter maximizes the speed operation range of the machine. Therefore, the maximum available power from the wind can be extracted within the rated limits. Also, the grid operators requirements can be satisfied considering a proper grid side converter control. These systems can give support to the grid voltage/frequency and they can incorporate ride-through strategies. Usually a DC chopper is included in the DC bus link to bypass the voltage sags during the operation.

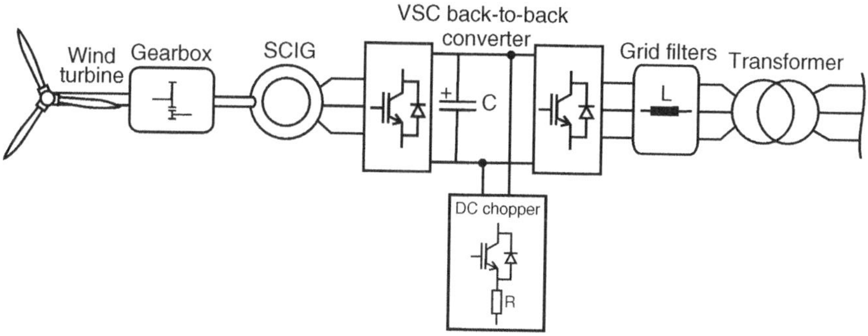

Figure 5.6 Variable speed—SCIG—Multiple-stage gearbox.

This concept can be implemented for different types of generator, namely, SCIG, permanent magnet synchronous generator (PMSG), and wound rotor synchronous generator (WRSG) [11]. Next, the most common proposals are described.

5.3.4.1 SCIG Generator Regarding the induction generators, the variable-speed common proposal is based on an SCIG generator and a full power converter (Figure 5.6) [12]. Through the converter, the generator is totally isolated from the grid, so the speed is totally controllable. Also, the reactive power needed for the machine operation can be supplied by the machine side converter. This concept is not available without gearbox due to the construction inconveniences of having multi-polar SCIG generators [13]. One of the main drawbacks could be the relatively low efficiency of the global system, considering the generator, the gearbox, and the full power converter losses [6].

5.3.4.2 Direct Drive WRSG This concept consists in a WRSG with DC rotor excitation connected to the grid by means of a back-to-back converter (Figure 5.7).

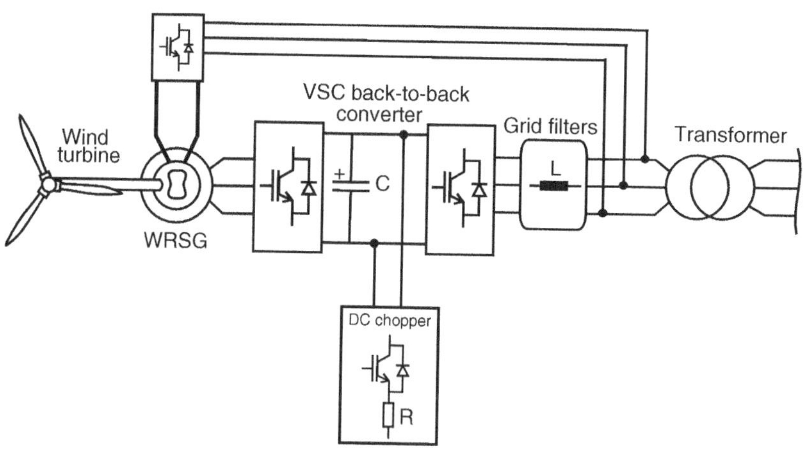

Figure 5.7 Variable speed—WRSG—Direct drive.

This type is usually implemented without gearbox through incrementing the number of poles of the rotor, obtaining a direct-driven system. The increment on the number of poles implies a large generator diameter, obtaining a machine of bigger dimensions [14]. On the one hand, the associated costs of the gearbox are eliminated. On the other hand, the generator costs are increased. Another interesting feature of this machine is the DC controllable excitation. Through the DC converter, the rotor flux can be controlled for different purposes: field-weakening strategies for high-speed operation, loss minimization techniques, and so on [15]. However, it must be considered that the excitation control is performed by a converter, which is located outside the rotor and connected to the machine windings commonly through slip rings. This system implies losses during the operation due to the current flowing the DC windings, and also regular maintenance must be considered if slip rings are used.

5.3.4.3 Direct-Drive PMSG Related to the previous WRSG concept, the PMSG direct-driven generator connected to the grid by means of a full-power converter is proposed (Figure 5.8). This type of generator comes from the idea of substituting the wounded rotor DC excitation by magnets, thus eliminating the excitation losses [15].

The PMSG possibilities of being isolated from the grid through the converter are equivalent to the WRSG. PMSG rotors can also be constructed with a large number of poles. However, PMSG are smaller and lighter when including a large number of poles than the equivalent WRSG [14], and therefore this topology achieves a higher power density.

On the other hand, working with permanent magnets can be difficult during the manufacturing and the operation stages. Also, the permanent magnet materials are quite expensive. However, in recent years its costs have been reduced considerably [6].

During the operation, the fact of having a constant flux in the rotor produced by the magnets may cause large voltage values at the machine terminals for high speeds, forcing the converter to apply larger voltages which could not be available. In the WRSG, the excitation could be weakened in order to reduce the rotor flux, thus reducing the DC excitation voltage. In a PMSG, the flux cannot be reduced because the excitation is permanent. Then, a flux weakening technique must be employed to reduce the terminal voltages at high speed [16]. This technique must

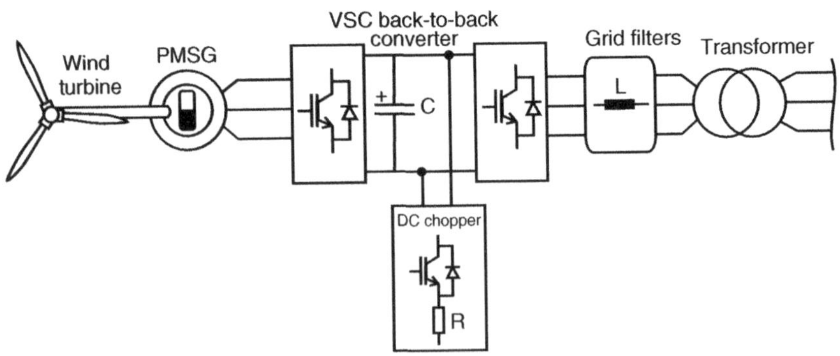

Figure 5.8 Variable speed—PMSG—Direct drive.

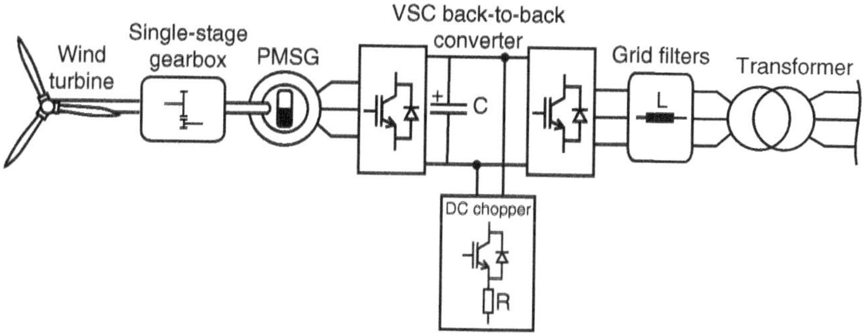

Figure 5.9 Variable speed—PMSG—Single-stage gearbox.

be done without too much effort in order not to demagnetize the magnets. Also, this demagnetization can be caused by certain possible faults in the machine. Also, these magnets are sensitive to the temperature, and therefore it should be controlled during the operation.

There are many different types of PMSG, which can be classified by its flux direction, the type of rotor, the type of magnets, and the winding layout.

5.3.4.4 PMSG with a Single/Multiple-Stage Gearbox This topology combines a PMSG with a single-stage or a multiple-stage gearbox system in order not to construct a bigger generator with a large number of poles (Figure 5.9) [14]. The single-stage topology does not achieve a high transmission ratio. It is interesting because it combines a small gearbox with a relatively small generator with a few number of poles. Therefore, the generator does not need to be too large and expensive and the gearbox complexity can also be reduced. The generator is operated with medium-speed and torque characteristics.

On the other hand, the PMSG with a multiple-stage gearbox can operate a conventional low-torque, high-speed generator (Figure 5.10) [6]. The PMSG used for this design is lighter than the other topologies, with a usual number of poles. The main drawbacks are its cost, weight, maintenance, and the complexity of the gearbox needed for the solution.

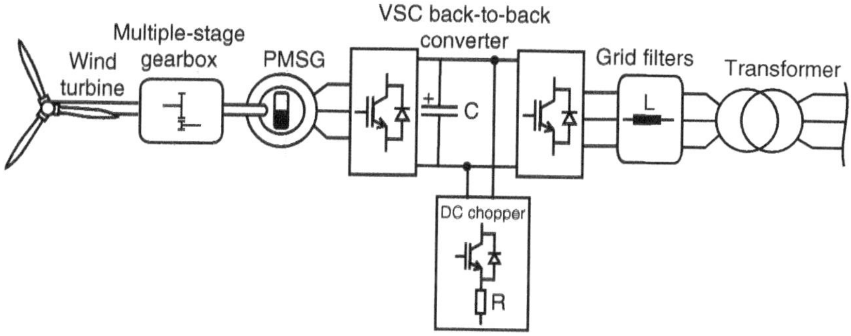

Figure 5.10 Variable speed—PMSG—Multiple-stage gearbox.

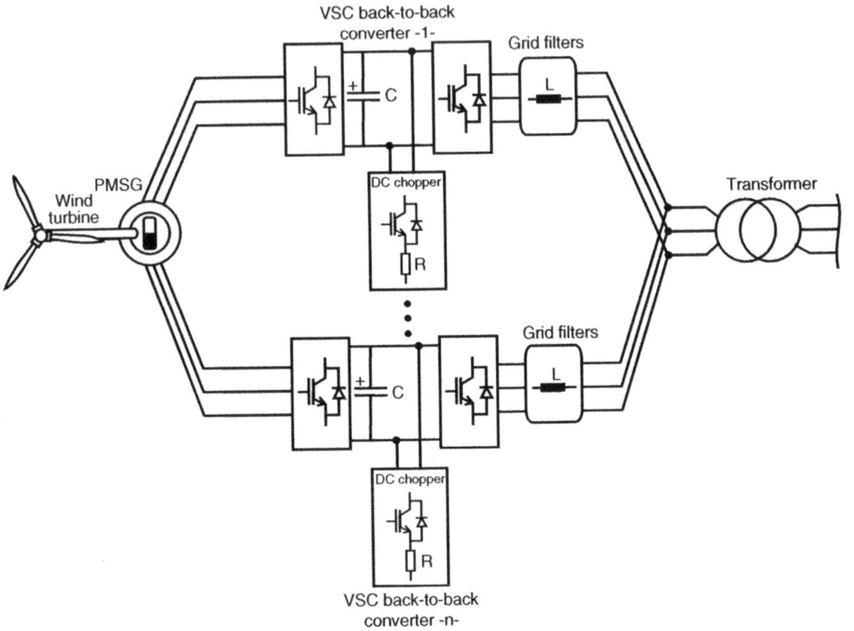

Figure 5.11 Variable speed—Multiphase PMSG—Direct drive.

5.3.4.5 Direct-Drive Multiple Three-Phase PMSG Recently, with the growing interest on the offshore wind power plants, factors such as reliability and efficiency gained importance. Three-phase PMSGs is one of the topologies which could be installed in these offshore plants (Figures 5.11 and 5.12). However, the back-to-back converter offers other possibilities which could be profitable for the offshore operation. Since the generator is isolated from the grid, the number of the generator phases is not limited to three, so redundant structures based on multiphase machines can be performed in order to increase the system reliability [11].

One of the proposed technologies is the direct-driven multipolar PMSG with multiple stators, which may include two or more three-phase winding sets with isolated neutral points [17]. Each stator is connected to a three-phase back-to-back converter in order to control the power flowing through each of them. This structure increases the reliability by means of redundancy, allowing to perform different fault tolerance strategies. Moreover, these redundant structures reduce the amount of power flowing through each converter, compared to a single three-phase one, decreasing the stress suffered by the power electronics.

5.3.4.6 Other Proposals In the literature, other new wind turbine topologies can be found. One of these topologies is the brushless DFIG (BLDFIG) [11], which consists in eliminating the slip rings of the DFIG concept which is one of the main drawbacks of the topology (see Figure 5.13). This topology can be obtained by utilizing a machine with a double stator, connecting the partial-scale converter to the second one, avoiding the slip rings of the rotor connection. This machine is operated similarly as a conventional DFIG machine.

Figure 5.12 Alstom Haliade-150 offshore wind turbine (Courtesy of Alstom).

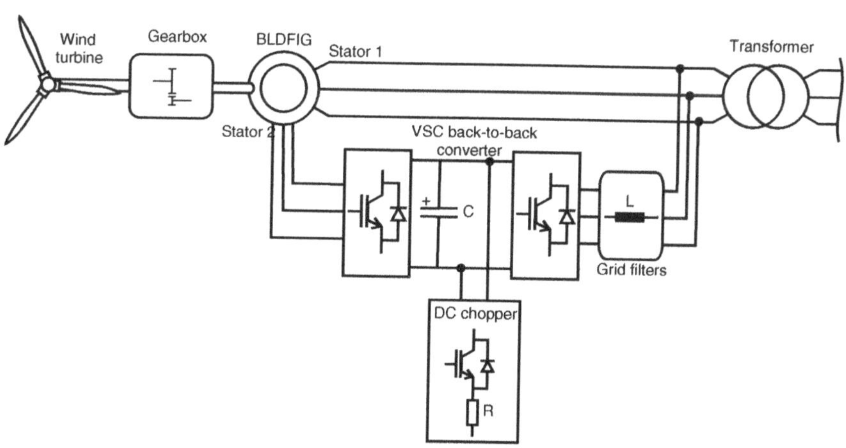

Figure 5.13 Brushless DFIG (BLDFIG).

5.4 CONCLUSIONS

Since the first steps of wind power generation, manufacturers have proposed a number of turbine concepts. Two of these types, fixed speed and variable speed with partial-scale converter wind turbines, have been the workhorse of the wind industry the past decades. However, the current trend is leading the wind technology to the full-scale converter wind turbines due to the possibilities offered by this concept, together with the price reduction of the power electronics and the increasing restrictions imposed by the transmission operator grid codes. To conclude, the evolution of the wind turbine topologies can be explained as a combination of the technology improvements and other factors surrounding the wind industry.

REFERENCES

1. E. Hau, *Wind Turbines: Fundamentals, Technologies, Application, Economics*. Berlin: Springer, 2005.
2. E. Golding, *The Generation of Electricity by Wind Power*. London: E. and F. N. Spon Ltd., 1977.
3. J. Slootweg, H. Polinder, and W. L. Kling, *Wind Power in Power Systems*, T. Ackermann, Ed. Hoboken, NJ: Wiley, 2005.
4. GWEC, Ed., *Global Wind Energy Outlook*. Washington, D.C.: GWEC, 2012.
5. EEA, Ed., *Europe's Onshore and Offshore Wind Energy Potential*. Boston, MA: EEA, 2009.
6. H. Li and Z. Chen, "Overview of different wind generator systems and their comparisons," *Renewable Power Generation, IET*, vol. 2, no. 2, pp. 123–138, 2008.
7. L. H. Hansen, L. Helle, F. Blaabjerg, E. Ritchie, S. MunkNielsen, H. Bindner, P. Sørensen, and B. Bak-Jensen, *Conceptual Survey of Generators and Power Electronics for Wind Turbines*. Roskilde, Denmark: Risø National Laboratory, 2001.
8. M. Khadraoui and M. Elleuch, "Comparison between optislip and fixed speed wind energy conversion systems," in *Systems, Signals and Devices, 2008. IEEE SSD 2008. 5th International Multi-Conference on*, 2008, pp. 1–6.
9. R. Pena, J. Clare, and G. Asher, "Doubly fed induction generator using back-to-back pwm converters and its application to variable-speed wind-energy generation," *Electric Power Applications, IEE Proceedings*, vol. 143, no. 3, pp. 231–241, 1996.
10. O. Gomis-Bellmunt, A. Junyent-Ferre, A. Sumper, and J. Bergas-Jan, "Ride-through control of a doubly fed induction generator under unbalanced voltage sags," *Energy Conversion, IEEE Transactions on*, vol. 23, no. 4, pp. 1036–1045, 2008.
11. J. H. Z.Q. Zhu, "Electrical machines and power electronic systems for high-power wind energy penetration applications: Part 1—market penetrationm, current technology and advanced machine systems," *COMPEL: The International Journal for Computation and Mathematics in Electrical and Electronic Engineering*, vol. 32, no. 4, pp. 7–33, 2012.
12. P. H. D. Slootweg, "Basic operation principles and electrical conversion systems of wind turbines," in *Electric Machines and Drives, 2005 IEEE International Conference on*, 2005, pp. 543–550.
13. M. R. Dubois, H. Polinder, and J. A. Ferreira (2000, June). "Comparison of generator topologies for direct-drive wind turbines," in *Proceedings of IEEE Nordic Workshop on Power and Industrial Electronics* (Norpie 2000), Aalborg, Denmark, pp. 22–26.
14. H. Polinder, F. Van der Pijl, G.-J. de Vilder, and P. Tavner, "Comparison of direct-drive and geared generator concepts for wind turbines," in *Electric Machines and Drives, 2005 IEEE International Conference on*, 2005, pp. 543–550.
15. A. Grauers, "Modelització i control d'un sistema de generació elèctrica turbina de vent," Master's thesis, ETSEIB-UPC, 1996.
16. A. EL-Refaie and T. Jahns, "Optimal flux weakening in surface pm machines using fractional-slot concentrated windings," *Industry Applications, IEEE Transactions on*, vol. 41, no. 3, pp. 790–800, 2005.
17. M. Duran, S. Kouro, B. Wu, E. Levi, F. Barrero, and S. Alepuz, "Six-phase pmsg wind energy conversion system based on medium-voltage multilevel converter," in *Power Electronics and Applications (EPE 2011), Proceedings of the 2011 14th European Conference on*, 2011, pp. 1–10.

CHAPTER 6

OFFSHORE WIND POWER PLANTS (OWPPS)

Mikel De Prada-Gil[1] Jose Luis Domínguez-García[1] Francisco Díaz-González[3], and Andreas Sumper[1,2]
[1]Research Group Wind Energy, Electrical Engineering Department, Catalonia Institute for Energy Research (IREC), Barcelona, Spain
[2]Research Group CITCEA, Electrical Engineering Department, Technical University of Catalonia (UPC), Barcelona, Spain
[3] Electrical Engineering Department, Technical University of Catalonia (UPC), Barcelona, Spain

6.1 INTRODUCTION

Offshore wind power is becoming more relevant and of interest during the last few years. It is mainly because the wind speed at sea is higher and steadier than on land and because of the lower installation restrictions, which allows the possibility of using larger wind turbines [1]. To date, most of the existing offshore wind power plants (OWPPs) are of a relatively small up to medium-sized rating (up to few hundreds MW) and are close enough to the shore that it is feasible to transmit the power through AC submarine cables. However, the fact that OWPPs are increasingly larger in size and located further away from shore, forced by environmental and social factors, is making it more interesting to be connected via HVDC technology. This trend is expected to continue over the coming years, since there are already several projects in operation in the North Sea [2].

The construction of an OWPP, as well as the O&M tasks performed throughout its lifetime (typically 20 years), entail some challenges and risks that should be taken into account in order to determine both its optimal location and the viability of the project itself. Thus, the design process of an OWPP should consider comprehensive studies based on the environmental conditions to be expected at a proposed site over the offshore wind project's lifetime. These environmental conditions mainly refer to the wind, wave, water depth, and soil and seabed characteristics.

Focusing on wind conditions, parameters such as the wind speed, the wind direction, or the turbulence intensity, among others, have a strong influence in determining the optimal OWPP layout which maximizes its energy generation. For example, the arrangement of the wind turbines within an OWPP will be conditioned by

HVDC Grids: For Offshore and Supergrid of the Future, First Edition. By Dirk Van Hertem, Oriol Gomis-Bellmunt, and Jun Liang.

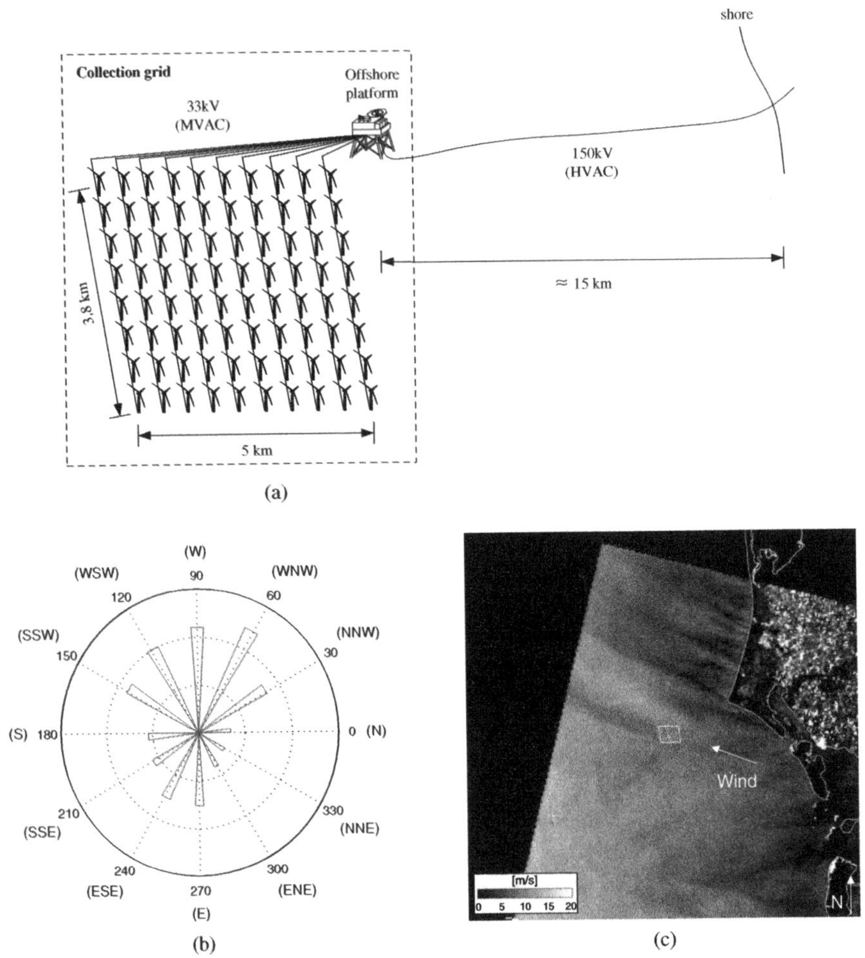

Figure 6.1 Horns Rev WPP layout, wind rose, and wind speed map. (a) Horns Rev offshore wind farm layout. (b) Wind rose distribution at Horns Rev offshore wind farm [3]. (c) Wind speed map based on the ERS-2 synthetic aperture radar (SAR) scene captured on 25 February 2003 over Horns Rev offshore wind farm (white trapezoid) [4].

the prevailing wind direction(s) at that site. Likewise, the wind speed reduction and the high turbulence levels in downstream wind turbines caused by the wake effect should be estimated to determine the spacing among wind turbines. Typically, wind turbines are spaced between 5 and 9 rotor diameters (D) away from each other in the prevailing wind direction and between about 3 and 5 rotor diameters for winds coming perpendicularly, in order to reduce the effects of wakes.

Figure 6.1 shows the layout of the Horns Rev offshore wind farm located in the North Sea (in Denmark), as well as, its wind rose and wind speed map based on remote sensing observations [3, 4]. As can be seen, it is laid out in a slightly oblique rectangle matrix of 8 rows and 10 columns. The spacing between two nearby wind turbines is 7 rotor diameters in both directions. The collection grid consists of 80 wind turbines of 2 MW each interconnected by buried 33-kV cable network (by adopting a radial design) and connected to an offshore substation platform. The power generated by the OWPP is transmitted to the main onshore grid by means of an HVAC-link at 150 kV.

This chapter aims to provide a general overview of different offshore wind power plant designs considering the use of both HVAC and HVDC transmission links to deliver the generated power. It is mainly focused on the conventional AC wind power plants, by introducing some possible wind power plant topologies and briefly describing the required technologies that encompasses the collection grid (except for the HVDC power converter and the wind turbines explained in detail in Chapters 3 and 5, respectively). This chapter also deals with the description of the electrical design methodology performed in the conventional offshore wind power plants. The methodology focuses on explaining in detail the cable selection process and how the technical assessment of an OWPP is performed, considering only the collection grid area. Likewise, the chapter shows the result of applying such methodology to a particular case study. Moreover, it presents a short presentation of future offshore wind power plants based on DC technologies (both the collection grid and the transmission system), as well as other proposals for AC wind power plants connected to the onshore grid through an HVDC transmission links.

6.2 AC OWPPs

In general, the design of the offshore wind power plants is broadly similar to those installed onshore—that is both the collection grid and the transmission link to the main electrical network—by using AC technology. However, HVDC transmission systems arose in recent years, posing an alternative to the conventional AC offshore transmission systems. The decision on what kind of transmission link is most suitable depends on several factors. Power losses in DC could be lower than in AC systems, but the capital costs of the DC transmission systems are intended to be higher than AC equivalents due to the youth of the technology. Other aspects have to be considered though, such as the need of installing reactive power compensators in AC systems, which also increases the capital costs of the system [5]. All these aspects lead to a clear trade-off analysis in which the length of the cables results decisive as it greatly affects the magnitude of the capital costs of the transmission facility.

Regardless of the type of transmission used (HVAC or HVDC), any wind farm whose capacity exceeds a certain minimum requires an offshore substation platform. In the following, the electrical design of an offshore substation platform is presented for AC OWPPs with both HVAC and HVDC transmission systems.

6.2.1 Topologies Overview of Offshore Wind Power Plant

Leaving aside the transmission system and focusing on the AC OWPP collection grid, there are mainly three different possible connection designs known as radial, ring, and star [6–10].

6.2.1.1 Radial In the radial collection system, the wind turbines included within the same feeder are installed in string configuration as shown in Figure 6.2. The

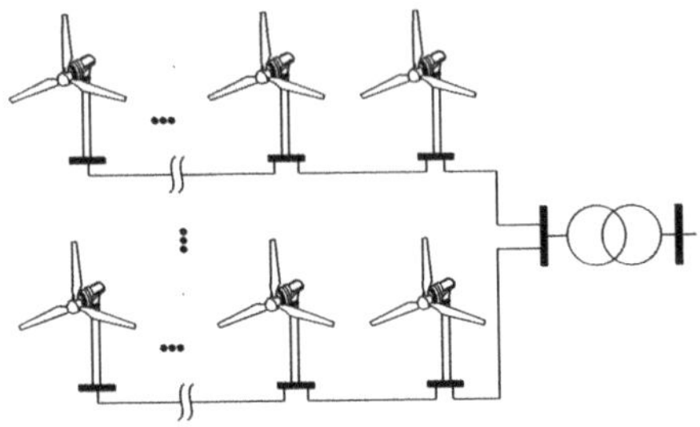

Figure 6.2 Radial collection configuration.

maximum number of wind turbines that can be connected to one feeder is determined by the cable ampacity and the rated power of the generators. It is the most common, cheapest, and simplest collection system but it presents some reliability issues, since if a failure occurs in the cable connecting the first turbine and the hub of a feeder, all the power generated by the downstream wind turbines in the string will be lost [6–9, 11].

6.2.1.2 ***Ring*** The ring collection system (shown in Figure 6.3) can be understood as an improved version of the radial design in terms of reliability, but it becomes costly. There are different ring designs defined depending on how the ring is formed; they are known as single-sided, double-sided, and multi-ring [7, 9, 11]. In all cases, redundant cables are added so that the power flow within a feeder have more options to be transmitted. Thus, a single-sided configuration (Figure 6.3a) consists in connecting a cable from the outermost turbine in the feeder to the collector hub whereas in the double-sided ring two feeders are connected together by means of a cable as it is shown in Figure 6.3b. One drawback of this scheme is that some cables must be oversized to allow bidirectional power flow in case of a cable failure.

6.2.1.3 ***Star*** The star collection system attempts to reduce the cable ratings of the cables which connect the wind turbines and the collector point. Such common

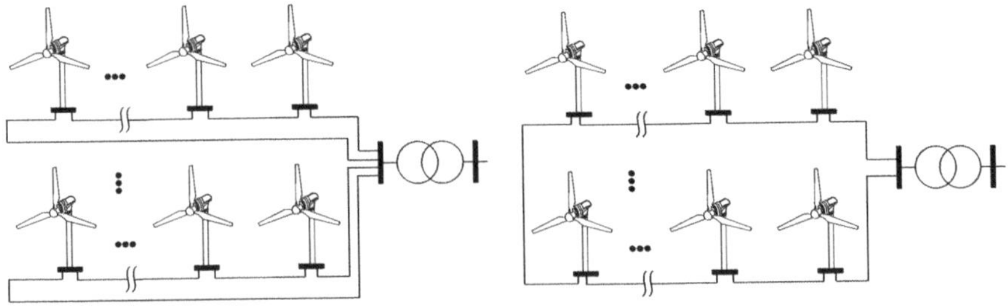

(a) (b)

Figure 6.3 Ring collection configuration. (a) Single-sided ring example. (b) Double-sided or multi-ring examples.

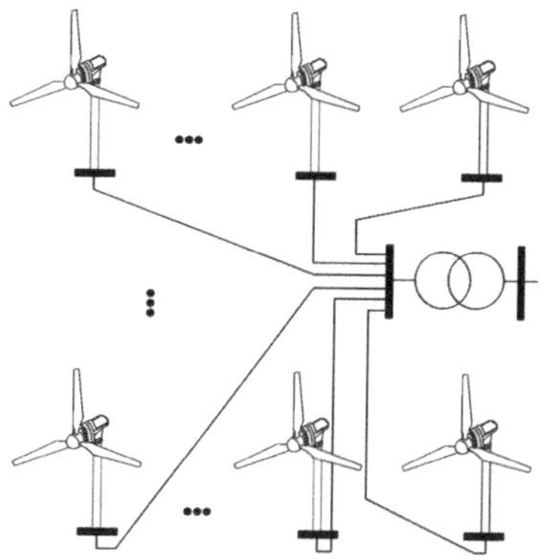

Figure 6.4 Star collection configuration.

connection point is usually located in the middle of all wind turbines disposition, as drawn in Figure 6.4.

The advantage of this topology is that the reliability of the system increases, since a cable failure causes the loss of only one machine. However, due to the longer cable lengths and the use of lower voltage ratings for the connection of wind turbines in this configuration, cable losses and their costs are significantly higher than in other WPP designs [7–9, 11, 12].

6.2.2 Wind Power Plant Components

After the presentation of different AC OWPP topologies, the required technologies to the development and implementation of those (such as cables, transformers, protections, control systems, and offshore platforms) are briefly described in the following subsection. It is worth remarking that the different existing concepts of wind turbine technologies as well as the VSC HVDC are not included since they are described in detail in Chapters 3 and 5. Likewise, the transformers, protections, and control systems presented below refer to those located at the offshore platforms.

6.2.2.1 Cables In general terms, AC cables can be classified in single-core and three-core cables (see Figure 6.5). The former refers to three single cables (one per

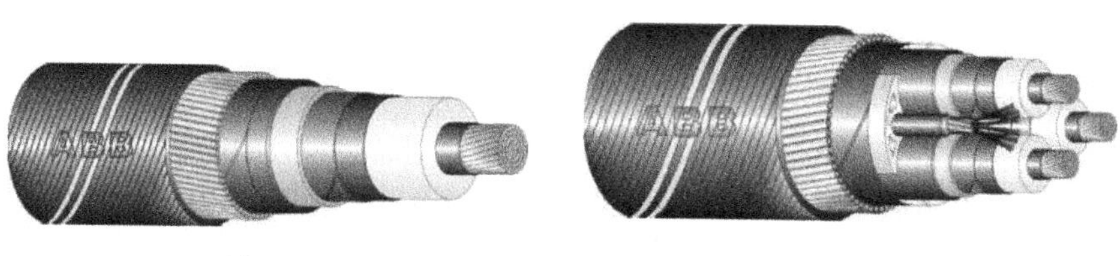

(a) (b)

Figure 6.5 Sea cables for offshore wind farms (courtesy of ABB). (a) Single-core cable with lead sheath, and wire armour. (b) Three-core cable with optic fibers, lead sheath, and wire armour.

Figure 6.6 Liquid-filled, transformer. (Courtesy of ABB.)

phase) separated by a certain distance from each other. The latter consist of three insulated cores bundled together in trefoil configuration, made up of copper (usually) or aluminium (for cost and weight reduction but for small power ratings), all sharing a common over sheath and armouring. Three core cables have the advantage of presenting reduced power losses due to the cancellation of magnetic fields between three cores. In addition, these are considerably cheaper to install compared to single cables, since the cable is laid in only one instance and it does not require extra operations for laying fiber optical cable as it is possible to produce the cable with integrated communications. However, and despite sharing some components, the handling of three core cables is more difficult than installing three single core cables due to their greater weight [13]. Furthermore, other drawbacks of three-core cable refer to the greater technical difficulty of making cable joints and the fact that a single failure occurring in a three-core cable means that we need to replace the damaged cable with an entire new one instead of having to only substitute the single-phase cable affected.

Generally speaking, the most used submarine cable technologies for power transmission and distribution systems are listed in the following:

- **Mass-Impregnated Cables:** These cables are characterized by the fact that the insulation consists of impregnated paper tapes with a high-viscosity compound. Also, laminated polypropylene can be used instead to further increase the power ratings. These insulations are good enough to avoid using pressured fluids inside the cable as further protection. Furthermore, the protection is enhanced with layers of polypropylene and galvanized steel wire armour [13]. The conductor is made up of copper, but considering high operating temperatures it can be also aluminum. Concentric conductor designs are also available to improve reliability in the case of a single cable fault. They are commercially suitable for voltages of up to 45 kV AC and 500 kV DC [14] and are able to transfer hundreds of MW per cable.

- **Extruded Insulation Cables:** As a difference with mass-impregnated cables, the insulation of extruded cables is based on laminated EPR and XLPE. This allows us to operate cables at high temperature (near 100°C) and to withstand high electrical stresses. Also, other sheets of lead sheath or aluminum are applied over each core for water isolation in case of using XLPE. This additional isolation is not needed in the case of using EPR, however [14]. Furthermore, other components such as additional layers based on polypropylene for corrosion protection and galvanized steel wire armour to increase tensile strength could be included for submarine applications.

6.2.2.2 ***Transformers*** Power transformers located at the offshore substations are needed to step up the voltage of the interarray cables (usually from 10–36 kV to 110–150 kV) in order to reduce export cable losses. Power transformers could provide voltage control services thanks to the inclusion of tap changers.

The design of power transformers for offshore applications does not differ much from those used in the electrical power system. However, corrosion issues are specially addressed for protection in marine environments. Transformers are usually made up of copper windings, which are wrapped around laminated iron cores. These components are immersed in oil for cooling purposes. Power transformers is a widely proven and mature technology field. This way, high levels of reliability and long life (more than 20 years) can be achieved. Nevertheless, offshore transformers require special attention as they are designed with the objective to minimize maintenance and adverse affects of the offshore environment. As such, different transformers are used, generally more costly than onshore transformers.

6.2.2.3 ***Switchgear*** Electrical and mechanical protection elements are necessary to protect all the electrical components within the wind farm from any type of electrical fault. The most relevant protection devices (switchgear) are the circuit breakers, relays, fuses, and so on. Those elements may need coordinated actions among them in order to ensure proper fault clearing, as well as both safe and correct equipment operation; for example, any element functioning correctly is disconnected by mistake [15].

The main purposes of the switchgear at medium voltage side (offshore grid collector) is to protect both wind turbine equipment and wind farm array components, connect various wind turbines in string or radial, and finally control the connection and disconnection of single wind turbines or strings. To be capable of driving such control remotely when needed, switchgears may use the Supervisory Control and Data Acquisition (SCADA) system to provide the control signals [16].

In general, existing switchgear can be classified into two main types according to their insulation technology. The first type is based on air, which is the one used in conventional substations, while the other uses the sulfur hexafluoride (SF_6) gas, which is a superior dielectric gas [17]. The latter is the technology mainly used for offshore locations due to its compact structure and low maintenance needs in comparison with the former.

It is important to remark that the switchgear must be as small, light, and reliable as possible due to the special conditions related to offshore locations. To be able to reduce offshore substations, the switchgear must fit in limited space when located at the wind turbine towers, and there must be tight limits on weight.

6.2.2.4 *Offshore Platforms* Offshore platforms are a crucial component of the entire offshore grid's connection to shore. They can be divided into two types: collector platform and HVDC platform. The former is an intermediate offshore platform located between the OWPP and HVDC substation whose main tasks are to gather the MVAC interarray cables from the wind turbines and to reduce electrical losses by increasing the voltage (from MV to HV) just before the transmission link to shore. The latter is usually located slightly away from the OWPP to facilitate others OWPPs connections and to transmit the combined power generated by all of them through a single HVDC-link.

Figure 6.7 shows the BorWin alpha HVDC platform that was installed in the German part of the North Sea in June 2009 [18]. As shown, the platform consists of a foundation structure and a topside.

The most relevant foundation structures for offshore platforms (both collector and HVDC platforms) can be a monopile (similar to the wind turbines), a hybrid or gravity-based (built up of a concrete caisson with a steel leg structure mounted on its top), or a jacket construction. The last of these has been widely used in the offshore

Figure 6.7 BorWin Alpha HVDC offshore platform located in the German North Sea, around 130 km from the coast. (Courtesy of ABB.)

oil and gas industry for many years and appears to be as a solid alternative for the construction of the future offshore platforms. It is made up of three or four main legs (depending on the seabed conditions and the platform weight) and supported by piles in each corner of the foundation structure. Moreover, it contains J-tubes to route the interarray cables from the offshore collection grid onto the platforms. Aside from being required as a steel frame supporting the topside, the jacket structure is designed to bear against multiple constraints such as the impact of the waves, corrosion, or the flow of the sea water streams and tides, among others.

The platform topside is equipped with different devices to make a substation operational such as the HVDC converter, transformers, switchgears, backup diesel generators, coolers for the ventilation, winch to hoist the subsea cables, pumps, fans, and so on. Furthermore, it accommodates other required necessities such as living quarters to host all the people on board, a crawler crane, a helipad, a boat landing, safety equipment, and a meteorological mast, among others.

In order to design the appropriate platform topsides dimensions and weights, key factors such as water depth and the size of the wind power plant are essential to be considered. Typically, platforms topsides can weigh about 2.000 tonnes and have an area of 800 m^2 and a height of 25 m above sea level. In the case of a large wind power plant, more than one offshore platform may be required.

Due to the large size of the offshore platforms and the extreme difficulty of constructing the installation at sea, all platforms are assembled on land and then transported out to the sea. Regarding their location, platforms should be situated somewhere which can be easily accessible in the future and without damaging any interarray or export cable.

To date, the only HVDC offshore platform constructed is the Borwin Alpha HVDC platform (see Figure 6.7) [18]. However, there are new projects, considering offshore wind power plants connected to the main grid through HVDC transmission links, under development—for example, Dolwin1 (ABB), Helwin 1 (Siemens), and Dolwin 3 (Alstom) [19].

6.2.2.5 Control Systems The global control and management of the offshore wind farm is done by the Supervisory Control and Data Acquisition (SCADA) system. The SCADA system is the responsible to provide real time visibility of the wind power plant operation by means of monitoring, and it has the ability to control the wind farm both centrally or remotely. The latter is the usual choice for offshore locations [20].

The SCADA introduces some advantages on the wind power plants, including: integration with wind turbine controls; detailed information for wind turbine and network diagnosis; reduction of the maintenance needs, which is really important offshore, and fulfillment of power generation requirements [21].

With the aim of wind power plant monitoring and control, the SCADA requires a communication network between the wind turbines and the HVDC converter, and the latter also with shore [16]. Such communications can be done through different technologies such as copper twister pair (RS485), radio telemetry, or fiber optic, which is the most common for plant interconnections due to its speed and bandwidth.

It is worth remarking that not only wind turbine manufacturers develop their own SCADA system—for example, WindAccess (Alstom Wind) [22], Wind Farm

SCADA (REpower) [23], VestasOnline Compact II (Vestas) [21]—but also other companies such as GH SCADA (Garrad Hassan) [24], among others.

6.2.3 Protections and Grounding of Offshore Collector Grids

Distribution network protective systems are designed to detect excess of current caused by insulation failures. The typical fault types are three-phase, two-phase with or without ground, single-phase to ground (depending on the grounding of the neutral), and double earth fault. These faulty conditions are only allowed in a short time (approximately 0.5 s) in order to limit the possible hazards associated with such high currents:

- Risk to life of persons by high step and touch voltages caused by the high currents flowing into the ground
- Risk to plant equipment caused by destructive heating and electromagnetic forces
- Risk to the stability of the power system

Protection systems for offshore collector grids are usually designed as established in well-known practice for protection of distribution level power systems and according to standards and national codes. However, there are two significant differences [25]:

- By definition, offshore collector grids are connected to weak power grids and therefore low-fault currents has to be expected. In such conditions, protective device may have difficulties to detect faulty conditions correctly. This can lead to more than less permanent faulty condition that may increase the risk to humans and equipment.
- Inverter interfaced wind turbines are not able to supply high short-circuit currents.

The medium-voltage networks connecting the individual wind turbines in an offshore wind farm have typically voltage levels between 11 and 33 kV. The particular case of extended undersea cable networks needs also special attention in the design stage [26]. For the proper design of such networks, the system earthing (i.e., an intentional connection of the neutral point of the electrical power system to earth) is especially important.

Solid grounding leads to larger single line ground fault currents that enable fast selective clearance [27]. The amplitude of the single line ground fault current is determined by the transformer zero sequence impedance and can be influenced by the number and location of the transformer neutrals used in the grid. A low short circuit during the faulty condition leads to overvoltages, and so a trade-off between the fault currents and overvoltage has to be done during the design. The isolated neutral arrangement is generally not recommended in such large cable networks due to the high capacitive earth fault current, greater than the standard allowable limit for isolated grids [28]. Additionally, excessive voltage stress with overvoltages reaching about 1.73 p.u. during fault would pose a danger to the whole grid. The well-known advantages of resonant grounding are not paying for in an undersea cable network [27].

Usually, the protection system will be organized within a zone protection system. On the machine level, HV fuses and MV breakers are used, depending on economic and space requirements. Circuit breakers that protect against faults in the transformer and in the cable to the next turbine can be operated from the shore via the SCADA system. Also, manually operated equipment to isolate and earth the cables and transformer can be found. Gas-insulated switchgear (GIS) is suitable for this type of application and is available from several manufacturers with demonstrated performance on offshore installations [29]. Special attention may be drawn to the transformer inrush currents to avoid unwanted tripping of the protection.

Single phase to ground faults should be cleared by means of neutral voltage displacement relays connected to HV-side voltage transformers as some transformer connections do not transfer zero-sequence currents. The export cable to the grid should also be protected by tripping the breakers on both ends [25]. Due to the low short-circuit currents, mostly overcurrent protection is not very efficient in such situations. One solution is to apply transfer tripping between upstream protection systems and the wind park. Tripping of protection relays can be transmitted by means of communication. Also the use rate-of-change of frequency relays (ROCOF) or loss-of-mains relays can be used for islanding protection. For transformer or reactor is sensitive to overvoltage in combination with underfrequency. This is called "volts per hertz" protection [30].

To guarantee safety of personnel and equipment in the case of electrical failures and to prevent fires, mechanical damage or malfunction resulting from lightning or static electricity, proper earthing, and equipotential bonding is necessary [31]. As the platform will be exposed to lightning, especial precautions shall be made to prevent damage to the structures and the equipment on the platform. The lightning protection systems shall be designed in order to ensure that no major equipment is exposed to a direct lightning flash [31]. The main functions of the grounding system for lightning of the wind turbine are:

- to ensure safe step and touch voltages for people staying close to the wind turbine tower footing during possible lightning strikes and
- to ensure a low grounding resistance for the lightning current and create a reference potential to each equipment can be connected.

Equipotential bonding shall be made locally in close proximity to the parts which require bonding. Every bonding connection to earth shall be of corrosion resistant and secured against becoming loose due to vibration, which is especially important in offshore environments [31].

6.2.4 Offshore Wind Power Plant Electrical Design

Offshore wind power is an emerging energy sector with a great untapped potential. Today, offshore wind is still a young industry that must overcome technical, economic, and political challenges to be fully competitive in the longer-term compared to other types of electricity generation. According to reference 32, the current levelized cost of energy (LCOE) for offshore wind power is estimated to be between 119 and

194 €/MWh, while for onshore wind it ranges from 45 and 107 €/MWh. These figures highlight the necessity for cost reduction, which can be achieved, inter alia, through a commitment from government and industry to encourage the planning and development tasks carried out in the early stage of any offshore wind project. An important task within all these activities corresponds to electrical offshore wind power plant (OWPP) design. This aims to reduce LCOE by improving system reliability and availability, reducing O&M costs and/or increasing energy generation.

6.2.4.1 ***Offshore Platform Electrical Design*** With the aim of providing an insight on equipment arrangements and layout within an offshore platform, two electrical designs based on National Grid's references offshore design arrangements that act as a standard block [13] are introduced below as illustrative examples.

These designs are represented as a single line network and refer to a 900-MW AC OWPP with an HVAC transmission (AC900) and a 1000-MW AC OWPP with an HVDC transmission (HVDC1000).

AC OWPPs Connected with HVAC The reference arrangement AC900 consisting of common electrical equipment such as transformers, circuit breakers, busbars, earthing transformers, and cables is displayed in Figure 6.8. As can be seen, it

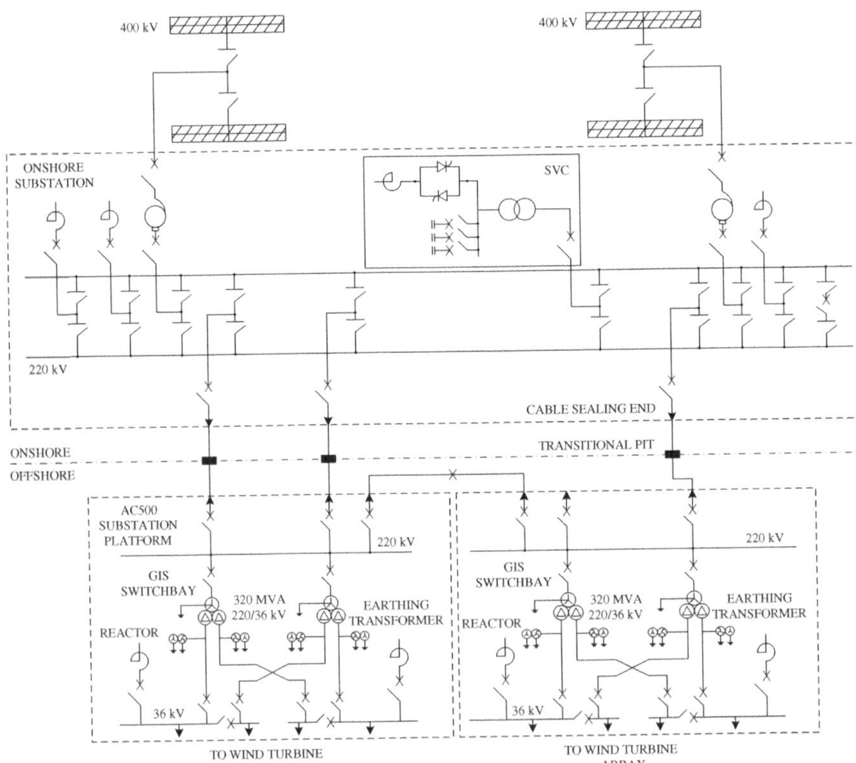

Figure 6.8 Single-line diagram based on the AC900 scheme proposal based on National Grid's reference offshore design arrangements [13].

is composed of two 600-MVA offshore substation platforms to which the radial feeders from the wind turbine array are connected. Each substation contains two 320-MVA (220/36-kV) tertiary transformers with a star/delta/delta configuration, with the primary winding solidly earthed. Two earthing transformers are installed on the secondary side to provide an earth on the delta winding of the transformer so that the protection system is enhanced. The low-voltage secondary side of the network has a number of busbars which can operate as one solid bus or as a sectionalized busbar in order to provide more redundancy to the system and, thus, improve its reliability. Additionally, two shunt reactors per platform are connected to the low-voltage busbar aiming to cater to reactive power support. As this system is oversized because it has four 320-MVA transformers and a wind power plant capacity of 900 MW, a fault or a maintenance outage on one of the tertiary transformers would not result in a loss of the power generated by the wind farm. In reality the transformers are operated at 70–90% of their rated value when wind speed is higher than its nominal value in order to optimize their lifetime.

The two offshore platforms have the possibility to be interconnected to form one 220-kV solid bus or to sectionalize or isolate one of them in case of a fault. The power generated by the offshore wind power plant is transmitted to shore through three 220-kV three-core AC submarine cables. The transition joint pit is an intermediate point where the onshore network meets the offshore network.

Regarding the onshore side, two feeders are connected to the main 400-kV onshore network providing redundancy to the system to improve its secure operation. Each feeder is connected in turn to an autotransformer which steps down the voltage from 400 kV to 220 kV and is earthed on the primary star point. In order to provide voltage support, an SVC and three shunt reactors are required to be connected to the 220-kV busbar.

AC OWPPs Connected with HVDC To illustrate the electrical system design enclosed between the main onshore network and an offshore substation platform that has been built to operate an AC OWPP with an HVDC-link transmission system, the reference HVDC1000 reported in reference 13 is described. Figure 6.9 depicts the detailed single-line scheme.

As shown in Figure 6.9, the HVDC1000 design consists of three offshore platforms (composed of one VSC converter and two 520-MVA substations) and one VSC onshore converter station platform. Thus, the radial feeders from the wind turbines array are directly connected to one of two 520 MVA AC intermediate collector platforms. The AC offshore platforms arrangements are similar to that presented above for the AC900 scheme, with the only exception that no shunt reactor is required. Therefore, two 320 MVA terciary transformers are installed in each platform increasing reliability. Both platforms are directly connected to the solid bus of the VSC offshore platform via 3-core AC submarine cables, which voltage is rated at 220 kV. Likewise, the two onshore and offshore VSC station platforms are interconnected by means of two ±300-kV submarine dipole cables. Finally, the HVDC VSC converts the ±300-kV DC voltage into 400-kV AC voltage for the main onshore grid. As in the previous case, the primary winding of the autotransformer is solidly earthed.

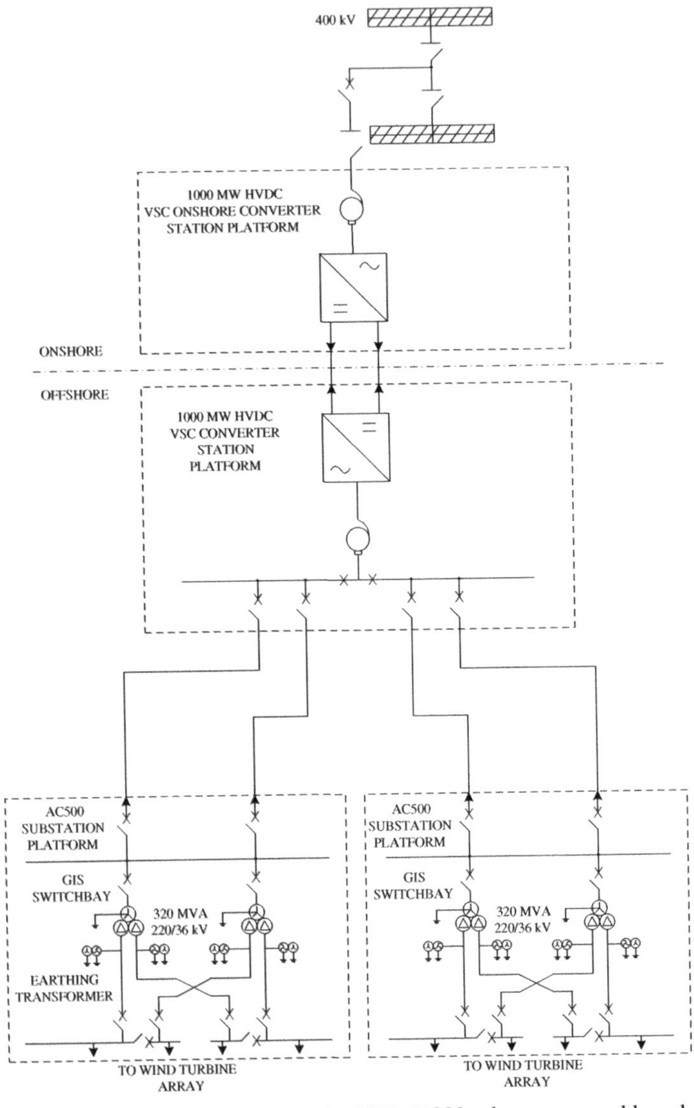

Figure 6.9 Single-line diagram based on the HVDC1000 scheme proposal based on National Grid's reference offshore design arrangements [13].

Theoretically, this type of design based on DC technology should be more economical than any AC scheme that involves large rated powers (over 1000 MW) and that is located further from the shore.

6.2.4.2 Design Methodology When designing an offshore wind power plant project, one of the tasks that is performed prior to technical analysis is the cable selection process. This activity is carried out once all the wind turbine positions

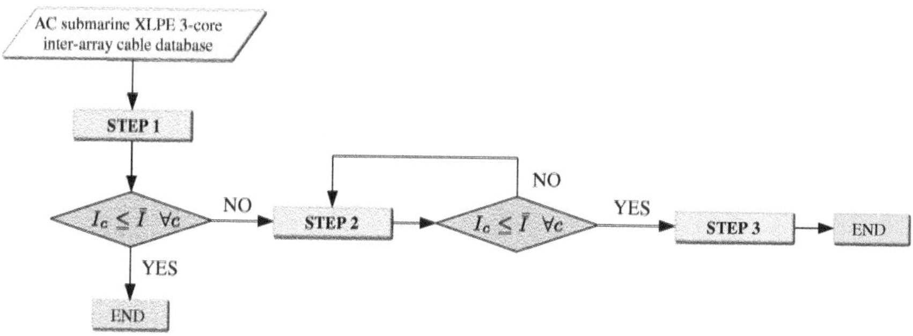

Figure 6.10 Flow chart of the cable selection process.

and locations of the offshore substations have been defined by optimizing the wind farm layout based on environmental, met station, and sea bed surveys. It aims at determining the optimal number of parallel lines to install and selecting their optimal cable cross sections to be used to interconnect the wind turbines, so that the capital expenditures of the cables and their installation costs are minimized, provided that the proper operation of the system under full load is ensured. Figure 6.10 shows a flow chart of the methodology used for the cable selection process.

From the proces of Figure 6.10, it follows that during an initial step the selection of the most economical available cables (i.e., minimum cross sections) is performed, ensuring that the maximum admissible loading under full load operation is not reached ($I_c \leq \bar{I}$, with I_c the maximum cable current and $\bar{I}$ the rating of the cable) (STEP 1). Next, if any of the interarray cables interconnecting wind turbines is overloaded (since none of the available cables in the database is capable of operating properly under full load conditions), STEP 2 is executed. Otherwise, the process ends. STEP 2 deals with the selection of the minimum number of parallel cables that are needed to work within the admissible conditions. This process is only carried out for those cables with the highest available cross section that are overloaded. Finally, once all the currents of the cables are below its allowable limit, STEP 3 is performed. This step aims to reduce the cross sections of those cables with two or more parallel cables in order to minimize the total cable costs of the offshore wind power plant.

Once the cable selection process is completed, the technical analysis of the offshore wind power plant is performed. It consists in estimating the total energy yield by the wind farm over its lifetime. This technical assessment may be more or less accurate depending on the degree of rigor required by the study being undertaken. For instance, in those studies aiming to get a rough idea of how much energy an offshore wind power plant can generate throughout its lifetime, a common assumption is based on computing the potential energy output generated by the OWPP (i.e., the hypothetical energy generated by the wind farm if it were continuously operating under full load conditions) and multiplying the resulting value by the capacity factor of the OWPP. This capacity ratio accounts to be approximately 40% for the existing offshore wind power plants [33]. Another possible approach is to consider that all wind turbines have the same wind speed and to calculate the energy produced by the wind farm using the average wind speed at the OWPP location during the time given by the available data. Other technical assessments, by contrast, estimate the energy generated by the OWPP by multiplying the electrical power produced by the wind turbines as a function of the wind speed and taking into account its probability of occurrence. Mathematically, this is expressed as

$$E^G = NT \int_{v_{min}}^{v_{max}} P^G(v) f(v)\, dv \tag{6.1}$$

where N is the number of wind turbines comprising the wind farm, T is the period of time considered for the analysis, v_{min} and v_{max} are the minimum and maximum wind speeds considered for the case under study, $P^G(v)$ is the electrical power generated by a wind turbine as a function of the wind speed, and $f(v)$ is the likelihood of occurrence of each wind speed (which is assumed to be given by the Weibull distribution function). Thus, $P^G(v)$ and $f(v)$ can be computed as follows:

$$P^G(v) = \frac{1}{2}\rho A C_P(\lambda, \beta) v^3 \tag{6.2}$$

$$f(v) = \frac{k}{c}\left(\frac{v}{c}\right)^{k-1} e^{\left[-\left(\frac{v}{c}\right)^k\right]} \tag{6.3}$$

where ρ is the air density, A is the area swept by the turbine blades, C_P is the power coefficient which is a function of the tip-speed ratio λ, and the pitch angle β, v is the wind speed, and k and c are the dimensionless shape (k) and scale (c) parameters characteristics of the Weibull distribution function.

Figure 6.11 shows the Weibull distribution function (a), the power generation curve (b), and the energy yield by a wind turbine as a function of the wind speed (c). It is worth noting that at high wind speeds (under full load operation) the total available power that can be extracted from the wind (dash gray line) is curtailed by the pitch control action with the aim of keeping the power output constant at rated power (solid black line) in order to avoid overloading. Therefore, the area enclosed by the solid black curve of Figure 6.11c corresponds to the total actual energy produced by a wind turbine during a certain period of time T.

Both studies previously mentioned estimate the total energy generated by the OWPP assuming an identical behavior for all wind turbines since all of them are considered to be subjected to the same wind regime. However, any wind power plant presents some wind speed variability among wind turbines caused by the wake effect given for each wind direction. Thus, Eq. (6.1) must be modified (if required) by including the wind direction distribution function (i.e., the wind rose distribution), and taking into account the wake effect among turbines. Thereby, the energy generated by a single wind turbine and the WPP can be computed as

$$E^G_{wt-i} = T \int_{0°}^{360°} \int_{v_{min}}^{v_{max}} P^G_{wt-i}(v, d) f_1(v) f_2(d)\, dv\, dd \tag{6.4}$$

$$E^G_{WPP} = \sum_{i=1}^{N} E^G_{wt-i} \tag{6.5}$$

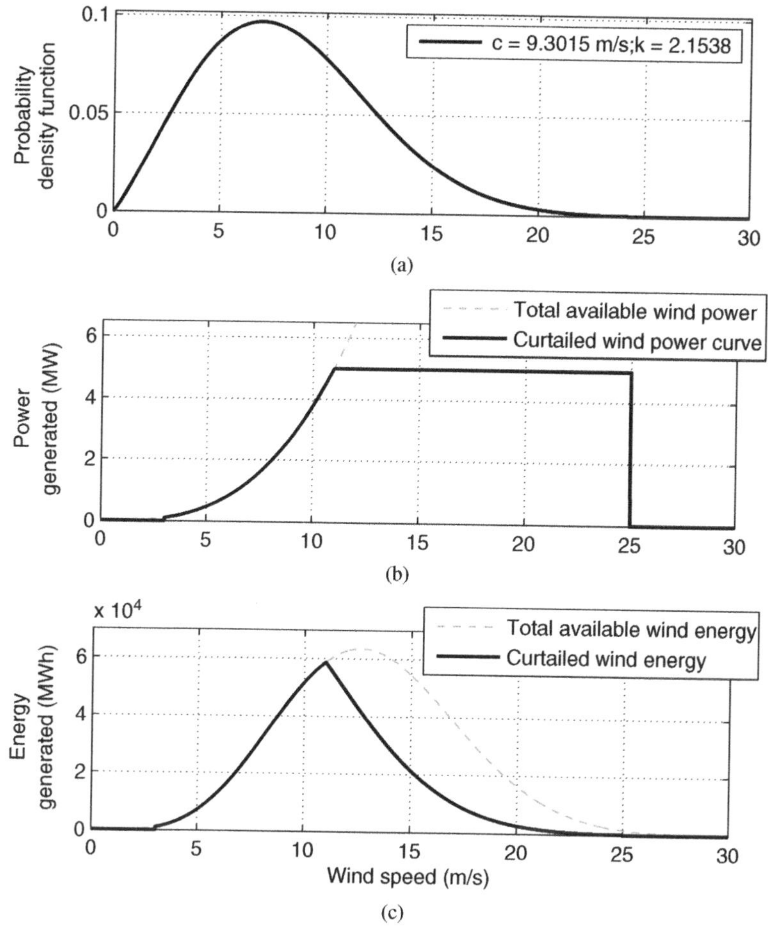

Figure 6.11 Energy yield as a function of wind profile and power capability function. (a) Weibull distribution function. (b) Power generation curve of a wind turbine. (c) Energy yield function by a wind turbine during a certain period of time T.

where $P^G_{wt-i}(v,d)$ is the power generated by each wind turbine-i for a particular average wind speed (v) and a certain wind direction (d) and $f_1(v)$ and $f_2(d)$ are the probabilities of occurrence according to the Weibull and wind rose distribution functions, respectively.

In practice, in order to obtain the most accurate computation of the total energy that can be produced by an OWPP over a certain period of time, a large number of factors should be taking into consideration, such as the aerodynamics of the wind turbines, the wind farm layout, the wind speed, and wind direction distributions and the turbulence levels.

It is also worth noting that such total energy that can be generated by an OWPP over a certain period of time for a particular wind farm layout and specific wind conditions does not correspond to the actual energy delivered to the grid through the point of common coupling (PCC) during this time. To do so, the total energy losses produced in the OWPP and its availability should be considered in the technical analysis. These losses can be classified into two main groups: the steady-state losses or power flow losses and the unavailability or maintenance losses.

The former type of losses refers to the electricity dissipated as heat within the OWPP as a result of the efficiency losses of its components (transformers, converters, cables, etc.). and it is computed as

$$E_{pf}^{L} = T \int_{0°}^{360°} \int_{v_{min}}^{v_{max}} [P^{G}(v,d) - P^{T}(v,d)] \cdot f_1(v) f_2(d)\, dv\, dd \qquad (6.6)$$

where $P^{G}(v,d)$ is the power generated by all wind turbines comprising the WPP and $P^{T}(v,d)$ is the net active power transferred to the grid at the Point of Common Coupling (PCC) obtained by means of power flow calculations.

The latter type of losses are more complex to calculate, since they occur due to the unforeseen equipment failure (corrective or unexpected maintenance losses) or because of partial or total outages of the installation during a fixed time for preventive maintenance purposes (preventive or planned maintenance losses).

With regard to the unplanned maintenance losses, it can be calculated based on the reliability multi-state models explained in detail in reference 34, in which the expected energy not supplied (EENS) depends on the failure rate (λ) and the mean time to repair (MTTR) of all the components encompassed within the OWPP. This model considers that each component has several states of service and has a probability of malfunction in each state per year. Concerning the preventive maintenance losses, they are planned to be performed during periods of low wind speeds where the energy yield is limited or nonexistent in order to maximize the availability of the offshore wind power plant. However, this is not always possible to be achieved due to the stochastic behavior of the wind, so some downtimes may occur in periods of high winds. Hence, the actual energy transferred to the grid by the OWPP can be computed as

$$E_{WPP}^{T} = \eta_{availability} \cdot E_{WPP}^{G} \qquad (6.7)$$

According to some experiences from Middelgrunden and Nysted OWPPs reported in references 35 and 36, respectively, the average availability of the existing OWPPs can be assumed to be 95–96% (covering both wind turbine and collection grid availability).

6.2.4.3 *Case Study* In this section, the results of applying the previously explained design methodology to a particular case study are shown. For this example, the OWPP analyzed consists of 80 wind turbines laid out in a rectangular matrix form of 10 columns and 8 rows. It is considered that the wind farm adopts a radial design. Likewise, the spacing between two nearby wind turbines is 7 rotor diameters (D) in both directions. The wind conditions (wind rose and Weibull distribution functions) used for the OWPP under study are presented in Table 6.1 [3]. Twelve 30° wind

TABLE 6.1 Wind Conditions (Wind Rose and Weibull Distribution Functions) at the OWPP Location [3]

Wind Direction	Probability %	Mean Wind Speed (m/s)	Scale Factor (c) (m/s)	Shape Factor (k)
N	8.2	7.34	11.15	1.27
NNW	4.5	6.39	9.04	0.55
WNW	3.7	6.89	10.02	0.53
W	3.9	7.59	10.98	1.26
WSW	9.5	9.13	10.95	1.71
SSW	13.5	8.72	10.93	1.62
S	12.7	7.93	11.22	1.73
SSE	12.3	8.34	10.90	1.69
ESE	10.1	8.41	10.82	1.70
E	6.4	8.36	10.61	1.48
ENE	6.5	8.01	10.55	1.97
NNE	8.7	8.08	10.86	2.14

direction sectors and a different set of Weibull parameters have been considered (mean wind speed, scale (c) and shape (k) parameters) for each direction. The electrical characteristics of the AC submarine 33-kV XLPE three-core cables database are shown in Table 6.2.

For the cable selection process, three examples are displayed in Figure 6.12 by varying the rated power of the wind turbines. As can be seen, when the rated power of each wind turbine is 2.5 MW (Figure 6.12a), only two different cross sections (95 mm^2 and 120 mm^2) are needed to ensure proper operation of the wind farm under full load conditions. Nevertheless, the number of cable cross sections needed to minimize the costs associated with cables and their installation while avoiding

TABLE 6.2 Electrical Characteristics of the AC Submarine 33-kV XLPE Three-Core Cables Database

Section (mm^2)	R(Ω/km)	L(mH/km)	C (μF/km)	Ampacity (A)
095	0.2478	0.420	0.161	358
120	0.1967	0.401	0.176	406
150	0.1597	0.387	0.188	452
185	0.1281	0.374	0.203	507
240	0.0981	0.358	0.228	582
300	0.0790	0.344	0.244	649
400	0.0629	0.331	0.270	713
500	0.0505	0.315	0.300	790
630	0.0409	0.305	0.328	861

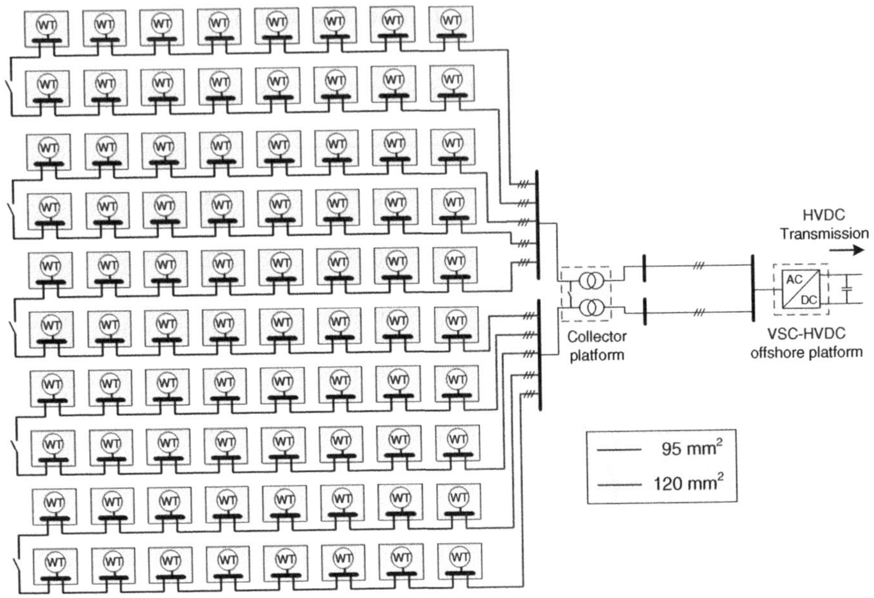

(a)

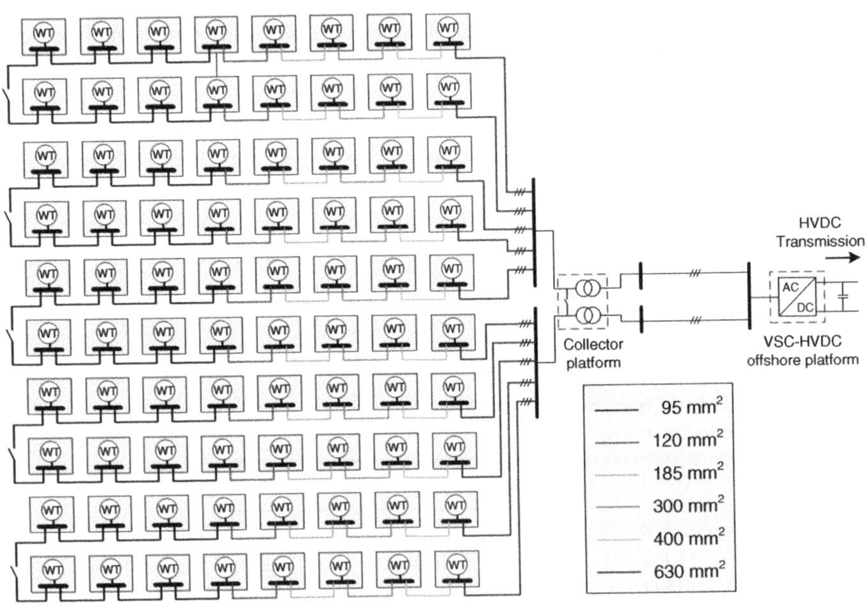

(b)

Figure 6.12 Result of applying the cable selection process to 3 OWPPs consisting of 80 wind turbines with different power ratings. (a) Rated power of each wind turbine = 2.5 MW. (b) Rated power of each wind turbine = 5 MW.

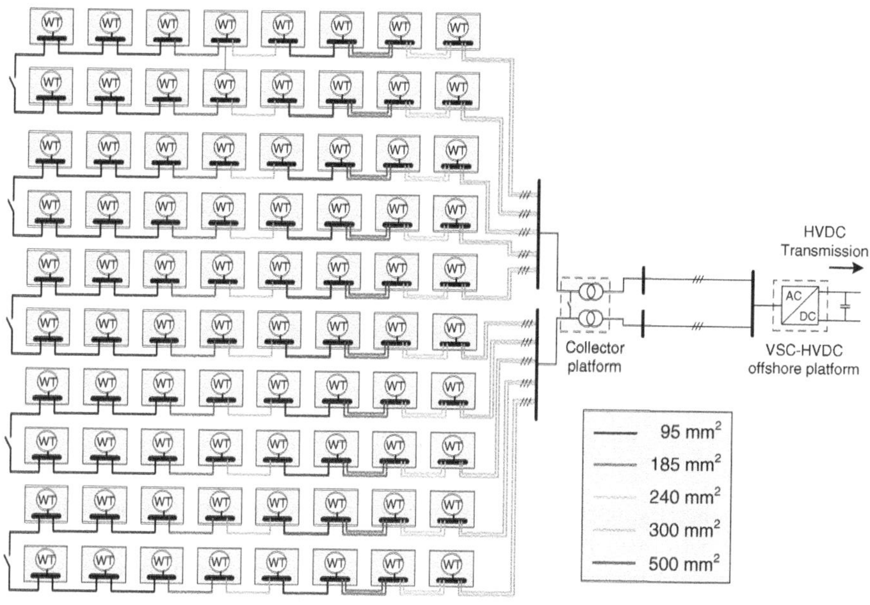

(c)

Figure 6.12 (Continued) (c) Rated power of each wind turbine = 7.5 MW.

overloading operation is increased if the rated power of the wind turbines amounts to 5 MW (Figure 6.12b). Both examples refer to STEP 1 of the cable selection process of Figure 6.10, since there is no need for adding parallel lines with the available cable database. Conversely, STEP 2 and STEP 3 of the cable selection methodology are required for the third example shown in Figure 6.12c, where the wind turbine rated power is 7.5 MW. Thus, the number of cables placed in the first three rows must be duplicated for guaranteeing the correct behavior of the system under full load operation. Thereby, these cables can be reduced in cross section to optimize the total cable cost.

In order to demonstrate that, for the given cable database, it is always more economical to install a single cable (albeit the cable has the highest available cross-section, worst case) than installing two parallel lines with the smallest cable cross sections (best case), Figure 6.13 is presented. As expected, the results indicate that as the cable length increases, the cost increment by installing two parallel cables is much greater. Therefore, it justifies carrying out STEP 2 of the cable selection process only if STEP 1 has not been successful.

Due to the currently limited vessels suitable for offshore wind power plant installations and the excessive cost that may lead to requiring one of them for a long period of time, it may be more feasible, depending on the cases, to simplify the results obtained in the cable selection process by reducing the number of cable cross sections to two or three different ones, so that the required number of trips is minimized.

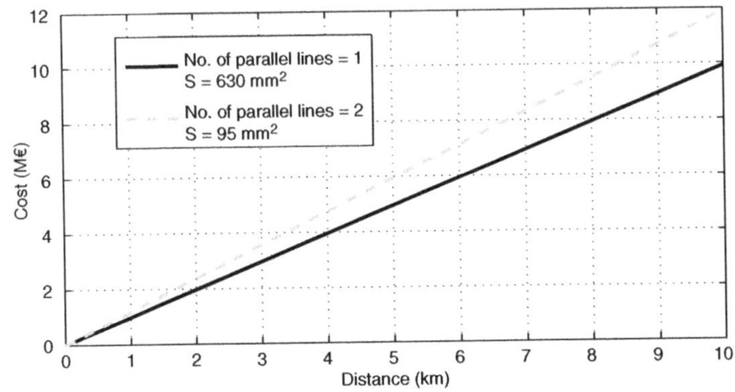

Figure 6.13 Demonstration that it is more economical to install a single 630-mm^2 cross-section cable than two parallel cables of 95 mm^2.

Finally, once the interarray AC submarine cables are selected, the technical analysis can be performed. Figure 6.14 shows an example of the results obtained from a load flow simulation considering wind turbines operating at rated power (2 MW). As can be seen, the power flow calculation determines the voltage magnitude (u) and the voltage angle ($phiu$) of the nodes, as well as the active (P) and reactive (Q) power flow on branches. Furthermore, the figure displays other results such as the apparent power (S) and the current (I), loading ($loading$) and power losses (P_{loss})of each branch. According to the results, the net active power transferred to the grid at the PCC ($P^T(v, d)$) for this particular example (rated wind speed and 0°) is 149.67 MW. Therefore, the total energy yield per year considering an availability of 96%, and taking into account the wind conditions presented in Table 6.1 amounts to 503.47 GWh/year.

6.3 DC OWPPs

The DC nature of the high voltage transmission can be extended to the collection grid and to consider the possibility of having an entire OWPP in DC [12, 37–41]. Thus, within this section, a brief introduction to possible topology designs as well as technological research and development needs for future DC OWPPs, are provided.

6.3.1 Topologies Overview

As it occurs in the conventional AC case, there are various topology proposals for DC OWPPs. These proposals are named regarding how the wind turbines are connected among them such as shunt, series and hybrid.

6.3.1.1 Parallel Topology In the shunt topology, the wind turbines are electrically connected in "shunt" as shown in Figure 6.15. In this case the output voltage of each wind turbine is maintained constant, while the current flowing through the

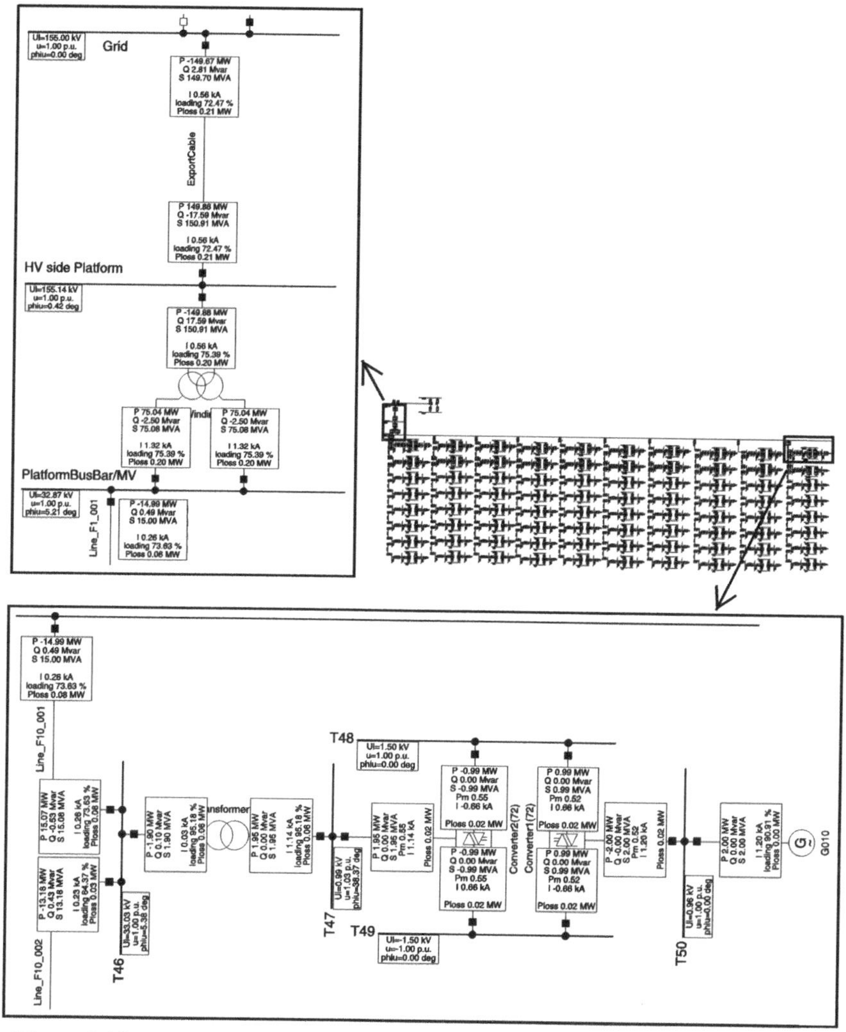

Figure 6.14 Load flow solution obtained with DIgSILENT Power Factory® software.

interarray cables depends on the number of turbines connected on it. Thus, the power delivered by each feeder can be expressed as

$$P_{feeder} = V_{wt} \sum i_{wt}$$

It is worth remarking that this topology is the most similar to the conventional AC case; therefore, it is the logical first step for DC OWPPs.

There are many possible OWPP alternatives based on shunt topology. The four most common proposals are briefly described as follows:

- **Configuration (a).** This shunt topology consists in connecting all the DC cables directly to the offshore HVDC converter platform. Thus, the output voltage of

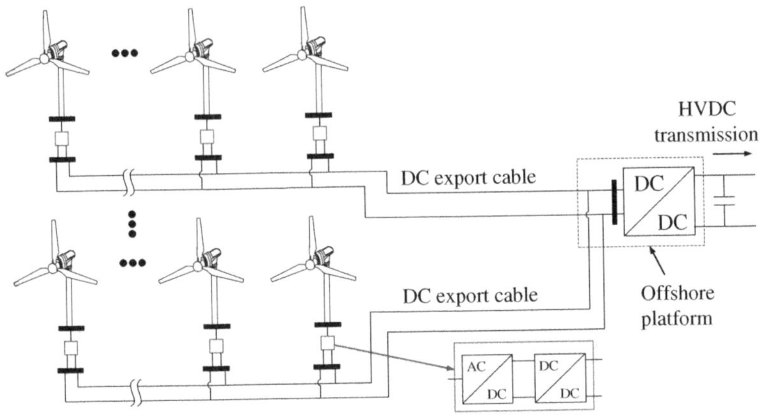

Figure 6.15 Proposal of the DC OWPP configuration (a).

each wind turbine is stepped up by a DC/DC power converter, and the power is transmitted at medium voltage DC.

- **Configuration (b).** The idea of this scheme (Figure 6.16) is very similar to the previous case except for gathering all the interarray cables into an offshore collector platform. Thereby, the wind turbines are connected to the collector platform by means of the interarray cables, while the export cable connects the collector platform with the HVDC offshore platforms. As in the previous case, the voltage is stepped up by means of a DC/DC converter, and the power is delivered to the onshore grid via an HVDC transmission link.
- **Configuration (c).** In order to reduce the export cable losses, this DC OWPP proposal is based on installing a DC/DC power converter in the intermediate

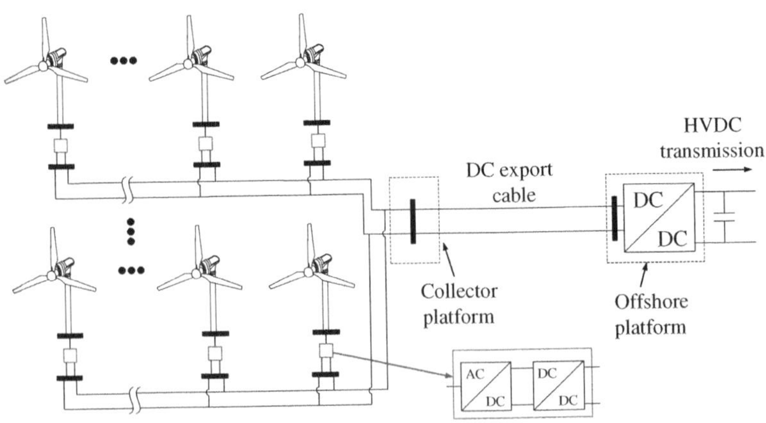

Figure 6.16 Proposal of the DC OWPP configuration (b).

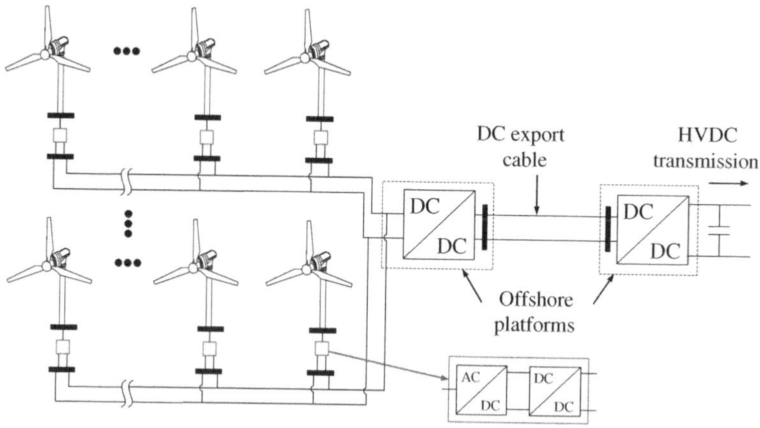

Figure 6.17 Proposal of the DC OWPP configuration (c).

offshore platform (Figure 6.17). Consequently, the output voltage is stepped up twice (both at the wind turbine and the collector platform level).

- **Configuration (d).** This configuration (Figure 6.18) consists in installing one DC/DC power converter per feeder at the intermediate collector platform in attempt to increase the reliability of the system. In this case, the intermediate offshore platform should be bigger than the others topologies considered.

6.3.1.2 Series Topology For the series topology case, the wind turbines are electrically connected in "series," as shown in Figure 6.19. In such topology, the current of each wind turbine is kept constant and the output voltage increases as the

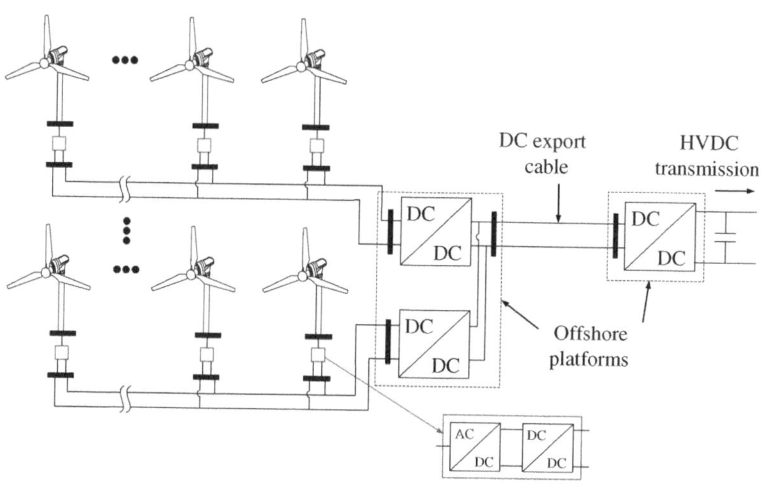

Figure 6.18 Proposal of the DC OWPP configuration (d).

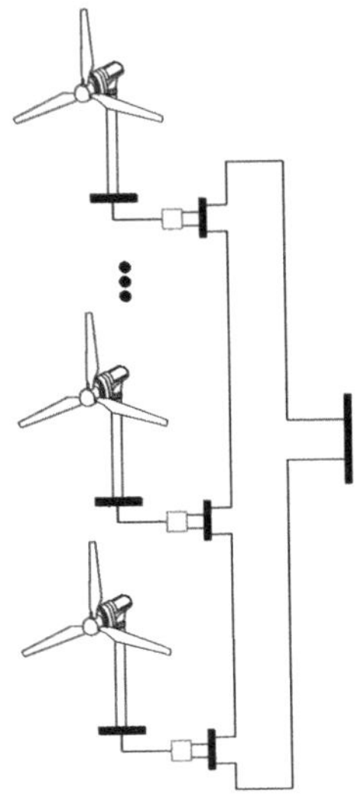

Figure 6.19 Example of a series collection configuration for DC offshore technology.

turbines are connected in series [41, 37]. Therefore, the power delivered by each feeder can be expressed as

$$P_{feeder} = i_{wt} \sum V_{wt}$$

Series topology present some main drawbacks such as the control concept modification (since it implies the regulation of the voltage instead of the current as in the conventional case) and the oversizing of some electrical components of the wind farm to the maximum power of the whole wind farm.

6.3.1.3 Hybrid Topology Finally, the hybrid topology is defined as a mix of both previous topologies. It is designed as a short number of wind turbines electrically connected in "series," but the feeders are connected in "shunt" between them.

Figure 6.20 represents a possible design of the hybrid topology. Nevertheless, in case that only one feeder is considered, it represents the series topology case; since all wind turbines are connected in "series."

As occurs with the previous case, hybrid topologies have some issues to deal with. For instance, they also imply the control of voltage trying to keep current as a constant, and some electrical components of the wind farm must be oversized, since there are only a few turbines connected in series. However, this oversizing is less pronounced compared to series connection.

6.3.2 Technology Development Needed

In order to make offshore wind power plants based on DC technology come true, further steps towards DC technology standardization must be done. Therefore, it is worth

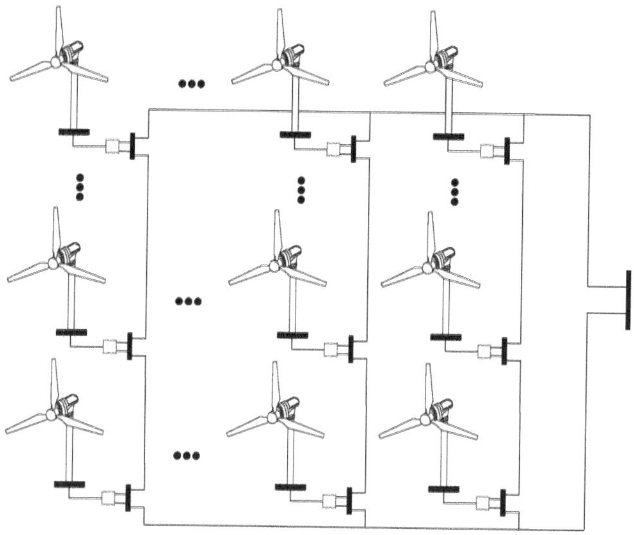

Figure 6.20 Example of a hybrid collection configuration for DC offshore technology.

remarking that extensive research and development activities must be performed to ensure the proper maturation of the required technologies.

In this regard, DC/DC converters and MVDC breakers are some of the most relevant devices to have into consideration during such development, since they could be limiting technologies. Other technical challenges concerning the electrical design of a DC WPP are the protection and earthing issues and the isolation system. Some of those are specially restrictive for the series or hybrid wind power plants cases, which require electrical devices oversizing.

6.4 OTHER OWPPs PROPOSALS

The ability of HVDC transmission to decouple the OWPP from the onshore power system leads us to propose new OWPP designs in order to find a more economical solution than the aforementioned OWPP topologies in this chapter. The fact that offshore HVDC substations accommodates a voltage source converter (VSC) allows us to operate the wind farm collection grid out of sync with the onshore electrical network (50 or 60 Hz). Thus, two possible options to be considered in order to optimize the total OWPP costs could be to vary the internal frequency or to work at a nonstandard frequency [42] within the wind farm.

In the following, the variable frequency operation proposal for AC wind power plants connected to the onshore grid through an HVDC transmission link is presented.

AC Variable Frequency Taking advantage of HVDC technology, several researchers have been investigating the suitability of a proposed OWPP topology based on variable frequency operation within the offshore collection grid.

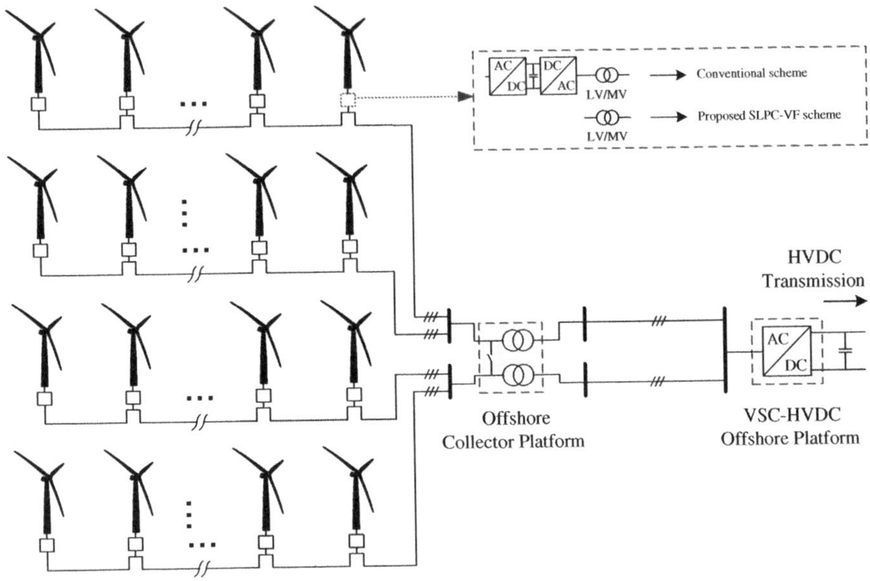

Figure 6.21 Layout of an OWPP collection grid considering both the conventional topology and the novel single large power converter–variable frequency (SLPC–VF) concept proposed in references [43–47].

This novel concept design for OWPPs can be applied to any wind farm regardless of the type of wind turbines installed. However, it is specially worthwhile for wind farm configurations composed by SCIG (squirrel cage induction generator) wind turbines, WRIG (wound rotor induction generator with rotor resistances) wind turbines, or DFIG wind turbines, since the common HVDC converter can regulate a cluster of wind turbines (or the entire wind farm), so that the individual power converter of each wind turbine can be reduced or removed. Figure 6.21 shows both the conventional and the proposed WPP layout for SCIG wind turbine case.

To date, some of the studies carried out are focused on the implementation and validation of different controls for this OWPP proposal [43, 45, 48–50], while others deal with the energy capture analysis of both WPP topologies (conventional and proposed) without taking into account any cost assessment [46, 47, 51, 52]. In references [45] and [12], costs are considered in order to analyze both the technical and economic benefits of the proposed scheme compared to the conventional case. However, the former only computes the investment costs without taking into consideration the energy losses produced during the lifetime of the installation, and the latter calculates both capital expenditures and costs associated with energy losses, but it provides neither corrective nor preventive maintenance energy losses. In general, the authors suggest a good potential for the proposed concept (based on a centralised control that removes the need for individual converters for each turbine) by concluding that, despite the fact that the optimum operation point of each turbine cannot be assured due to the inherent configuration of the novel scheme, the absence of dedicated power converters for each turbine and the variable frequency operation within the wind farm result in a significant saving of the total WPP costs.

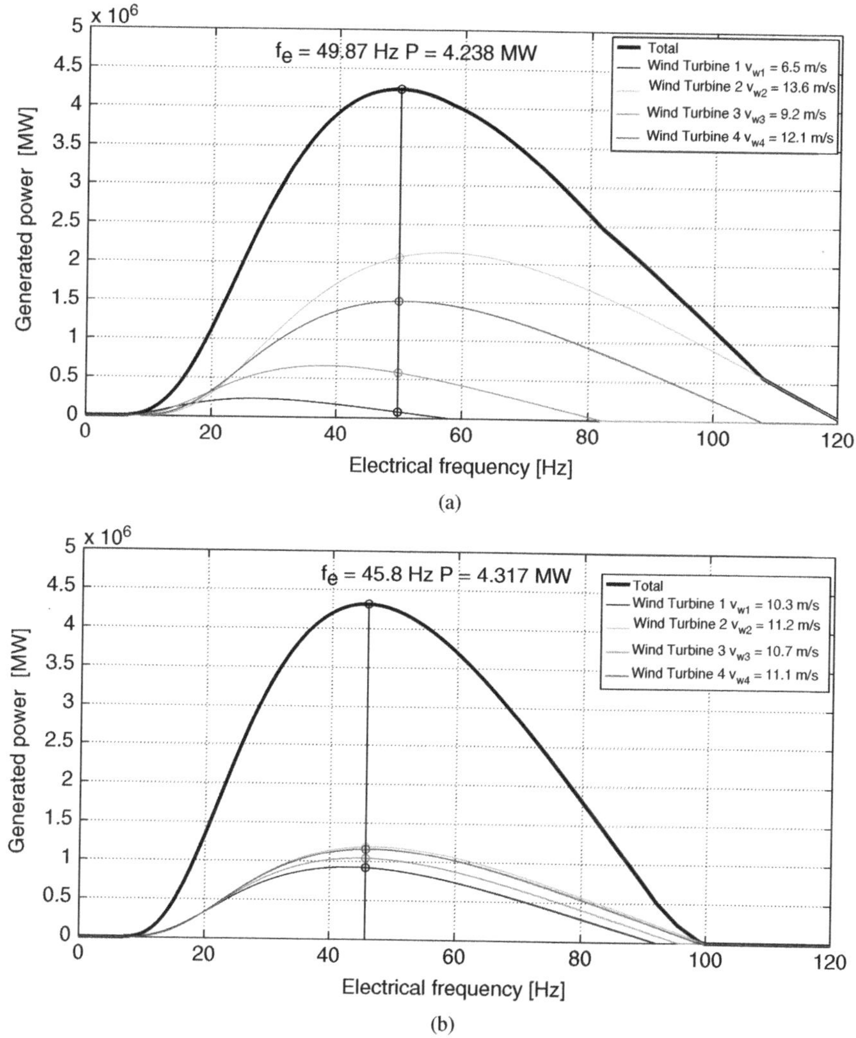

Figure 6.22 Generated power depending on the electrical frequency in a wind turbine cluster of four wind turbines with different wind speed variabilities among turbines. (a) $\sigma = 5$ m/s (b) $\sigma = 0.5$ m/s.

This new WPP topology is considered for future offshore wind power plants (with HVDC transmission link) where the wind flow is assumed to be more uniform compared to onshore and where maintenance issues have a greater relevance. The fewer the number of components that are working in the wind farm, the lower the probability of failure of one of them, which can be of crucial importance when the OWPP is inaccessible by boat or helicopter because of harsh weather conditions.

Figure 6.22 shows the optimum electrical frequency that maximizes the total generated power of a cluster composed of four wind turbines considering different wind speed variabilities among them. The wind speed of each individual turbine has been randomly generated by means of a normal distribution function, $\mathcal{N}(\mu, \sigma^2)$. As can be seen, the larger wind speed diversity causes larger deviations of the operating point of each turbine from its optimum, and therefore the energy wasted by the WPP or wind turbine cluster is greater.

6.5 CONCLUSIONS

This chapter has given some insight on AC offshore wind power plants, by presenting the electrical methodology design within an offshore collection grid, as well as describing some possible topology designs and introducing some of the technologies required to construct them. As an illustrative example, the result of applying such methodology to a particular case study has been shown. It is worth remarking that those presented concepts start from the assumption of at least one HVDC-link existence. Moreover, the option of wind power plants based on DC technology, their collector grid inclusive, has been presented. Different wind turbine connection alternatives (i.e., shunt, series, and hybrid) have been shown giving some advantages and drawbacks of each, and some technology development needs to achieve this aim have been highlighted. Finally, a novel concept for AC wind power plants connected to the onshore grid through an HVDC transmission link, based on optimizing the OWPP total costs by operating the collection grid at variable frequency, has been presented.

REFERENCES

1. J. Wilkes and J. Moccia, "Wind in power: 2012 European statistics," EWEA, Technical Report, 2013.
2. R. Berger, "Offshore wind toward 2020: On the pathway to cost competitiveness," RolandBerger, Strategy Consultants., Technical Report, 2013.
3. C. B. Hasager, A. Pena, M. B. Christiansen, P. Astrup, M. Nielsen, and F. Monaldo, "Remote sensing observation used in offshore wind energy," *IEEE Journal of Selected Topics in Applied Earth Observations and Remote Sensing*, vol. 1, no. 1, pp. 67–79, Oct. 2008.
4. M. B. Christiansen and C. B. Hasager, "Wake effects of large offshore wind farms identified from satellite SAR," *Remote Sensing of Environment*, vol. 98, pp. 251–268, 2005.
5. N. Barberis Negra, J. Todorovic, and T. Ackermann, "Loss evaluation of HVAC and HVDC transmission solution for large offshore wind farms," *Electric Power Systems Research*, vol. 76, pp. 916–927, 2006.
6. S. Lumbreras and A. Ramos, "Offshore wind farm electrical design: A Review," *Wind Energy*, vol. 16, no. 3, pp. 459–473, 2013.
7. T. H. Hansen, "Offshore wind farm layouts. Performance comparison for a 540 MW offshore wind farm," Master's thesis, Norwegian University of Science and Technology. Department of Electric Power Engineering, July 2009.
8. P. K. Steimer, O. Apeldoorn, L. Dalessandro, R. Grinberg, and S. Ebner, "Efficient medium voltage power collection windparks." Copenhagen, Denmark: EWEA, 16–19 April 2012.
9. A. Sannino, H. Breder, and E. K. Nielsen, "Reliability of collection grids for large offshore wind parks," in *9th International Conference on Probabilistic Methods Applied to Power Systems*, KTH, Stockholm, Sweden, June 11–15 2006.
10. T. Ackermann, *Wind Power in Power Systems*. Hoboken, NJ: Wiley, 2005.
11. J. Pierik, M. Damen, P. Bauer, and S. de Haan, "Steady state electrical design, power performance and economic modeling of offshore wind farms," *EPE Journal*, vol. 16, pp. 44–49, 2006.

12. M. A. Parker and O. Anaya-Lara, "Cost and losses associated with offshore wind farm collection networks which centralise the turbine power electronic converters," *IET Renewable Power Generation*, vol. 7, no. 4, pp. 390–400, 2013.
13. "Offshore development information statement, 2010 annual review. Appendix 1," National Grid, Technical Report.
14. P. G. Cables and Systems, "Submarine energy systems" [Online]. Available: http://prysmiangroup.com/en/business_markets/markets/hv-and-submarine/downloads/datasheets/LOW-A4-Leaflet-Submarine.pdf
15. R. MacLean, "Electrical system design for the proposed one gigawatt beatrice offshore wind farm," Master's thesis, University of Strathclyde, 2004.
16. P. Gardner, L. Craig, and G. Smith, "Electrical systems for offshore wind farms," in *Twentieth BWEA Wind Energy Conference Wind Energy*, 1998.
17. J. D. McDonald, Ed., *Electric Power Substations Engineering*. CRC, 2003.
18. "ABB HVDC Reference Projects in Europe: BorWin1" [Online]. Available: http://www.abb.com/industries/ap/db0003db004333/a8e328849ac67b66c125774a00243367.aspx
19. "Tennet offshore wind projects" [Online]. Available: http://www.tennet.eu/de/netz-und-projekte/offshore-projekte.html
20. IEEE Wind and Solar Plant Collector Design Working Group, "Wind Power Plants SCADA and Controls" [Online]. Available: http://grouper.ieee.org/groups/td/wind/Wind_Power_Plant_SCADA_and_Controls.pdf
21. C. Nicolae, "SCADA—Supervisory control and data acquisition," *Journal of Sustainable Energy*, vol. 2, no. 4, pp. 1–5, 2011.
22. "WindAcWind—Alstom's SCADA for onshore and offshore applications" [Online]. Available: http://www.alstom.com/Global/Power/Resources/Documents/Brochures/windaccess-scada-control-system-onshore-offshore-wind.pdf
23. "The REpower SCADA system" [Online]. Available: http://www.repower.de/fileadmin/download/produkte/RE_Br_Scada_EN_low.pdf
24. "GH SCADA" [Online]. Available: http://www.gl_garradhassan.com/en/software/scada.php
25. "Course on Electrical protection in wind turbines, Aalborg University."
26. S. G. Johansson, L. Liljestrand, J. Krogh, F. Karlstrand, and J. Hanson, "AC cable solutions for offshore wind energy." Copenhagen Offshore Wind Conference, October 2005.
27. R. van de Sandt, J. Löwen, J. Paetzold, and I. Erlich, "Neutral earthing in off-shore wind farm grids." IEEE Bucharest Power Tech Conference, Bucharest, Romania, June 28th–July 2nd 2009.
28. *Manahmen bei Beeinflussung von Fernmeldeanlagen durch Starkstromanlagen: Beeinflussung durch Gleichstrom-Bahnanlagen*. Berlin: DIN, 1987.
29. P. P Gardner, L. M. Craig, and G. J. Smith, *Electrical Systems for Offshore Wind Farms*. Glasgow, UK: Garrad Hassan & Partners, June 2003.
30. P. M. Anderson, *Power System Protection*, Wiley-IEEE Press, 1998, Ed, no. ISBN 0-7803-3427-2.
31. *Offshore Transformer Platform: Earthing, Bonding and Lightning Protection*, Energinet.dk Std.
32. "Levelized cost of electricity renewable energy technologies," Fraunhofer ISE, Technical Report, Nov. 2013.
33. "The economics of wind energy," European Wind Energy Association (EWEA), Technical Report, Mar. 2009.
34. L. Anatoly, F. Ilia, and D. Yi, *Multi-state System Reliability Analysis and Optimization for Engineers and Industrial Managers*, Berlin: 1st edition. Springer, 2010.
35. J. H. M. Larsen, H. C. Soerensen, E. Christiansen, S. Naef, and Vølund. Experiences from Middelgrunden 40 MW Offshore Wind Farm. In Copenhagen Offshore Wind Conference, October 2005.
36. P. Volund, P. H. Pedersen, and P. E. Ter-Borch, "165 MW Nysted Offshore Wind Farm. First year of operation - performance as planned," Technical Report, 2004, http://www.offshorecenter.dk/log/bibliotek/165 MW Nysted Offshore Wind Farm.pdf.
37. N. Holtsmark, H. Bahirat, M. Molinas, B. Mork, and H. K. Hoidalen, "An all-DC offshore wind farm with series-connected turbines: An alternative to the classical parallel AC model?" *IEEE Transactions on industrial Electronics*, vol. 60, pp. 2420–2428, 2013.
38. D. Valcan, P. Kjaer, L. Helle, S. Sahukari, M. Haj-Maharsi, and S. Singh, "Cost of energy assessment methodology for offshore AC and DC wind power plants," in *13th International Conference on Optimization of Electrical and Electronic Equipment (OPTIM)*, 2012, pp. 919–928.

39. M. De Prada-Gil, J. L. Domínguez-García, F. Mancilla-David, E. Muljadi, M. Singh, O. Gomis-Bellmunt, and A. Sumper, "Type-2 wind turbine with additional subsynchronous resonance damping," in *Fifth Annual Green Technologies Conference (IEEE GreenTech 2013)*, 2013.
40. G. Stamatiou, "Techno-economical analysis of DC collection grid for offshore wind parks," Master's thesis, University of Nottingham, 2010.
41. G. Stamatiou, K. Srivastava, M. Reza, and P. Zanchetta, "Economics of dc wind collection grid as affected by cost of key components," in *World Renewable Energy Congress*, 2011.
42. J. L. Domínguez-García, D. Rogers, C. Ugalde-Loo, J. Liang, and O. Gomis-Bellmunt, "Effect of non-standard operating frequencies on the economic cost of offshore AC networks," *Renewable Energy*, vol. 44, pp. 267–280, 2012.
43. L. Trilla, O. Gomis-Bellmunt, A. Sudrià-Andreu, and J. Liang, "Control of SCIG wind farm using a single VSC." Power Electronics and Applications (EPE 2011), Proceedings of the 2011–14th European Conference, 2011, pp. 1–9.
44. O. Gomis-Bellmunt, J. Liang, J. Ekanayake, R. King, and N. Jenkins, "Topologies of multiterminal HVDC-VSC transmission for large offshore wind farms," *Electric Power Systems Research*, vol. 81, pp. 271–281, 2011.
45. D. Jovcic and N. Strachan, "Offshore wind farm with centralised power conversion and DC interconnection," *IET Generation, Transmission & Distribution*, vol. 3, no. 6, pp. 586–595, 2009.
46. V. Gevorgian, M. Singh, and E. Muljadi, "Variable frequency operations of an offshore wind power plant with HVDC-VSC." San Diego, California: IEEE Power and Energy Society General Meeting (PES), July 22-26 2012.
47. M. De Prada-Gil, O. Gomis-Bellmunt, A. Sumper, and J. Bergas-Jane, "Power generation efficiency analysis of offshore wind farms connected to a SLPC (single large power converter) operated with variable frequencies considering wake effects," *Energy*, vol. 37, no. 1, pp. 155–168, 2012.
48. O. Gomis-Bellmunt, A. Junyent-Ferré, A. Sumper, and J. Bergas-Jané, "Control of a wind farm based on synchronous generators with a central HVDC-VSC converter," *IEEE Transactions on power systems*, vol. 26, no. 3, pp. 1632–1640, 2011.
49. D. Jovcic, "Offshore wind farm with a series multiterminal CSI HVDC," *Electric Power Systems Research*, vol. 78, pp. 747–755, 2008.
50. T. Vrionis, X. Koutiva, N. Vovos, and G. Giannakopoulos, "Control of an HVDC link connecting a wind farm to the grid for fault ride-through enhancement," *IEEE Transactions on Power Systems*, vol. 22, no. 4, pp. 2039–2047, 2007.
51. M. De-Prada-Gil, O. Gomis-Bellmunt, A. Sumper, and J. Bergas-Jané, "Analysis of a multi turbine offshore wind farm connected to a single large power converter operated with variable frequency," *Energy*, vol. 36, pp. 3272–3281, 2011.
52. O. Gomis-Bellmunt, A. Junyent-Ferré, A. Sumper, and S. Galceran-Arellano, "Maximum generation power evaluation of variable frequency offshore wind farms when connected to a single power converter," *Applied Energy*, vol. 87, pp. 3103–3109, 2010.

PART III

PLANNING AND OPERATION OF HVDC GRIDS

PART III ADDRESSES the planning and operation of HVDC grids. During the planning phase of transmission systems, some challenging questions need to be answered which can be summarized as the four W's of the transmission expansion problem: Where to invest? What type of investment? When to invest? Who pays for the investment? The approach towards planning new transmission assets and some examples on how to to perform such studies are shown. Their applications are illustrated using case studies. A transmission expension plan is a result of an optimal routing, rating, and technology problem which is time-dependent.

The second chapter in this part focuses on how an HVDC grid will look like. The different possible HVDC grid layouts are presented, distinguishing between radial and meshed topologies with special focus on offshore grid layouts. The different HVDC converter station configurations are introduced as well as the substation configuration.

An aspect, which is sometimes neglected in technical discussions, is the governance model which will be applied to the future HVDC grids. Uncertainty on such a model may bring serious difficulties for the development of HVDC grids. Governance models are discussed, including issues such as the ownership, financing, risk, and regulation of HVDC grids.

The last two chapters in this part address the operation of power systems with HVDC grids and the operation of offshore wind

HVDC Grids: For Offshore and Supergrid of the Future, First Edition. By Dirk Van Hertem, Oriol Gomis-Bellmunt, and Jun Liang.

power plants. The operation of hybrid AC/DC power systems requires special attention to the management of power system reliability. Furthermore, new ancillary services between the different AC and DC system are introduced, including some references for their implementation in future grid codes. The operation of offshore wind power plants connected to HVDC grids is presented, with special attention to support of the AC grid from offshore wind power plants.

CHAPTER 7

HVDC GRID PLANNING

Hakan Ergun and Dirk Van Hertem
Research Group Electa, Department of Electrical Engineering, University of Leuven, Leuven, Belgium

7.1 CONTEXT OF TRANSMISSION SYSTEM PLANNING

System development can be defined as the process of determining the optimal investment policy for the development of generation facilities, transmission systems, and distribution systems to meet the future (electrical) energy demand.

Transmission systems are developed to transmit bulk power (>1 GW) over larger distances to ensure that electrical energy that is generated is transmitted in a cost-effective and reliable manner to the larger consumers or to the distribution system. Transmission system planning aims at developing an optimal scheme for the extension of the transmission system. The word "optimal" in this case should be understood as the "best possible solution in the frame of the given investment policies." The transmission system is developed considering only the partial energy system rather than performing a global optimization. There are actually numerous interactions between the three subsystems (generation, transmission, and distribution) which are usually not taken fully into account. Because of the complexity, the investment strategies in each of the systems were determined separately even before the liberalization. Interactions with other systems were taken into account in a simplified manner to that end. Taking account of these interactions after the liberalization becomes even more difficult, because the subsystems are managed by different actors, each with his own policy and agenda. Furthermore, power generation and transmission systems must be planned in a global system of the energy use of a country or region, taking into account the energy provision and security of supply in all its aspects. Electricity is here one type of energy use that is set in competition with others, yet the development of a future transmission system greatly depends on it. Furthermore, the investments have up till now been generally weighted on the basis of what are known as "internal" or "private" costs, but what are known as "external costs" should also be taken into account during their life cycle.

Overall, transmission system planning is a very challenging and complex task. This is especially so for long-term planning, where the planning horizon covers one or several decades. The uncertainties in load and generation which need to be considered while designing the transmission system are considerable. This section

HVDC Grids: For Offshore and Supergrid of the Future, First Edition. By Dirk Van Hertem, Oriol Gomis-Bellmunt, and Jun Liang.

aims to provide insights in the complex transmission expansion planning problem by explaining the aspects to be considered and by showing the difficulties during transmission system planning.

Transmission investments can be characterized as follows [1]:

- A natural monopoly, with only one responsible firm for the construction of grids within a given area.
- A capital intensive business, using expensive equipment which occupies considerable space. The recurring costs for grid investments are normally substantially more important than the operational costs.
- Transmission assets have a long life, with expected life ranging from 20 to 40 years, or even longer for transmission lines.
- Transmission investments are irreversible, meaning that it is virtually impossible to move transmission assets from their original site. This means that investment decisions made now remain present in the system and might become stranded if the energy system does not develop as anticipated.
- Transmission investments are lumpy, meaning that equipment is normally offered in a limited set of distinct ratings for voltage and power.
- Transmission networks are strongly subjected to economies of scale, meaning that it is comparatively cheap to build infrastructure with high power capacity.

7.1.1 Functionality of Grids

In theory, each consumer can install his own local generator next to the load. In that case, there is no need for the transmission of electrical energy. Yet, the system is developed such that electrical energy is transported over thousands of kilometers, even though that history shows that electrification originated from isolated centres, with only local use of the energy produced. Later, these centers were interconnected to form grids, mainly to provide aid in the case of an outage of a production unit. This is known as the *interconnection function* of the grid. As a traditional generation unit is available on average for only 85% of the time (downtime or maintenance), an individual user would require an auxiliary supply to cover in case of unavailability of the primary source. Interconnecting significantly reduces the generation reserve coefficient. A second important advantage of the interconnection function is the common supply of all individual grid users, evening out the individual peaks in consumption which are not simultaneous.

The network can of course also operate to transmit the energy: the *transmission function*. In this way, energy is transported from possibly distant sources to load centers. This will allow the access to all energy resources and load centers, also if these are remote (e.g., hydro or wind generation).

A final role of the grid is the *market facilitator function*. It must make commercial transactions of the different market parties (generators/suppliers and clients) possible. Increased transmission capacity reduces the congestion in the system and increases the social welfare at a system level which in turn enhances the competitiveness of the market. It should be pointed out here that not every commercial transaction leads to a physical power flow. A distinction must be drawn between commercial transactions (and their financial flows) and physical flows. The market actually influences the grid only to the degree that it brings a physical power flow.

The system developers have the responsibility to ensure a reliable energy provision taking into account the future energy energy scenarios (see Chapter 2). They have to prepare the system for the energy flows that can occur in the future while taking credible contingencies into account. On the other hand, transmission system investments are costly and a balance must be sought between investment costs and long-term system-wide benefits [1].

7.1.2 The Four W's of the Transmission Expansion Problem

The goal of transmission system expansion planning is to obtain the optimal transmission expansion plan under a given (uncertain) future. This is done by maximizing the social welfare or through a cost–benefit analysis. While maximizing the social welfare, a number of constraints have to be satisfied which are defined to keep a certain level of security of supply. In its most general definition, the transmission expansion plan has to provide an answer to the following four main questions:

- Where to invest?
- What type of investment?
- When to invest?
- Who is going to pay for the investment?

7.1.3 Where to Invest?

The first question to be answered is where the future transmission system investments must be realized: Which transmission corridors or regions require new or upgraded transmission infrastructure? The answer to this question depends on the motivation of the transmission expansion. If the motivation of the expansion is the maximization of revenues (e.g. for a merchant line), the optimal location of the line might conflict with the optimal location from a TSO perspective.

The determination of the location of new transmission infrastructure is a problem with different geographical scales. Firstly, it is important to determine which transmission zones (e.g., countries) to connect, in order to achieve the maximum social benefit in a multi-zonal perspective. Optimization of investment needs between multiple zones is mostly carried out using a market model and simplifying the transmission grid typically to one or two nodes per zone as depicted in Figure 7.1. This way the optimization problem is simplified, as these calculations are carried out with a large number of generators and loads over a time period of one or several years on an hourly basis.

After determination of transmission investment needs between the different zones, the system can be designed to fulfill the investment needs. To achieve an optimal system design, the following question has to be answered: where to place

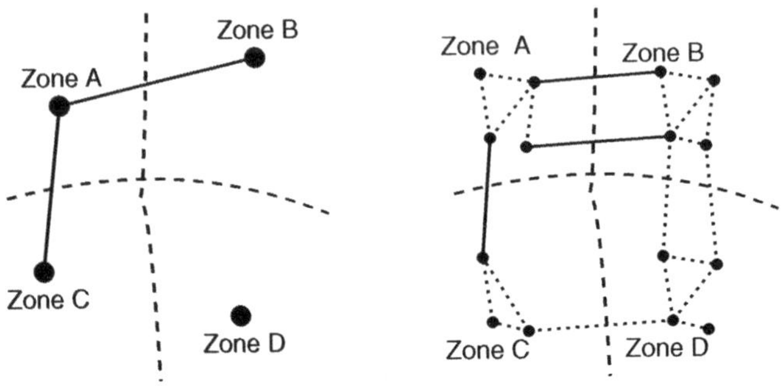

Figure 7.1 Determination of investment needs in a multi-zonal (left) and on substation level (right).

actual lines and which substations to connect with each other considering a number of technical, social, and environmental constraints (Figure 7.1).

From the pure technical perspective, the optimal location depends on the spatial and temporal distribution of generation and demand. These distributions can accurately be predicted in the operational domain or short-term planning. In long-term planning, the prediction of these distributions is a very difficult task as many uncertainties occur.

In the case of demand, the spatial distribution depends mainly on the population density. Therefore, the future spatial distribution can be predicted rather easily. The temporal distribution of demand is mainly determined by the consumption behavior of the population. On short and medium term, this consumption behavior is unlikely to change. Nevertheless, there is a possibility of extreme events (e.g., the mortgage crisis of 2008), causing a sudden drops or increase in consumption. In a time frame of several decades, the consumption behavior can change due to increased usage of demand response and demand side management or increased use of electrical vehicles.

The spatial and temporal distribution of generation is somehow connected. The spatial distribution of generation depends on the availability of energy resources. Conventional power plants are ideally located there, where the infrastructure allows easy transportation of primary energy resources. In case of renewable generation (e.g., wind and solar energy), the favorable locations are determined by the wind speeds and solar resources. Although the best location from the geographic point of view is known, the actual location of future generation infrastructure is still difficult to predict. Generation assets may be forced to be built on different locations due to public opposition. Generation assets may even not be built as foreseen due to lack of capital or unexpected changes in the capital market.

The choice of the primary energy resources is driven strongly by (geo-) political, economic, and environmental drivers. The temporal distribution of generation depends as such on the primary energy resources. Conventional generators can deliver power whenever needed. In contrast, renewable power generation depends on the availability of intermittent sources. Another uncertainty in the temporal distribution is the market price for electricity as well as the market prices for the primary energy sources. Eventually, these prices determine, together with the demand, the

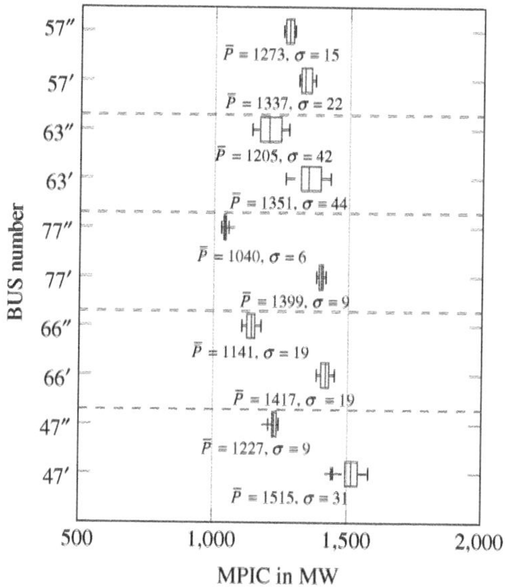

Figure 7.2 Maximum power injection capabilities (MPIC) of selected nodes of the IEEE 118 bus test system. $\bar{P}$ indicates average power injection capability, and σ indicates the standard deviation of the injection capability. ′ indicates MPIC calculated without $N-1$ constraint and ″ indicates MPIC calculated with $N-1$ constraint [2].

commitment of different power plants. Summarizing, the temporal distribution of generation affects the commitment of power plants, which further affects the spatial distribution of power injections into the system.

Security constraints play a major role during the selection of the location for new investments. Depending on the system layout and the desired security level, different criteria such as $N-1$, $N-1-1$, $N-2$, or diverse probabilistic security criteria can be used in the planning process. Figure 7.2 illustrates the effect of the $N-1$ constraint. The figure shows the amount of power which can be injected in a grid, without causing any overloads [2]. These capabilities are been calculated taking into account different spatial and temporal distributions of generation and demand as well as the cost of generation. The figure clearly shows that the maximum power injection capability decreases when the $N-1$ security constraint is taken into account during the calculation. The figure also shows that the consideration of the security constraint changes the ranking of favourable nodes.

The routing of the lines is another decision criterion on the actual location of transmission system investments. The route selection of new transmission corridors is directly coupled to other types of infrastructure such as railways, roads, shipping routes, and gas and oil pipelines as well as prohibited zones such as natural reserves, water reservoirs, and military zones. Additionally, social acceptance of transmission system expansion is one of the major influence factors for the selection of the location for investments. Such decision criteria are very case-specific and therefore very difficult to be considered in generic transmission planning tools.

7.1.4 What Type of Investment?

The second question to be answered is the technology or type of investment that is selected. The system developer needs to make two fundamental choices: AC or DC technology, and overhead or underground transmission. Also the desired voltage level and transmission capacity needs to be determined.

In classic transmission system planning, high-voltage AC overhead lines was the technology that was considered first. However, underground cabling and HVDC technology have gained a high importance in today's power system. In Europe, HVDC links were mainly used for long-distance submarine connections or connection of nonsynchronous areas. Current projects such as the Swedish south west-link, the Spain–France connection, the Belgium–Germany connection, or the British west-link show that embedded HVDC links are an attractive option for new transmission system investments.

HVDC technology is vital to integrate large-scale offshore wind power into the electricity system. As elaborated in Chapter 4, HVDC technology has virtually no length limitation which enables the connection of distant offshore wind farms with high power ratings. Beyond that, with the development of VSC technology and the possibility of multi-terminal operation, HVDC connections can be operated in a meshed grid, allowing the connection of large offshore wind farms and several countries as a first step of a super grid.

The choice of the type of investment depends mainly on the transmission distance and the transmitted power. The transmission distance is a limiting factor for high-voltage AC transmission. Especially in underground transmission, high cable capacitance and the resulting charging currents limit the feasible transmission distance 4.

Besides the transmission distance, also the installation site can affect the technology selection. Installation costs of transmission equipment are a considerable part of the total investment costs. The installation costs for transmission equipment depend on the spatial properties of the area of installation as well as land acquisition costs for the right of way. Installation costs can be considered as fixed in a specific installation area, whereas the investment costs depend on the rated power of the equipment. If the total costs consisting of installation and investment costs are illustrated as a function of the rating of the transmission system, an inverse dependence as shown in Figure 7.3 is observed. The cost function has discontinuities at the maximum power rating of one circuit of the transmission equipment. If two different technologies are considered, the location of the discontinuities varies as different technologies have typically different maximum power ratings for one circuit. Depending on the installation costs for both technologies and the power ratings per circuit, different technologies can be economically more feasible in a certain range of power as shown in Figure 7.3. This can also result in multiple breakeven points between two technologies. Therefore, the optimization of location, technology, and power rating should be performed together to avoid suboptimal transmission grid layouts. This also means that any optimization result can be challenged as small deviations from the initial assumptions can lead to different results. Therefore it is necessary to vary initial assumptions and perform sensitivity analysis in order to obtain the most suitable result.

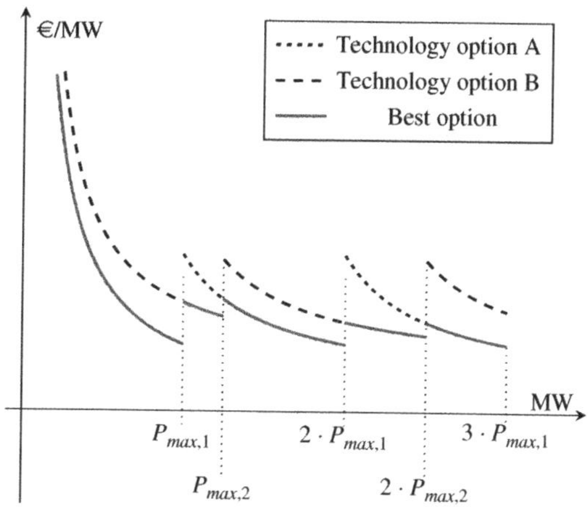

Figure 7.3 Transmission system investment costs in € per MW for two different technologies on the same transmission route [3].

7.1.5 When to Invest?

Every time a new transmission investment is done, the grid layout change and the power flows through the system change as well. As a consequence, the part of the transmission system that is most congested or which is the most prone to reliability disturbances shifts within the system. As such, every transmission system investment has an impact on the future investments.

A second effect is the economic benefit of delaying investments. On the other hand, the process of investing in a new transmission asset can take multiple years, which makes delaying investments a very difficult task given the uncertainties of the future scenarios. As such it is important to optimize the timing of transmission system investments, especially in the long term, taking into account a number of uncertainties. Some of the scenario uncertainties can be listed as follows:

- Uncertainty in generation
  - Location of generation
  - Type of generation
  - Generation profile
- Uncertainty in demand
  - Location of demand
  - Demand growth
  - Demand profile
  - Demand flexibility
- Uncertainty in technological developments

As the location and type of investment can change due to the timing of the investment, it is important to optimize timing, location, and type of investment at the same time.

7.1.6 Who Pays for the Investment?

The remuneration scheme for transmission system investments is an important decision criterion as each investor expects a given rate of return over the investment. This means a correct financial incentive to invest in the required transmission infrastructure. In case of investment projects where more parties with different remuneration schemes are involved, it can be difficult to find a compromise which can satisfy all involved parties. Also, cases where the benefits of the investment are not directly linked to the investor (e.g., when a given investment mainly benefits a third country) require the correct incentives in order to be built.

A clear example can be seen in the number of offshore investments in a given country when comparing the connection charging regime. Depending on the regulatory regime, it is the generator company (deep connection charges) or the system developer (shallow or super shallow connection charges) that is responsible to pay for the grid reinforcements. Depending on the "depth" of the regulatory schemes, different grid layouts are obtained as the optimal solution. In case of interconnecting lines, an harmonization of the regulatory schemes is needed to facilitate such investments.

Chapter 9 covers the topic of financing and regulation of grids in much more detail.

7.1.7 Other Influential Factors

Besides the four "W's" of transmission system planning, there are a number of influential factors putting additional constraints on the transmission investment problem.

7.1.7.1 Security Constraints There are two classes of security constraints, which have to be taken into account in transmission grid planning. The first class of constraints deals with long-term security of supply. Long-term security of supply constraints are related to transmission system investments to secure the use of energy sources of a specific area in long term. The Desertec project is such an example, where the integration of large-scale solar power in North Africa in the European transmission system is investigated [4–6]. The connection of north African solar power plants to Europe via Sub-Mediterranean HVDC links could help to supply Europe with renewable energy on long term and reduce its dependency on fossil fuels. In the same way, the connection of large-scale wind power from the North Sea can reduce Europe's dependence on fossil fuel imports [7]. Such major investment projects have to planned following a long-term vision.

The second class of security constraints concern operational security. The $N-1$ security criterion is widely used to ensure operational security in transmission systems. As such, a planned extension to the existing grid also has to comply with the

$N-1$ security criterion. In future grids with highly volatile injections from renewable power sources, stochastic security criteria can be applied to avoid over investments in the transmission grid.

7.1.7.2 Soft Constraints One of the biggest hurdles for new transmission infrastructure projects is the public opposition against such projects. The public opposition mostly relates to the visual impact and electromagnetic field exposure of infrastructure. Due to public opposition, in many cases TSOs are forced to put new transmission assets underground, which increases the costs significantly. Besides cost increase, high-voltage underground transmission results in selection of HVDC technology instead of HVAC due to high charging currents of HVAC underground cables.

It is very important to estimate the public opposition against transmission infrastructure projects in the initial phase of the planning where the technology selection and the line routing are determined. Belated changes in transmission technology or line routing delay the commissioning process significantly which eventually can result in the failure of the whole project.

7.1.8 Transmission Grid Planning in the Liberalized Electricity Market

In classic transmission expansion planning, the generation and transmission system planning is carried out by the central planner. In the classic planning scheme, the objective of the future planning is the minimization of operational costs while fulfilling security constraints.

In the liberalized market, where generation and transmission companies are different entities, the generation and transmission planning are separated. This puts additional complexity and uncertainty in the transmission planning. The main difference is that the objective of the planning is the maximization of profits as opposed to minimization of costs. Because generation companies and transmission companies want to maximize their own profits, the objectives of both can be different, which complicates the decision-making process.

Besides the different objectives of different parties, there is additional uncertainty for generation companies as well as transmission companies. Transmission companies are obliged to connect new generation units in their zone. The transmission system operator can have a stranded investment, in case the generation project is severely delayed or not realized. The same risk exists for the generation company, in case the commissioning process for the transmission system is delayed. Therefore, in current transmission system planning, risk assessment gains increasing importance.

7.1.9 Cross Border Investments—The Investment Paradox

Cross border investments deserve special attention in transmission expansion planning. The fact that different TSOs of different countries are involved in the investment decision complicates the decision-making process. In principle, the goal of such an investment is to maximize the total social welfare. In most cases, the maximum total social welfare is not the same as the sum of the maximum welfare of both countries. Therefore, most investment decisions in interconnectors result in a suboptimum in terms of total social welfare.

Another interesting issue with interconnectors is the so-called investment paradox. The justification for an investment in an interconnection link is the electricity price difference between the two zones. In the case of a high price difference, the commercial and physical flow on the interconnector will be high and the owner (a TSO or a merchant) of the line will get high revenues according network charges. After the interconnection is realized, the congestion between the zones decreases and the electricity price difference between both countries decreases accordingly. In the case of total price convergence between the two zones, there is no reason for the commercial flow between both countries. This also affects the physical flows and the network charges. This means that there will be a decrease in the revenues of the owner of the line, who provided the interconnection between the countries in the first place. In extreme cases, this can result in nonrecovered investment costs for the owner of the line.

7.2 TRANSMISSION EXPANSION OPTIMIZATION METHODOLOGIES

The transmission expansion planning is an optimization problem which can be formulated in different ways. In general, the optimization problem is defined as a two-step problem. The upper level is the investment problem, where the decision of building lines is made with the objective of minimizing the investment costs. In the lower level called the operational problem, the power ratings of the lines are optimized with the objective of minimizing operational costs. In the operational problem, constraints such as angle limits and security constraints are taken into account.

The formulation of the transmission expansion optimization problem depends on the desired objective and the solution methodology. As such, the transmission expansion problem can be classified according their solution methods, the planning time horizon, and the structure of the analyzed power system [8]. Each transmission expansion problem can be formulated in a deterministic or a nondeterministic way.

7.2.1 Classification by Solution Method

In general the transmission expansion problem can be solved using mathematical optimization or heuristic optimization. The combination of mathematical optimization and heuristics are the so-called meta-heuristics which are frequently used for transmission expansion optimization.

7.2.1.1 Mathematical Optimization Methods Mathematical solution methods require the description of the optimization problem with a closed set of equations. The formulation consists of an objective function and a set of constraints, which reflect technical, economical, and reliability imposed criteria. The problem is solved using classic methods of optimization such as linear programming (LP) [9–14], dynamic programming (DP) [15], nonlinear programming (NLP) [16], mixed-integer programming (MIP) [17–24], Benders' decomposition [25–28], hierarchical decomposition methods [29], and branch and bound algorithms, which are combinations of decomposition methods with other algorithms.

The level of abstraction and simplification depends on the chosen solution method. An LP optimization is chosen whether or not a transportation model is chosen [9] or the power flow equations are linearized. To represent voltage magnitue limits, reactive power limits and transmission losses in the operational problem, the use of non-linear optimization methodologies are required.

Transmission system investments are lumpy in their nature. To represent the lumpiness of transmission investments, integer optimization variables have to be used. Because of convergence problems and higher computation time due to integer variables, continuous formulation of circuit additions is still a widely used approach.

Nonlinear mathematical solution methods have often convergence problems due to the nonconvex nature of the problem. To limit the high computational burden, the problems have to simplified very often. Therefore, the obtained solutions have to be checked for feasibility to verify the applicability of the solutions.

7.2.1.2 Heuristics Heuristic methods are based on the step-by-step solving of the transmission expansion problem using a predefined set of rules. To do so, new expansion alternatives are generated, evaluated, and selected. This can be done with or without user interaction. These heuristic methods use local searches based on logical or empirical rules and sensitivities [8]. The search is performed until no better solutions can be found. Optimization methods based on heuristics mostly consider investment costs and operational costs and not served energy in the objective function.

There are several heuristic methods, such as adjoin network [30], sensitivity analysis [31–37], overload networks [9, 10, 12], decision trees [38], genetic algorithms [39–45], simulated annealing [46, 47], expert systems [48–50], fuzzy set theory [51], greedy randomized adaptive search [25], tabu search [52–54], particle swarm optimization [55], and multiobjective meta-heuristics [56, 57]. Game theory [58–61] based methods can be seen as heuristics. In game-theory-based methods, the interaction of different agents based on predefined rules are mimicked.

7.2.2 Classification by Planning Horizon

Transmission expansion problems can generally be separated into two classes regarding the planning horizon, namely static and dynamic problems. Static problems are solved only for one specific point in time. Intermediate time horizons are discarded. All aforementioned methods can be used for this kind of optimization problem. Referring to Section 7.1.2, these kind of expansion problems only answer the questions where to invest and what type of investments to choose.

The expansion problem gets dynamic if multiple time steps are considered. In this case, next to sizing of new investments also their timing along the whole planning horizon is taken into account has to be optimized [8]. Methods to solve such problems are currently underdeveloped and often require significant simplifications or restrictions. In references 14–16, 22, 23, 62, and 63, such problems are described.

To simplify dynamic problems, the algorithm can be performed for multiple sequential (and discrete) steps. This way, a pseudo-dynamic problem is created. Such pseudo-dynamic problems can be solved with two natural decompositions, namely with forward and backward methods. In the forward method [64], the problem is sequentially solved, where the starting point is the first year of considered planning horizon. In the backward methods, first an optimization is performed for the last year of the planning horizon. To solve the problem in the intermediate years, the information of the "future" investments are used. The backward method delivers, in general, better results [8]. The forward and backward method can be combined where the problem at every time step is solved in both directions.

Dynamic and pseudo-dynamic problems require large computational effort due to the large number of variables, which grows with the length of the planning horizon. In general, heuristic methods are used to solve dynamic and pseudo-dynamic problems.

7.2.3 Classification by Power System Structure

Before the liberalization of the electricity markets, the objective of the transmission expansion planning has been to minimize the life cycle costs of the system by considering security constraints. After liberalization, the electricity market structure has changed, causing multiple potentially competing objectives.

As mentioned in the previous section, generation and transmission planning can have contradicting objectives, which complicates the decision-making process. In that context, different entities can have different viewpoints in the following issues [8].

- Definition of the objective function (minimum operational costs, maximum cost–benefit ratio, global welfare, etc.)
- Interaction between transmission and generation expansion planning
- Definition of criteria to choose a specific expansion plan
- Assessment of different expansion options
- Flexibility requirements of the transmission planning process
- Relation between investments and pricing
- Optimization of existing network utilization
- Introduction of flexible transmission devices
- Definition of the greatest uncertainty level for the planning
- Review of reliability, security and quality criteria

7.2.4 Deterministic versus Nondeterministic Expansion Problems

The classic transmission expansion planning is performed in a deterministic way. In general, some worst-case scenarios are created and the transmission system planning is carried out such that security criteria are fulfilled during occurrence of the worst case. Such approaches might lead to over investments as the the probability of occurrence of such a scenario is very low. On the other hand, the quality of the expansion plan depends heavily on the quality of the worst-case scenario. The negation of important scenarios may lead to under investments and to curtailment of load in case these cases occur.

Due to increasing number of uncertainties such as renewable generation and flexible loads, probabilistic transmission expansion planning gains more popularity. Instead of using fixed load and generation values in certain scenarios, the load and generation are modeled with probability density functions. Using probabilistic load flow or probabilistic optimal load flow calculations, the probability of exceeding line limits in certain expansion plans can be determined and monetized.

Using probabilistic techniques, also security criteria can be redefined in a probabilistic way to replace the $N-1$ security criterion used in transmission grid planning.

Monte Carlo simulation is the current standard for probabilistic planning approaches. Monte Carlo simulation requires large computational effort due to large number of sample cases to be calculated. Due to that computational effort it is mostly applied to conceptual study cases in small grids. Especially for large grids with a high number of nodes, the definition of the probability density functions for load and generation as well as the definition of security criteria play a major role and deserve research priority to develop more efficient modelling techniques.

7.3 SPECIALTIES OF GRID PLANNING WITH HVDC TECHNOLOGY

Technical properties of HVDC transmission systems differ significantly from HVAC systems. These properties should be taken into account during the planning stage in order to optimize the layout of the transmission system. In this section, some aspects are listed to show the differences of considering HVAC or HVDC lines for transmission grid expansion.

7.3.1 Controllabillity

One of the major differences between HVAC and HVDC links is the controllability of power flows. During the classic transmission expansion planning, the power flows are distributed according to the admittance matrix of the system and can be calculated using Kirchhoff's current and voltage laws. If controllable power flow devices such as HVDC links are used, they can contribute to optimize the power flows in the entire system. Therefore, they can be modeled as optimization variables rather than lines with fixed impedance, which is a major difference for the optimization algorithm.

Figure 7.4 shows an example of how the controllability of HVDC links can influence the power injection capabilities in transmission grids. The figure compares the maximum possible power injection in selected nodes for the case that only HVAC lines are used against a case where three lines are replaced by HVDC lines. During the calculation the spatial and temporal distribution of generation and load in the entire system is considered [2]. The calculation is carried out using probabilistic methods

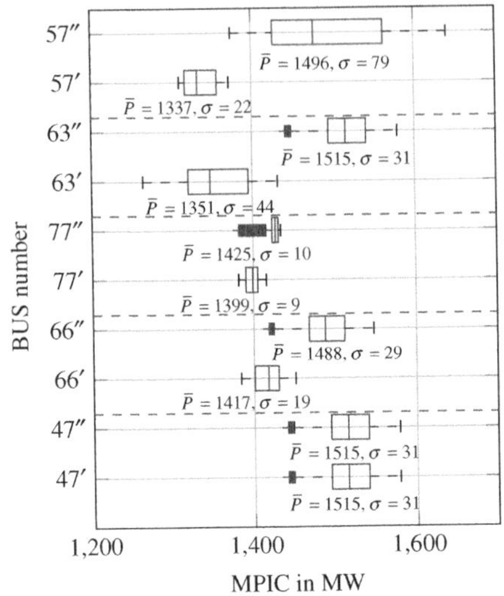

Figure 7.4 Effect of existing HVDC lines on the maximum power injection capability, ′ indicates the base case and ″ indicates case with HVDC lines [2].

and respects voltage angle and power flow limits. The calculations are performed on an IEEE 118 bus test system.

To investigate the effect of controllable HVDC lines on the maximum power injection capabilities, three lines has been replaced by HVDC lines with the power rating. The figure clearly shows that the use of HVDC lines increases the power injection capability in the investigated nodes. If the maximum power injection capability in certain nodes is increased, lines or generators with higher power ratings can be connected to these nodes.

Another interesting aspect of the controllability concerns the remuneration of HVDC assets, if flow-based cost allocation schemes are used to recover the costs of transmission assets. Additional flows caused by the controllability of HVDC links cause flow-based cost allocation schemes to under-recover the total costs of transmission [65]. Especially in a multinational context, the use of HVDC lines in one zone can significantly affect the allocation of transmission system costs to generators and loads in another zone.

7.3.2 Offshore Applications

For offshore applications, submarine cables have to be used. With increasing transmission distance and power rating, the use of HVAC technology becomes infeasible so that HVDC technology has to be used. Hence, classic transmission expansion optimization methodologies cannot be used for offshore applications. Additional constraints and optimization variables have to be included in the optimization algorithm to ensure a feasible system design if HVDC technology is used.

Although currently not done, in the near future meshed operation of HVDC links will be possible. In transmission expansion optimization algorithms, the possibility of meshed HVDC grids has to be modeled with an additional constraint. If multiple time steps are used (dynamic optimization), this constraint can be relaxed depending on the technology forecast for availability of meshed HVDC transmission systems.

As mentioned above, the controllability of HVDC lines causes variation in power flows which have to be modeled as optimization variables. This only holds if HVDC converters are used for point-to-point connections with AC nodes in between. For a pure meshed HVDC grid, the power flows will distribute according to the resistance of HVDC lines and the converters will be modeled as equivalent PV nodes in classic power flow.

These aspects complicate the development of robust optimization algorithms as the definition of the optimization problem is affected such that optimization variables (flows on DC lines) become state variables depending on the technology development.

7.4 ILLUSTRATIVE EXAMPLES

This section shows the complexity of transmission expansion planning using quantitative, illustrative examples. The examples will show the dependence of the "4Ws" of transmission expansion planning to give the reader an idea of how to define an optimization problem to solve the transmission expansion problem.

7.4.1 Optimal Routing, Rating, and Technology

The total investment costs C_{tot} for transmission lines consist of equipment costs C'_{eq} and installation costs C'_{inst} as depicted in Eq. (7.1a). In general, both cost components depend linearly on the transmission system length l. The equipment costs C'_{eq} per km depend linearly on the rated power P_{nom} of the equipment as shown in Eq. (7.1b), where C_1 is a power-rating-related cost and C_2 is a fixed cost which is associated with costs during the manufacturing and transport. The installation costs per kilometer depend on the geographic properties of the area of installation and the number of circuits to be installed, which is an integer number depending on the nominal power rating [Eq. (7.1c)].

$$C_{tot} = (C'_{eq} + C'_{inst}) \cdot l \tag{7.1a}$$

$$C'_{eq} = C_1 \cdot P_{nom} + C_2 \tag{7.1b}$$

$$C'_{inst} = C_{inst} \cdot n_{circ}(P_{nom}) \tag{7.1c}$$

Using Eq. (7.1), the total investment cost per MW of installed equipment for a transmission path with a given length l can be written as shown in Eq. (7.2). This discontinuous function is illustrated in Figure 7.3 for two different technology options.

$$C'_{tot} = C_1 \cdot l + \frac{C_2 + C_{inst} \cdot n_{circ}(P_{nom})}{P_{nom}} \cdot l \tag{7.2}$$

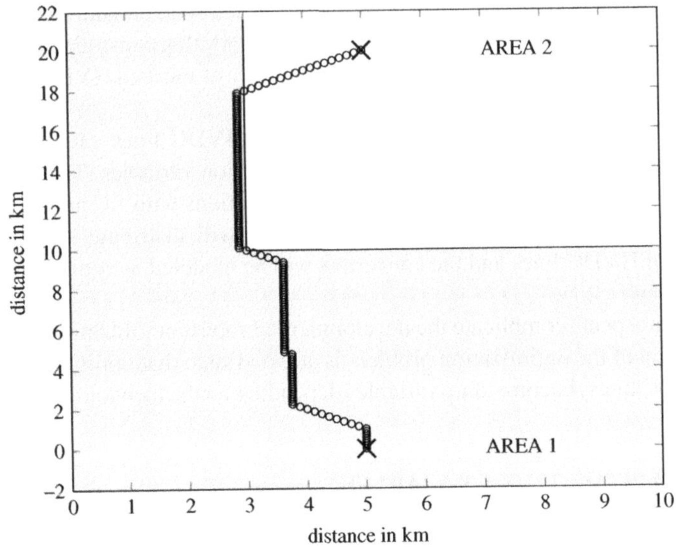

Figure 7.5 Optimal route of a transmission system with given rating using different spatial installation costs.

7.4.1.1 Illustrative Example for Optimal Routing Let's consider an optimization problem where the objective is the minimization of the total investment costs C_{tot}. In the simple case where the power rating is given, the optimal technology can be selected using the relationships in Eq. (7.1). Unfortunately, the factor C_{inst} depends on the geographic properties of the area of installation. If several areas with different C_{inst} exist, an optimal routing algorithm must be used to calculate the best transmission route and the according costs.

Figure 7.5 shows such an example using the methodology developed in reference 3. The area of focus is divided into two areas with different installation costs. The investment and installation costs are given in Table 7.1 for two different technologies.

If only the first technology is considered, the optimal transmission route is as illustrated in Figure 7.5 with total costs of 58.64 €. The optimal transmission route and the selected technology change if both technologies are considered in the optimization (Figure 7.6). In this case the total costs are 55 €.

The example shows that the transmission route can be determined easily if the nominal power is known. In the case where the nominal power is an optimization

TABLE 7.1 Costs

	Technology 1	Technology 2
Investment costs	1 €/km	1 €/km
Installation costs Area 1	1.5 €/km	2 €/km
Installation costs Area 2	3 €/km	2 €/km

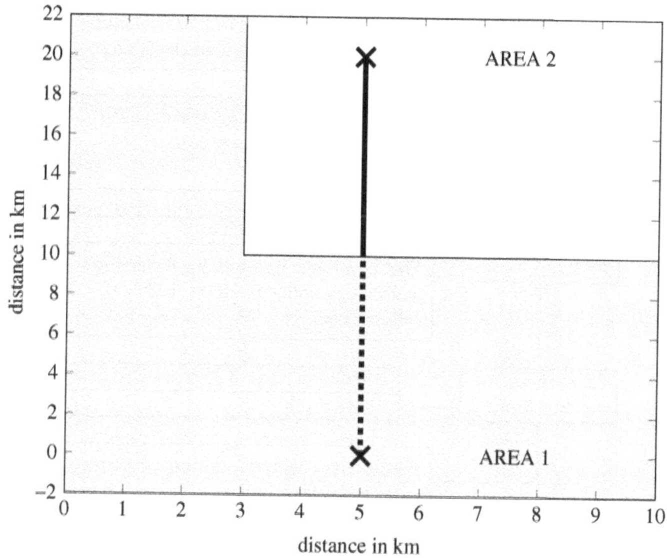

Figure 7.6 Optimal route of a transmission system with given rating using different spatial installation costs and two technologies. Dashed line, technology 1; solid line, technology 2.

variable, the determination of the optimal routing and rating will have to be combined. This complicates the optimization problem significantly. In such a case, the objective is to determine the optimal rating of multiple possible transmission paths. There are two principle methods of combining the optimization of route and rate.

The first method calculates for each possible transmission path and power rating the optimal transmission route and technology. This way the minimum costs of each path and rating are determined. The optimal topology of the system with optimal power ratings per path are determined using the minimum costs of paths per power rating. This approach requires large computational effort as all possible combinations of paths and power ratings are considered.

Another method is to use both approaches iteratively as shown in Figure 7.7. Therefore, first initial costs for each transmission path and rating are approximated. Using the approximated costs, the rating of possible paths are determined. Sequentially, the optimal routing algorithm is applied to determine the actual minimum costs for the obtained topology and power ratings. Knowing the actual minimum costs, the approximated costs are updated and the procedure is repeated until there is no difference between the costs obtained by the optimal routing algorithm and the updated average costs.

7.4.2 Effect of Security Constraints

In transmission planning, traditionally the $N-1$ security constraint is used. This constraint states that the transmission system must be operable if one major grid element fails. As simple as it sounds, the definition of the $N-1$ security constraint

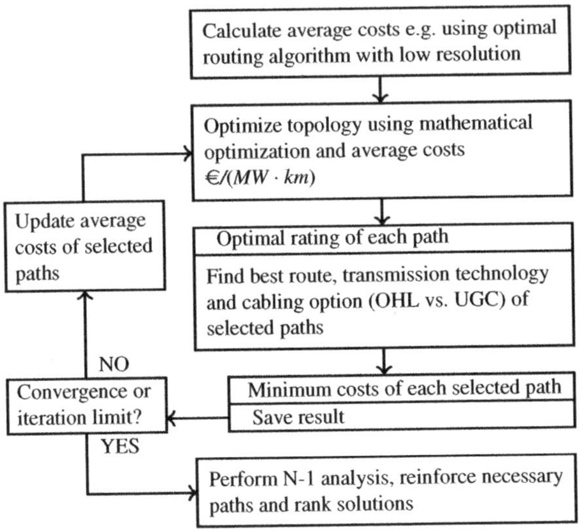

Figure 7.7 Iterative optimization of power rating, technology, and routing of transmission systems [3].

is not that straightforward. For overhead lines, the $N - 1$ security constraint might be interpreted as the failure of one system, or the failure of a transmission tower which can contain multiple circuits.

7.4.2.1 Illustrative Example In this section, the effect of considering $N - 1$ security on the optimal transmission topology is illustrated. Therefore, the optimization methodology depicted in Figure 7.7 is used. More information about the optimization algorithm can be found in reference 3. The objective function and the constraints are illustrated in Eqs. (7.3)–(7.6).

$$\min_{\substack{\forall U^+, U^- \\ \in B^{N_b \times N_k}}} C_{cap} = (U^+ + U^-) : (\vec{1} \otimes \mathbf{k}) \cdot \tilde{C} : L \tag{7.3}$$

$$= \sum_i \sum_j \left(U^+_{ij} + U^-_{ij} \right) \cdot K_i \cdot \tilde{C}_{ij} L_{ij}$$

$$s.t.$$

$$P^{inter} \leq (U^+ + U^-) : \mathbf{t} \otimes \mathbf{k} \tag{7.4}$$

$$P_{MPIC} \geq \Delta_{MPIC} \cdot (U^+ + U^-) \cdot \mathbf{k} \tag{7.5}$$

$$P^{max}, \quad \forall i \in [1, N_b] \geq P_i \tag{7.6}$$

The objective of the optimization is to find the economically best solution for additional transmission capacity between two zones. U^+_{ij} and U^-_{ij} are integer decision variables for each possible branch out of N_b branches and available power ratings out of N_k power ratings. The decision variables are defined bidirectional to account for asymmetry and flow cancellation. $\tilde{C}_{ij}$ are the investment costs for each possible branch and power rating in €/$(MW \cdot km)$ and L_{ij} is the length of each branch. Note

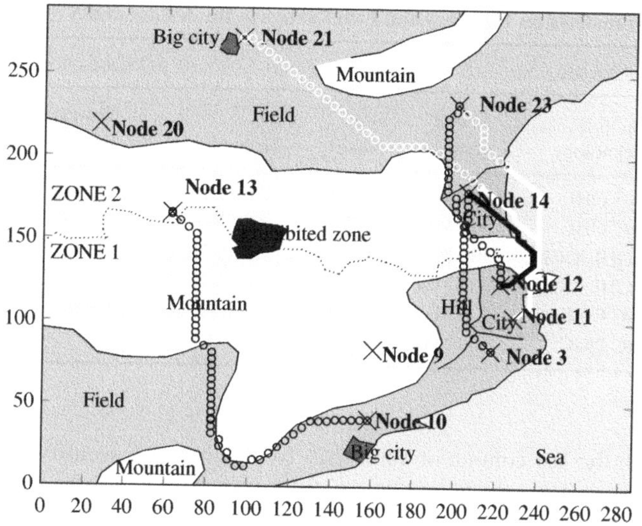

Figure 7.8 Spatial map of area of focus showing optimal transmission topology for 7-GW interconnection capacity. Dark areas indicate HVAC connections. White areas indicate HVDC connections. Solid lines indicate underground cables. Circles indicate overhead lines.

that the length of the one branch can change due to its power rating, as different technologies might be used for different ratings, resulting in different transmission routes, as shown in the previous section.

Constraint (7.4) states that a certain interconnection power P^{inter} between both zones have to established. Constraint (7.5) ensures that the maximum power injection capabilities in the candidate nodes are not exceeded, taking into account the change in the power flow capabilities due to flows on the new built connections (Δ_{MPIC}). Constraint (7.6) restricts the maximum rating of a transmission path to P^{max}.

An example with zones are illustrated in Figure 7.8. To take the spatial properties of the area of focus into account, an optimal routing algorithm is used. During the optimization, AC and DC technology as well as overhead lines and underground cables can be used as possible technologies. The desired interconnection power is 7 GW with a maximum of 2 GW per transmission path.

Figure 7.8 shows the optimal transmission topology obtained for the N case. The figure shows that the additional investments are realized using AC and DC technology as well as both underground cables and overhead lines, depending on the equipment installation costs in the different areas. The power ratings of the transmission paths are illustrated in Table 7.2.

The optimal transmission topology for the $N-1$ case is depicted in Figure 7.9. As the figure shows, the consideration of $N-1$ security criterion changes the grid topology significantly. The topology is affected by the change in power ratings of the transmission paths depending on the technology selection and routing. Table 7.2 shows the transmission paths and their ratings for the optimal solution of the $N-1$ case. The power rating of the necessary circuit additions to obtain $N-1$ security

TABLE 7.2 Transmission Topology and Path Ratings in the *N* Case and *N* – 1 Cases

N Case			*N* – 1 Case			
Nodes	Technology	Power Rating	Nodes	Technology	Power Rating	Additional Power Rating
10–13	AC OHL	1500 MW	10–14	AC OHL	1500 MW	1500 MW
3–14	AC OHL	1500 MW	3–21	AC OHL	1400 MW	1400 MW
11–21	DC OHL/UGC	1400 MW	11–20	DC OHL/UGC	1500 MW	1500 MW
11–23	DC OHL/UGC	600 MW	11–14	DC UGC	600 MW	—
12–23	AC OHL	1500 MW	12–23	AC OHL/UGC	2000 MW	200 MW
12–14	AC UGC	500 MW				

are shown in the last column of Table 7.2. To obtain $N-1$ security for the shown transmission topology, the power rating of three transmission paths have to be doubled. This is also reflected in the total costs of the transmission system. In the $N-1$ secure case, the total transmission system costs are 71.1% higher in comparison to the optimal solution in the *N* case.

7.4.3 Optimal Timing and Investment Coordination

Each investment decision has an effect on the following investment decisions as the topology of the transmission grid is changed. Due to topology change, the impedance

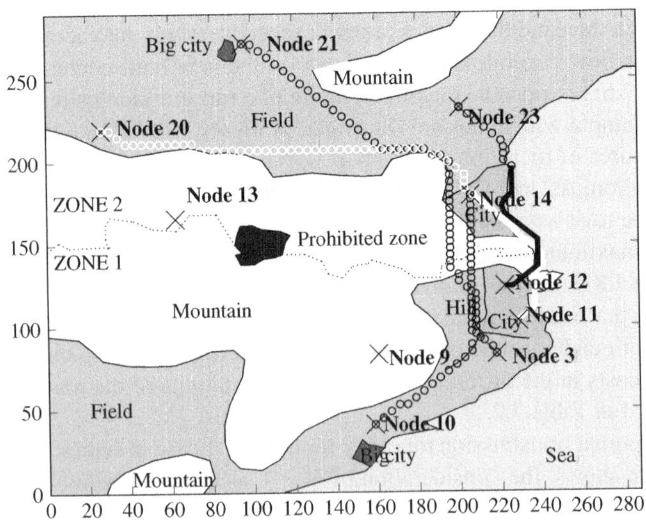

Figure 7.9 Transmission topology for 7-GW interconnection capacity considering $N-1$ criterion. Dark areas indicate HVAC connections. White areas indicate HVDC connections. Solid lines indicate underground cables. Circles indicate overhead lines.

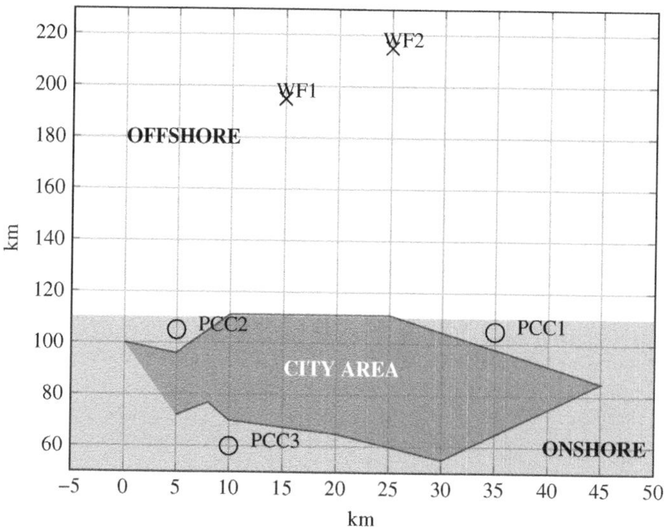

Figure 7.10 Wind farms and points of common coupling of study case.

and therefore the flow pattern in the transmission changes. The timing and coordination of transmission system investment is of major importance due to the lumpiness of transmission system investments.

The project cycle from planning to construction of a transmission system investment can take several years up to decades due to a number of technical and nontechnical reasons. Therefore, it is important to foresee future developments in the transmission grid and coordinate and optimize investments accordingly.

In optimization problems, each investment decision can be seen as a control variable or a state variable as it changes the transmission system structure. For investment problems, where the time horizon is divided in two or three steps, the optimization of the time point and coordination can be carried out using scenario techniques or calculating all possibilities in each time step as the number of control variables is limited. In case that more time steps are used, the optimization of the investment time points should be carried out using dynamic programming approach.

7.4.3.1 Illustrative Example The following paragraphs present a simple case study to show the importance of the optimal timing of investment and coordinated investment planning [66]. The case study is illustrated in Figure 7.10 and consists of two offshore wind farms (*WF*) with a rated capacity of 550 MW each. There are three potential points of common coupling (*PCC*). It is assumed that WF_2 is connected five years later than WF_1 to the main grid. The optimization of the offshore grid layout is carried out using a rule-based genetic algorithm [67].

The objective of the optimization is the minimization of life cycle system costs, consisting of investments, costs of losses, costs of maintenance, and costs of not harvested energy due to unavailability of the transmission equipment [Eq. (7.7)].

$$C_{lcsc} = C_{inv} + \sum_{i=1}^{\text{lifetime}} (C_{loss,i} + C_{main,i} + C_{nte,i}) \tag{7.7}$$

The optimization constraints are the Kirchhoff laws for current and voltage [Eq. (7.8a)–(7.8e)], the maximum power per transmission path [Eq. (7.8f)] and the maximum power injection capabilities of the *PCC* [Eq. (7.8g)].

$$\mathbf{0} = 3\underline{\mathbf{U}} \cdot \underline{\mathbf{Y}}^* \cdot \underline{\mathbf{u}}^* - (\mathbf{p} + j\mathbf{q}) \tag{7.8a}$$

with

$$\underline{\mathbf{U}} = diag(\underline{U}_1 \underline{U}_2 \cdots \underline{U}_N) \tag{7.8b}$$
$$\underline{\mathbf{u}} = [\underline{U}_1 \underline{U}_2 \cdots \underline{U}_N]^T \tag{7.8c}$$
$$\mathbf{p} = [P_1 P_2 \dots P_N]^T \tag{7.8d}$$
$$\mathbf{q} = [Q_1 Q_2 \dots Q_N]^T \tag{7.8e}$$
$$\mathbf{P}_i \leq \mathbf{P}_{max} \tag{7.8f}$$
$$\mathbf{P}_{inj} \leq \mathbf{P}_{mpic} \tag{7.8g}$$

Two different planning approaches are compared using the optimization algorithm. The first approach is the integrated approach, where the optimal connection topology is determined for both wind farms at the same time. In the separated approach, the connection topologies of both wind farms is optimized independently.

In the integrated approach the delayed connection of the second wind farm is considered. While calculating the optimal offshore grid layout, both wind farms are treated as they are connected to the transmission system at the same time. The share of the costs caused by the second wind farm is discounted using the interest rate in Table 7.3.

In the case where both wind farms are treated separately, first, the connection layout of WF_1 is optimized. Thereafter, WF_1 is treated as a new *PCC* and the transmission layout for WF_2 is optimized. Also with this approach, the obtained costs for the transmission investments for WF_2 are discounted accordingly. The costs for investments, maintenance, losses, and not supplied energy over the lifetime of the

TABLE 7.3 Assumptions for Transmission System Optimization

Capacity $PCC_{1,2}$:	200 MW
Capacity PCC_3:	1200 MW
Voltage level $PCC_{1,2,3}$:	400 kV
Voltage level $WF_{1,2}$:	30 kV
Lifetime:	25 *ae*
Energy price:	50 €/MWh
Interest rate:	5%

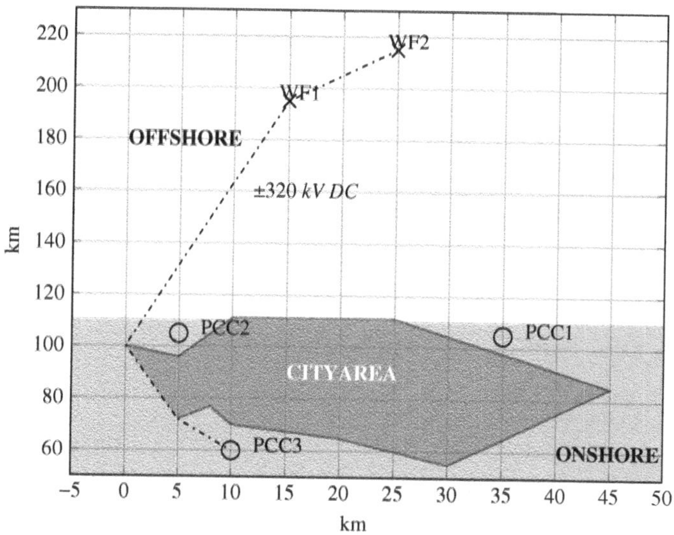

Figure 7.11 Optimal transmission layout for integrated approach.

offshore transmission system are taken into account. A further assumption is that the onshore transmission is realized with underground cables (AC or DC) [66].

Figure 7.11 shows the optimal offshore transmission topology using integrated planning. The transmission system consists of a multi-terminal HVDC system connecting both wind farms and PCC_3. The ideal transmission voltage equals ±320 kV, and one pair of HVDC cables is used to transmit the power. The cable between WF_1 and PCC_3 is rated for the total amount of wind power, namely 1100 MW. This means that the cable is overrated, before WF_2 is connected to transmission system. The total transmission system costs equal 669.4 M€. The total transmission system length equals 160 km, whereas the length of offshore connections equals 109.4 km.

Figure 7.12 shows the transmission system layout if the optimization is carried out separately for both wind farms. The optimal transmission system for WF_1 consists of HVAC connections. The optimal transmission voltage equals 220 kV. WF_1 is first connected to PCC_2 via a pair of three-core HVAC submarine cables. Two hundred megawatts of the generated wind power are injected in PCC_2 to the main transmission system, whereas the rest, namely 350 MW, are transmitted to PCC_3 via one system of 220 kV single core underground land cables. The system costs of the proposed transmission system equal 375.3 M€. The length of the transmission system is 139.1 km, whereas 85.5 km of the transmission cables are situated offshore.

During the determination of the optimal transmission system layout of the second wind farm, it is assumed that the load level in the city has been increased with time and that an additional 100 MW can be injected in both PCC_1 and PCC_2. Due to the used crosssection of the submarine and underground cables for the connection of the first wind farm, an additional 60 MW can be injected in the offshore substation of WF_1. Using these values as input, the optimal transmission system for the second wind farm is determined and shown in Figure 7.12.

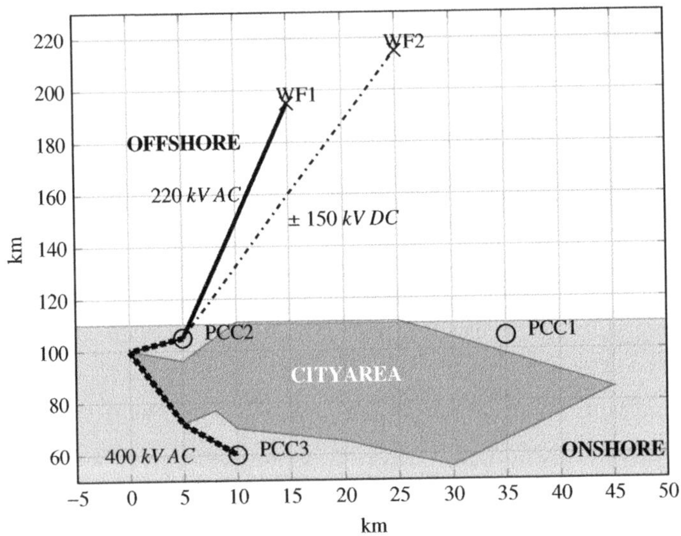

Figure 7.12 Optimal connection of WF_1 and WF_2 with nonintegrated approach.

The optimal transmission system for WF_2 consists of an HVDC link between WF_2 and PCC_2 and an HVAC connection between PCC_2 and PCC_3. Here, ±150 kV is chosen as the best transmission voltage for the HVDC link using one pair of HVDC submarine cables. In this case, 100 MW are injected into PCC_2 and the rest of the generated wind power is transmitted to PCC_3 via one system of 400-kV single core HVAC cables. The transmission system length is in total 160.3 km, with 106.7 km of transmission cables situated offshore. The costs for the transmission system equal 430.3 M€.

If the costs of the transmission system of the second wind farm are discounted and added to the costs of the first system, the obtained total costs equal 708, 3 M€. Comparing the system costs to the integrated approach, this is a cost increase of 38.9 M€, which equals 5.5%.

The breakeven time between the integrated and the nonintegrated planning can be calculated using the obtained cost information. The breakeven time states how many years the construction of the second wind farm may be delayed, so that the overrating of the HVDC cable from the integrated solution is still feasible. The breakeven time for different interest rates is listed in Table 7.4. In the given study case using an interest rate of 5%, the breakeven time equals 9.3 years. This means that if the connection of the second wind farm takes place 9.3 years after the first wind farm, it is still more cost effective to overrate the HVDC cable connecting WF_1 and PCC_3 (see Figure 7.11) and foresee transmission capacities for the second wind farm.

Another advantage of investment coordination besides cost saving is that the transmission system consists of one single voltage level and transmission technology whereas the optimal solution requires three different voltage levels and both HVAC and HVDC technology. The use of different technologies complicates the operation of the transmission system.

TABLE 7.4 Breakeven Time Between Integrated and Not Integrated Approach in Dependence of Interest Rate

Interest Rate (%)	Breakeven Time (*ae*)
5	9.3
10	4.5
15	3
20	2.1
25	1.7

The most important benefit of the investment coordination is that the onshore transmission system has to be commissioned only once as the cable is overrated. The offshore connections for the second wind farm and the extension of the converter station onshore has to be realized after five years. If the investments are not coordinated, the transmission links will be commissioned separately as different voltage levels and transmission technologies are used. In that case also two substations (PCC_2 and PCC_3) are affected by the connection of the wind farms.

7.4.4 Conclusions

Transmission grid planning is a difficult task due to a large number of variables, constraints, and uncertainties. Until recently, transmission grid planning was performed using expert knowledge. Currently, optimization tools are being developed and used which consider different aspects of this complex problem. Nevertheless, these optimization tools need improvement. In the future, more renewable energy sources will be connected to the grid whereas the electricity grid and the electricity demand will become more flexible. This will add higher uncertainties and control variables in this complex problem. Therefore, research in probabilistic transmission grid planning is needed to provide more efficient planning tools in the future.

REFERENCES

1. D. Kirschen and G. Strbac, *Fundamentals of Power System Economics*. Wiley Online Library, 2004.
2. H. Ergun, D. Van Hertem, and R. Belmans, "Identification of power injection capabilities for transmission system investment optimization," *10th IEEE PowerTech Conference*, June 2013, Grenoble, France.
3. H. Ergun, B. Rawn, R. Belmans, and D. Van Hertem, "Technology and Topology Optimization for Multizonal Transmission Systems." *IEEE Transactions on Power Systems. Institute of Electrical and Electronics Engineers nr. 29*, pp. 2469–2477, ISSN 0885–8950.
4. Desertec Industrial Initiative (Dii), "Industrial initiatives dii and medgrid join forces for the development of industrial-scale renewable energy," 2011 [Online]. Available: http://www.dii-eumena.com
5. L. Moreno, "Concentrated solar power (csp) in desertec—Analysis of technologies to secure and affordable energy supply," in *Intelligent Data Acquisition and Advanced Computing Systems (IDAACS), 2011 IEEE 6th International Conference on*, vol. 2, Sept. 2011, pp. 923–926.

6. Desertec Foundation, "Tunisian sun will light European homes by 2016," 2012 [Online]. Available: www.desertec.org
7. J. De Decker and P. Kreutzkamp, "Offshore electricity grid infrastructure in Europe—Final report," Technical Report, Oct. 2011.
8. G. Latorre, R. Cruz, J. Areiza, and A. Villegas, "Classification of publications and models on transmission expansion planning," *Power Systems, IEEE Transactions on*, vol. 18, no. 2, pp. 938–946, May 2003.
9. L. Garver, "Transmission network estimation using linear programming," *Power Apparatus and Systems, IEEE Transactions on*, vol. PAS-89, no. 7, pp. 1688–1697, Sept. 1970.
10. R. Villasana, L. L. Garver, and S. J. Salon, "Transmission network planning using linear programming," *Power Engineering Review, IEEE*, vol. PER-5, no. 2, pp. 36–37, Feb. 1985.
11. V. Levi and M. Calovic, "Linear-programming-based decomposition method for optimal planning of transmission network investments," *Generation, Transmission and Distribution, IEE Proceedings C*, vol. 140, no. 6, pp. 516–522, Nov. 1993.
12. ——, "A new decomposition based method for optimal expansion planning of large transmission networks," *Power Systems, IEEE Transactions on*, vol. 6, no. 3, pp. 937–943, Aug. 1991.
13. R. Chanda and P. Bhattacharjee, "Application of computer software in transmission expansion planning using variable load structure," *Electric Power Systems Research*, vol. 31, no. 1, pp. 13–20, 1994.
14. K. Kim, Y. Park, and K. Lee, "Optimal long term transmission expansion planning based on maximum principle," *Power Systems, IEEE Transactions on*, vol. 3, no. 4, pp. 1494–1501, Nov. 1988.
15. Y. Dusonchet and A. El-Abiad, "Transmission planning using discrete dynamic optimizing," *Power Apparatus and Systems, IEEE Transactions on*, vol. PAS-92, no. 4, pp. 1358–1371, July 1973.
16. H. Youssef and R. Hackam, "New transmission planning model," *Power Systems, IEEE Transactions on*, vol. 4, no. 1, pp. 9–18, Feb. 1989.
17. V. A. Levi, "A new mixed-integer methodology for optimal transmission expansion planning," *Electric Power Systems Research*, vol. 32, no. 3, pp. 227–238, 1995.
18. L. Bahiense, G. Oliveira, M. Pereira, and S. Granville, "A mixed integer disjunctive model for transmission network expansion," *Power Systems, IEEE Transactions on*, vol. 16, no. 3, pp. 560–565, Aug. 2001.
19. S. Haffner, A. Monticelli, A. Garcia, J. Mantovani, and R. Romero, "Branch and bound algorithm for transmission system expansion planning using a transportation model," *Generation, Transmission and Distribution, IEE Proceedings*, vol. 147, no. 3, pp. 149–156, May 2000.
20. A. Seifu, S. Salon, and G. List, "Optimization of transmission line planning including security constraints," *Power Systems, IEEE Transactions on*, vol. 4, no. 4, pp. 1507–1513, Nov. 1989.
21. J. dos Santos, A., P. Franca, and A. Said, "An optimization model for long-range transmission expansion planning," *Power Systems, IEEE Transactions on*, vol. 4, no. 1, pp. 94–101, Feb. 1989.
22. A. Sharifnia and H. Aashtiani, "Transmission network planning: A method for synthesis of minimum-cost secure networks," *Power Apparatus and Systems, IEEE Transactions on*, vol. PAS-104, no. 8, pp. 2025–2034, Aug. 1985.
23. A. Meliopoulos, R. Webb, R. Bennon, and J. Juves, "Optimal Long Range Transmission Planning with AC Load Flow," *Power Apparatus and Systems, IEEE Transactions on*, vol. PAS-101, no. 10, pp. 4156–4163, Oct. 1982.
24. S. Lee, K. Hicks, and E. Hnyilicza, "Transmission expansion by branch-and-bound integer programming with optimal cost–capacity curves," *Power Apparatus and Systems, IEEE Transactions on*, vol. PAS-93, no. 5, pp. 1390–1400, Sept. 1974.
25. S. Binato, M. Pereira, and S. Granville, "A new Benders decomposition approach to solve power transmission network design problems," *Power Systems, IEEE Transactions on*, vol. 16, no. 2, pp. 235–240, May 2001.
26. S. Siddiqi and M. Baughman, "Value-based transmission planning and the effects of network models," *Power Systems, IEEE Transactions on*, vol. 10, no. 4, pp. 1835–1842, Nov. 1995.
27. R. Romero and A. Monticelli, "A zero-one implicit enumeration method for optimizing investments in transmission expansion planning," *Power Systems, IEEE Transactions on*, vol. 9, no. 3, pp. 1385–1391, Aug. 1994.

28. M. Pereira, L. Pinto, S. Cunha, and G. Oliveira, "A decomposition approach to automated generation/transmission expansion planning," *Power Apparatus and Systems, IEEE Transactions on*, vol. PAS-104, no. 11, pp. 3074–3083, Nov. 1985.
29. R. Romero and A. Monticelli, "A hierarchical decomposition approach for transmission network expansion planning," *Power Systems, IEEE Transactions on*, vol. 9, no. 1, pp. 373–380, Feb. 1994.
30. W. Puntel and R. Fischl, "Computer aided design of electric power transmission networks," *IEEE Winter Power Meeting*, Jan. 1972.
31. M. Pereira and L. Pinto, "Application of sensitivity analysis of load supplying capability to interactive transmission expansion planning," *Power Apparatus and Systems, IEEE Transactions on*, vol. PAS-104, no. 2, pp. 381–389, Feb. 1985.
32. A. Ekwue and B. Cory, "Transmission system expansion planning by interactive methods," *Power Apparatus and Systems, IEEE Transactions on*, vol. PAS-103, no. 7, pp. 1583–1591, July 1984.
33. A. Ekwue, "Investigations of the transmission system expansion problem," *International Journal of Electrical Power; Energy Systems*, vol. 6, no. 3, pp. 139–142, 1984.
34. A. Monticelli, A. Santos, M. Pereira, S. Cunha, B. Parker, and J. Praca, "Interactive transmission network planning using a least-effort criterion," *Power Apparatus and Systems, IEEE Transactions on*, vol. PAS-101, no. 10, pp. 3919–3925, Oct. 1982.
35. R. Bennon, J. Juves, and A. Meliopoulos, "Use of sensitivity analysis in automated transmission planning," *Power Apparatus and Systems, IEEE Transactions on*, vol. PAS-101, no. 1, pp. 53–59, Jan. 1982.
36. C. Dechamps and E. Jamoulle, "Interactive computer program for planning the expansion of meshed transmission networks," *International Journal of Electrical Power; Energy Systems*, vol. 2, no. 2, pp. 103–108, 1980.
37. C. Serna, J. Duran, and A. Camargo, "A model for expansion planning of transmission systems—A practical application example," *Power Apparatus and Systems, IEEE Transactions on*, vol. PAS-97, no. 2, pp. 610–615, Mar. 1978.
38. G. Latorre-Bayona and I. Perez-Arriaga, "CHOPIN, a heuristic model for long term transmission expansion planning," *Power Systems, IEEE Transactions on*, vol. 9, no. 4, pp. 1886–1894, Nov. 1994.
39. E. da Silva, H. Gil, and J. Areiza, "Transmission network expansion planning under an improved genetic algorithm," in *Power Industry Computer Applications, 1999. PICA '99. Proceedings of the 21st 1999 IEEE International Conference*, pp. 315–321, July 1999.
40. E. Da Silva, H. Gil, and J. Areiza, "Transmission network expansion planning under an improved genetic algorithm," *Power Systems, IEEE Transactions on*, vol. 15, no. 3, pp. 1168–1174, Aug. 2000.
41. K. Yoshimoto, K. Yasuda, and R. Yokoyama, "Transmission expansion planning using neuro-computing hybridized with genetic algorithm," in *Evolutionary Computation, 1995., IEEE International Conference on*, vol. 1, Nov.–Dec. 1995, p. 126.
42. R. Gallego, A. Monticelli, and R. Romero, "Comparative studies on nonconvex optimization methods for transmission network expansion planning," *Power Systems, IEEE Transactions on*, vol. 13, no. 3, pp. 822–828, Aug. 1998.
43. ———, "Transmission system expansion planning by an extended genetic algorithm," *Generation, Transmission and Distribution, IEE Proceedings*, vol. 145, no. 3, pp. 329–335, May 1998.
44. A. Escobar, R. A. Gallego, R. L. Romero, and S. A. de Oliveira, "Hybrid model analysis and constructive algorithms in the performance of a genetic algorithm for the planning of an electric power transmission system," *Proceedings of the 17th International Conference on CARS and FOF*, 2001.
45. H. Gil and E. Da Silva, "A reliable approach for solving the transmission network expansion planning problem using genetic algorithms," *Electric Power Systems Research*, vol. 58, no. 1, pp. 45–51, 2001.
46. R. Romero, R. Gallego, and A. Monticelli, "Transmission system expansion planning by simulated annealing," in *Power Industry Computer Application Conference, 1995. Conference Proceedings, 1995 IEEE*, pp. 278–283, May 1995.
47. R. Gallego, A. Alves, A. Monticelli, and R. Romero, "Parallel simulated annealing applied to long term transmission network expansion planning," *Power Systems, IEEE Transactions on*, vol. 12, no. 1, pp. 181–188, Feb. 1997.

48. R. Teive, E. Silva, and L. Fonseca, "A cooperative expert system for transmission expansion planning of electrical power systems," *Power Systems, IEEE Transactions on*, vol. 13, no. 2, pp. 636–642, May 1998.
49. R. Teive, A. Hawken, and M. Laugthon, "Knowledge-based system for electrical power networks transmission expansion planning," in *Transmission and Distribution Conference and Exposition: Latin America, 2004 IEEE/PES*, pp. 435–440, Nov. 2004.
50. J.-R. Shin and Y.-M. Park, "Optimal long-term Transmission Planning by Expert System Approach," in *TENCON '93. Proceedings. Computer, Communication, Control and Power Engineering.1993 IEEE Region 10 Conference on*, no. 0, oct 1993, pp. 713–717 vol. 2.
51. H. Sun and D. Yu, "A multiple-objective optimization model of transmission enhancement planning for independent transmission company (ITC)," in *Power Engineering Society Summer Meeting, 2000. IEEE*, vol. 4, 2000, pp. 2033–2038.
52. R. Gallego, R. Romero, and A. Monticelli, "Tabu search algorithm for network synthesis," *Power Systems, IEEE Transactions on*, vol. 15, no. 2, pp. 490–495, May 2000.
53. E. Da Silva, J. Ortiz, G. De Oliveira, and S. Binato, "Transmission network expansion planning under a Tabu Search approach," *Power Systems, IEEE Transactions on*, vol. 16, no. 1, pp. 62–68, Feb. 2001.
54. "Transmission network optimal planning using the tabu search method," *Electric Power Systems Research*, vol. 42, no. 2, pp. 153–163, 1997.
55. P. Sensarma, M. Rahmani, and A. Carvalho, "A comprehensive method for optimal expansion planning using particle swarm optimization," in *Power Engineering Society Winter Meeting, 2002. IEEE*, vol. 2, 2002, pp. 1317–1322.
56. P. Maghouli, S. Hosseini, M. Buygi, and M. Shahidehpour, "A multi-objective framework for transmission expansion planning in deregulated environments," *Power Systems, IEEE Transactions on*, vol. 24, no. 2, pp. 1051–1061, May 2009.
57. H. Kakuta and H. Mori, "A multi-objective memetic algorithm for probabilistic transmission network expansion planning," in *Systems Man and Cybernetics (SMC), 2010 IEEE International Conference on*, pp. 1414–1419, Oct. 2010.
58. J. Contreras and F. Wu, "A kernel-oriented algorithm for transmission expansion planning," *Power Systems, IEEE Transactions on*, vol. 15, no. 4, pp. 1434–1440, Nov. 2000.
59. ———, "Coalition formation in transmission expansion planning," *Power Systems, IEEE Transactions on*, vol. 14, no. 3, pp. 1144–1152, Aug. 1999.
60. J. Contreras, "A cooperative game theory approach to transmission planning in power systems," Ph.D. thesis, University of California Berkeley, 1997.
61. A. Orths and Z. Styczynski, "Game theoretical approach to power network planning," in *Power Tech Proceedings, 2001 IEEE Porto*, vol. 1, 2001, p. 6.
62. J.-C. Kaltenbach, J. Peschon, and E. Gehrig, "A mathematical optimization technique for the expansion of electric power transmission systems," *Power Apparatus and Systems, IEEE Transactions on*, vol. PAS-89, no. 1, pp. 113–119, Jan. 1970.
63. J. Dodu and A. Merlin, "Dynamic model for long-term expansion planning studies of power transmission systems: The Ortie Model," *International Journal of Electrical Power; Energy Systems*, vol. 3, no. 1, pp. 2–16, 1981.
64. S. Binato and G. C. Oliveira, "A heuristic procedure to cope with multi-year transmission expansion planning," *IEEE/KTH Stockholm Power Technical Conference*, June 1995.
65. M. Mekonnen and R. Belmans, "The influence of phase shifting transformers on the results of flow-based market coupling," in *European Energy Market (EEM), 2012 9th International Conference on the*, 2012, pp. 1–7.
66. H. Ergun, D. Van Hertem, and R. Belmans, "Building a new overlay grid for Europe," in *IEEE PES General Meeting*, July 2012.
67. ———, "Transmission system topology optimization for large-scale offshore wind integration," *Sustainable Energy, IEEE Transactions on*, vol. 3, no. 4, pp. 908–917, 2012.

CHAPTER 8

HVDC GRID LAYOUTS

Jun Liang,[1] Oriol Gomis-Bellmunt,[2] and Dirk Van Hertem[3]

[1]Institute of Energy, School of Engineering, Cardiff University, Cardiff, Wales

[2]Research Group CITCEA, Electrical Engineering Department, Technical University of Catalonia (UPC), Barcelona, Spain; and Research Group Wind Energy, Electrical Engineering Department, Catalonia Institute for Energy Research (IREC), Barcelona, Spain

[3]Research Group Electa, Department of Electrical Engineering, University of Leuven, Leuven, Belgium

SO FAR, most HVDC transmission uses a point-to-point link between two AC systems. With the development of HVDC transmission systems, future HVDC grids will be formed through the evolution of existing HVDC links which will be connected to each other, or through systematic plans and designs of HVDC networks.

When considering this HVDC grid, different design choices need to be made which fix the layout of the grid, but also its performance and capabilities. Three levels of layout need to be distinguished: the overall system topology, the HVDC converter station configuration, and the substation arrangement. In each of these three categories, a choice must be made between functionality or reliability and costs [1].

Topologies of HVDC grids are determined by the locations of sources (or sending AC systems) and loads (or receiving AC systems), redundancy and reliability of power transmission, and costs of transmission circuits and devices which include converters configurations, DC cables/lines, and DC switchgear.

The HVDC converter station configuration determines how the converters are connected inside a substation, as well as the grounding scheme and the grounding point(s).

The substation arrangement determines how the different feeders and converters are connected within a DC substation.

This chapter starts with a discussion on what a DC grid is. Next the design considerations of HVDC grids for offshore applications are analyzed in greater detail. The chapter continues with the discussion on the configurations of VSC stations including grounding and the different substation arrangements.

HVDC Grids: For Offshore and Supergrid of the Future, First Edition. By Dirk Van Hertem, Oriol Gomis-Bellmunt, and Jun Liang.

8.1 WHAT IS AN HVDC GRID?

An interesting question, not only from a philosophical point of view, is what the minimum requirement is to talk about an HVDC grid. A first consideration is that we consider systems with more than two nodes. When looking at possible multi-terminal HVDC configurations, different topologies are possible (Figure 8.1) [2].

A first potential topology (Figure 8.1a) is a simple *multi-terminal system*, which can be described as a DC bus with several tappings. While being the simplest form of a multi-terminal HVDC system, there are no meshes and no redundancy in the DC system itself. It is not really a "grid" as it offers no redundancy. Such a topology would typically be useful as an alternative to a single AC line, as a connection between two asynchronous zones with an additional connection—for example, for an offshore wind farm.

A second possible configuration (Figure 8.1b) is to have a *grid of DC lines*, where all the buses are AC buses, and the classic transmission line is replaced by a DC circuit with two converter stations. In this topology, all DC lines are fully controllable. It can consist of a mixture of LCC and VSC HVDC lines, and different lines can operate at different voltages. It is also easier to incorporate existing HVDC lines in the new DC grid. However, it needs a more complex control of the flows to keep the mains frequency in case of different isolated AC grids or nodes. The protection in this topology can also be done using existing AC protection systems and equipment.

The third topology (Figure 8.1c) proposes a *meshed DC system*, with a number of connections between the AC and DC system. Connections within the DC system without converter are possible, and there is redundancy in the DC system.

A major issue with the second topology is the amount of converter stations that are needed. When considering "normal" large grids, a rule of thumb says that the number of branches is usually 1.5 times the number of nodes. This would require $2 \times 1.5 \times \#DC_{nodes}$ converters. A grid using the DC topology of Figure 8.1c requires only $\#DC_{nodes}$ converters. This is especially important because converters are the most expensive, sensitive, and lossy components of the DC grid [2]. Nevertheless, such a system does not require harmonized voltage levels and can even consist of both LCC and VSC systems. Furthermore, such systems already exist, for instance when considering the connections between synchronous zones in Europe (there are currently seven HVDC lines between Scandinavia and continental Europe).

Only Figure 8.1c is a "real" DC grid that can provide inherent redundancy. If it is required that the HVDC system provide redundancy within the HVDC grid itself rather than it using the alternative path through the AC grid, this is the only viable option. The first option can be seen as the first step towards a grid and has considerable advantages towards protection.

8.2 HVDC GRID TOPOLOGIES

When looking at the grid structure or layout, we can identify two basic topologies of HVDC networks, namely, radial and meshed topologies. Radial topologies consist of string connections and star connections.

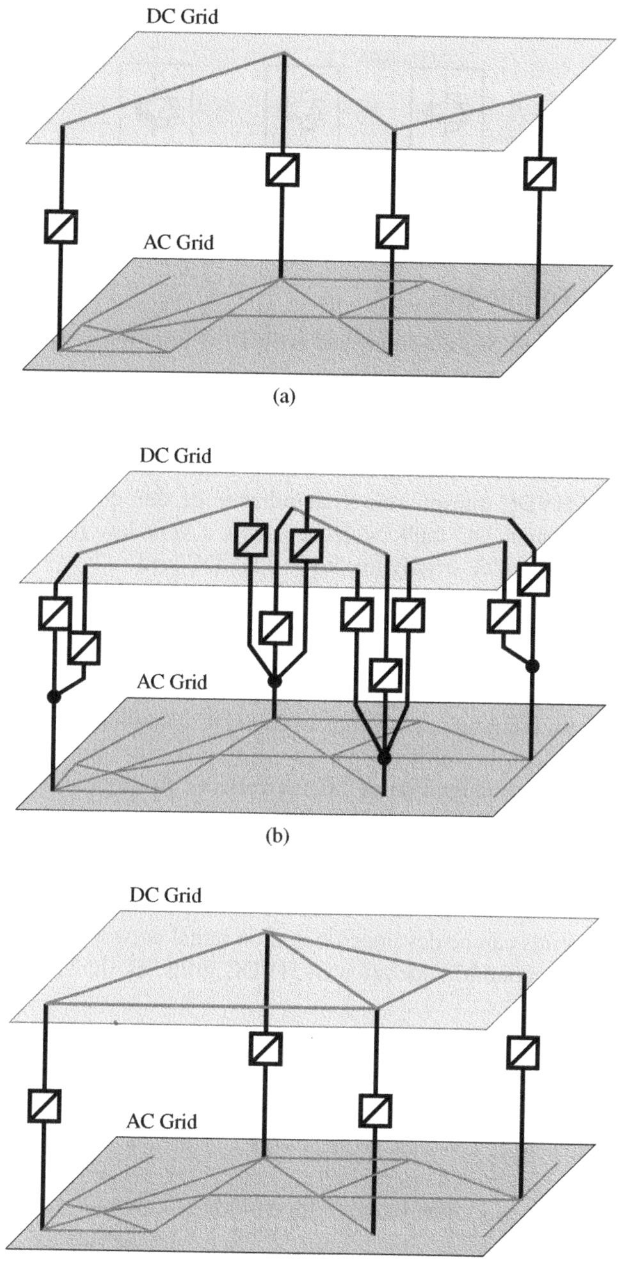

Figure 8.1 Different potential topologies for DC grids [2]. (a) Multi-terminal grid with tappings. (b) Grid of independent DC lines. (c) Meshed DC grid.

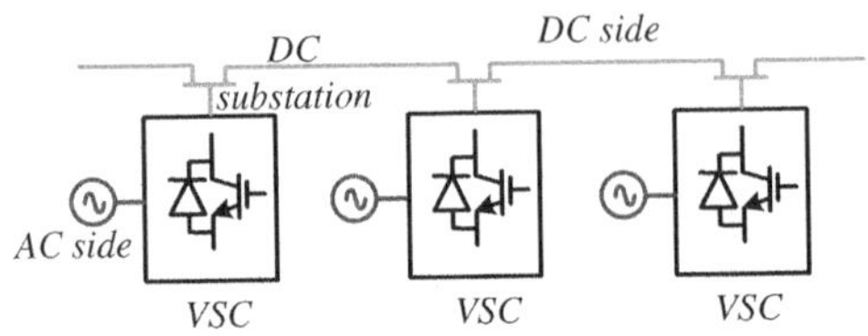

Figure 8.2 Single line diagram of a string connection of DC substations.

8.2.1 Radial Topologies

In Figure 8.2, multiple VSCs are connected at the DC side to form a string connection. This configuration can be extended from the existing point-to-point links. DC circuit breakers may be required at the both ends of each DC cable or overhead line between converters. DC circuit breakers can also be installed at selected circuits to separate the system into two or more parts in case of a DC fault. Without DC circuit breakers, a multi-terminal HVDC transmission system has to be shut down completely when subject to a permanent DC fault, which presents a very low reliability of power transmission and cannot be strictly treated as a HVDC grid.

In Figure 8.3, VSCs are connected to a central DC switching station through DC line/cable to form a star connection. A DC switching station in a HVDC grid is a node where no converter is installed, and only DC lines/cables, DC circuit breakers, and isolators are connected. A DC substation is a node where converters are installed. DC circuit breakers are required at both ends of each DC cable/line between converters. This configuration can be formed from an existing point-to-point link by tapping other terminals for the integration of AC networks or AC sources such as offshore renewable power.

8.2.2 Meshed Topologies

Meshed HVDC grids can be developed based on radial networks to further enhance the reliability and economic operation of HVDC grids, as shown in Figure 8.4 as

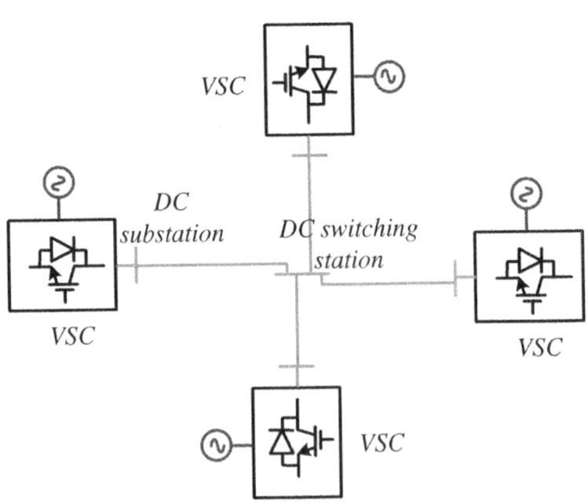

Figure 8.3 Single line diagram of a star connection of converters.

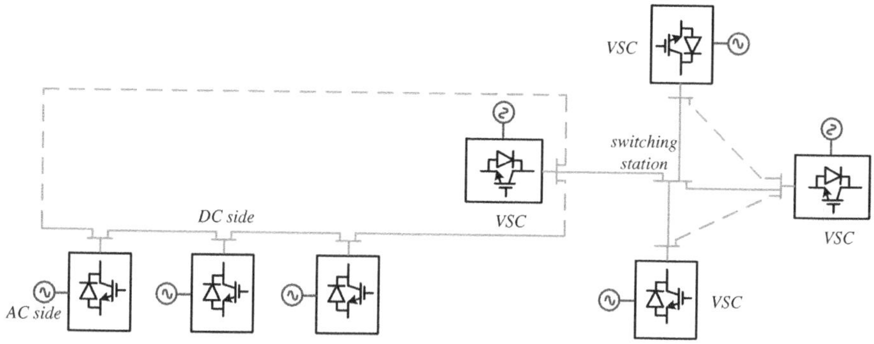

Figure 8.4 A meshed HVDC grid which is developed from radial networks.

an example. The dashed lines indicate the development from radial networks to a meshed HVDC grid.

A general configuration of a meshed HVDC grid is shown in Figure 8.5, which consists of meshed networks of DC lines/cables, DC substation nodes, and DC switching station nodes. A special node could be found in a HVDC grid, which interconnects a number of DC links together with wind parks via an islanded AC network. This equivalent node is actually a VSC-based AC hub in HVDC transmission, also referred as a Supernode, which is an interim stage for building a Supergrid [3].

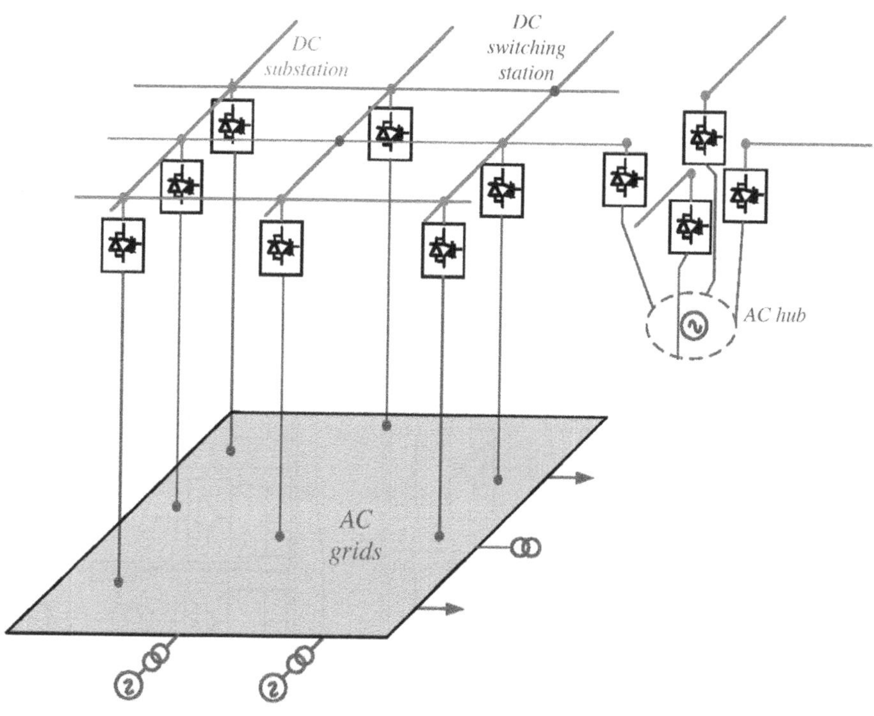

Figure 8.5 A general configuration of a meshed HVDC grid.

The main benefit of the AC hub is that DC circuit breakers are not required. However, more VSCs are required for the AC hub and the power flow through an AC hub is constrained by the power ratings of the VSCs. This concept is addressed in more detail in Chapter 22.

8.3 TOPOLOGIES OF HVDC GRIDS FOR OFFSHORE WIND POWER TRANSMISSION

Point-to-point (PPT) HVDC links have been used to transfer the offshore wind power to the onshore grids. No HVDC circuit breaker is required for the PPT topology. In the event of a converter or HVDC circuit failure, the appropriate action is to disconnect the faulted line by opening the AC circuit breakers of the grid side converter and let the turbines trip off on overspeed or DC-link overvoltages. As a consequence, the wind farms connected the link will be lost. This configuration lacks redundancy.

HVDC grids are likely to be first used for offshore wind power transmission. The purpose of the offshore HVDC grids is to integrate several large offshore wind farms into the onshore AC grids, as shown in Figure 8.6 in which WF represents an offshore wind farm side converter and SS represents an onshore substation side converter. Various topologies of the offshore HVDC grids are analyzed and compared according to a number of criteria: number and capacity of HVDC circuits, number of HVDC circuit breakers, maximum power loss, flexibility, redundancy, lines utilization, and need of offshore switching platforms and fast communications [4].

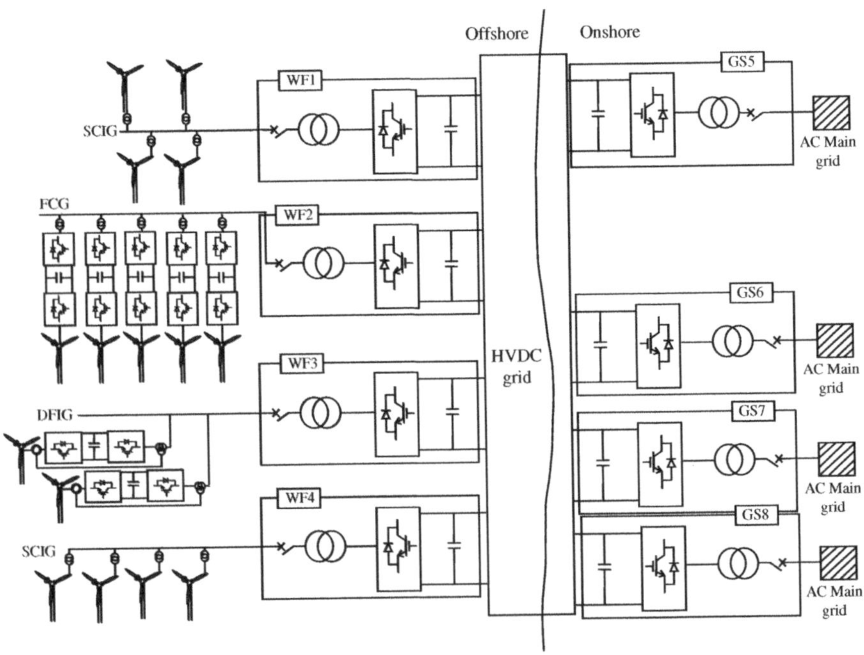

Figure 8.6 HVDC grid for integration of offshore wind farms.

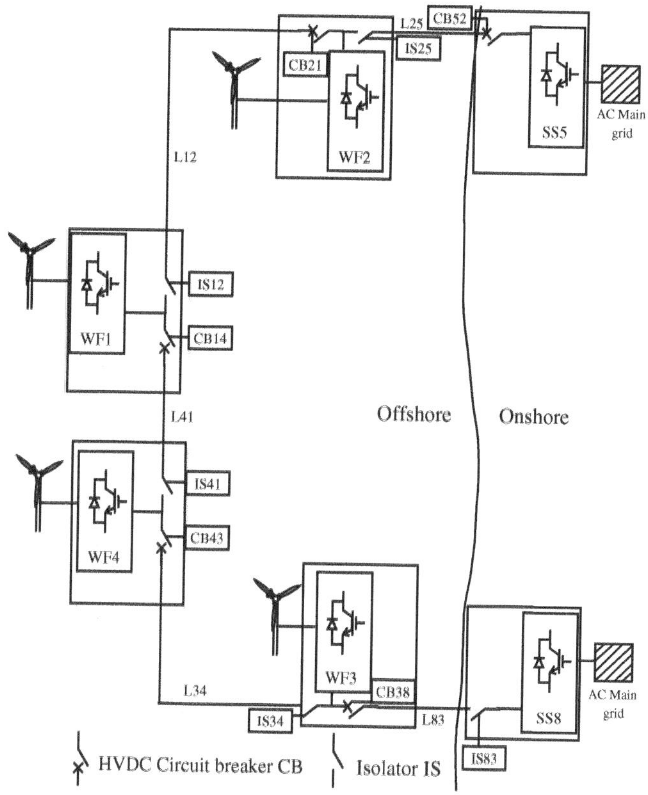

Figure 8.7 Ring topology HVDC grid for offshore wind integration.

8.3.1 General Ring Topology

The general ring topology (GRT) is shown in Figure 8.7. The general ring topology is an HVDC grid with the lines connected to all the nodes composing a ring. The ring can be operated in closed loop, having all the circuit breakers and isolators closed in normal operation, or in open loop, having a circuit breaker or an isolator in the ring opened. In the case of a converter or DC grid fault, the first action is to open the two circuit breakers connected to the fault, leaving the system in open ring. When the fault current reaches zero, the isolators can isolate the faulted area and the appropriate circuit breakers can be connected again. Therefore the power ratings of cables must be determined to meet both normal operation and abnormal operation when the ring is opened.

This configuration has flexibility at the cost of increasing the power ratings of some cables in order to transfer all the wind power. For permanent DC faults or maintenance, the system can be operated in open ring. During normal operation, open ring may also be used to simplify the protection and limit the fault currents.

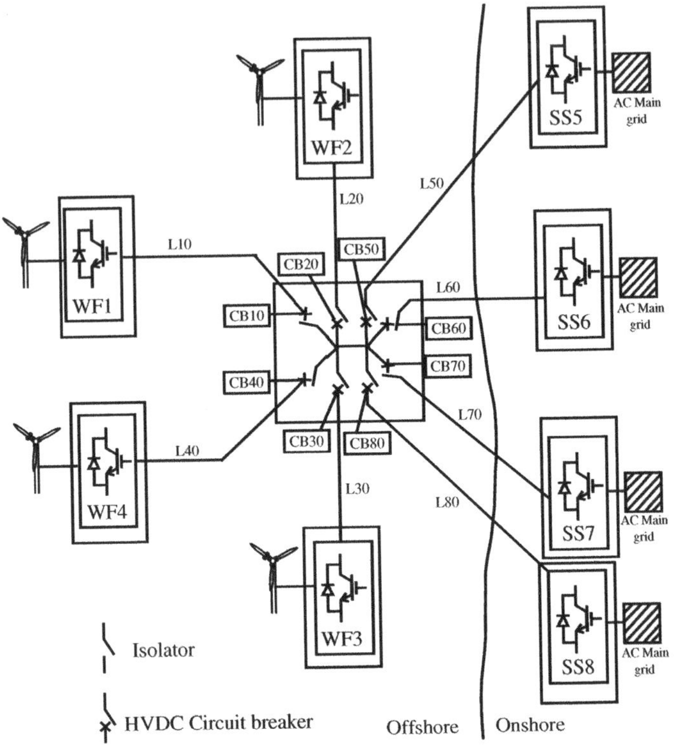

Figure 8.8 Star topology.

8.3.2 Star Topology (ST)

The star topology (ST), shown in Figure 8.8, is an HVDC grid where each line that is connected to a wind farm or an onshore substation is connected to a central star node. In this topology, the rating of each line corresponds to the rating of the wind farm or substation. The main drawback of this configuration is that a fault at the central node can cause the entire system to go offline. Therefore, in spite of the various advantages offered by ST, it is not a feasible topology for multiple wind farm connections. Disregarding a fault on the central node, DC and converter faults can be handled by disconnecting the corresponding line using the HVDC circuit breakers. The star topology needs an offshore platform (or submarine installation) at the central node on which to place all the circuit breakers and star-point connections. The flexibility of this configuration is not as good as GRT because for a permanent fault in a line from the central node to a wind farm, the full wind farm is permanently lost.

8.3.3 Star with a General Ring Topology (SGRT)

Ring and star topologies can be combined to form a hybrid topology star with a general ring topology (SGRT) (Figure 8.9), which is essentially a star configuration with a central switching ring. This configuration offers the advantages of GRT and

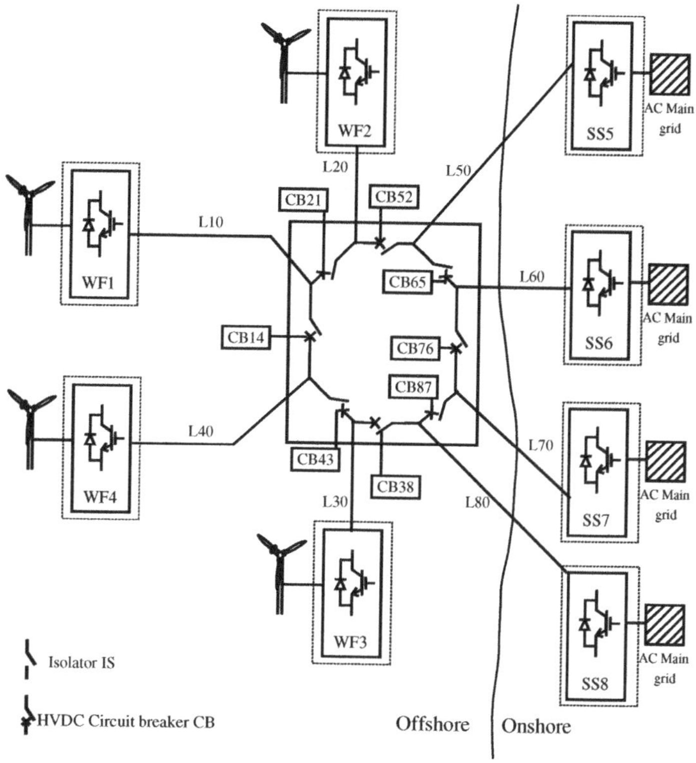

Figure 8.9 Star with a general ring topology.

ST: It needs only full power rating in the central ring, circuit lengths are kept to a minimum (wind farms or substations to central ring), and it can isolate a fault while meeting the maximum loss of power criterion. The central ring in the Figure shows one example arrangement of the DC circuit breakers. Other arrangements such as double busbar and breaker-and-a-half can also be used to increase the reliability of power transmission. One of the main drawbacks of this configuration is the need of an offshore platform on which all circuit breakers in a ring are installed. The capacity of the central ring lines should be equal to the total system power. SGRT shows flexibility for short- and long-term HVDC faults. Regarding flexibility, SGRT has the same problem as ST: A full wind farm is lost for a permanent fault in a line between the central node and the wind farm.

8.3.4 Wind Farm Ring Topology (WFRT)

Wind farm ring topology (WFRT), shown in Figure 8.10, can minimize the number of HVDC circuit breakers. Fast line isolators are used along with HVDC circuit breakers. This topology comprises a wind farm ring having a number of HVDC circuit breakers equal to the number of wind farms and circuits connected to the substations. The topology is similar to the PPT but with the added flexibility of controlling the

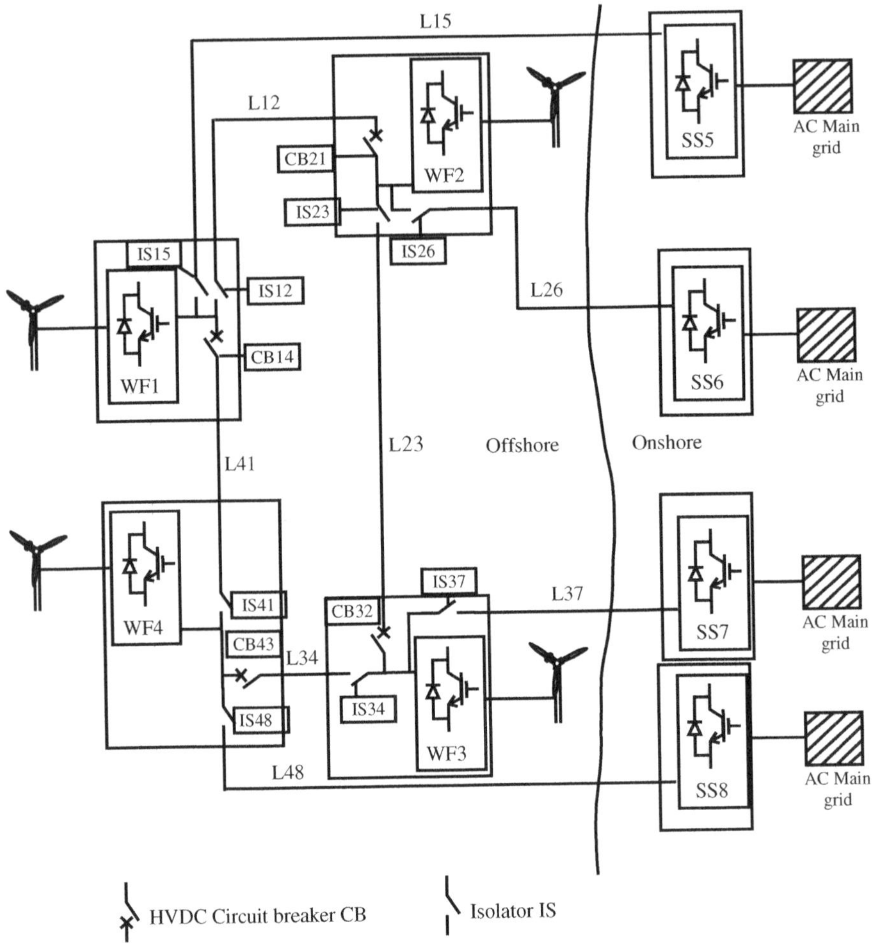

Figure 8.10 Wind farm ring topology.

power flow between wind farms and onshore substations. In the event of a DC fault, the faulted line can be isolated as if it were a PPT thus temporally disconnecting a wind farm. However, the disconnected wind farm can be reconnected to another substation. Depending on distance and costs, the WFRT can be reconfigured as a large ring or a concentrated switching ring as in the case of SGRT.

In the event of a circuit fault, the two HVDC circuit breakers in either side of the faulty location are opened, thus disconnecting the wind farm and the substation. Once there is no current in the disconnected line, the line isolator can be opened and the HVDC circuit breakers can be reconnected isolating the faulted line.

The substation circuit capacity will be equal to the rated substation power. For the wind farm ring, power rating of the line between two wind farms can be the sum of the rated power of the two wind farms. WFRT features excellent flexibility as it can deal with faults in any line without permanently losing power from the wind farms.

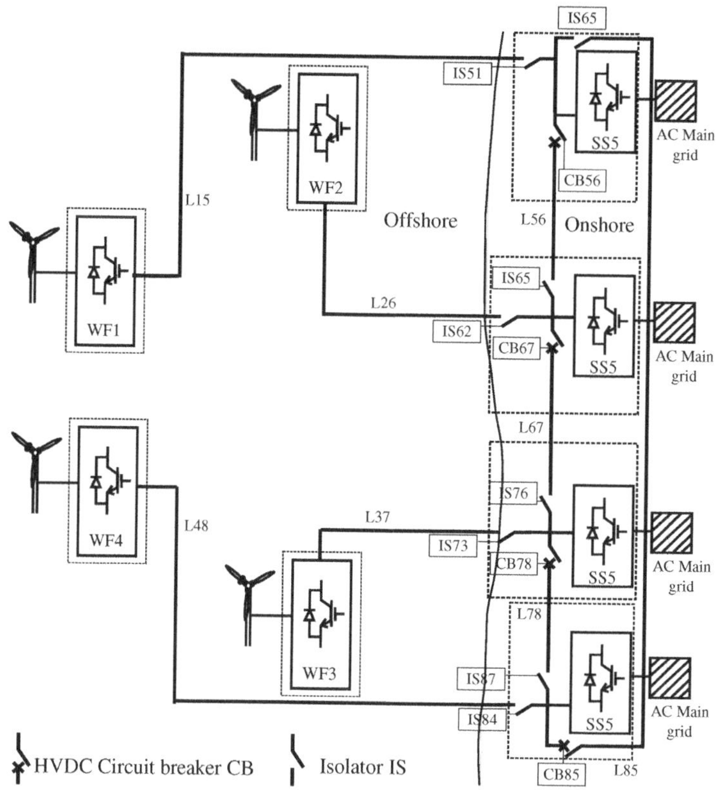

Figure 8.11 Onshore substation ring topology.

8.3.5 Substations Ring Topology (SSRT)

In a manner similar to that of the WFRT, an onshore substation ring topology (SSRT) can be devised as illustrated in Figure 8.11. This configuration performs similarly to WFRT. The difference is that when there is a fault in a HVDC line, the isolated converter will be the wind farm VSC (instead of the VSC in the onshore AC grids). This allows more flexibility in the main grid side at the cost of losing flexibility in the wind farm side both in faulty operation and during long-term failures and maintenance. In most applications it will be more convenient to use WFRT, since faults in the HVDC circuits will allow continued delivery of all the power from the wind farms.

More complex topologies can be formed at the cost of adding more lines and HVDC circuit breakers with different power ratings according to the topologies, such as mixed ring with HVAC and HVDC, double ring topology (WFRT plus SSRT), double whole system rings (doubling the GRT), double star topology (with different central nodes), and star plus ring topology (superposing a star and a ring topology). These configurations can provide higher performance in certain cases at higher costs.

8.3.6 Comparison of the Topologies

PPT does not need HVDC circuit breakers. GRT, ST, and SGRT need $(N_{wf} + N_{ss})$ pairs of HVDC circuit breakers, one pair for each converter station considering both positive and negative poles. Here N_{wf} denotes the number of wind farm side converters and N_{ss} denotes the number of onshore substation converters. The partial ring topologies WFRT and SSRT need only N_{wf} or N_{ss} pairs of circuit breakers respectively.

Circuits of PPT, ST, and SGRT can be rated at the same power of the converter where they are connected. The lines of GRT circuits have to be rated at full system power, thus are economically not attractive. WFRT and SSRT circuits can be rated to the rated power of a single station for the offshore to onshore circuits and to the double of the rated power of a station for the ring circuits. Except PPT, all the HVDC grids can be considered flexible since they allow redirecting the power flow in the event of a fault. The only exception is a fault in the central node with ST, where the maximum power loss criterion is not met.

In the event of a long-term failure or a circuit disconnection, flexibility becomes crucial. With PPT the single wind farm is lost permanently. For GRT, if the fault is in the ring, the faulted circuit can be disconnected, and remaining lines and converters can be in operation. For faults in the converter, the opened ring can still be in service. For ST and SGRT, only the faulted line can be disconnected, resulting in a converter lost. For WFRT a substation will be disconnected for faults in the offshore-to-onshore line, and the opened ring can be in service for faults in the ring. Similarly, for SSRT a wind farm will be disconnected for faults in the offshore-to-onshore line, and the opened ring can be in service for faults in the ring. The topologies which employ a central switching node or ring (ST, SGRT) need an offshore platform. The others can be implemented without an offshore platform. GRT, WFRT, and SSRT need fast communication in order to coordinate the protections. The most relevant features of each topology are summarized in Table 8.1.

It is suggested to consider the following conditions when selecting a topology for the integration of offshore wind farms.

- Flexibility for wind farms
- Fast fault isolation
- Maximum power loss criterion
- Numbers of HVDC circuit breakers
- Length of DC cables

Selection of a topology for a given application depends not only on operation and robustness requirements but also on the geographical location of the substations and wind farms and the eventual cost of HVDC circuit breakers and cables. With the current high cost of solid-state HVDC circuit breakers, it seems that the most appropriate configuration would be the WFRT [4], since it can meet all the required criteria with a reduced number of solid-state HVDC circuit breakers and without needing an offshore switching platform. However, detailed analysis is required before selecting the best topology for a given application because factors such as distances between

TABLE 8.1 Comparison of Topologies of HVDC Grids for Large Offshore Wind Farms

Topology		Offshore Platform	Communication	Flexibility	Redundancy	Comments
PPT		No	No	No	No	Simple but it lacks flexibility
GRT		No	Yes	Good	Good	Flexible but some circuits have to be rated for the full system power
ST	SS side	Yes	No	Good	Yes	The circuits' rating equal to the rating of the wind farm or substation to which the circuit is connected, but it has a weak point at the central node
	WF side	Yes	No	Bad	Yes	
	Overall	Yes	No	Poor	Yes	
SGRT		Yes	No	Poor	Yes	Has the advantages of both GRT and ST, but still needs full power in the central ring.
WFRT		No	Yes	Good	Good	Allow the isolation of a faulty circuit as in the case of point-to-point topology and without needing full system power rating of the ring circuits
SSRT	Ring	No	Yes	Good	Yes	
	Line	No	Yes	Bad	Yes	
	Overall	No	Yes	Poor	Yes	

nodes, power rating and cost of cables, circuit breaker costs, and communication system availability will vary from project to project.

8.4 HVDC CONVERTER STATION CONFIGURATION

There are three basic configurations of VSC stations in a HVDC grid, which are asymmetrical monopole, symmetrical monopole, and bipolar configurations [5] (see also Chapter 3).

8.4.1 Asymmetrical Monopole

Figure 8.12 shows the arrangement of an asymmetrical monopolar point-to-point link. A three-phase connection is to the DC converter at the midpoint. At the DC side, one pole is grounded.

This topology requires only one high voltage conductor (overhead line or cable) and one neutral conductor, which is grounded at one point. The neutral conductor or metallic return is fully rated for the load current but only lightly insulated and thus

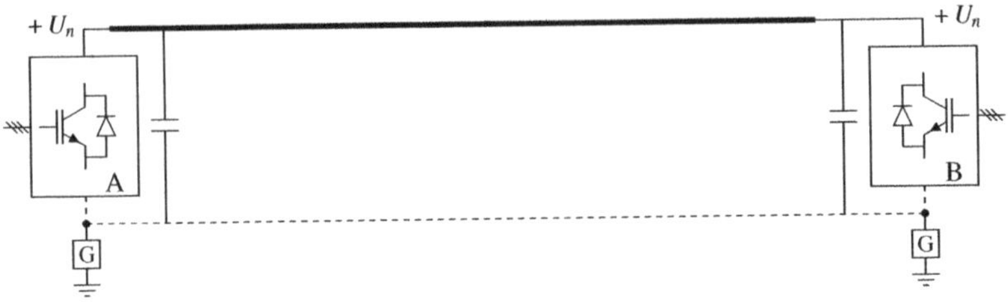

Figure 8.12 Asymmetrical monopolar configuration [6].

lower cost and lower probability of dielectric failure compared to the high voltage conductor. This wire can even be omitted in case the system is grounded at both sides and when substantial ground currents are allowed. The disadvantage is that a HVDC transformer is needed to withstand continual DC stress, at half the DC voltage, on its secondary windings. The loss of any major component of the converter, either due to a fault or during maintenance, leads to the complete disruption of power transfer through this converter.

The symbol G in Figure 8.12 refers to the possible locations for grounding the HVDC system and can represent any of the basic grounding options, shown in Figure 8.15 (see also Section 8.4.4).

8.4.2 Symmetrical Monopolar Configuration

Figure 8.13 shows the arrangement of a symmetrical monopole. As with the asymmetrical monopole, the AC connection is made to the midpoint of the converter. The two DC terminals connect to the positive and negative pole, which operate at an equal but opposite DC voltage. As the DC voltage is symmetrical, the transformer connection experiences no steady-state DC voltage stress. This system can be grounded through a high impedance grounding, for instance by connecting the midpoint of shunt capacitors. Two high-voltage conductors are required: each fully rated for the current duty and fully insulated. It is important to note that in case of a pole-to-ground fault, the voltage on the healthy pole will increase to a (theoretic) maximum of $2 \cdot U_n$. Like an asymmetrical monopole, the loss of any major component leads to the complete loss of power transfer. Symmetrical monopoles have been widely used for VSC HVDC schemes.

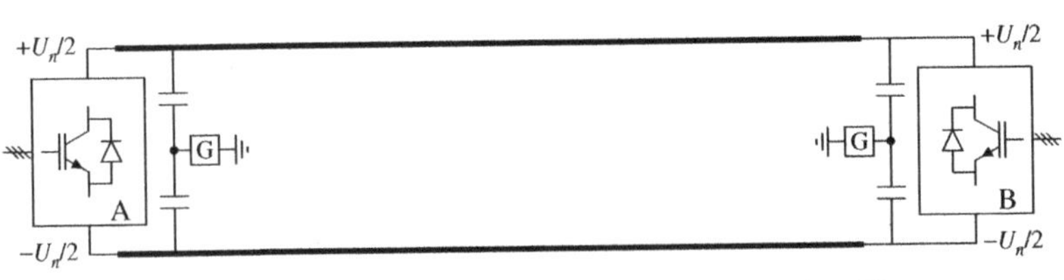

Figure 8.13 Symmetrical monopolar configuration [6].

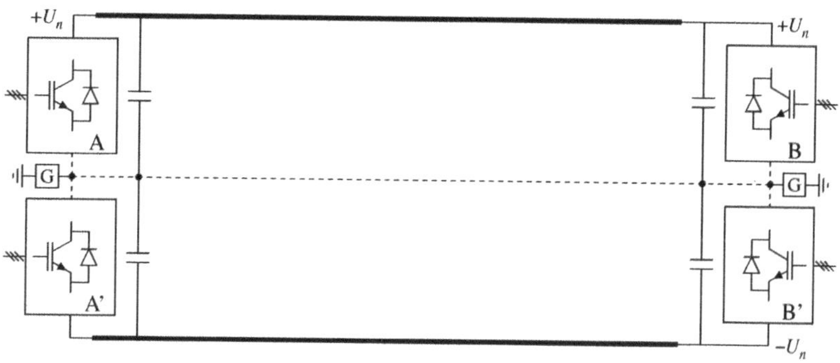

Figure 8.14 Bipolar configuration [6].

8.4.3 Bipolar Configurations

Figure 8.14 shows a bipolar configuration for voltage source converter HVDC point-to-point link. Two asymmetrical monopoles are connected to each other by sharing the ground connected pole. At the AC side, the two converters are powered in parallel. Each pole has its own transformer connection to the AC grid. One converter is connected to the $+U_n$ pole and another converter is connected to the $-U_n$ pole. A neutral conductor can be used to connect the midpoints of the two poles.

In normal operation the two poles are operated in a balanced mode to ensure that the DC current circulates between the $+U_n$ and U_n conductors, with zero current flowing in the ground connection. This is important if the ground (or sea) is used as the return current path, as for environmental reasons and to avoid depletion of the ground (or sea) electrodes, long-term use of the ground path is normally avoided and a metallic return cable is foreseen to carry the return current.

The bipolar configuration requires the use of converter transformers as they experience a DC shift in voltage. Unlike the 12-pulse LCC HVDC system, in which the winding types of the two power transformers are different (one is star–star and the other is star–delta to get a smooth output voltage), the two transformers can be the same in a VSC HVDC system, because there is no need to superpose the output DC voltage of each converter to avoid harmonics. Actually, the two converters in bipolar VSC HVDC system can operate independently. Therefore, the benefit of a bipolar arrangement is that the loss of any major element of a converter entails only the loss of 50% of the transmission capacity of the scheme, provided that a metallic return is installed or the system is grounded at both sides and the earth can be used as a return conductor. In this case, the remaining converter can continue to operate as an asymmetrical monopole, maintaining operating capacity and reliability.

8.4.4 HVDC Grid Grounding

Grounding forms an essential part of the electric circuit, and it determines how the system is designed, operated, and protected. For HVDC systems one can

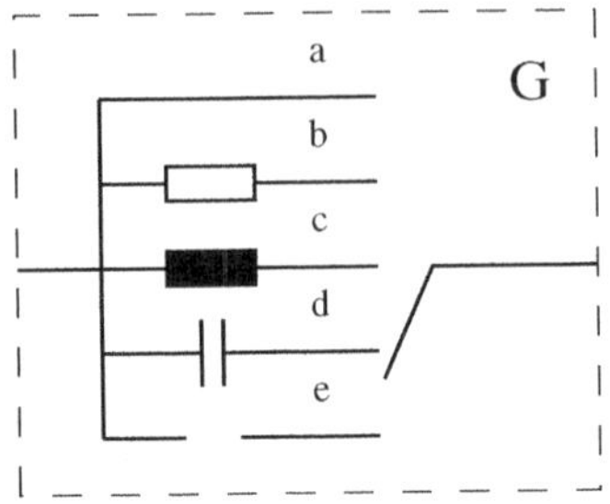

Figure 8.15 Grounding options [6].

consider multiple options for the grounding. The most common options are indicated in Figure 8.15: (a) direct grounding, (b) resistive grounding, (c) inductive grounding, (d) capacitive grounding, (e) high-impedance grounding. Out of these five options, the inductive and capacitive grounding are frequency-dependent: The inductive ground forms a low-impedance path for low frequencies and a high-impedance path for high frequencies while for capacitive grounding it is the reverse.

With a low-impedance grounding, the voltages on the system will retain a fixed point during normal operation and will keep the voltage at the healthy pole low. However, the low-impedance results in a low impedance current path in case of pole to ground fault, which causes high short-circuit currents.

High-impedance grounding results in (near) zero short-circuit currents, yet the voltage on the poles is floating, and in case of a pole-to-ground fault, the voltage doubles on the "healthy" pole, which might influence the cost of the transmission line significantly.

Next to the type of grounding, also the number of grounding points and the location of the ground connections is of interest. A single grounding point suffices to keep the system from floating; however, it is also a single point of failure. Furthermore, all (capacitive) pole-to-ground currents will flow through the ground connection. As such, it also influences the dimension of the earth electrodes [7]. The fault current caused by a pole-to-ground fault also returns through the single ground point, influencing a large part of the system.

Instead of a single grounding point, one can also opt for multiple points, potentially even at all converter stations. This way, the voltage throughout the system is kept closer to zero. However, multiple grounding points cause ground stray currents during normal operation and significant currents during asymmetric operation. In the case where some, but not all, DC nodes are grounded, an additional imbalance might be introduced, depending on the location of the grounding points, the earth return resistance, and the imbalance in the DC grid.

In an European overlay grid, earth currents might need to be avoided as they can endanger human safety, enter the AC grid [8], or have environmental impacts [9].

It is clear that the location, number, and type of grounding points affects the system configuration. At this moment it is not clear how such a system will be implemented.

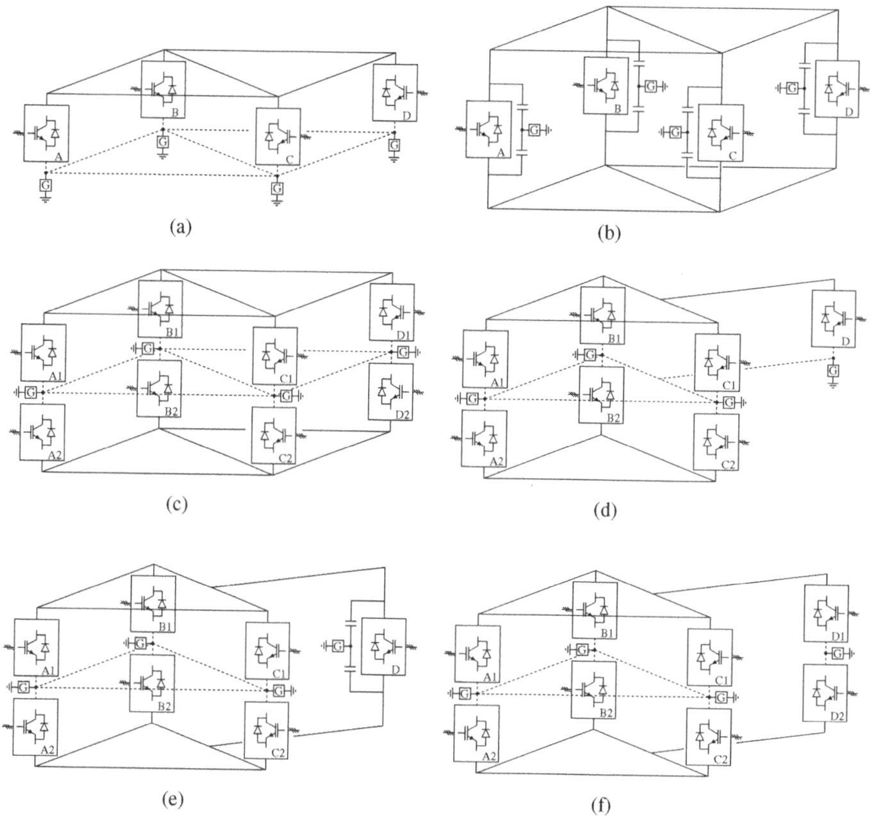

Figure 8.16 Possible VSC-HVDC grid configurations [6]. (a) Asymmetric monopolar grid with earth return. (b) Symmetric monopolar grid. (c) Bipolar grid with metallic return. (d) Bipolar grid with metallic return and asymmetric monopolar tappings. (e) Bipolar grid with metallic return and symmetric monopolar tappings. (f) Bipolar grid with metallic return and bipolar tappings with earth return.

8.4.5 HVDC Configurations Extended to HVDC Grids

When building an HVDC grid, the same configurations as mentioned before can be considered, extended over multiple converter stations. Figures 8.16a, 8.16b, and 8.16c–f show possible extensions of the asymmetric monopolar, symmetric monopolar, and bipolar grid into a four-node HVDC grid.

As an HVDC grid may evolve from point-to-point systems, comprising both monopolar systems and bipolar systems, operation in an unbalanced mode may become quite common, forcing the need for neutral metallic conductors. The converter station arrangement will be different in either due to historical reasons, to connect existing infrastructure, or because specific topologies provide the optimum solution. An example could be the connection of an intermediate offshore wind farm to a large interconnector.

As a result, the end configuration of a HVDC grid might be quite different. These configurations are compared in terms of extensibility and flexibility for post-fault operation.

In an asymmetric monopolar grid (Figure 8.16a), the metallic return conductor can have a lower voltage rating because of low-impedance grounding. For this type of grids, a fast-acting fault clearance is needed because of high fault currents. In the case of multiple grounding points in the system, earth currents will flow in normal operation. These currents can be restricted by using resistance grounding. Alternatively, the grid can be low-impedance grounded at only one single point and high-impedance grounded elsewhere (e.g., as proposed in references 10 and 11). Several backup grounding points with an active device are required in case of an outage of the effectively grounded point. Moreover, the grounding points need to be dimensioned to sustain the total fault current supplied by all converters. A drawback is different fault behavior in the system depending on the location of the low-impedance grounding.

A symmetric monopolar grid, as shown in Figure 8.16b, is effectively high impedance grounded. No steady-state earth currents flow in normal operation when the system is grounded at multiple points. For standardization of, for example, converter insulation requirements or protection settings, it is conceivable that each converter is grounded in the same manner. A symmetric monopolar grid is extensible with high-impedance grounded bipoles. Every extension must have cables and converters rated to carry the nominal converter voltage. This cost can be high in comparison to the power rating of the extension.

Figure 8.16c shows a bipolar grid configuration. A low-impedance grounded bipolar grid is similar to an asymmetric monopolar grid regarding voltage rating of the poles and the need for fast acting fault clearance. Compared to the asymmetric monopolar grid, the power that can be transported over a link is doubled at the cost of only one extra cable. The bipolar grid can be low-impedance grounded at multiple locations. In balanced steady-state operation, no (or only small) earth current flows. Similar to the asymmetric monopolar grid, earth currents in unbalanced operation can be limited by resistance grounding or low-impedance grounding at only one point.

A low-impedance grounded bipolar grid is extensible with other low-impedance grounded bipoles or asymmetric tappings between the metallic return conductor and one pole (Figure 8.16d). The latter option provides a possibility to extend the grid by smaller rated systems at a reduced cost with respect to a symmetric grid. However, grid control can be more difficult as the system will be continuously operated in unbalanced conditions.

A high-impedance grounded bipolar grid is similar to the symmetric monopolar grid in terms of cable voltage ratings, protection requirements, and grounding locations. The metallic return must be rated at the nominal converter voltage. If the cable voltage ratings are the same as used in the monopolar grids, the nominal converter voltage must be halved compared to monopolar grids. The high-impedance grounded bipolar grid is extensible with other high-impedance grounded bipoles, symmetric monopoles (Figure 8.16d), and asymmetric tappings (Figure 8.16e). However, insulation requirements of these tappings will be higher compared to asymmetric tappings on a low-impedance grounded bipolar grid.

A bipolar grid with metallic return cables can also be extended by bipolar tappings with earth return (Figure 8.16f). In the case of unbalanced operation, earth currents are possible because of the low-impedance grounded tapping.

A more comprehensive overview and comparison of the different HVDC configurations can be found in [6].

8.5 SUBSTATION CONFIGURATION

Within a grid, substations are formed at places where different grid assets are connected. In an AC system, this usually entails a number of overhead lines or cables, transformers, and switchgear, together with the necessary auxiliaries. Also, within a DC system we can expect a similar concept of substations connecting to multiple DC assets.[1]

Within an AC system, busbars are used to collect the different infeeds. In order to protect the system, breakers are installed to open faulted connections, and disconnectors are used to be able to do maintenance to the breakers and to visually assure that a line is disconnected. The actual configuration of the busbars depends on the importance of the substation, the number of connections, and specific local demands such as minimal short-circuit power. The number of busbars and the number of breakers per feeder determine redundancy, associated expected failure/outage rates, complexity, and the costs [12]. Some of the most common systems are listed below, although different variations exist:

- Single busbar (possibly with separation) (Figure 8.17a)
- Double busbar, single breaker (Figure 8.17b)
- Double busbar with breaker-and-a-half scheme (Figure 8.17c)
- Double busbar, double breaker (Figure 8.17d)
- Ring substation

For DC grids we expect the same types of busbar arrangements based on similar decisions. Figure 8.17 gives a single line representation of such a system. Note that, as with AC systems, the busbar configurations are mirrored for all poles and the neutral if present. However, the operation of individual breakers in the different poles of the same connection can be different, contrary to AC systems.

8.6 CONCLUSIONS

This chapter discusses the different layouts that can be encountered when discussing future DC grids. Different terms can be understood by "layout." Therefore it is important to distinguish between the different sizes: from system layout or topology over converter configuration and busbar arrangement. Different choices need to be

[1] A substation that contains both an AC and a DC side can be denoted as a conversion station and usually contains an AC switchyard, a DC switchyard, and one or more converters.

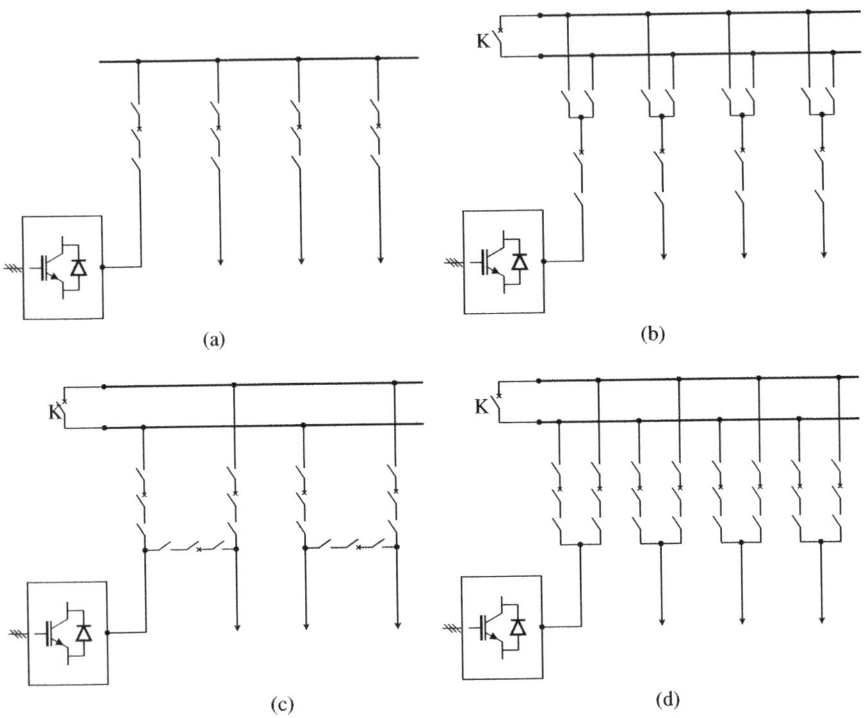

Figure 8.17 Possible HVDC busbar arrangements in a substation with one HVDC converter and three outgoing feeders (single pole representation). (a) Single busbar, single breaker configuration. (b) Double busbar, single breaker configuration. (c) Breaker and a half configuration. (d) Double busbar, double breaker configuration.

made when developling a HVDC grid, depending on the required reliability and cost as well as the application.

REFERENCES

1. G. Asplund, C. Barker, U. Baur, J. Beerten, P. Christensen, S. Cole, D. Van Hertem, W. Jialiang, D. Jovcic, L. P., K. Lindén, A. Marzin, N. Pahalawaththa, M. Rashwan, J. Rittiger, K. Søgaard, D. Westermann, E. Wilkening, and C. Yue, "Working group b4-52—HVDC grid feasibility study," Cigré, Technical Report 533, Apr. 2013.
2. D. Van Hertem and M. Ghandhari, "Multi-terminal VSC HVDC for the European supergrid: Obstacles," *Renewable and Sustainable Energy Reviews*, vol. 14, no. 9, pp. 3156–3163, 2010.
3. Friends of the Supergrid, "Response to the UK parliament energy and climate change committee inquiry into supergrid," Friends of the Supergrid, Technical Report, 2011.
4. O. Gomis-Bellmunt, J. Liang, J. Ekanayake, R. King, and N. Jenkins, "Topologies of multiterminal HVDC-VSC transmission for large offshore wind farms," *Electric Power Systems Research*, vol. 81, no. 2, pp. 271–281, 2011.
5. J. Arrillaga, Y. H. Liu, and N. R. Watson, *Flexible Power Transmission: The HVDC Options*. Hoboken, NJ: John Wiley & Sons, 2007.

6. W. Leterme, P. Tielens, S. De Boeck, and D. Van Hertem, "Overview of Grounding and Configuration Options for Meshed HVDC Grids," in *Power Delivery, IEEE Transactions on*, vol. 29, no. 6, pp. 2467–2475, Dec. 2014 doi: 10.1109/TPWRD.2014.2331106
7. E. W. Kimbark, *Direct Current Transmission*. New York: Wiley-Interscience, 1971, vol. 1.
8. R. Zeng, Z. Yu, J. He, B. Zhang, and B. Niu, "Study on restraining DC neutral current of transformer during HVDC monopolar operation," *Power Delivery, IEEE Transactions on*, vol. 26, no. 4, pp. 2785–2791, 2011.
9. T. Magg, H. Mutschler, S. Nyberg, J. Wasborg, H. Thunehed, and B. Sandberg, "Caprivi link HVDC interconnector: Site selection, geophysical investigations, interference impacts and design of the earth electrodes," *CIGRE Session Paper B4-302, Paris*, 2010.
10. N. Gibo, K. Takenaka, S. Verma, S. Sugimoto, and S. Ogawa, "Protection scheme of voltage sourced converters based HVDC system under DC fault," in *Transmission and Distribution Conference and Exhibition 2002: Asia Pacific. IEEE/PES*, vol. 2. New York: IEEE, 2002, pp. 1320–1325.
11. M. Takasaki, N. Gibo, K. Takenaka, T. Hayashi, H. Konishi, S. Tanaka, and H. Ito, "Control and protection scheme of HVDC system with self-commutated converter in system fault conditions," *Electrical Engineering in Japan*, vol. 132, no. 2, pp. 6–18, 2000.
12. J. Schlabbach and K.-H. Rofalski, *Power System Engineering: Planning, Design, and Operation of Power Systems and Equipment*. Hoboken, NJ: John Wiley & Sons, 2008.

CHAPTER 9

GOVERNANCE MODELS FOR FUTURE GRIDS

Muhajir Tadesse Mekonnen, Diyun Huang, and Kristof De Vos
Research Group Electa, Department of Electrical Engineering, University of Leuven, Leuven, Belgium

9.1 INTRODUCTION

The previous chapters discuss how the development of HVDC networks becomes feasible from a technical point of view. However, the actual realization of the future HVDC grid does not only face technical and economic barriers, but is also challenged by the regulatory framework of transmission systems. This chapter discusses the main regulatory principles impacting the development of future transmission networks.

Transmission of electricity has the characteristics of a natural monopoly. Where the liberalization process has been completed, network activities are unbundled from generation and retail and are treated as a regulated monopoly. The transmission system has become the responsibility of the transmission system operator (TSO), with the task to ensure a reliable transmission of electricity from generating units to the distribution networks and consumers. A regulatory framework is put in place to provide correct incentives for maintaining the quality of service at an acceptable cost. This framework impacts the planning, financing, ownership, pricing, and operation of transmission systems. The combination of these five elements is referred to as the governance model of electricity transmission. Compliance with the regulatory framework is monitored by National Regulation Authorities (NRAs).

The last decades have shown a trend towards the development of supranational and regional electricity markets, with the European Internal Electricity Market (IEM) as an extensive example of international cooperation. The IEM consists of directives and regulations which are enacted by the European institutions and afterwards adopted in national legislation. In the European Union, a trilateral relationship between the European political institutions, the Agency for Cooperation of the Energy Regulators (ACER), and the European Network Transmission System Operators (ENTSO-E) has been established to address the challenges of operating an integrated market while maintaining a cost-efficient and reliable system based on sustainable electricity supply [1].

HVDC Grids: For Offshore and Supergrid of the Future, First Edition. By Dirk Van Hertem, Oriol Gomis-Bellmunt, and Jun Liang.

Today, new technologies such as HVDC and corresponding long-distance grid architectures give rise to complex issues relating to the transnational aspects. For instance, asymmetric benefits for different countries may result in insufficient support for transmission investments, even though such investments may benefit the European market as a whole. The regulatory framework has to evolve in a way which ensures the identification of required investment projects (planning), the possibility to acquire the necessary funds (financing), the ability to decide upon the property rights of the assets (ownership), and the possibility for system operators or market players to recoup their investments (pricing). Finally, the rules are to be defined to ensure a reliable and efficient operation. This chapter aims to introduce the main regulatory principles and challenges in view of the integration of HVDC technologies in five sections, representing planning, financing, ownership, pricing, and operation of transmission systems, respectively.

9.2 TRANSMISSION GRID PLANNING

The growth in demand for electricity, the installation of new generation capacity, and the depreciation of existing transmission assets trigger the need for transmission network expansion. Transmission investments are costly, irreversible, and long-term and should only be undertaken if justified. On the one hand, overinvestment in transmission assets reduces the cost-efficiency of electricity supply. On the other hand, underinvestment not only decreases system reliability, but also limits the ability to integrate large shares of renewable energy into the system. A balance that achieves the right level of investment requires a detailed planning procedure to determine optimal and feasible transmission expansion scenarios. Therefore, interaction between several stakeholders such as TSOs, generation companies, regulators, consumers, and environmental groups is required before transmission investments can be approved and realized.

In Europe, regulated TSOs are responsible for building, maintaining, and operating national transmission facilities. The NRAs strive to incentivize national transmission network development towards efficient expansion and operation. The planning process of HVDC transmission projects is not fundamentally different from the conventional grid planning procedures. Depending on the stakeholders impacted by the transmission projects, different approaches are followed, as described in the following subsections.

9.2.1 Planning of National Transmission Network Expansions

At national level, the local TSO is responsible for developing transmission network expansion plans. Generally, the TSOs prepare periodic transmission expansion plan with the objective of improving national system reliability or economic efficiency. This is conducted by taking into account demographic and economic projections, as well as expected generation capacity expansion plans. Public consultations are conducted to enable participation by relevant and interested stakeholders during the planning process. The transmission expansion plan, which consists of both new and reinforcement transmission projects, is then proposed to the NRA. The plan should meet a predefined set of reliability and economic criteria to justify the proposed investments. Based on these criteria and an assessment of social and environmental impacts of the proposed expansion plan, the NRA decides which facilities can be built.

9.2.2 Planning of Regional Transmission Network Expansions

The integration of the European electricity market results in an increasing need for cross-border interconnection capacity. National network expansion planning focuses on national benefits that will not necessarily coincide with adequate cross-border investments to support increasing volumes of trade. Therefore, a regional transmission planning is required to achieve European policy goals.

A periodical Ten-Year Network Development Plan (TYNDP) is identified as an important step to address the need for a long-term European network development plan. It takes into account national plans (bottom-up approach) and also incorporates regional aspects (top-down approach) [2]. ENTSO-E publishes a biannual TYNDP taking into account the European generation forecast. ACER assesses the consistency of the TYNDP with the national network expansion plans. It also provides opinion and recommends amendments to both planning levels in case of inconsistency. However, the TYNDP is nonbinding and neither institution, ENTSO-E and ACER, has the power to dictate to national planners.

9.2.3 Planning Procedures of Third Party (Merchant) Transmission Projects

Prompted by concerns about underinvestment in cross-border electricity transmission, and the impacts this has on competition within the European market, third-party projects are allowed on interconnectors.

Similar to the national planning process, the procedures of third-party transmission projects also involve three stages. First, project promoters—which can be TSOs, third-party investors or a combination of both—identify the need for a new transmission facility. The objective for this new facility can be to improve system reliability, to enhance economic efficiency, or both. Then, the project along with its techno-economic and environmental feasibility study is proposed to the responsible NRAs, which evaluate the validity of the economic, technical, social, and environmental impact assessment of the proposed project. Based on their assessment the NRAs approve the project, suggest modifications or reject the project proposal. The difference from national planning is that this is slightly complicated as it involves several stakeholders including more than one TSOs and NRAs.

9.2.4 Projects of Common Interest (PCIs)

In order to further promote cross-border investments, transmission projects that are in line with the TYNDP are put forward to be considered as Projects of Common Interest (PCIs). These are approved and prioritized by the European Commission

(EC). Projects which are classified as PCIs can get financial assistance, accelerated permitting process, and facilitated regulatory measures. For a project proposal to be labeled as a project of common interest, it has to fulfill at least the following criteria [3].

- The project is necessary for the implementation of the energy infrastructure priority corridors and areas.
- The project displays economic, social, and environmental viability.
- The project involves at least two member states, either by directly crossing the border of one or more member states or by being located on the territory of one member state and having a significant cross-border impact.

PCIs are characterized by multiple jurisdictions and therefore multiple market and regulatory frameworks, as well as the influence of multiple stakeholders. This calls for a clear and agreed-upon method for cost and benefit assessment as well as the cost and benefit allocation of these projects.

9.2.5 Coordination of Regional Transmission Plans

Different studies show how the integration of high shares of renewable energy sources for electricity (RES-E) in the European generation system would benefit from a transmission overlay network [4–6]. This long-distance network would not only access renewable energy sources, but would also allow balancing of their short-term and seasonal variability. This is, for instance, done by tapping both massive wind power production in Central Western Europe and the solar production in Southern Europe. This way, power from RES-E technologies would be transmitted across the EU, and beyond, by what is envisioned as the "supergrid." The "supergrid" widens the geographical footprint from where energy can be transported by linking up across countries. Due to its technological advantages over HVAC for long-distance transmission, HVDC technology is expected to serve as the backbone of this supergrid (see also Chapter 4).

Deploying a European overlay transmission grid is difficult in the current institutional setup. When transmission planning involves multiple countries, it is not self-evident that all the national and European interests coincide. As a responsible national transmission planner, TSOs plan with the objective of enhancing the security and economic efficiency in their own control area, in the best case, in collaboration with their neighboring TSOs. Hence, current transnational projects usually remain limited to a maximum of two countries.

Transmission planners at different levels can have different priorities. Allowing national planners to look at possible investments beyond their jurisdiction can result in a better solution [7]. The European supergrid is perceived to encompass a wide geographical area and a European planning is thus crucial. This is important to achieve optimal investments at the European level as well as to avoid short-sighted investments.

9.3 TRANSMISSION GRID OWNERSHIP

The ownership model answers the question of who owns and operates transmission assets. It clarifies the incentives, responsibilities and liabilities of transmission investments. This section deals with the regulation on ownership of transmission assets, and corresponding remuneration schemes. Consequently, these aspects have a direct impact on investments and operational decisions. There are three main models to organize the relation between the ownership and operation of transmission assets, which are discussed: ISO/TO, TSO, and ITO, where:

- TO is a transmission network owner,
- SO is a system operator,
- ISO is a fully unbundled system operator without the network assets,
- TSO is a transmission system operator that owns the system assets, and ownership of the network is fully unbundled from generation,
- ITO is a transmission system operator owning the network assets and belonging to a vertically integrated company.

At this moment, the most common ownership model applied in Europe is the TSO model. However, in other places such as the United States and parts of the United Kingdom, the ISO/TO scheme is applied.

There are two main types of asset ownership: the regulated and the merchant model. Regulated assets are subject to traditional regulatory schemes such as cost-based regulation and performance-based regulation. In contrast, merchant assets do not follow these traditional regulations and can be seen as an alternative to regulated transmission investments. Sometimes, merchant transmission investments are adopted to bridge non-coordinated regulatory regimes that are interconnected by the network.

9.3.1 Remuneration Schemes for Regulated Assets

There are two main approaches to provide incentives to a regulated TSO to pursue an efficient level of investments: cost-based and incentive-based regulation.

9.3.1.1 Cost-Based Regulation Cost-based regulation in its pure form, i.e. without ex-post regulatory tests to reopen evaluation of either efficiency or necessity of the sunk investment—is characterized by a fixed rate of return based on incurred costs (Figure 9.1). Under this design, costs are fully passed to grid users (generation and demand) and the TSO does not face the risks following a changing demand or generation system [8]. Consequently, lower risk exposure level results in a lower cost of capital for the TSO. On the other hand, since its revenue depends on the capital spending, the TSO has less incentive to control its costs which results in a tendency to overinvest [9]. A typical cost-based regulation can be represented by

$$R_t = C_t + D_t + T_t + RAB_t \times r_t \tag{9.1}$$

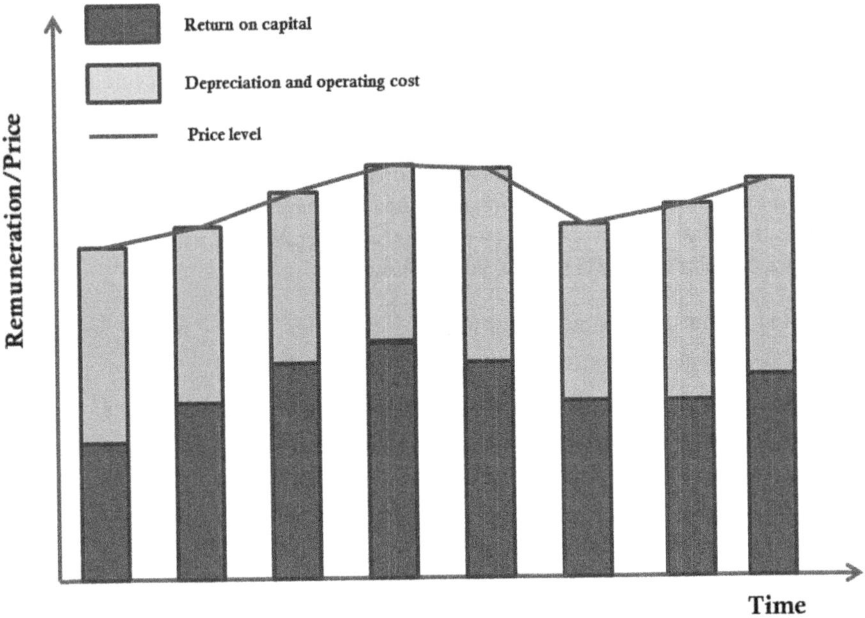

Figure 9.1 Cost-based regulation.

where

R_t	is the required revenue in year t,
C_t	is the operating cost in year t,
D_t	is the depreciation in year t,
T_t	is the taxes in year t,
RAB_t	is the value of the regulated asset base in year t,[1]
r_t	is the allowed rate of return in year t,
t	is the year index.

9.3.1.2 *Incentive-Based Regulation* In general, under incentive-based regulation, the revenue of the regulated TSO is indexed to one or more indicators of its performance during a regulatory period. The partial or total detachment of revenue from incurred cost exposes the TSO to a higher level of business risk opposed to cost-based regulation.

Price/Revenue Cap Regulation. Under a price or revenue cap, an upper limit on the price or revenue is set prior to a predefined regulation period, as illustrated in Figure 9.2. In this case, the price or revenue of the regulated TSO is completely or partially decoupled from the incurred cost [10]. Since the TSO profits from the economic gain by outperforming the efficiency improvement, often represented by productivity growth factor set by

[1] In order to remunerate transmission network investors for their sunk capital commitments on existing assets, regulatory entities usually include efficiently conducted network investments into a regulated asset base (RAB).

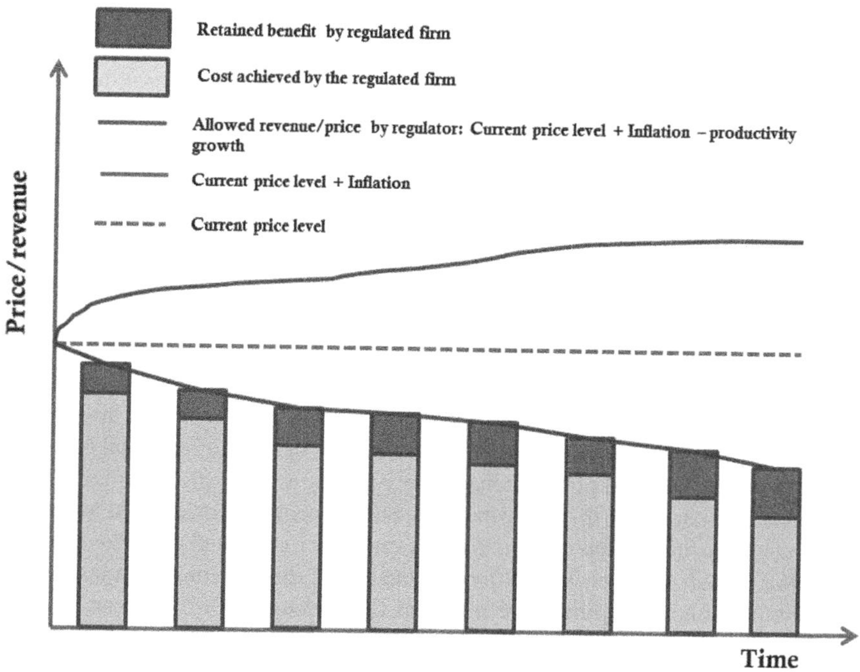

Figure 9.2 Price/revenue cap regulation.

regulators, price cap regulation provides a direct incentive for the TSO to reduce its costs. A typical price or revenue cap design can be represented by

$$R_t = R_{t-1}(1 + RPI_t - X) \tag{9.2}$$

where

R_t	is the limit on price or on revenue,
RPI_t	is the the retail price index,
X	is the productivity growth factor that reflects the assumed improvements in rate of efficiency and other relevant factors,
t	is the year index.

Yardstick Competition. The idea of yardstick competition is to expect the same efficiency gains observed at other TSOs, which is to be matched by the regulated TSO. The objective is to implement a mechanism where firms have incentives to compete for efficiency. Under this regulation approach, the price or revenue allowed for the TSO is usually linked to the average or best performing industry peer firms.

In practice, a mixture of elements from both cost-based and incentive-based regulation is applied on transmission investments. For instance in some parts of the USA, a cost-based regulation is implemented in combination with an ex post assessment in order to check whether the investment is conducted efficiently. Return on investment is then remunerated only based on efficiently incurred cost. Therefore, investors face the risk in which the sunk cost is not recognized by the regulator to be included in the regulated asset base.

Even within the same company, different regulatory scheme for different services can be imposed by the NRA. Splitting regulatory treatment on CAPEX and OPEX is adopted by some NRAs to allow different regulations on cost items of distinct characteristics. Cost-based regulation on CAPEX and incentive-based regulation on OPEX have, for instance, been implemented in France, Italy, and Portugal [11].

9.3.1.3 Merchant Investment Merchant investments can be described as market-driven investments. This model provides an alternative means to attract investments in addition to regulated investment. The salient attributes of the merchant transmission investment model are decentralized ownership and a market-based revenue mechanism [12]. Decentralized ownership makes it possible for investors to collect local information, such as open solicitation for demand of potential transmission customers, and then plan merchant line capacity accordingly. This tackles the problems associated with disjoint generation and transmission investment according to the rules of an unbundled market. When revenues are generated by investors based on market-based approaches, such as an auction of the incremental transmission capacity, investment incentives are provided for investors. In a European context, revenues result from the congestion that results from the price difference between the interconnected zones. However, these prices are uncertain and result in exposure to a high risk to be borne by the investors, which in turn translates into a higher cost of capital.

EU legislation states that new DC interconnectors may be exempted, from regulated third-party access and the use of the collected congestion rent for a limited period of time [2]. In exceptional cases, AC interconnectors can also be granted exemption when the costs and risks of the investment are particularly high when compared with those which are normally incurred when connecting two neighboring national transmission systems.

9.3.1.4 Considerations for HVDC Grid Investments Table 9.1 summarizes the characteristics of the three models of asset regulation: cost-based, incentive-based, and merchant. The extent to which cost pass-through elements are incorporated in the regulation is the most important consideration for investors.

The implication of regulation for technology differs in CAPEX or OPEX. In terms of capital expenditure, for investors who select new technologies such as multi-terminal HVDC, a cost-based regulation with a strong cost pass-through mechanism in the planning and construction phase offers an important advantage by reducing uncontrollable technology risks such as delivery delay and unexpected additional construction costs. For the operational phase when the incurred investments are included in the RAB, investors have the incentive to favor higher cost technologies. Under cost-based regulation, higher investment costs establish a higher base to generate stable revenue in the agreed depreciation period. Concerning operational costs, incentive-based regulation with benchmarking mechanisms has an adverse effect on newer technology compared with cost-based regulation. New technologies might encounter higher reliability uncertainties in the operation phase compared with more mature

TABLE 9.1 Investment Characteristics of Three Types of Asset Regulation [13, 14]

Type of Regulation	Cost-Based	Incentive-Based	Merchant
Regulatory framework	Cost of service regulation	Price/revenue cap regulation, Yardstick	Locational market price
Rate of return on investment	Fixed (nominal or real)	Adjusted for performance	Determined by market price
Investment process	Central planning by TSOs	Approval of investment by regulator	Decentralized planning by private firms
Investment incentive for a private investor	Increase regulated asset base	Increase regulated asset base and maximize performance payment	Maximize market value
Main financing method	Corporate finance	Corporate finance	Project finance or corporate finance channeled into project company

ones. System operators who bear more risks under an incentive-based regulation for operation might therefore favor mature technologies.

9.4 TRANSMISSION GRID FINANCING

Financing transmission network investments is based on two main pillars. The first pillar concerns financing sources, which provide the financial means with a combination of debt and equity, and the second pillar concerns the cost of capital. First, an overview of different financing sources and their pros and cons from an investor's perspective is given. Then, a detailed analysis is unfolded to reveal cost of capital considerations for new transmission network investments, taking into account the potential technology impact. In particular, the investment risk is a key factor and is strongly impacted by the regulatory framework of transmission networks (Section 9.3). In contrast, this section focuses on the relation between investor risk exposure and its implication for cost of capital.

The financing considerations are presented for transmission network investments in general, as they do not fundamentally differ for HVDC investments. However, specific aspects of HVDC investments require additional attention, as they remain a relatively new technology for which the costs are still expected to evolve.

9.4.1 Financing Sources

The main sources of the debt and equity in electricity transmission network investments are listed below [13]:

- Debt:
  - International financing institutions: loans from international financing institutions such as European Investment Bank (EIB) are important financing

sources. The main advantage of this type of loan is the favorable conditions such as long maturities, i.e. from 10 to 25 years, and low interest rates.

- Corporate bonds: corporate bonds are a popular form of financing for large network companies with credit ratings. A commercial bond typically has a maturity of 10 years. The rate of bonds is linked with the ratings of the issuance company.
- Commercial bank loans: loans from commercial banks are featured with short maturities, approximately 5 to 10 years.

- Equity:
  - Government investment: government investments are an essential source of equity injections in network companies where the government is a major shareholder.
  - Internal equity: this financing source origins from the system operator's own cash flows.
  - External equity: external equity could be injected by investors such as pension funds, infrastructure funds from investment banks, or insurance companies.

Government grants play an important role to send a positive signal to financing parties. In the case where TSO debt level in total asset is very high at the corporate level, grants could act as replacement or supplement of equity.

9.4.2 Cost of Capital

The cost of capital is the rate of return that is required on capital expenditures by the transmission network investors. It is typically represented by

$$r_a = r_f + \beta_a(r_m - r_f) \tag{9.3}$$

where

r_a	is the required return on an investment asset,
r_f	is the risk free rate, which often uses long term government bonds as a proxy,
β_a	is the asset beta, which measures how the return on a particular investment co-varies with the return of on the market, usually represented by a market risk premium,
$r_m - r_f$	is the market risk premium, which is the expected additional return for holding higher risk assets, in addition to that of a risk-free asset.

The allowed cost of capital level is set by the regulator. From the perspective of a regulator, it is a result of a trade-off between setting investor incentives and achieving cost-efficiency. On the one hand, it should not be set too low in order to allow sufficient remuneration for network investors with respect to the risks they face. This aims to ensure an adequate network investment level and a healthy financial structure to keep credit rating at investment grade. On the other hand, it should not be too high as this directly leads to higher network capital costs, and creates welfare losses resulting in an increasing electricity bill for the consumers. There is an important interaction between risk assessment from the investor perspective and the cost of capital. From the investor perspective, the expected rate of return depends on the risk allocated to the investor and the associated cost of capital.

9.4.3 Risk and Cost of Capital

The relation between the regulatory framework, risk, and cost of capital is fundamental when financing transmission assets. The regulatory framework determines the risk, and the risk determines the cost of capital. Financing theory shows that the cost of capital required by the market depends on the systematic risk exposure in the business. Systematic risks refers to the risk which investors are not able to eliminate by investment portfolio diversification, which are usually linked to exogenous macro-economic factors such as interest rate or the political and regulatory uncertainties. The latter could exert impact to transmission network investor risk in a number of ways, such as shift of energy policy or ex-post expropriation on investors. There are multiple sources of systematic risks that are identified for network investments.

Scenario Risks: this relates to the development of drivers concerning the demand for electrical energy, as well as long-term generation risk in terms of technology mix and location. These risks are driven by energy policy uncertainties and demand evolution resulting from unexpected economic growth or downturn, or demand side management.

Change to Input Cost: this relates to the different project phases. Firstly, during the construction phase, additional project cost resulting from uncertainties of new technologies, such as supply chain bottlenecks, might not be automatically passed through to consumers. Secondly, in the operational phase, the value of the RAB can be determined by means of a replacement cost valuation method, where the RAB value is determined by means of the estimated cost of replacing the existing assets, instead of the actual incurred costs. Consequently, evolutions in market price of transmission technologies can affect the value of the RAB.

Financial Risks: these arise when the cashflow of transmission network investments is adversely affected by the broader macro-economic context, and can include changes in inflation or interest rate.

For transmission network investments adopting relative new technologies such as HVDC, the uncertainty associated with the RAB value is conceived to be higher when using the replacement cost valuation method. The uncertainty of future technology cost plays an important role in this type of RAB evaluation method. If an investor would choose a new technology for which the replacement cost valuation method is used on a yearly basis, and the cost to replace the original asset decreases with 50% in the year after the investment, this would result in a reduction of the revenues from that asset in the RAB, in the years to come. The investor is thus exposed to undesired high risk. On the other hand, if the future cost for realizing the same service increases due to the advent of technology breakthrough, this cost rise could lead to a higher reevaluation of the existing assets and exert positive impact on investor cash flows. In this case, consumers usually bear the risk of change to input cost with rising tariffs. For instance, experience in New Zealand shows that since the introduction of the replacement cost method in its distributed network, asset value rose up and consumer price increased [15].

9.5 TRANSMISSION GRID PRICING

The costs of transmission investments are recovered from grid users through transmission pricing. Transmission pricing implies determining who should pay for transmission investment and how much. For regulated transmission investments, it involves the allocation of cost of transmission facilities, including normal rate of return, among the users of the transmission grid using a certain cost allocation approach. For third-party (merchant) investments, transmission pricing might involve selling the right to use the transmission line to market participants. The general pricing principles for an HVDC grid is not different from the conventional principles which are discussed in this section.

In general, transmission cost allocation must obey some basic principles or guidelines that result from the combination of micro-economic theory and power systems engineering [16]. These principles are cost causality, economic efficiency, stability, and predictability. According to the cost causality principle, the allocation of costs to grid users (both generators and consumers) should reflect the costs they cause in the network. The cost caused by a grid user is usually assumed to be equivalent to its benefits. Transmission facilities are built when aggregated benefits exceed incurred costs and hence cost is allocated in proportion to estimated benefits. When benefits are difficult to estimate, some proxy to benefits such as network utilization can be used.

The economic efficiency principle mandates that cost allocation should be chosen to maximize the global surplus of market agents both in short term (operational decisions) and long term (investment decisions). Therefore, transmission pricing should not be based on who trades with whom, rather it should be based on where and when the network is used.

The third principle suggests that stable and predictable economic signals should be sent for investors so that they may choose the most convenient site with low financial risk. Any possibility of future updates to the pricing or allocation mechanism introduces uncertainty and thus raises capital costs. Hence, it is recommended that cost allocation should be determined ex ante, i.e., before the transmission project becomes fully operational, on the basis of anticipated benefits and should also be left in place for the life of the project or at least a long period (e.g., 10 years).

9.5.1 Pricing of National Transmission Projects

National transmission projects can be part of either *connection facilities* or *network facilities*. The former are facilities that are used to connect network users (i.e., power plants or consumers) to the main transmission network. The latter, however, are

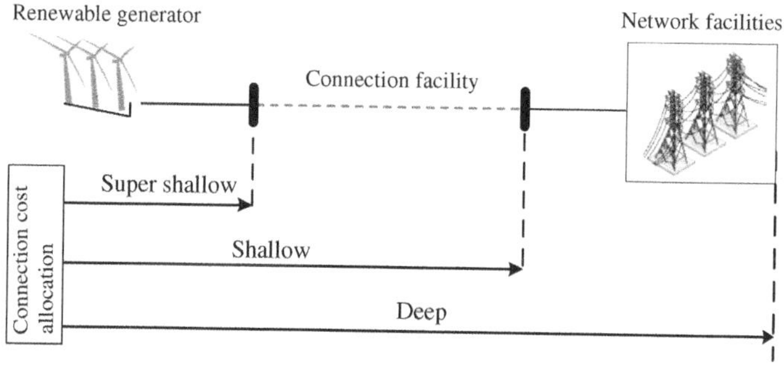

Figure 9.3 Connection charging practices.

facilities that are used to transport electricity from one place to another place in the network.

9.5.1.1 Pricing of Connection Facilities Transmission investments that connect power plant modules (e.g., offshore wind parks or long-distance hydro) to the main transmission grid are usually labeled as generation connection projects. The pricing of such investments depends on the connection charging practices of their respective country which can be distinguished as deep, shallow, or super-shallow, (Figure 9.3).

In a deep connection charging policy, generators are responsible for the costs of their internal grid, their connection to the main grid, and any required reinforcement on the main grid caused by their connection. In a shallow connection charging practice, generators bear the cost of their internal grid and their connection to the main grid. In a super-shallow charging policy, generators pay only for their internal grid. The cost of connection facilities and cost of reinforcing the the main grid that is not included in the connection charges is considered as part of the shared network facilities and TSOs become responsible for the pricing.

Although deep and shallow practices are justifiable based on the cost causality principle as they attribute connection costs to grid users who cause these costs, they hinder generation investments, which is particularly relevant for deep offshore wind farms. Furthermore, non-aligned national connection charging practices can hinder cooperative grid planning among generation companies and TSOs.

This has been demonstrated in the case of the Kriegers Flak project in the Baltic region (Figure 9.4), where initially a joint feasibility study by TSOs of three countries, (i.e., Denmark, Germany, and Sweden) showed a higher benefit-to-cost ratio (Figure 9.4a). However, TSOs in Denmark and Germany are responsible for connecting offshore wind farms at their expense (super-shallow connection practice), while in Sweden the wind farm developers are responsible for their own connection (deep connection practice) [17]. This was one of the main reasons for the Swedish TSO to decide to withdraw from its part of the joint venture, while the Danish and the German counterparts are still pursuing this innovation (Figure 9.4b).

9.5.1.2 Pricing of Shared Transmission Projects (Network Facilities) Shared transmission network facilities are used by several grid users for electricity

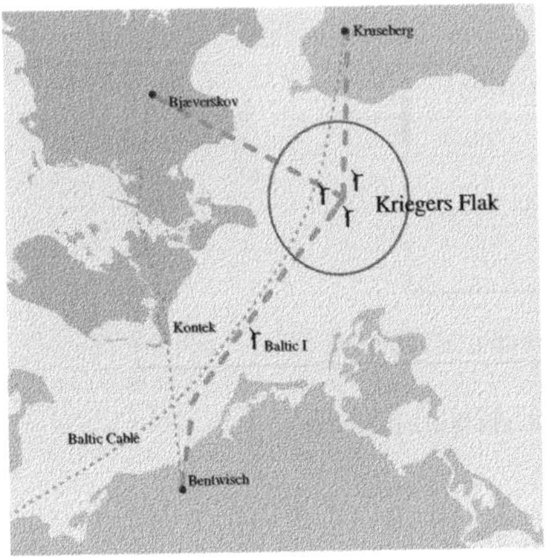

(a)

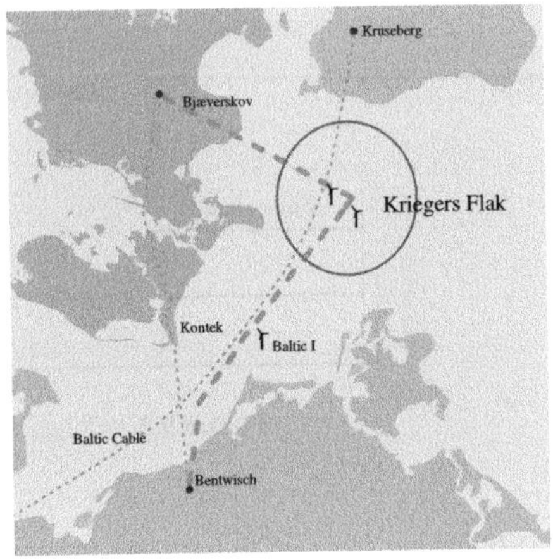

(b)

Figure 9.4 The Kriegers Flak project [3, 18]. (a) Initial proposal. (b) Current status.

transport. Pricing of such facilities should be non-discriminatory as well as economically efficient, which can be achieved in theory by a nodal transmission pricing model. A nodal price is the marginal cost of injecting and offtake electricity at a given node and time. A nodal price is a charge for energy, but it internalizes the impact of losses and congestion of the network [19]. Congestion in the network creates different nodal prices. As a result, differences of nodal prices between two nodes at a certain time can be seen as the short-term marginal cost of transporting electricity between them.

The current system in Europe is a zonal system where groups of nodes under similar market organization are aggregated into market areas, i.e. bidding zones which usually correspond to countries with no internal transmission constraints. Zonal price differences can be used as a short-term marginal cost of electricity transport across cross-border interconnections.

Although nodal transmission pricing promotes economic efficiency by sending efficient short-term signals, it does not fully recover the network investment costs [20]. Ideally, full cost recovery is possible if the transmission design is optimal in a strict economic sense. However, in actual networks there is a tendency to overinvest in the transmission network due to the discrete nature of transmission facilities, the economies of scale of the transmission business, and the need for reliability as well as other financial, environmental, or technical restrictions.

9.5.1.3 Transmission Tariffs Zonal or nodal transmission pricing should be complemented with some additional charges, usually known as *transmission tariffs*, and can be used to provide long-term signals for the network users. These complementary charges are calculated by allocating the remaining cost of the transmission facilities that is not recovered through nodal or zonal transmission pricing to grid users based on their extent of grid use or based on the estimated benefits of the grid users [21, 22]. Several methods are suggested in the literature for this cost allocation. The methods range from simple cost socialization to more complex game theoretic approaches. Generally, they can be categorized as non-flow-based and flow-based approaches.

Non-flow-based approaches allocate transmission costs to grid users based on their injection, withdrawal, and distance between the injection and withdrawal points of the transaction. However, the actual power flow is dictated by technical characteristics of the network and does not follow transaction paths. Therefore, cost allocated to users using these methods does not reflect the actual use of the network by the users or the value they derive from being connected. In many cases, some users cross-subsidize others. Methods under this category include postage stamp, contract path, and the distance-based MW-mile [23].

Flow-based methods allocate transmission costs to grid users based on the real network utilization of each user. Power flow analysis methods are used to estimate the flow caused by each user on the network as a proxy for network utilization. Methods such as flow-based MW mile, average participation, marginal participation, and Aumann–Shapley fall under this category [24, 25]. These approaches are generally perceived to send correct economic signals to grid users as they take the actual system operation into account. However, since they are based on power flow calculations, their results can be affected by the settings of power flow controlling devices such as HVDC lines that parallel AC network and FACTS [26]. The flow-based models described in the literature and also in practice do not take into account HVDC lines. These methods can be adapted for pricing HVDC lines by calculating equivalent impedance of the HVDC lines based on assumed flow through those lines. This approach is proposed in the United Kingdom via Project Transmit [27].

In the zonal pricing model of Europe, the remaining cost of cross-border lines and network enhancements within a zone are recovered through transmission tariffs. Most European countries use a uniform (postage stamp) approach to calculate their transmission tariffs. Only a few countries such as the United Kingdom, Sweden, and Italy apply locational transmission tariffs.

Transactions do not take place only within countries (zones) but also between zones involving generation companies and consumers (agents) from different zones. Such transactions cause flows not only between the agents but also throughout the system, making use of the transmission infrastructure of zones that are not directly involved in the transaction. Operators hosting such transits incur additional costs on their local network. Charging such grid users for the use of the system charge of each system they cross according to the TSO's transmission tariff results in pancaking, which is often criticized to distort wholesale competition. However, the intermediate countries must still be compensated for the external use of their grid. The Inter-TSO-Compensation (ITC) mechanism is a mechanism where TSOs are compensated for costs incurred as a result of hosting cross-border flows of electricity on their network [28, 29].

9.5.2 Cost Sharing of Transnational Transmission Projects

When investments occur between two TSOs, the cost sharing of transnational projects usually involves equal sharing of the costs among the TSOs. Each TSO passes these costs to its grid users through the national transmission tariff. Recently, however, innovative transmission projects integrating offshore wind farms with cross-border interconnectors are being considered. Such projects involve not only TSOs but also other stakeholders such as offshore wind farm generators who want to share the interconnector instead of individually connecting to the national grid. Different cost-sharing alternatives could affect the level of cooperation among the stakeholders and hence the final grid layout [30].

9.5.3 Pricing of Third-Party Transmission Projects

As an alternative to regulated transmission investments by TSOs, there are third-party investment projects (also known as merchant investments) (Section 9.3.1.3). These projects are usually market-driven and can be either "tariff-financed" or "contract-financed." In the former, the project costs are recovered via regulated transmission tariffs. In the latter, project costs are recovered via the sale of transmission rights, (i.e., explicit or implicit capacity auctioning) to competitive market participants [31].

9.6 TRANSMISSION GRID OPERATION

The components of the governance model that affect the operation of a transmission system are important when trying to achieve acceptable reliability and efficiency standards. A regulatory framework usually contains a various set of rules and mechanisms to ensure security, i.e. the short term aspects of reliability or security of supply. The framework defines reliability standards, congestion management mechanisms, network access rules, defense plans, and the mechanisms to ensure the availability of grid supporting services such as balancing reserve capacity, reactive power, and black-start capabilities.

In addition, the regulatory framework ensures that this level of quality is met at a reasonable cost. It determines the way that transmission capacity is allocated to the market and therefore has a substantial impact on the way electricity markets operate. It can also affect the design of specific market rules such as nomination and balancing responsibility, as well as the procurement of such grid supporting services discussed above by market players. In order to achieve regional power system integration, it is important to align market and network operation between the different systems. In the European power system, common rules for market and system operation are under development by means of a trilateral interaction between the European Commission (requesting the initiative following the third legislative package), ACER (development of the vision on market and operation subjects), and ENTSO-E (drafting of the network codes). After iterations between the involved parties and stakeholders, these are to be adopted in the European regulatory framework.

Network codes have been structured in a series of topics which are being developed over a period of several years. An overview is given in Table 9.2. Numbers 4–7 focus particularly on operational aspects, while 8–10 deal with market aspects.

TABLE 9.2 Network Code Areas [32]

1	Requirements for Generators (RfG)	Sets requirements that new generators connecting to the network (both distribution and transmission) and existing generators (in very limited cases) will need to meet, as well as responsibilities on TSOs and distribution network operators.
2	Demand Connection Code (DCC)	Sets requirements for new demand users and distribution network connections to the network, outlines basic demand side response capabilities for meeting transmission system needs, and outlines responsibilities on TSOs and distribution network operators.
3	HVDC Connection Code (HVDC)	Sets requirements for HVDC connections and offshore DC connected generation.
4	Operational Security Network (OS)	Sets common rules for ensuring the operational security of the pan-European power system.
5	Operational Planning and Scheduling (OPS)	Sets requirements, ranging from the year-ahead time frame to real time, for assessing the adequacy and operational security of the interconnected power system and for planning outages required by TSOs and grid users when they have cross-border impacts on power flows.
6	Load Frequency Control and Reserves (LFCR)	Provides for the coordination and technical specification of load frequency control processes and specifies the levels of reserves (back-up) which TSOs need to hold and specifies where they need to be held.
7	Emergency and Restoration (ER)	Deals with the procedures and remedial actions to be applied in the emergency, blackout, and restoration states. This involves preparation of system defense, system restoration, and resynchronization plans in advance, dealing with information exchange, procedures for operating when a system enters into one of these states and ad hoc analysis of such incidents.
8	Capacity Allocation and Congestion Management (CACM)	Creates the rules for operating pan-European day-ahead and intraday markets, explains how capacity is calculated, and explains how bidding zones will be defined.
9	Forward Capacity Allocation (FCA)	Sets rules for calculating and buying capacity in timescales before day ahead and for hedging price risk between bidding zones in long-term time frames.
10	Balancing Network Code (EB)	Sets rules to define the roles and responsibilities of TSOs and market parties to procure and exchange balancing products to balance the system from day ahead to real time in the most efficient way. It also includes financial principles for the payment of these services.

9.7 CONCLUSIONS

This chapter discusses the governance model for HVDC transmission network developments. The main aspects of this governance model are represented by means of five building blocks representing the regulatory framework of electricity transmission networks, i.e. planning, financing, ownership, pricing, and operation. This governance model is a crucial element when reviewing the planning and operation of new HVDC transmission assets, which are to be implemented under an existing regulatory framework. Understanding the elementary aspects of this framework is an essential step towards (a) assessing the interaction with future grid evolutions such as HVDC and (b) the justification of potential adaptations of the regulatory framework.

Transmission network planning has proved to be much more difficult when two or more countries or states are involved. This is thus particularly relevant for long-distance HVDC network components, which are anticipated as crucial pieces of regional networks such as the European supergrid. Five crucial building blocks are needed to support a transmission infrastructure evolution that supports continental renewable integration goals, ensures capacity for unfettered trade, and remains reliable while transparently and fairly allocating costs and benefits. In general, the governance model has to incentivize an efficient development and operation of the transmission system, ensuring cost-efficiency, reliability, and sustainability.

First, planning has to be done on a regional level favoring the investments which benefit the region as a whole. Second, ownership structure and remuneration schemes must provide transmission network investors with appropriate means to recoup their investment cost. This reduces the investment risks and therefore reduces the financial needs of the required investment projects. Third, stable, transparent, and predictable pricing schemes should be deployed to reduce investment risk and to allow a fair allocation of the transmission network costs. Fourth, efficient operation of cross-border transmission capacity needs to be achieved by common rules, such as the European network guidelines which are currently in development.

All of these aspects are strongly interrelated and should be analyzed and developed as a whole when pursuing European energy policy goals such as competitiveness, reliability, and sustainability while ensuring transparency and fairness. The introduction of a clear and well-chosen governance model will help to keep the complexity and risk of transmission network infrastructure development as low as possible.

REFERENCES

1. I. Pérez-Arriaga, *Regulation of the Power Sector*. Berlin: Springer, 2013.
2. European Parliament and Council, "Regulation (EC) No 714/2009 of the European parliament and of the council of 13 July 2009 on conditions for access to the network for cross-border exchanges in electricity and repealing Regulation (EC) No 1228/2003 (text with EEA relevance)," *Official Journal of European Union*, vol. 211, pp. 15–35, 2009.
3. European Commission, "Energy infrastructure—Projects of common interest," 2014. [Online]. Available: http://ec.europa.eu/energy/infrastructure/pci/pci_en.htm

4. Desertec Foundation, "Clean power from deserts The DESERTEC Concept for energy, water and Climate security," 2007 [Online]. Available: http://www.desertec.org/fileadmin/downloads/DESERTEC-WhiteBookensmall.pdf
5. J. De Decker, A. Woyte, B. Schödwell, J. Völker, and C. Srikandam, "Directory of offshore grid initiatives, studies & organisations," Technical Report, Sept. 2009.
6. Woyte, A. and De Decker, J. and Vu Van, T., "A North Sea electricity grid [R]evolution-Electricity output of interconnected offshore wind power: a vision of offshore wind power integration.Greenpeace 3E," 2008 [Online]. Available: http://www.greenpeace.org/raw/content/eu-unit/press-centre/reports/A-North-Sea-electricity-grid-(r)evolution.pdf
7. P. Buijs, "Transmission invesments: concepts for European collaboration in planning and financing," Ph.D. dissertation, KU Leuven, 2011.
8. R. Baldwin, and M. Cave, "Understanding regulation: theory, strategy, and practice." Oxford; New York: Oxford University Press, 1999.
9. H. Averch and L. Johnson, "Behavior of the firm under regulatory constraint," *The American Economic Review*, vol. 52, no. 5, pp. 1051–1069, 1962.
10. K. Petrov, V. Ajodhia, and D. Resnjanskij, "Regulatory incentives for investments in electricity networks," *Proceedings of the third annual conference on Competition and Regulation in Network Industries*, pp. 1–21, November 19, 2010.
11. ACER, "On incentives for projects of common interest and on a common methodology for risk evaluation," Technical Report, 2014.
12. P. Joskow and J. Tirole, "Merchant transmission investment," *Journal of Industrial Economics*, vol. 53, p. 233264, 2005.
13. Roland Berger strategy consultants, "The structuring and financing of energy infrastructure projects, financing gaps and recommendations regarding the new TEN-E financial instrument," Technical Report, 2011.
14. P. Rabensteiner, "Multi-dimensional risk and investment return in the energy sector: The case of electric transmission networks," PhD dissertation, University of St. Gallen, 2013.
15. G. Bertram and D. Twaddle, "Price-cost margins and profit rates in New Zealand electricity distribution networks since 1994: the cost of light handed regulation," *Journal of Regulatory Economics*, vol. 27, no. 3, pp. 281–307, 2005.
16. I. J. Pérez-Arriaga and Y. Smeers, "Guidelines on tariff setting," *Transport Pricing of Electricity Networks*, (ed F. Lévéque), Kluwer Acadamic Publishers, pp. 175–204, 2003.
17. L. Meeus and M. Saguan, "Innovating grid regulation to regulate grid innovation: From the Orkney Isles to Kriegers Flak via Italy," *Renewable Energy*, vol. 36, no. 6, pp. 1761–1765, June 2011.
18. Energinet.dk, Svenska Kraftnat, Vattenfall Europe Transmission, "An analysis of offshore grid connection at Kriegers Flak in the Baltic Sea—Joint pre-feasibility study," Technical Report, 2009.
19. R. E. Bohn, M. C. Caramanis, and F. C. Schweppe, "Optimal pricing in electrical networks over space and time," *Rand Journal of Economics*, vol. 15, no. 3, pp. 360–376, 1984.
20. I. Perez-Arriaga, F. Rubio, J. Puerta, J. Arceluz, and J. Marin, "Marginal pricing of transmission services: An analysis of cost recovery," *IEEE Transactions on Power Systems*, vol. 10, no. 1, pp. 546–553, 1995.
21. H. Rudnick, R. Palma, and J. E. Fernibdez, "Marginal pricing and supplement cost allocation in transmission open access," *IEEE Transactions on Power Systems*, vol. 10, no. 2, pp. 1125–1142, 1995.
22. R. E. Rotoras, T. Lefevre, and R. B. Pacudan, "Marginal transmission pricing and supplemental cost allocation method: A case of Philippines," *Electric Power Systems Research*, vol. 63, pp. 213–227, 2002.
23. D. Shirmohammadi, X. Filho, B. Gorenstin, and M. Pereira, "Some fundamental, technical concepts about cost based transmission pricing," *IEEE Transactions on Power Systems*, vol. 11, no. 2, pp. 1002–1008, May 1996.
24. J. Marangon Lima, "Allocation of transmission fixed charges: An overview," *IEEE Transactions on Power Systems*, vol. 11, no. 3, pp. 1409–1418, 1996.

25. F. Rubio-Oderiz and I. Perez-Arriaga, "Marginal pricing of transmission services: A comparative analysis of network cost allocation methods," *IEEE Transactions on Power Systems*, vol. I, pp. 448–454, 2000.
26. M. T. Mekonnen, C. De Jonghe, B. Rawn, D. Van Hertem, and R. Belmans, "Power flow control and its effect on flow-based transmission cost allocation," *10th International Conference on the European Energy Market (EEM)*, pp. 1–8, May 2013.
27. National Grid, "CMP213 Project TransmiT TNUoS Developments: Final CUSC Modification Report," Technical Report, June 2013.
28. European Commission, "Commision regulation no. 774/2010 of 2 September 2010 on laying down guidelines relating to the inter-transmission system operator compensation mechanisms and a common regulatory approach to transmission charging," *Official journal of the European Union*, vol. 233, no. 774, pp. 1–6, 2010.
29. European Commission, "Commision regulation no. 838/2010 of 23 September 2010 on laying down guidelines relating to the Inter-transmission system operator compensationmechanisms and a common regulatory approach to transmission charging," *Official Journal of the European Union*, vol. 250, no. 838, pp. 5–11, 2010.
30. S. Hadush, C. De Jonghe, and R. Belmans, "The effect of welfare distribution and cost allocation on offshore grid design," *IEEE LTS*, 2014.
31. R. Coxe and L. Meeus, "Survey of non-traditional transmission development," *IEEE PES General Meeting*, pp. 1–6, July 2010.
32. ENTSO-E, "European Network Code Development—The Importance of network codes in delivering a secure, competitive and low carbon European electricity market," 2013.

CHAPTER 10

POWER SYSTEM OPERATIONS WITH HVDC GRIDS

Dirk Van Hertem,[1] Robert H. Renner,[1] and Johan Rimez[2]

[1]Research Group Electa, Department of Electrical Engineering, University of Leuven, Leuven, Belgium

[2]Division TGX, Elia Engineering, Brussels, Belgium

10.1 INTRODUCTION

The future HVDC grid will be connected to the existing AC power system. The resulting system will form an integrated energy system. This integrated system is expected to be operated in a similar manner as the current international AC power system. This means that an HVDC grid will be integrated in the current system operator business, and this within the different operational time frames and the operational states. This requires that the HVDC grid(s) be connected according to a given grid code, operate within the market and regulatory framework, and participate in the delivery of ancillary services.

Current HVDC lines are also connected to the existing power system, however, they are often not fully integrated in the overreaching system. Up to now, they also form a relative small share of the total transmission infrastructure. When HVDC grids are developed, it is expected that they will form a more important part of the overall transmission system, through their high power ratings and geographical size, and the larger impact on overal system operations will play a larger role. The controllability of the DC converters influences the operations of the system significantly and needs to be incorporated in the operational procedures.

One important difference in operation between an HVDC link and an HVDC grid is that the former is very often operated at full capacity, making maximum use of the (expensive) equipment. However, this is not the case for a meshed HVDC grid, where the rated capacity of the transmission lines is not directly linked to the rated capacity of the converter stations. Rather, these converter stations can be compared to the transformers in AC systems, interlinking different layers (see Figure 10.1 [1]). The flows in the DC grid are distributed following the laws of Kirchhoff and depend on the power injections by each converter station (see also Chapter 13). The capacity of the individual branches are fixed and clearly do not

HVDC Grids: For Offshore and Supergrid of the Future, First Edition. By Dirk Van Hertem, Oriol Gomis-Bellmunt, and Jun Liang.

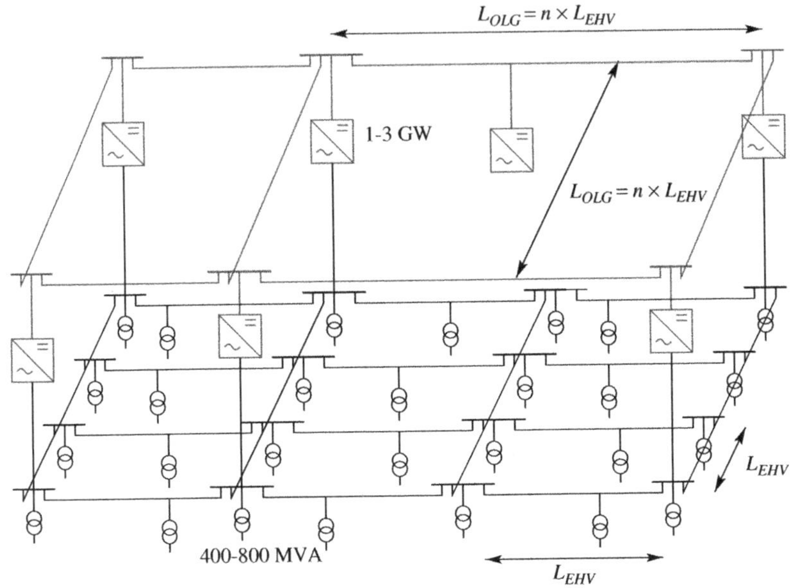

Figure 10.1 Grid layout with an additional overlay grid. Mesh sizes will increase and a limited number of converter stations will connect to the existing extra-high-voltage (EHV) AC grid with the new overlay grid in DC. The converter stations connect the HVDC layer with the AC system as the AC transformers do between voltage levels [1].

follow these injections. Furthermore, the grid operations need reserve capacity to take variations in injections and outages into account.

This chapter gives an overview of the key aspects of power system operations and highlights the possible influence or changes that can be expected with HVDC grids. Special attention is paid to the interface between stakeholders and systems. The choices that need to be made in the future to guarantee a successful and efficient system operation in the presence of both AC and DC systems are indicated. The principles discussed in this chapter are not necessarily only valid for power systems with a HVDC grid, but are also valid for systems with a significant amount of point-to-point HVDC links embedded in the system.

10.2 WHO OPERATES THE HVDC LINK OR GRID?

The manner in which the HVDC converters are controlled is mainly dependent on who operates the DC system, and the requirements under which the link is operated (the grid code, see also Section 10.7).

In some cases it is important to make a distinction between the entity that performs the operations and the entity that dictates the operations. An example is a link where the transmitted power is set through auctioning on a market, while the actual control room operators are "only" responsible for assuring that the system is operated as such. For the remainder of this chapter, the operator refers to the decision maker.

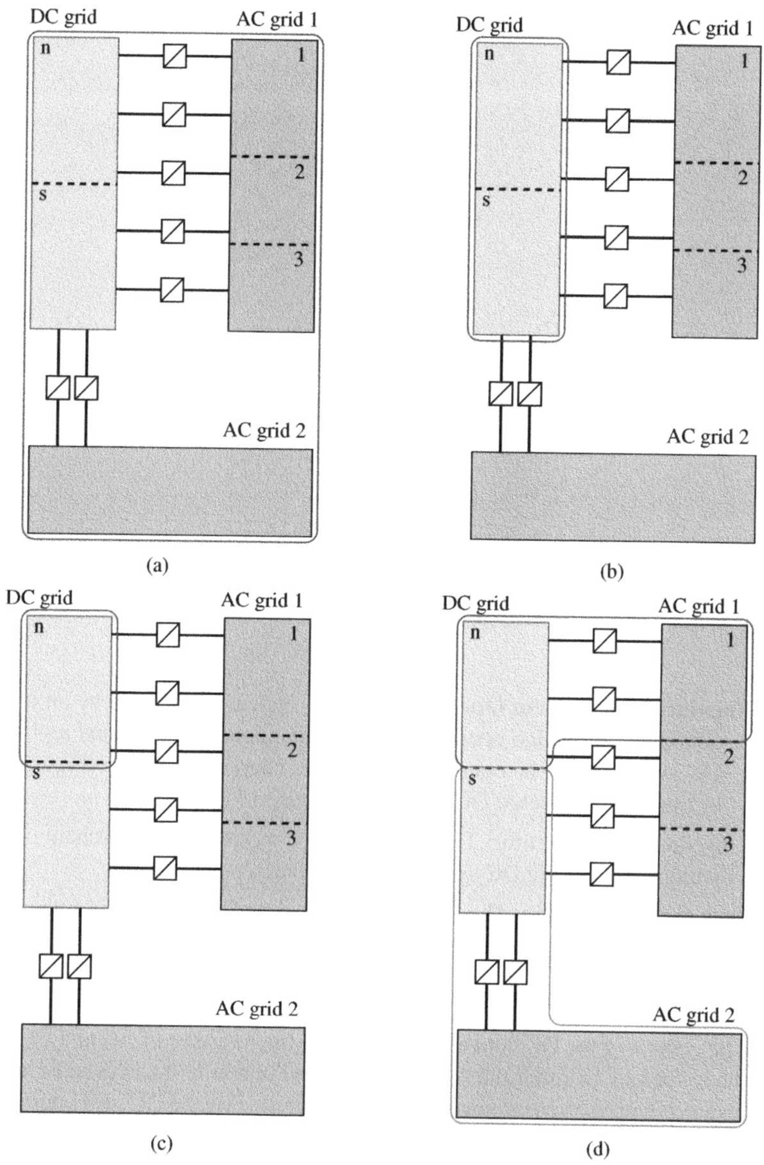

Figure 10.2 Different operator configurations for HVDC systems. (a) Single system operator. (b) DC grid-operated separately. (c) Multiple independent DC operators. (d) System operator boundaries determined by geographic area.

A hybrid AC/DC grid can have the following operator options:

Single Operator. In the single operator configuration, the combined AC and DC system is operated by a single entity, which can optimize the entire system (Figure 10.2a);

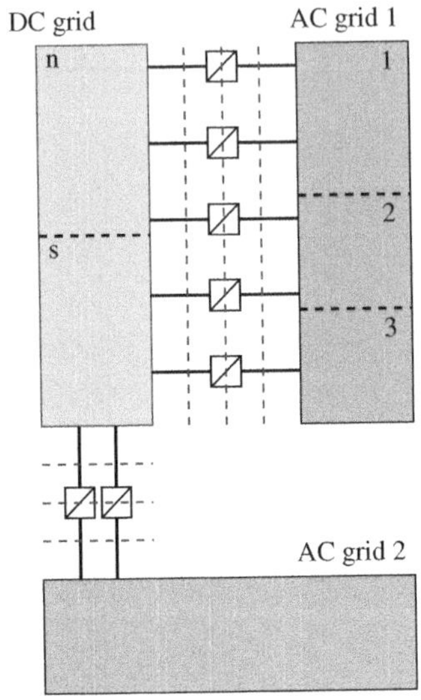

Figure 10.3 The boundaries of different power systems need to be well established, with a clear definition of the point of common coupling (PCC): at the DC side, at the AC side, or somewhere inside the HVDC converter.

Regulated DC System Operator. The DC system is operated by an independent operator, which operates the DC system under a regulated regime. The DC system could be operated as a single independent entity (Figure 10.2b) or multiple connected DC operators (Figure 10.2c).

Merchant DC Operator. The DC system is operated by a merchant operator, which operates the DC system independently.

Territorial Operator. The system operation and ownership is linked to the location (e.g., Figure 10.2d).

Two additional notes must be made to the aforementioned options:

- The control of the DC converter, which is at the interface of AC and DC, can in theory belong to either the AC or the DC grid operator. It can even be split (for instance with the active power control is governed by the DC grid operator, and the reactive power or voltage controlled by the AC operator) (see Figure 10.3).
- The control of the power exchanges between AC and DC systems can have an effect on a neighboring system, even if there are no direct injections in that system.

An illustrative example of how the ownership can influence the operation of the combined AC and DC system is given with the help of Figure 10.4. Consider an AC and a DC system that are connected with five converter stations. In the north of the DC system, a large power injection is present, which needs to be transferred to the south of the AC zone where there is a large load. As all links between the AC and DC system are fully controllable, the transfer can to a large extent be controllable

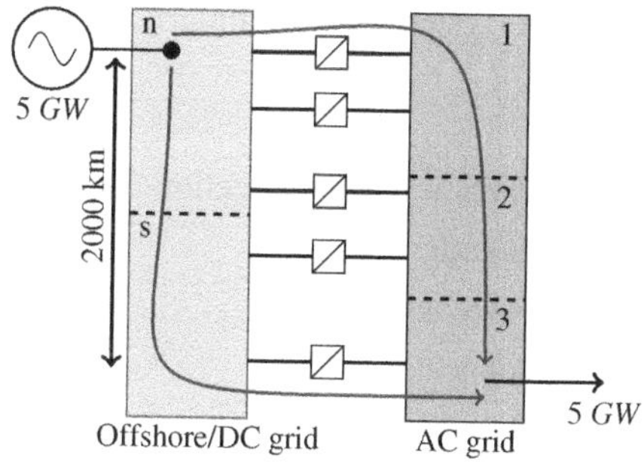

Figure 10.4 DC and AC system flow.

as well. A first option is to use the DC grid to transmit the power from the north to the south and then inject this power as near as possible to the load center. A second option is to directly transfer the electric power towards the AC system by using the northern interconnections. In a third option, the power transfer is distributed over the different links.

Consider now the case where the objective is to minimize the losses in the system. In the case of the DC operator, the second option is the most beneficial, while in the case of the AC grid operator, it is more beneficial to use option one, and transfer the power over the DC grid. In a system-wide approach, an optimization over both systems is needed.

The matter is even more complicated when we consider that both the AC system and the DC system could be split into multiple system operators. In this case, the operation involves coordination over multiple borders. For instance, the flow through AC zone "2" is strongly influenced by the control of the different converters.

In real power systems, the operator of the power system has to ensure that there is sufficient:

- Capacity on each converter
- AC transmission capacity
- DC transmission capacity
- Reserves in the case of an outage in the AC system, the DC system, or a converter station

These requirements are part of the ancillary services between AC and DC systems (see Section 10.6.1).

10.3 RELIABILITY CONSIDERATIONS IN SYSTEMS WITH HVDC

10.3.1 Reliability at the DC Grid Side

In order to assure a reliable operation, the power in each branch should remain below the rated value and the voltage between the minimum and maximum limit.

Contingencies in the DC system occur, similarly to traditional power systems. These contingencies could be:

- Converter station malfunction or disconnection
- A line opening (all poles) after fault
- A pole opening in a bipolar system
- Metallic conductor disconnection
- Busbar disconnection

In each of these cases, the original flow pattern will be disturbed and the power flow will be redistributed according to Kirchhoff's laws. This change in flows could create overloads in the system which can be alleviated by changing the injections into the DC grid (DC redispatch) or through switching actions. Potentially also the use of DC power flow controlling devices is possible (see Chapter 21).

The manner in which the operator manages the possible contingencies in the system depends on the reliability criterion that is used by the operator (e.g., "$N-0$," "$N-1$," or a risk-based approach) and the interaction with the AC system.

10.3.2 Reliability of the Hybrid AC/DC Grid

When considering the hybrid AC/DC grid, contingencies in both systems need to be taken into account as they influence the other system. Line openings at the DC side might cause overloads in the AC system, or a short circuit at the AC system might cause a converter station to disconnect.

The hybrid AC/DC grid consists of a number of AC and DC systems operating in parallel. When considering the hybrid AC/DC grid reliability, it is important to specify on which part of the system the reliability criterion is calculated. To use the example of the $N-1$ criterion, one needs to decide whether the AC and the DC system(s) need to be "$N-1$" secure individually, or that the combined system(s) is seen as a single meshed system. In the first case, a multi-terminal HVDC system without meshes can never attain "$N-1$" secure operation as there is no inherent redundancy in that system. However, the same system might be "$N-1$" secure when the AC system to which it is connected is considered in the reliability calculation.

Similarly as with the DC grid reliability, when a line contingency occurs at either the AC or the DC side, a redispatch of the converter injections might alleviate the overload. The controllability of the DC converters can be used to make optimal use of the installed assets. How this is done depends largely on expected level of reliability, the number of parallel paths (in both AC and DC, as well as in the combined system) and the involved stakeholders.

In case a converter outage occurs, the energy exchange between the AC and DC system gets disturbed and a balancing action is required. This is dealt with in more details in Section 10.4.

10.3.3 Operating the Power System under Different Time Frames

Power system operations is a term which encompasses an entire range of activities performed or influenced by different stakeholders. The activities of the operation fall within a time frame of several weeks, days, or hours in advance, up to real-time (see Figure 10.5). Within these time frames, all processes and short-term decisions which prepare the system to be *ready* for the actual moment of energy delivery happen. The system is *ready* when at any instance a reliable and cost-effective energy supply can be guaranteed, without hampering the market.

Months to days in advance, the operational planning of the power system focuses on maintenance, generation scheduling, and assessing the grid capacity between zones. Closer to actual operations, the energy is traded on the day-ahead market, taking into account longer-term contracts (e.g., year contracts) and the available capacity between different bidding zones. This available market capacity is calculated using either an ATC (available transmission capacity)-based methodology or according to a flow-based mechanism [3]. Based on this input, market participants make offers to the market. The bids for the actual day come in the day before the actual operation ($D - 1$), before gate closure. Based on bidding that occurs on the market, the different system operators perform the day-ahead congestion forecast (DACF) to determine whether the provided generation schedule can be maintained or whether adjustments are needed. The DACF forecasts the system flows for each individual hour, while taking into account security constraints ("$N - 1$"). The DACF also includes the forecasted generation from various renewable energy sources. Based on this data, a security assessement is performed.

In the case where certain security issues arise, the system operators take appropriate action and anticipate. This can be in the form of preventive measures which avoid the system to experience those security issues. Alternatively, the operator can choose to take the risk and prepare for corrective measures in case this situation occurs. These measures can be in the form of market actions (e.g., redispatch or reserve activation) or through TSO actions (topology changes, changing the settings of power flow controlling devices such as the HVDC controllers, etc.). TSO actions are preferred as they are normally cheaper to implement.

On the day of actual operation itself, the TSO monitors the grid behavior, which in normal operation differs within acceptable limits from the predicted state due to normal deviations in generation, load, and the grid itself. These normal deviations can be addressed by standard operations. The operator is required to intervene using corrective actions and even load curtailment when more serious contingencies occur in the system, such as large unforeseen generation shifts, sudden outages of important transmission assets, or adverse weather.

10.3.4 Steady-State Operations of HVDC Connections and Grids

Point-to-point HVDC lines behave fundamentally different than traditional AC power lines. The main difference lies in the controllability of the power flow through the converters. This ability makes it on the one hand impossible to overload the converter

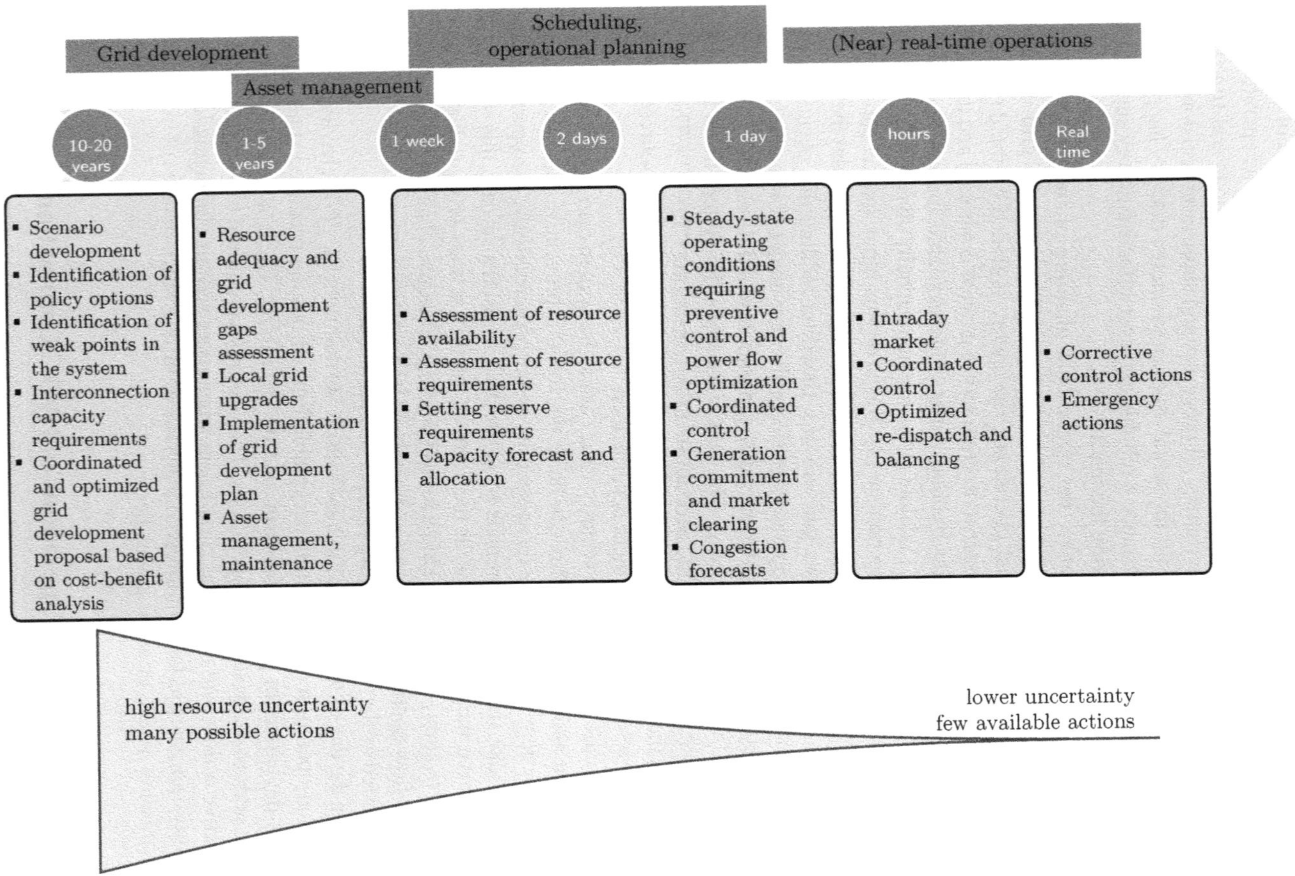

Figure 10.5 Smart operation and planning time frames [2].

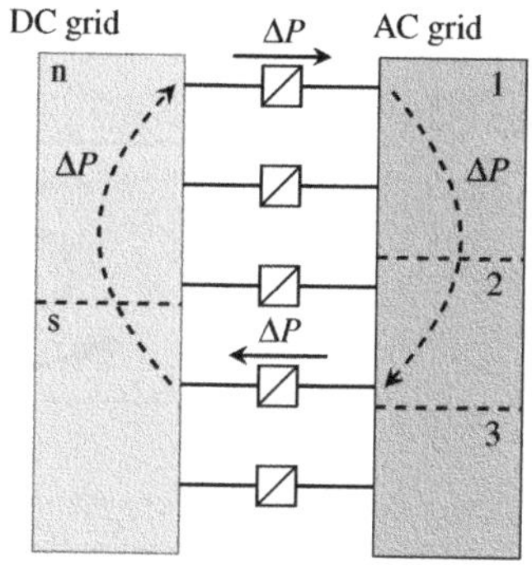

Figure 10.6 Shift of power through the use of an HVDC connection or grid. A power transfer of ΔP from one converter to another causes a corresponding flow in both the AC and the DC system.

stations (and as a consequence the DC lines). On the other hand, the DC system can be used up to its limits as the system without actual danger for overloads. The DC grid also behaves differently than an equivalent AC systems as all the interfacing flows can be fully controlled.

By controlling the converter active power at each node, the power flow throughout the entire transmission system is influenced, both AC and DC. A shift in power injection between nodes of the HVDC grid can be considered as a redispatch of power and creates a compensating transfer between the AC and DC system (see Figure 10.6). Such a redispatch is possible by keeping the sum of the injections in each synchronous zone[1] equal to zero.

$$\sum_{synchr\ zone} \Delta P_i = 0 \tag{10.1}$$

This inherent capability to manage the power exchanges influences the operational procedures of the transmission system in the different time frames:

- By changing the injection pattern, the available transmission capacity on different transmission corridors can be directly influenced. The transmission capacity between zones can be directly influenced by altering the flow over a DC interconnector, or indirectly by alleviating congestion elsewhere. At the same time, it is also possible to reduce the capacity between (other) zones.
- During the security assessment, redispatching the HVDC injections can be used by the system operator to appropriately shift power between transmission lines.
- During the actual operation (real time), the fast controls of the HVDC converters allow to apply corrective actions, bringing the system back to safety or increasing the reliability.

[1]Although that it is only required that the redispatch happens within one synchronous zone, it is likely that in practice this will be limited to a control area.

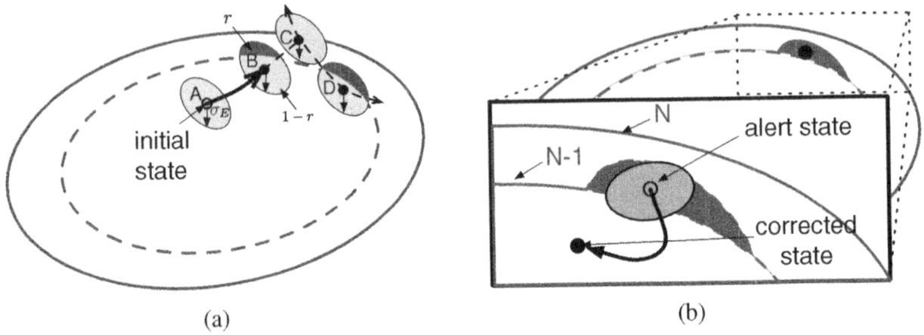

Figure 10.7 Controlling the power flow in the power system using HVDC can manage uncertainty in the system. (a) Choosing the operating point through flexible energy power flow management. (b) Uncertainty over the power flows changes over the different time frames.

It is important to note that most likely not all control actions on each converter station are available to manage the system capacity and system security.

10.3.4.1 Operation under Uncertainty Power system operations have always operated under significant uncertainty: Load, generation, and the exact location of the power injections are subjected to variations. With the massive introduction of renewable energy sources, the uncertainties increased even further. The uncertainty in the actual operating point causes the system operators to either (a) take larger security margins, at a cost for the energy market, or (b) operate closer to the operational limits, at a higher risk [4].

The use of controllable devices can aid in managing the operating point in such a way that the system is used up to its limits, yet with acceptable reliability. Figure 10.7a illustrates conceptually how the operating point can change depending on a power transfer between two zones and its effect on the power system security. By increasing the power transfer, the operating point moves from the initial operating point A up to the security boundary ($N - 1$ in this case). The operating point itself is subjected to an uncertainty σ_E, indicated by the gray zone. Operating state B represents the system with power transfer P_T, resulting in a probability or risk of being outside the reliability limits of r. With a higher power transfer ($P_T + \Delta P_T$), the operating state moves up to point C, with a higher risk. Using the controllability of the HVDC system, the operating point can be shifted so that the system can handle the increased power transfer while retaining an equal risk r of surpassing the security boundaries. Figure 10.7 b illustrates the use of corrective actions in the case where uncertainty indeed causes a potential security violation.

10.3.4.2 Coordination Between Operators The set-point change of DC converters comes at a low direct cost, given that there is sufficient capacity available in the system. At least, this is the case when DC converter stations can take part freely in the reliability management and are under the control of the grid operator performing the redispatch. In the case where the set-point changes are not freely available—for instance, if they are set by the market or set by another operator—the actual possibilities to use HVDC converters to aid systems operations depend on the framework and the coordination between system operators. It can also be expected that the level of cross-zonal coordination will depend on the severity of the event. Specifically in the case of large events that potentially may lead to blackouts, "force majeur" might be used.

In a general system where several interconnected systems contain different independent HVDC lines or independently controlled converters of an HVDC grid, the coordination will play an increasingly important role, as the effect of the DC control actions can reach far across interconnected AC systems [5, 6].

10.4 MANAGING ENERGY UNBALANCES IN THE SYSTEM

In any closed system, the energy balance has to be kept in order to retain stable operations. This is also true for the DC grid. In AC power systems, the frequency, generally considered the heartbeat of the system, is kept constant due to the shared duty of all generators involved. The rate of rise of the AC frequency is determined by the energy stored in the rotating mass of the synchronous generators ($E \propto J \cdot \omega_m^2$, with J the moment of inertia and ω_m the mechanical speed of the generators). The voltage amplitude at the different nodes in the system is generally controlled locally and is linked with the reactive power management in the system.

In a DC grid, the energy stored is located in the grid capacitance ($E \propto C \cdot U^2$, with C the grid capacitance and U the DC voltage). The DC voltage fulfills a role similar to the frequency in an AC system. Every outage or contingency will be reflected by an increase or decrease of the voltage. Compared to an AC grid, where the frequency acts as an indicator for the system imbalance, the change of the voltage in the case of a power imbalance is much faster, because in a DC grid the energy is not buffered in the rotating mass like in AC grids, but mainly in the converter and cable capacitances. A state-of-the-art multilevel modular voltage source converter (MMC) has a time constant of around 10–100 ms, which is a few orders of magnitude faster than AC generators. The result is a fast change of voltage, even by a small deviation of the power balance. For DC grids the time constant will increase, because the total capacitance increases with every converter, but will always be small in comparison to AC power systems.

In order to manage such quick variations, the system might require additional devices such as DC choppers to be installed, or might require the coordinated control of many individual wind turbines in order to reduce power fast enough.

The control action can be performed using either a DC slack bus or voltage droop control, where the consensus in literature seems to prefer the droop-controlled DC voltage (see Chapter 15) and much like the primary control actions which are currently used in AC systems. Note, the droop setting in different converter stations might differ, resulting in different power exhanges with the connecting system. The DC slack bus control is a special case of the droop control where one converter takes the complete deviation, while the others do not take part in the droop regulation.

Uncontrollable systems (such as AC systems with only loads or offshore wind power plants which do not take provide reserve power) will not partake in the droop action.

10.4.1 Unbalances in HVDC Grids

Unbalances in DC grids can originate in a similar manner as in AC systems: either through a fault disconnecting a converter station which imports or exports electrical power or through an unforeseen power injection—for example, from a large uncontrollable power injection such as an offshore wind farm. During the unbalance, the energy flowing in or out the system is not zero, which means that energy is accumulated in the DC grid. This means that the capacitors in the DC circuit are charged or discharged and the DC voltage will increase or decrease.

When an energy unbalance occurs (e.g., when a converter station is disconnected), the control of the DC voltage by the different converters will adjust the power injections immediately, distributing the deficit according to the droop settings. The overall voltage level in the HVDC grid is slightly lower or higher than the reference value due to this control action. This behavior is similar to the primary frequency control. After this action, two further actions are needed. The DC voltage needs to be restored and the change in power exchanges with the different neighboring systems need to return to the original values. As such, there is a need to adjust the power injection set-points to meet the scheduled exchanges, especially so if multiple zones or synchronous zones are connected to the DC grid. This requirement is very much in line with the secondary and tertiary control which is used in the AC system.

Figure 10.8 illustrates how the unbalances in a DC grid connected to multiple zones and multiple AC systems can be addressed in a system that utilizes droop control. Consider a simplied system where all the droop characteristics are equal, and for the sake of simplicity we neglect the effect of the voltage drop on the droop and assume that each converter is operating sufficiently far from the limits. A DC converter station outage causes an unbalance $-\Delta P$ in the DC system (Figure 10.8a). As a result, the voltages in the system will drop throughout the DC grid, and the droop causes the healthy converters to change their power. Each converter will contribute for $\Delta P/6$ as the droop settings are equal (Figure 10.8b). At this moment the DC system is balanced, yet there is still an unbalance in both AC systems. This unbalance is corrected in the third phase (Figure 10.8c), where contribution from AC grid 2 is canceled and taken over by AC grid 1. In a final phase, the correction is made per zone, and the entire ΔP is handled by the converter(s) connected to the same AC zone 1 to which the outaged converter is connected (Figure 10.8d).

Every power imbalance in the DC grid is directly shifted to the connected AC grids and not necessarily in an expected or desired way. With the expansion of the DC grid, this could become a problem, because if the imbalances reach a level where it influences the AC frequency beyond the control deadband for primary reserve (in continental Europe $\pm 20\,mHz$ [7], North America $\pm 36\,mHz$ [8]), it decreases the primary reserves of the AC grids in normal operation. In order to manage the HVDC grid power injections in the different nodes, a higher level control is needed which balances the entire system [9].

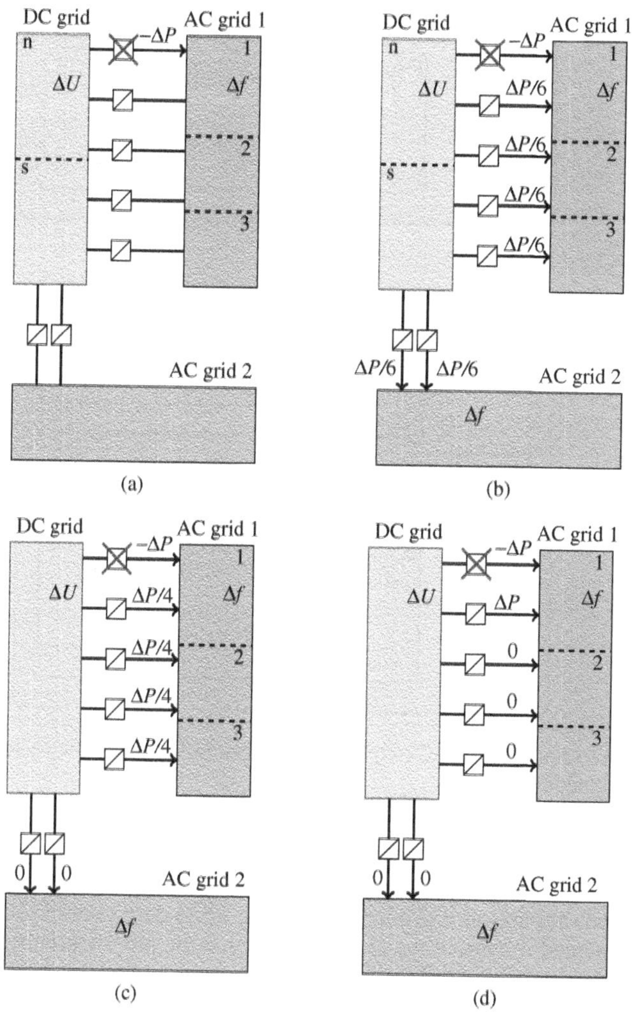

Figure 10.8 Addressing an unbalance in the DC grid through adjustment of the power injections. (a) Outage of a converter station connecting the DC grid with AC grid 1, zone 1. (b) Equal droop reaction causes all converters connected to the DC grid to contribute. (c) The schedule with AC grid 2 is corrected, only AC grid 1 contributes. (d) Control zone 1 of AC grid 1 takes the full unbalance over from the other systems.

10.4.2 Integrated Scheme for the Hybrid AC/DC Grid

As stated before, the management of AC and DC system unbalances can be organized in a very similar manner; furthermore, they interact due to the nature of the control actions (see Figure 10.8). In [9], a hierarchical control structure is introduced which provides an upper-level controller to the HVDC voltage control which can be compared to classic AC grid controls, consisting out of a primary, secondary, and tertiary control. This structure permits us to create an easy interaction between

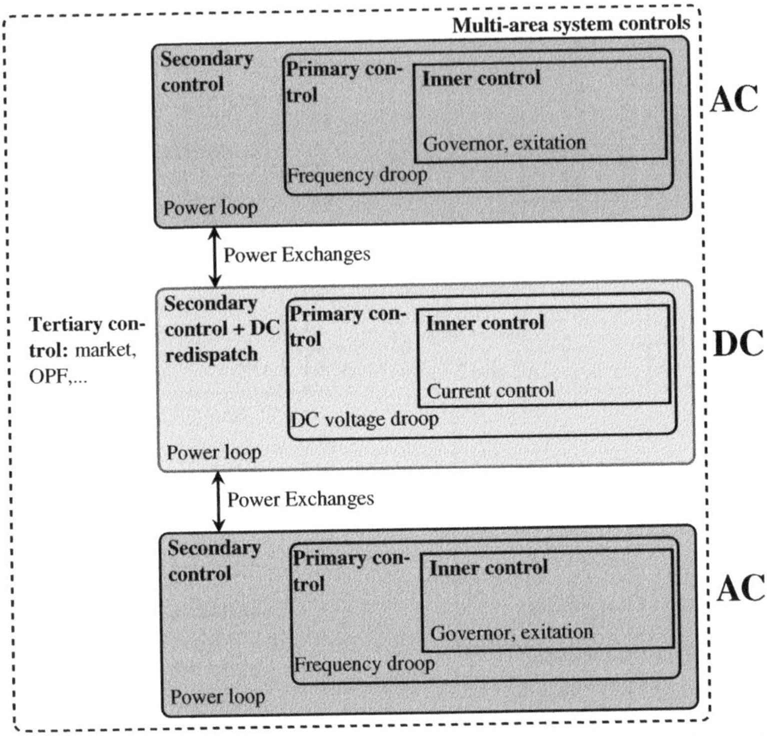

Figure 10.9 Integration of the AC frequency control and DC voltage control into a single hierarchical methodology [9].

power dispatch and the lower controllers using a methodology which is well accepted and known in the power sector. A voltage controller is implemented in each VSC power converter and acts as a primary controller. Secondary control, or power control, is implemented to correct the exchanges so that the pre-disturbance conditions are restored. The practical implementation is a change of set-points in the affected VSC controllers. The new set-points are provided by a system-wide DC redispatch. Finally a tertiary control mechanism calculates the power reference according a more optimal operation power system: market, losses, security, and so on. Tertiary control can be based on OPF (optimal power flow) algorithms considering AC and DC grids. The internal current control loop of the power converters has the same function as the governor and the excitation of a synchronous generators. Figure 10.9 shows how the AC and DC control approach are alike and how they can be integrated in a single framework. Note that the time constants in the different control loops differ.

10.5 ACTIVE AND REACTIVE POWER CONTROL

One of the key properties to power system operations is the ability to control the active power flow through the converter and the reactive power exchanged at each converter. This section investigates the active and reactive power exchange at the point of common coupling and its resulting power capability curve.

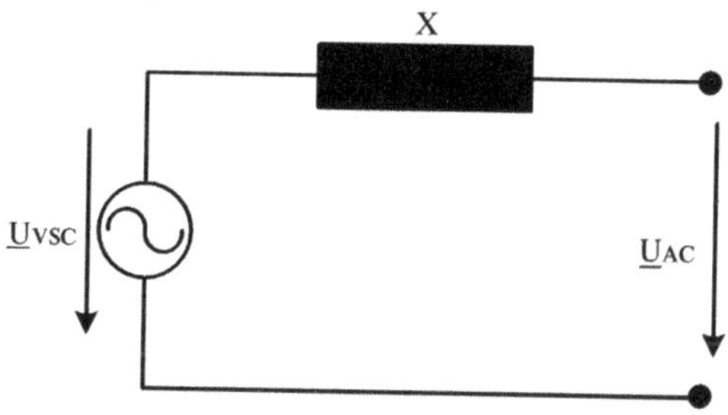

Figure 10.10 Equivalent circuit of a VSC-HVDC terminal.

10.5.1 Active and Reactive Power Exchange

A simplified representation of VSC-HVDC converter station which can be used for steady-state analysis consists of an ideal voltage source behind an inductance (Figure 10.10), which is similar to the model of a synchronous generator. The power electronic converter allows us to control the voltage amplitude ($|\underline{U}_{VSC}| = U_{VSC}$) by changing the modulation index and the phase angle between the converter and the AC voltage ($\angle(\underline{U}_{VSC}, \underline{U}_{AC}) = \delta$) by altering the switching pattern. The well-known simplified equations for the power transfer through an inductive line (see (10.2a) and (10.2b)) give the relation between the converter voltage amplitude and phase and the active and reactive power through an inductive line and hence through the converter. The transformer and phase reactor represent the line inductance.

$$P_{AC} = \frac{U_{AC} \cdot U_{VSC}}{X} \cdot \sin(\delta) \tag{10.2a}$$

$$Q_{AC} = \frac{U_{AC}^2}{X} - \frac{U_{AC} \cdot U_{VSC}}{X} \cdot \cos(\delta) \tag{10.2b}$$

The active power exchanged (10.2a) between the DC converter and the AC transmission system can be altered by modifying the phase angle between the AC voltage and the VSC voltage. Changing the rms value of the VSC voltage has only a minor influence on the active power flow. The control of the AC system voltage is therefore nearly independent from the active power flow control.

As shown by (10.2b), the reactive power exchange is more directly linked to the voltage amplitudes and less to the angle between them (when considering the voltage angles to remain relatively low, and the voltage amplitudes "around" p.u.).

Equations (10.2a) and (10.2b) can be rewritten to describe a circle in the PQ-plane ((10.3), Figure 10.11).

$$\begin{aligned} P^2 + Q^2 &= \left(\frac{U_{AC} \cdot U_{VSC}}{X} \cdot \sin(\delta)\right)^2 + \left(\frac{U_{AC}^2}{X} - \frac{U_{AC} \cdot U_{VSC}}{X} \cdot \cos(\delta)\right)^2 \\ &= \frac{U_{VSC}^2}{X^2} \cdot \left(U_{VSC}^2 + U_{AC}^2 - 2 \cdot U_{AC} \cdot U_{VSC}\right) \\ &= (U \cdot I)^2 = S^2 \end{aligned} \tag{10.3}$$

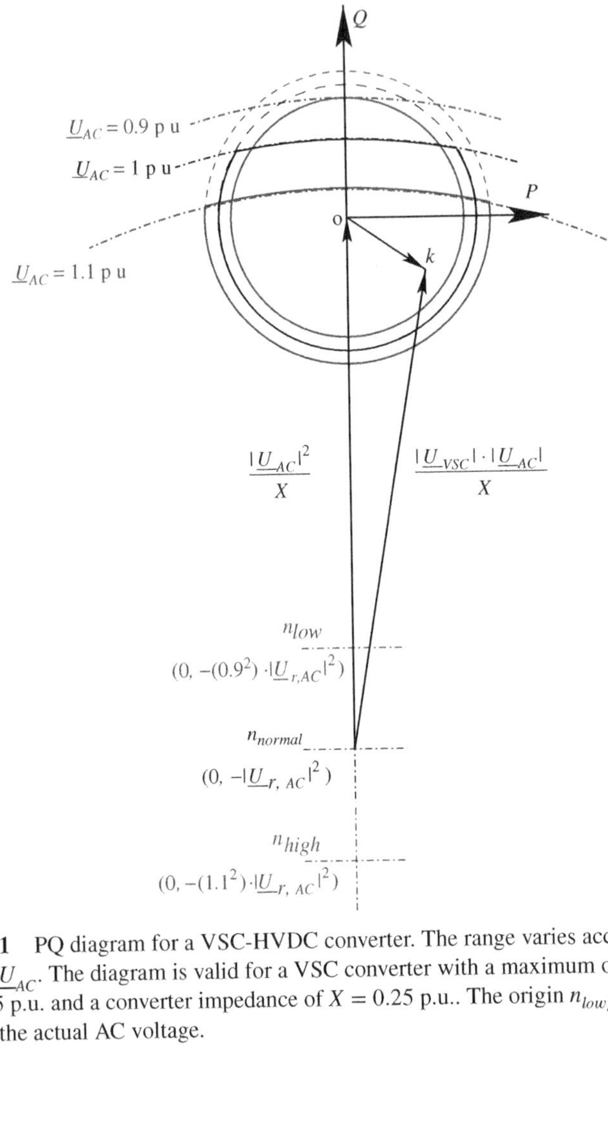

Figure 10.11 PQ diagram for a VSC-HVDC converter. The range varies according to the AC voltage $\underline{U}_{AC}$. The diagram is valid for a VSC converter with a maximum output voltage $\underline{U}_{VSC} = 1.15$ p.u. and a converter impedance of $X = 0.25$ p.u.. The origin $n_{low/normal/high}$ depends on the actual AC voltage.

10.5.2 Power Capability Curve

The VSC-HVDC system can deliver the desired active and reactive power whenever they are within the curve representing the largest circle, when $S = S_{max}$. The VSC-HVDC terminal can operate in all four quadrants. In quadrant 1, it delivers active and reactive power (behaving as a capacitor) to the system. In quadrant 2 it absorbs active and delivers reactive power. In quadrants 3 and 4, the terminal absorbs reactive power (behaving as an inductance), while it withdraws, respectively delivers, active power from/to the grid. The active power exchanges at the terminals are linked: $\sum_C P = P_{Loss}$, with C the set of converters connected to the DC grid. However, the reactive power outputs at both sides of the DC link are independent.

The ability to control the reactive power independently at each individual converter station can be seen as a significant advantage of the VSC-HVDC technology. This reactive power support can be used to manage the steady-state voltage in the vicinity of the converter stations.

Equations (10.2a) and (10.2b) are represented in the phasor diagram of Figure 10.11. The control range of the VSC converters is limited by the following practical converter design properties:

- The maximum current through the converter. With the impedance between converter and AC side being constant, this results in a maximum voltage $\Delta \underline{U} = \underline{U}_{VSC} - \underline{U}_{AC}$ across this component. In the PQ diagram of Figure 10.11, the maximum current is represented by a circle around O. As the AC voltage increases, the maximum power flow increases, for an equal voltage drop across the converter impedance.
- The maximum voltage across the converter IGBTs, and consequently the maximum AC voltage at converter side $\underline{U}_{VSC}$.
- The maximum DC voltage and current the DC cable can withstand.

Next to the design properties, also the voltage at the grid side $\underline{U}_{AC}$ has an influence on the power control range. In a normal operating state, the voltage can vary between 0.9 p.u. and 1.1 p.u..

The maximum voltage across the converter stack determines the cost of the converter station significantly as it depends on the number of components placed in series. This converter voltage can be higher (or lower) than the AC voltage, depending on whether the converter delivers (or absorbs) reactive power. In the case where the AC voltage is higher than the nominal voltage (yet within limits), the converter voltage needs to exceed the AC voltage as well when delivering reactive power. The converter voltage needs to be restricted in that case and the full circle cannot be maintained.

As a clarifying example, the separate PQ diagrams under different AC voltages for a VSC-HVDC converter delivering 0.6 p.u. active power, with a fixed converter voltage amplitude $\underline{U}_{VSC} = 1.12$ p.u. are shown in Figure 10.12.

Although it is technically possible to increase the converter voltage, this parameter is limited during design for two reasons:

- Increasing the reactive output would result in a further increase of the AC voltage. In normal operation, this would be an undesirable operation as the AC voltage is expected to be ≈ 1 p.u.
- A higher converter voltage would require more components in series and thus a more expensive installation or the utilization of special modulation schemes which might reduce the maximum current.

10.5.3 Active and Reactive Power Dynamics

The power electronic converter is able to react very rapidly to changes of the converter set-points. The reaction speed is limited by the dynamic characteristics of the

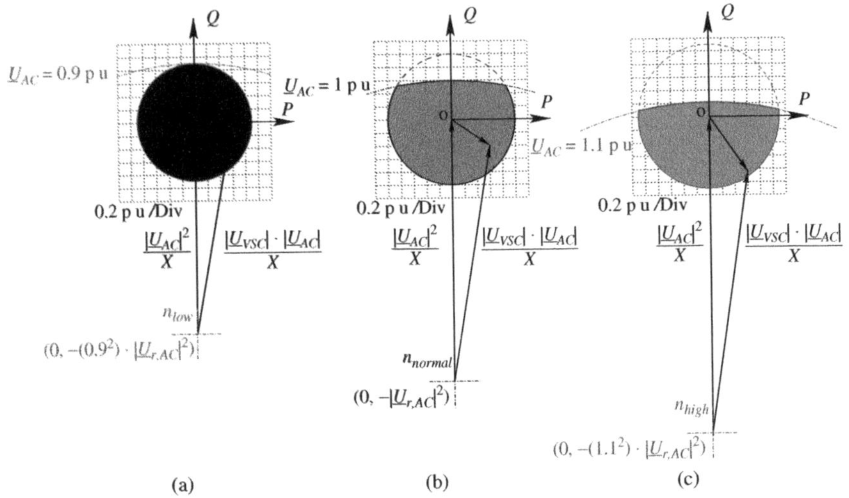

Figure 10.12 Vector diagram of a VSC-HVDC link delivering 0.6 p.u. active power with a fixed converter voltage. As the converter output voltage is limited, the VSC HVDC link cannot deliver active and reactive power in a "full power" circle. At high AC voltages, the converter cannot deliver high amounts of reactive power (i.e., act as a capacitor). (a) PQ diagram at low AC voltage (0.9 p.u.). (b) PQ diagram at rated AC voltage. (c) PQ diagram at high AC voltage (1.1 p.u.).

converter switching only. Both the active and reactive power set-point can be changed independently.

The fast-acting equipment can be used to manage dynamic events in the power system. For instance, the controllability of the DC converter can be used to mitigate the effect of AC short circuits or it can be used to damp oscillations in the power system.

Altering the reactive power or voltage set-point can be done without influencing the DC side power or voltage. This allows the use of the VSC converters to enhance the system stability locally, without impact on the DC side.

The active power can also be used to dampen oscillations in the AC system. Contrary to reactive power control, it will affect the energy balance in the DC grid and the oscillations are transferred to other parts in the AC system. As such, solving a problem at one location might cause problems at another site. When multiple AC systems are connected via a DC system, the active power control might be used to dampen the dynamic response of the combined system. Clearly, the use of active power damping requires in-depth study.

The most effective control, through active or reactive power, depends on the connected system(s) and the influence of active and reactive power modulation on the damping. Chapter 19 goes into more detail on power oscillation damping using VSC-HVDC systems.

10.6 ANCILLARY SERVICES

Ancillary services are all services required by the system operator to guarantee a certain level of continuity of energy supply, but which he cannot provide exclusively by himself or that others might be able to deliver at a lower cost. They are necessary to support the main service of power transmission from sellers to purchasers in a reliable and secure manner, compliant with defined power quality standards. They are commonly provided to a stakeholder, usually the grid operator, by other stakeholders (often generator companies or possibly loads). As such, ancillary services are a direct consequence of the unbundling of the energy system. The procurement can be mandatory or market-based, dependent on the regulatory framework.

The definitions of ancillary services are formulated in a general manner, so that everything which is valuable for a grid could be traded as an ancillary service. The following ancillary service are commonly defined for an AC system[2]:

- Active power/frequency control
  - Primary control or frequency containment reserves (FCR)
  - Secondary control or frequency restoration reserves (FRR)
  - Tertiary control or replacement reserves (RR)
- Reactive power/voltage control
- Loss compensation
- Black start and restoration

Based on the definitions of ancillary services [10, 11], it is possible to identify a number of services which are required in systems with DC grids. The DC grid needs at least energy balance, loss compensation, and black start and restoration. Furthermore, the controllable converters can alter the power flows in the AC and DC grids through redispatch. This could result in an extra ancillary service: DC transmission capacity reserve and power flow control.

As such, the minimum set of ancillary services that need to be defined for DC grids are:

- Energy balance and reserves
- DC transmission capacity reserve and power flow control
- DC loss compensation
- Energizing of DC subsystems, DC black start and restoration

Additionally, other ancillary services can be defined such as limitation of fault and inrush currents, power oscillation damping, fast voltage support and limitation, and so on, but they have to be defined by each responsible regulator dependent on grid needs.

Although this book focuses on DC grids, the general provision of ancillary services will change strongly if DC grids arise. The reason is an additional dimension of needs and provisions for ancillary services. Currently, ancillary services are well-defined for AC systems. ENTSO-E, the European network of transmission system operators, is developing a network code on "HVDC connections and DC connected power park modules" [12]. Ancillary services are described, but the focus lies on ancillary services from DC equipment for AC grids. However, more ancillary services

[2]The naming and definition of different ancillary services may vary depending on the system that is considered.

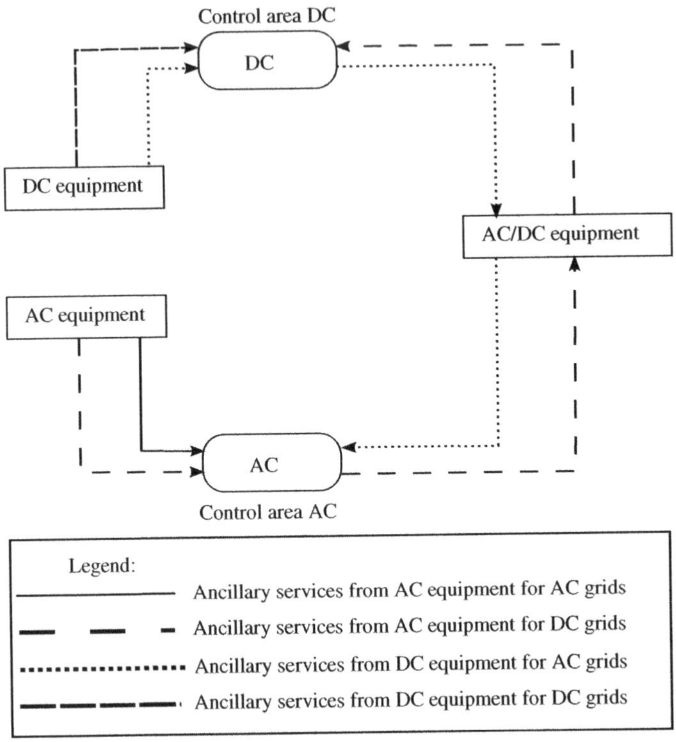

Figure 10.13 Overview of the four ancillary services/grid combinations in a power system with HVDC grids [13].

will be required for the DC grid. They can be defined from the DC system to the AC system, from the AC system to the DC system, and between different DC systems as well (see Figure 10.13).

10.6.1 Different Ancillary Services Defined

The ancillary services that were previously listed for AC systems can be defined as well with respect to DC grid operations. While some services are limited to either AC or DC systems, often similar services exist for both AC and DC systems.

Active Power Balancing. Received the most attention so far in the literature, and was explained in greater detail in Section 10.4, while the implementation of the control is described in Chapter 15. The active power balancing will require different types of reserves in order to work successfully:

Primary Reserves. Fast acting and restores the energy balance at the DC grid side.

Secondary Reserves. Replacement of the primary reserve and increases or decreases the energy balance indicator back to the set-point after large outages, and during normal conditions they compensate small or slow deviations from the power set-points.

Transmission Reserve. Services which retain a margin on the system in order to provide services from other grids—for instance, power reserves. If primary reserve and/or secondary reserve is provided via another grid, this reserve has to pass through it.

Power Flow Control or Redispatch. Service which the operator can use to shift power between converter stations. This can be used to alleviate congestion or to increase reliability (Section 10.3). The correct set-points can be an optimal power flow (OPF) in the AC, DC, or both systems by controlling the converters power injection. A method to calculate the optimal power flow for combined AC and DC grids is presented in references 14 and 15 and in Chapter 14. The power flow control service can be defined for both the DC system and the AC system, depending on the control modes and responsibilities of the converter stations.

Reactive Power/Voltage Control. Service of the DC grid to other systems as the converters have the ability to provide a wide range of reactive power at any node in the system.

Rotor Angle Stability-Related Control. DC grids can aid in managing dynamic interactions in the connected systems through, for instance, power oscillation damping (POD) and even between systems that are operating asynchronously, connected via the DC grid.

Loss Compensation. All losses which result in the transmission of electrical power have to be compensated by the responsible system operator. In a DC grid, those losses are the transmission losses over the lines and from other DC equipment, as well as the losses of the converter if they fall within the responsibility of the DC operator. If the power which is needed to compensate the losses originates from power plants which are located in the AC grid, additional losses in the AC grid and the converter stations may have to be taken into account.

Black Start and Restoration. There are two possible definitions for black start capability of DC grids: (1) the ability of a DC grid to re-energize the DC system after it has been disconnected and de-energized or (2) the ability of the DC grid to re-energize an AC system which has been de-energized and is connected to the DC grid which has remained healthy. Although the first option is the only “true” black start capability, it is the second one that is often referred in relation to DC systems.

In any case, it should be possible to restore the energy provision within the DC grid after a DC blackout, possibly assisted by an AC power system. The power source to perform the restart and the number of converters which are able to perform such actions strongly depends on the size of the DC grid. For instance, the requirements of a small offshore DC grid collecting power from some wind farms might differ significantly from those of a full-fledged DC grid.

In the following sections, the different ancillary services for the different transactions are listed. The actual implementation will depend on the regulatory framework and the respective grid codes that will be developed for the DC grids.

10.6.2 Ancillary Services from DC Equipment for AC Grids

Unlike the ancillary services for AC grids, ancillary services from DC equipment for AC grids are not formally defined yet. For existing point-to-point HVDC connections, ancillary services are usually defined on a case-by-case base. These ancillary services have also been introduced in literature. In reference 16, a number of ancillary services are identified. In reference 12, requirements for HVDC connections are listed, which can be considered to be specifying ancillary services. Additionally, a DC transmission reserve and DC power flow control can be defined [13].

A nonlimited set of services that DC equipment can deliver to AC grids are:

- Active power/frequency control
  - Primary control or frequency containment reserve (FCR)
  - Secondary control or frequency restoration reserve (FRR)
- Reactive power/voltage control
- Rotor angle stability-related control
  - Avoiding loss of synchronism
  - Damping electromechanical oscillations
- Black start and restoration
- DC transmission reserve
- DC power flow control

10.6.3 Ancillary Services from AC Equipment for DC Grids

Likewise, a DC grid requires at least services which manage the following:

- Active power/voltage control
- DC loss compensation
- DC black start and restoration
- AC transmission reserve
- AC power flow control

10.6.4 Ancillary Services from DC Equipment for DC Grids

Interconnected DC grids, or rather DC grids that are split into multiple operational zones, will require services from each other, potentially indirectly from equipment placed in the AC grid. Also, equipment placed in the DC grid can provide such services: the DC voltage control of the converters, DC choppers, DC storage, and other equipment directly connected to the DC grid.

- Active power/voltage control
- DC loss compensation
- DC power flow control
- DC black start and restoration

10.7 GRID CODES

Grid codes or network codes are sets of rules which apply to one or more parts of the energy sector and define the boundary conditions for the interactions between stakeholders. The grid code were traditionally developed on a national level. Currently, the increase in international trade is pushing a more internationally harmonized approach to grid codes. In Europe, the EU regulation foresees the development of such codes by ENTSO-E [17]. These codes are approved by ACER and the national regulatory agencies.

The grid codes can be subdivided into requirements for the connection of new equipment to the system, rules for the planning and operation of the system, and rules for the manner in which the market is operated. At this moment, HVDC connections are, in general, subjected to the AC grid code, with custom extensions to the code representing the connection and operation agreements of the AC and DC system. ENTSO-E is currently developing a network code that aims at harmonizing the connection of HVDC systems to the AC grid [12].

When HVDC grids are developed, the need will arise to extend the existing grid codes with HVDC grid-related articles and even to develop new grid codes that are specifically aimed at the HVDC grid, addressing the connection and operational rules from the HVDC grid perspective. The combined set of grid codes will determine which equipment can be connected to a particular HVDC grid and how it should operate (under different conditions). Cigré working group B4-56 is drafting a first set of guidelines for the development of such grid codes [18].

10.8 CONCLUSIONS

This chapter provides an overview of the operation of a hybrid AC/DC system. It is important to recognize that both systems influence the flows in both systems. The hybrid AC/DC grid offers significantly more flexibility to the operator as the active and reactive power flows in the system can be managed through adjusting the controls of the converters. The available system assets can therefore be used closer up to the limits while retaining a similar level or higher level of reliability and retaining the scheduled energy transfer.

The manner in which the system is operated depends on who operates the system. A system-wide optimal operation is possible when considering a single system operator. However, in systems with multiple operators at AC and DC it is increasingly important to coordinate the operations in the different time frames as the controllability can be both a positive and negative contribution to system reliability. When optimally operated, the HVDC grid can help to manage uncertainties in the system.

Balancing the energy flows in the system is more difficult in HVDC grids and therefore will require rapid control actions which influence the power flows in the system.

This chapter paid special attention to the voltage support from HVDC terminals to the AC system and the corresponding power capability curve. Voltage support can be used at any converter station for steady-state operations, and the fast control actions of the converter stations can introduce additional dynamic support to the AC system. Also, active power can be used to dampen power oscillations. However, those power swings will be seen in the DC grid and throughout the connected grids.

The introduction of HVDC grids will lead to the definition of ancillary services. Next to the traditional ancillary services, new services are needed between DC equipment and the AC grid, between AC equipment and DC grid, and between DC grids. These ancillary services are likely introduced into new grid codes for systems with HVDC grids, together with rules that govern the operation, connection requirements, and market operation.

REFERENCES

1. H. Ergun, J. Beerten, and D. Van Hertem, "Building a new overlay grid for Europe," in *Power and Energy Society General Meeting, 2012 IEEE*. New York: IEEE, 2012, pp. 1–8.
2. L. Vanfretti, D. Van Hertem, and J. O. Gjerde, "A smart transmission grid for Europe: Challenges in developing grid enabling technologies," in *Smart Grids and Sustainable Energy Transformation*. Berlin: Springer, 2014.
3. M. Aguado, R. Bourgeois, J. Bourmaud, J. Van Casteren, M. A. Ceratto, M. Jäkel, B. Malfliet, C. Mestdag, P. Noury, M. Pool, W. Van Den Reek, M. Rohleder, P. Schavemaker, S. Scolari, O. Weis, and J. Wolpert, "Flow-based market coupling in the Central Western European region—On the eve of implementation," in *Cigré sessions*, no. C5-204. Cigré, Aug. 2012.
4. P. Panciatici, G. Bareux, and L. Wehenkel, "Operating in the fog: Security management under uncertainty," *Power and Energy Magazine, IEEE*, vol. 10, no. 5, pp. 40–49, 2012.
5. D. Van Hertem, R. Eriksson, L. Söder, and M. Ghandhari, "Coordination of multiple power flow controlling devices in transmission systems," in *AC and DC Power Transmission, 2010. ACDC. 9th IET International Conference on*. London: IET, Oct. 2010.
6. D. Van Hertem, J. Rimez, and R. Belmans, "Power flow controlling devices as a smart and independent grid investment for flexible grid operations: Belgian case study," *IEEE Transactions on Smart Grids*, vol. 4, no. 3, Nov. 2013.
7. "Operation Handbook," 2012 [Online]. Available: https://www.entsoe.eu/publications/system-operations-reports/operation-handbook
8. NERC, "BALANCING AND FREQUENCY CONTROL," North American Electric Reliability Corporation, Technical Report, Jan. 2011 [Online]. Available: http://www.nerc.com/comm/oc/rs; landing pagedl/relatedfiles/nercbalancingandfrequencycontrol040520111.pdf

9. A. Egea-Alvarez, J. Beerten, D. Van Hertem, and O. Gomis-Bellmunt, "Primary and secondary power control of multiterminal HVDC grids," in *AC and DC Power Transmission (ACDC 2012), 10th IET International Conference on*, Dec. 2012, pp. 1–6.
10. R. Elsen, B. Aouini, Z. Barócsi, T. Bonnet, A. Bouzoualegh, H. J. Compter, J. Galindo, M. Gomes, C. Ioanitescu, R. Jacquet, J. Jenko, G. Käck, J. Kasparu, D. Lozek, A. Mizan, M. O'Mahony, A. M. Papadopoulos, G. Sica, S. Spasojevic, K. Szynol, J. Tauschitz, J. Vaha-Piikkio, K. W. Hansen, and C. Zhang, "Ancillary Services; *Unbundling Electricity Products—An Emerging Market*," Union of the Electricity Industry—EURELECTRIC, Technical Report, 2003.
11. "Glossary of Terms Used in NERC Reliability Standards," Jan 2014 [Online]. Available: http://www.nerc.com/files/glossary_of_terms.pdf
12. "NC HVDC—PRELIMINARY SCOPE ," Aug. 2013 [Online]. Available: https://www.entsoe.eu/major-projects/network-code-development/high-voltage-direct-current
13. R. Renner and D. Van Hertem, "Ancillary services in electric power systems with HVDC grids," *IET Generation, Transmission and Distribution*, vol. 9, no. 11, pp. 1179–1185, 2015.
14. J. Rimez and R. Belmans, "A combined AC/DC Optimal Power Flow Algorithm for meshed AC and DC Networks linked by VSC Converters," *International Transactions on Electrical Energy Systems*, 2014.
15. R. Wiget and G. Andersson, "Optimal Power Flow for Combined AC and Multi-Terminal HVDC Grids based on VSC Converters," in *Power and Energy Society General Meeting, 2012 IEEE* . IEEE-PES, 2012.
16. Y. Phulpin and D. Ernst, "Ancillary services and operation of multi-terminal HVdc systems," in *International Workshop on Transmission Networks for Offshore Wind Power Plants*, Oct. 2011.
17. European Network of transmission system operators for electricity (ENTSO-E). (2014) Network codes for electricity [Online]. Available: http://networkcodes.entsoe.eu/
18. Cigré WG B4-56 (Convenor: Philippe Adam & Kamran Sharifabadi), "Guidelines for the preparation of "connection agreements" or "grid codes" for multi-terminal HVDC schemes and HVDC grids," Cigré, Technical Report, 2015, in preparation.

CHAPTER 11

OPERATION AND CONTROL OF OFFSHORE WIND POWER PLANTS

Oriol Gomis-Bellmunt[1,2] *and Mònica Aragüés-Peñalba*[1]
[1]Research Group CITCEA, Electrical Engineering Department, Technical University of Catalonia (UPC), Barcelona, Spain
[2]Research Group Wind Energy, Electrical Engineering Department, Catalonia Institute for Energy Research (IREC), Barcelona, Spain

ONE OF the main triggers for the development of offshore grids is the need for transmission systems for the increasing number of offshore wind power plants (WPP). When connected to HVDC transmission systems, offshore wind power plants are not synchronous with the main terrestrial grid and therefore the operation and control requirement differ from those of AC connected WPP. The offshore grid must be established and maintained by means of the VSC HVDC rectifier in close coordination with the wind turbines. Furthermore, the WPP are expected to provide support to both the HVDC transmission system and the terrestrial grid or grids according to the relevant grid codes [1–5]. The present chapter introduces the main functional requirements for HVDC connected WPP and describes some application examples.

11.1 INTRODUCTION

The operation of offshore wind power plants addresses the coordination and integration of WPP- and VSC-HVDC converters. The coordination of controllers and protections is a fundamental issue for VSC HVDC connected offshore wind power plants. An appropriate integration will often deal with different manufacturer technologies, and therefore it is extremely important to clearly specify how this integration can be conducted in a given installation [6].

The present chapter considers point-to-point VSC-HVDC connections integrated with wind power plants based on any wind turbine electrical concept, typically doubly fed induction generators (DFIG) or full power converter (FPC). The

HVDC Grids: For Offshore and Supergrid of the Future, First Edition. By Dirk Van Hertem, Oriol Gomis-Bellmunt, and Jun Liang.

considered scenarios include normal and fault operation. As far as the requirements are concerned, they include the start-up and the normal operation of the WPP, the eventual active power curtailment, the fault ride through capability, and the onshore grid support. The onshore grid support will be mainly implemented in the grid side VSC-HVDC converter, but it will require actions from the WPP.

11.2 SYSTEM UNDER ANALYSIS

The considered system is sketched in Figure 11.1. Detailed discussions on possible WPP arrangements are included in Chapter 6. The analyzed system considers a VSC-HVDC converter which is installed in a platform and connected to two transformers located in two platforms where they collect the power from several strings of wind turbines. Each transformer collects the corresponding nominal power in normal operation; however, the transformers will be typically overrated to allow to operate the WPP at partial load in certain circumstances with only one transformer. Reactive power compensation equipment may be considered depending on the configuration. Reactive power is also provided by the VSC-HVDC converter and the wind turbines. DC choppers will be considered in all the wind turbine converters and the VSC HVDC.

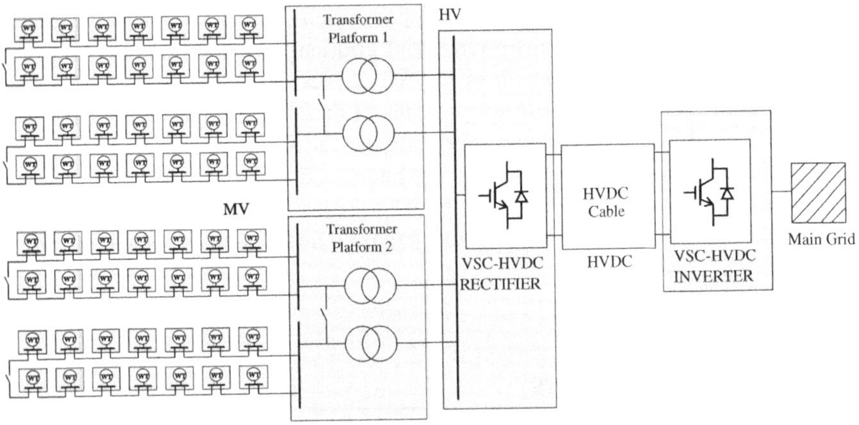

Figure 11.1 Offshore wind power plant connected to a VSC-HVDC unit.

11.3 CONTROL AND PROTECTION REQUIREMENTS

11.3.1 Requirements

The control and protection integration of the WPP and the VSC HVDC will be required to provide the following functions:

- The offshore grid frequency must be controlled at the desired value. This task is performed by the VSC-HVDC rectifier which imposes the offshore grid frequency.

- Reactive power management. The management of reactive power is strongly linked to the voltage control in both the wind turbines and the VSC HVDC. The WPP dispatches reactive power references to the wind turbines. The offshore grid voltage must be maintained in appropriate levels in all the wind turbines and in the VSC-HVDC converter.
- An appropriate start or stop sequence must be applied considering issues such as the transformers inrush current and the possible voltage transients. Auxiliary power and communication system are needed for the start-up and shut-down sequences.
- Active power management. Active power will be managed in order to extract the maximum possible power for low wind conditions, extract the nominal power in high wind conditions, and reduce power when needed or required by the grid operator.
- Fault ride-through capability.
- Main grid support. VSC-HVDC connections are expected to provide support to the grid where they are connected:
  - Voltage support will depend mainly on the grid side VSC-HVDC inverter.
  - Frequency support requires coordinating the VSC-HVDC inverter, the VSC HVDC rectifier, and the WPP. The WPP VSC-HVDC rectifier will be demanded to reduce or increase power for onshore grid frequency support. In this case, one possibility is to change via communications the active power command or to mimic the onshore frequency on the offshore grid in order to obtain the same response from the WPP as it would perform in an onshore system. In any case, communications are needed.
- Power system stabilizer capability.
- Operation under communication failure.
- Provide auxiliary power to the WPP (1–2%) when there is no wind.

11.3.2 Possible Scenarios

In order to be able to specify the control and protection functions required, the following situations are considered[1]:

- **Normal Operation.** The WPP is operated without any electrical restriction. In this case all the generated power should flow through the WPP grid and be injected in the VSC-HVDC cable. This situation is shown in Figure 11.2, where the green lines show the power flow.
- **Restricted Operation.** Due to eventual problems or maintenance operation, part of the electrical system might not be available. For example, one of the two transformers of the platform may be unavailable. Coordinated control will

[1]To illustrate the different situations, the following color code is used: green in used for systems operating in normal conditions, yellow is used for systems operated in restricted conditions, and red is used for systems which are faulted or are the problem origin.

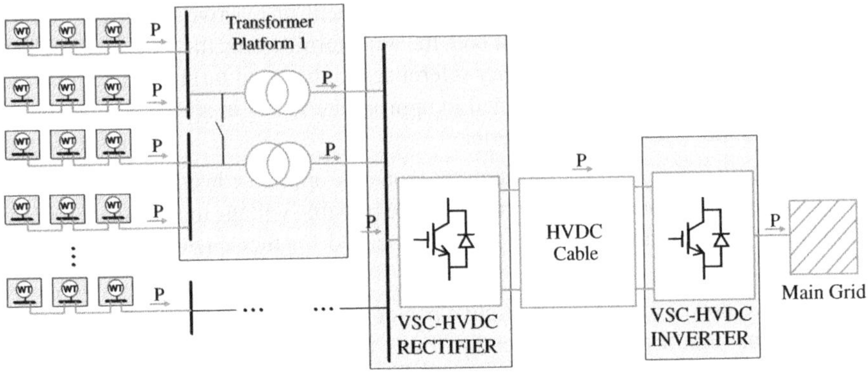

Figure 11.2 Normal operation.

be required to limit the generated wind power. As shown in Figure 11.3, the generated power will have to be reduced, but the operation of the system is possible.

- **Main AC Grid Fault.** A fault in the main AC grid reduces drastically the AC voltage in the VSC-HVDC inverter and therefore reduces the power extraction capability. These faults can be addressed providing fault ride-through capability by an appropriate use of a DC chopper in the grid-side VSC converter as shown in Figure 11.4.
- **Reduced Power Operation.** The grid operator may send power references to the WPP in order to reduce power, mainly due to congestion management issues. In this case a direct communication will be sent from the grid operator to the WPP which will dispatch reduced power references to the wind turbines (Figure 11.5).
- **Frequency Response.** In the case of frequency variations, the active power from the WPP may have to be changed according to the grid-code requirements.

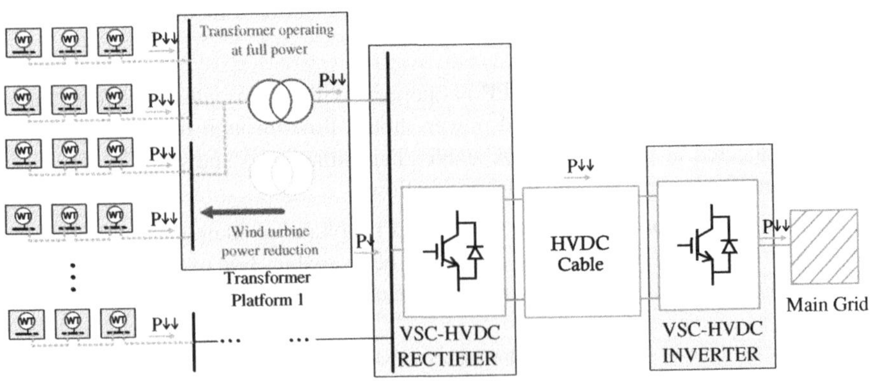

Figure 11.3 Restricted operation.

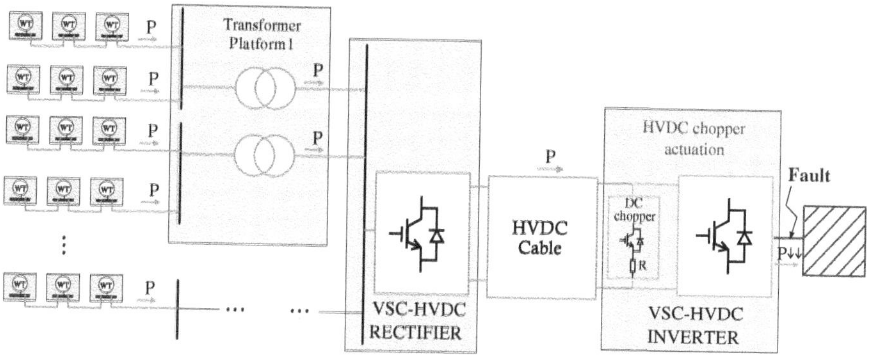

Figure 11.4 Fault in the main AC grid.

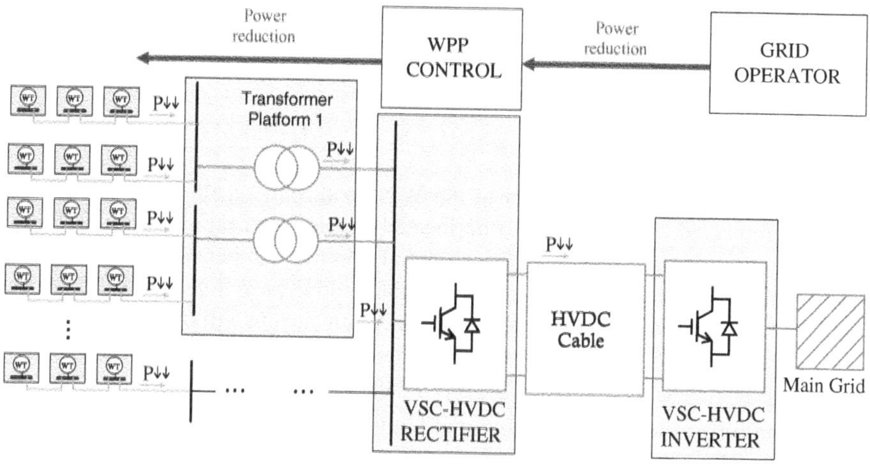

Figure 11.5 Reduced power operation.

In this case there are two possibilities (both assuming that the grid-side VSC HVDC is continuously measuring the onshore grid frequency):

- A direct communication is sent from the grid-side VSC HVDC to the WPP (containing the onshore grid frequency or the required WPP active power). The WPP control dispatches power references to the wind turbines (Figure 11.6).
- A direct communication is sent from the grid-side VSC HVDC to the wind power plant VSC HVDC rectifier which changes the offshore grid frequency to mimic the onshore grid frequency. The wind turbines react accordingly modifying the active power injected (Figure 11.7).

- **DC Fault.** A fault in the VSC HVDC or one of the VSC HVDC converters will imply disconnection of the entire system (in a point-to-point scheme[2]).

[2]For more complex topologies, see Section 8.2.

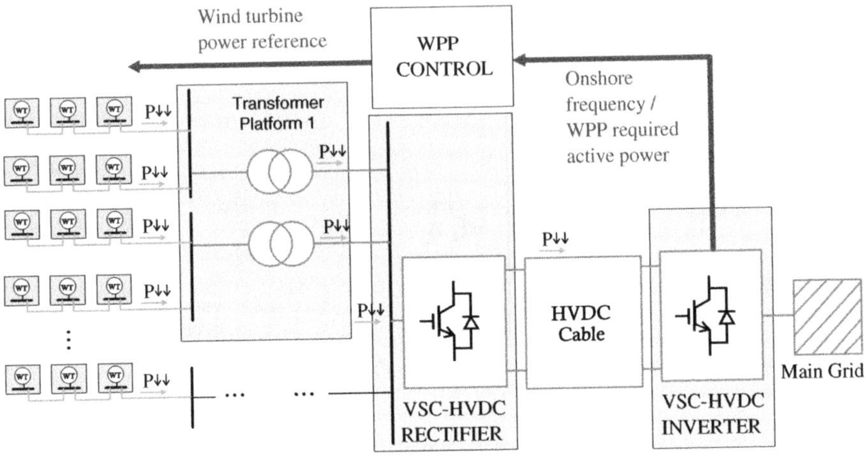

Figure 11.6 Frequency response: Direct communication from the grid-side VSC HVDC to the WPP.

- **WPP Grid Fault.** For faults in the WPP grid, part or all the WPP will be disconnected, the fault has to be located and isolated rapidly (Figure 11.8). The disconnected part will depend mainly on the number of circuit breakers available in the WPP grid and the available redundancy. Since the WPP grid is an only power electronics grid, the short circuit current will be provided by the several VSC converters, but mainly by the HVDC unit. It must be ensured that the relay protections operate appropriately.
- Wind turbine internal fault: Internal faults in the WT will imply the WT disconnection.

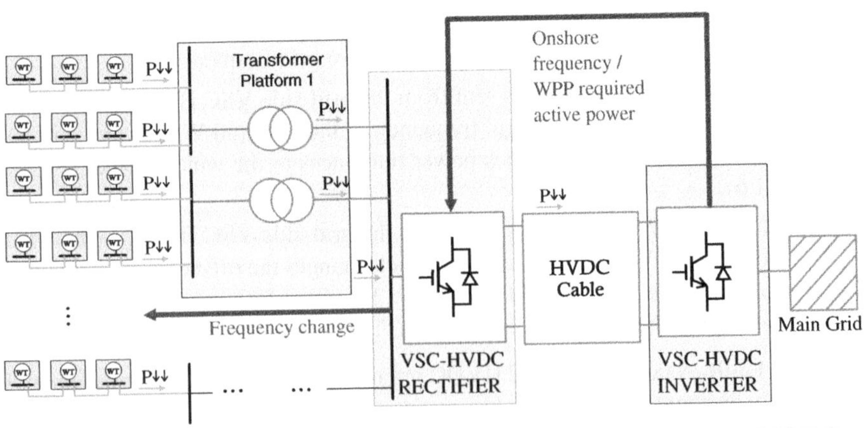

Figure 11.7 Frequency response: Direct communication from the grid-side VSC HVDC to the wind power plant VSC-HVDC rectifier.

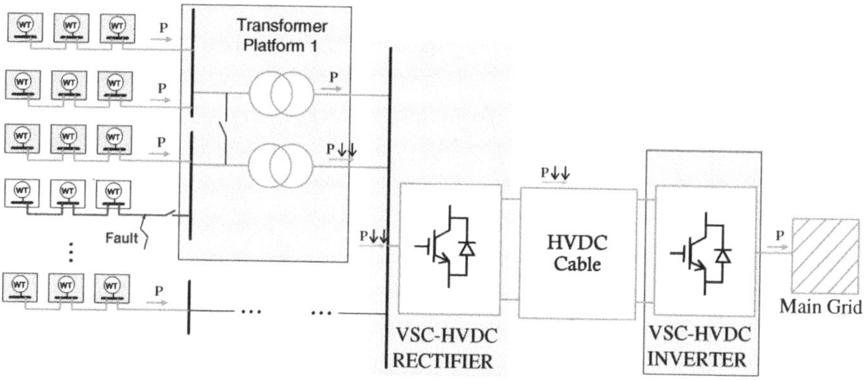

Figure 11.8 Fault in the WPP grid.

11.4 WIND POWER PLANT CONTROL STRUCTURE

The typical control structure for a wind power plant is hierarchical and can be divided into two levels: the control applied inside each wind turbine (local control or control at wind turbine level) and an outer control, applied to the whole wind power plant (global control or control at wind power plant level) [7]. The interaction between the wind power plant control and each wind turbine control is possible thanks to SCADA systems [8]. The SCADA ensures the supervision and control of the wind power plant as well as the acquisition of the data of its different components. Their specific functions include the real-time data capture, recording and management, data transmission among different devices, and data display through human machine interface (HMI). The following sections explain the control present in each generating unit of the wind power plant (wind turbine control) and the coordinated control of all the wind turbines (wind power plant control). Finally, an example in a 90-MW wind power plant is shown through simulations in DIgSILENT Power Factory®.

11.4.1 Wind Turbine Control

When a variable speed wind turbine is connected to the grid, operation at partial load and full load operation can be distinguished. Between these two operating zones (indicated by *I* and *III*, respectively, in Figure 11.9), there exists a transition (indicated by *II*, in Figure 11.9). The main characteristics of each operating mode are:

- **Partial Load (*I*).** Electrical power is being produced below the rated power. The objective is to maximize the power generation by means of MPPT (maximum power point tracking) algorithms. The pitch angle is adjusted to an optimum value and the generator control adapts the rotating speed.
- **Transition Between Partial Load and Full Load (*II*).**
- **Full Load (*III*).** Power is being limited through pitch control and at constant rotating speed. In this operating region, wind speed is above rated value and the generator is working at full load. The aerodynamic control (pitch control) and

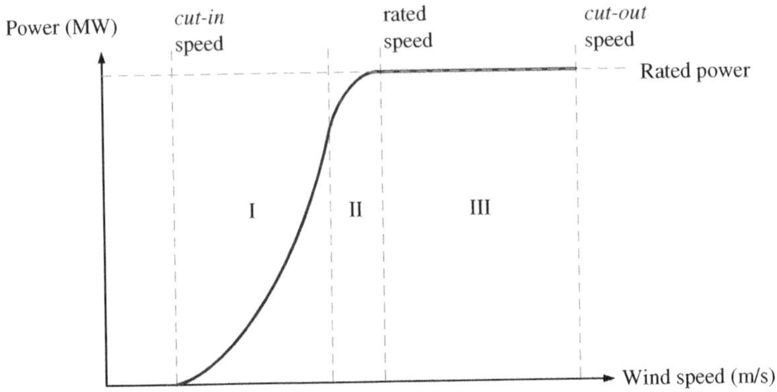

Figure 11.9 Main operating regions of a variable speed wind turbine.

the generator control which controls the rotating speed ensure the tracking of rotating speed and power values, set by the supervision system while limiting them to maximum values that could damage the wind turbine.

The wind turbine controls guarantee its correct performance along the before mentioned operating zones, and they can be classified according to the hierarchy reflected in Figure 11.10. The high control level includes controls of the mechanical and aerodynamics quantities, and a low-level control deals with the power converter and generator control.

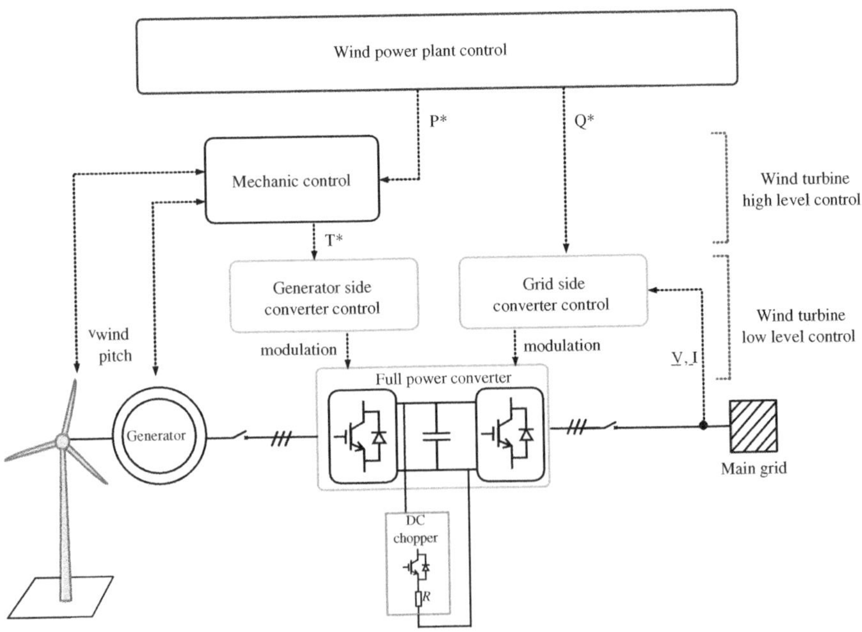

Figure 11.10 Control of a wind turbine equipped with full power converter.

11.4.1.1 Mechanic Control Mechanic (and aerodynamic) control optimizes the wind power caption through MPPT algorithms, reduces the overall wind turbine mechanical loading, orients the wind turbine nacelle in the wind speed direction using the yaw mechanism, and is in charge of reducing the power capture when necessary by means of pitch or stall systems. Thanks to the power control, the wind turbine generates electric power above the cut-in speed (around 3 m/s) and stops generating when the wind speed is too high and can damage the stability of the wind turbine (cut-out speed, close to 25 m/s). The main outputs of the mechanic control are (a) the generator reference torque which is sent to the low level control and (b) the pitch and yaw angles.

11.4.1.2 Converter Control Converter control includes the generator side converter control and the grid side converter control. The machine side converter regulates the current through the generator in order to obtain the needed torque demanded by the mechanical control, as well as the reactive power exchanged with the generator. The grid side converter control regulates the DC bus voltage to a desired value and the reactive power to a defined set-point. To do so, each converter control applies the necessary voltages at the AC side by means of a proper modulation strategy [9].

11.4.2 Wind Power Plant Control

Being at a higher hierarchy level than the wind turbine control, the control at wind power plant level presents slower dynamics. It permits the wind power plant integration to the main grid according to the TSO requirements and, consequently, leads to the participation of the wind power plant on primary and secondary controls, approaching its behaviour to the conventional power plant. The main objectives of the wind power plant control are the control of active power and reactive power at the PCC (point of common coupling).

11.4.2.1 Active Power Control The balance between active power generation and demand in a system ensures that its electrical frequency is kept within certain limits. Conventional power plants contribute to frequency regulation through primary and secondary strategies by varying the power they produce. Taking into consideration the dimension that wind power plants are acquiring, they are also required to participate in frequency regulation [5, 10–13]. When the WPP are connected to offshore grids, they also have to participate in the frequency control of the terrestrial system where they are connected. According to recent grid codes, wind power plants should also operate at a lower production so as to have a reserve to increase power output in case of underfrequency operation. Similarly to the behavior of conventional power plants, the power variation in WPP generally is required to respond to a frequency droop function, which changes the power output according to the grid frequency. On the other hand, even if the grid frequency does not change, the operator can ask for a specific active power set-point in the point of connection. The evolution from the WPP operating point to new operator set-points is usually required to follow a ramp rate (imposed by the TSO) which defines the temporary increase

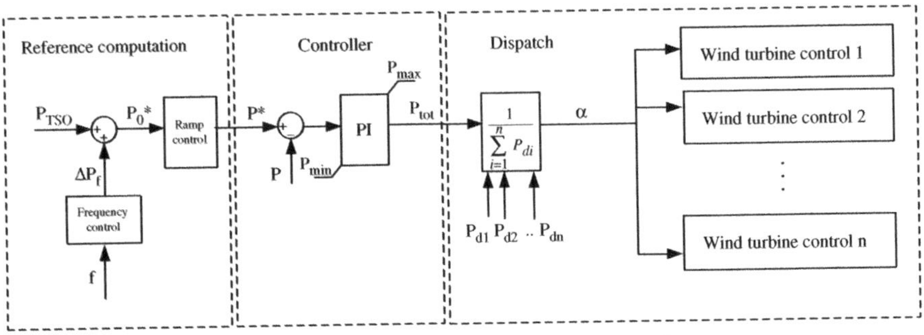

Figure 11.11 Active power plant control.

or decrease of power (generally MW/min). Therefore, the active wind power plant control must ensure the active power regulation in wind power plants to accomplish power curtailment, ramp rates, and frequency response. Further requirements related to active power include virtual (or synthetic) inertia emulation or power oscillation damping capability. Both requirements imply a fast response from the wind turbines and therefore require an appropriate control implementation in the wind turbine and wind power plant controllers.

An example of a possible interaction between the active power control, the AC grid, and the wind turbines is reflected in Figure 11.11. The operator sends an active power reference (P_{TSO}) to the active power control, which also receives the value of active power being produced by the power plant (P). Additionally, the grid electrical frequency must be known ($f_{onshore}$). Based on this data, the active power control computes an active power reference signal (α) to be sent to the wind turbine control.

In order to implement this control, a communication system is needed between the power plant controller and each generating unit. In case communications are bidirectional, the power plant controller can receive information of the operating point of each wind turbine, P_{di}. This allows to know the total active power available at the power plant instantaneously (by adding the capabilities of all the wind turbines) and to adjust the reference sent to each wind turbine accordingly.

11.4.2.2 Reactive Power Control For HVDC-connected wind power plants, the TSO reactive power requirements are ensured by the VSC-HVDC inverter/s. The reactive power plant control main goal is to maintain the voltage at the point of connection of the wind power plant and in all the wind turbines. A possible implementation of reactive power plant control is sketched in Figure 11.12, based on [14]. The control receives a reactive power reference or voltage reference to accomplish at the point of connection of the wind power plant and the reactive power being produced by the wind power plant (Q), measured at the wind power plant substation. The output of the PI controller is the new reactive power reference, Q_{tot}, needed to produce from the wind turbines, corrected in case reactive power compensations have been connected.

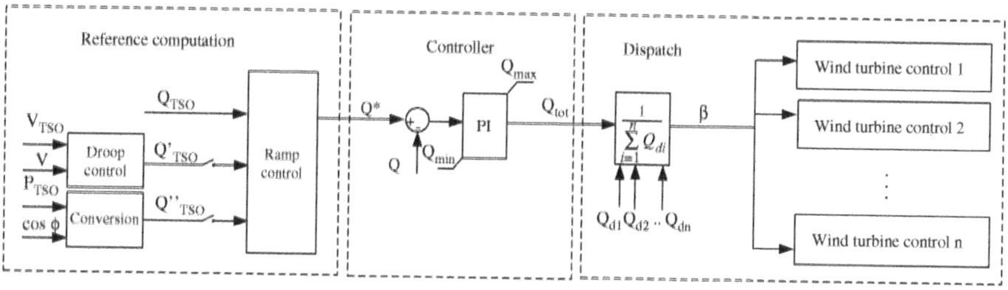

Figure 11.12 Reactive power plant control.

This new reactive power reference is divided by the total available reactive power in the wind power plant, so as to obtain a percentage to be sent to each wind turbine control, β. Similarly to the strategy implemented for active power control, the signal sent to each wind turbine is the same and each wind turbine control adjusts this percentage to a reactive power reference taking into consideration the available reactive power (Q_{di}).

11.5 DYNAMIC SIMULATION OF A SIMPLIFIED EXAMPLE

In order to show an example of active power control, a simplified system has been implemented in DIgSILENT Power Factory® for a 90-MW wind power plant, based on 18 PMSG, rated to 5 MW each. The wind power plant grid voltage level is 33 kV and it is connected to the main grid through a point-to-point HVDC of ± 150 kV. The operation scenario simulated is the one presented in Figure 11.5: reduced power operation. The wind farm is sketched in Figure 11.13 and the load flow for one of its wind turbines and the HVDC transmission system are shown in FIgures 11.14 and 11.15, respectively.

As shown in Figure 11.16, the wind power plant is generating nominal power until time instant 10 s. At this moment, the TSO asks for a lower active power set-point at the PCC: 85.5 MW. The active power ramp has a slope of ± 10 MW/min. Thus, the active power plant controller finds the appropriate active power signal to be sent to the wind turbines to reach the new set-point (represented in Figure 11.17), following a decrease of 10 MW/min. At time instant 50 s the TSO sends an active power set-point of 90 MW. The active power plant controller recalculates the new power signal needed and sends it to the wind turbines, guaranteeing a ramp rate increase of 10 MW/min. In Figure 11.18 It can been noted how the wind turbine control reduces the active power output until the TSO reference is reached, keeping it constant until a new set-point is required. This is also reflected at transmission level, in Figure 11.19, where the active power measurement on the AC side of the wind power plant converter and grid side converter evolves as the wind power plant power output does, ensuring the tracking of the TSO requests.

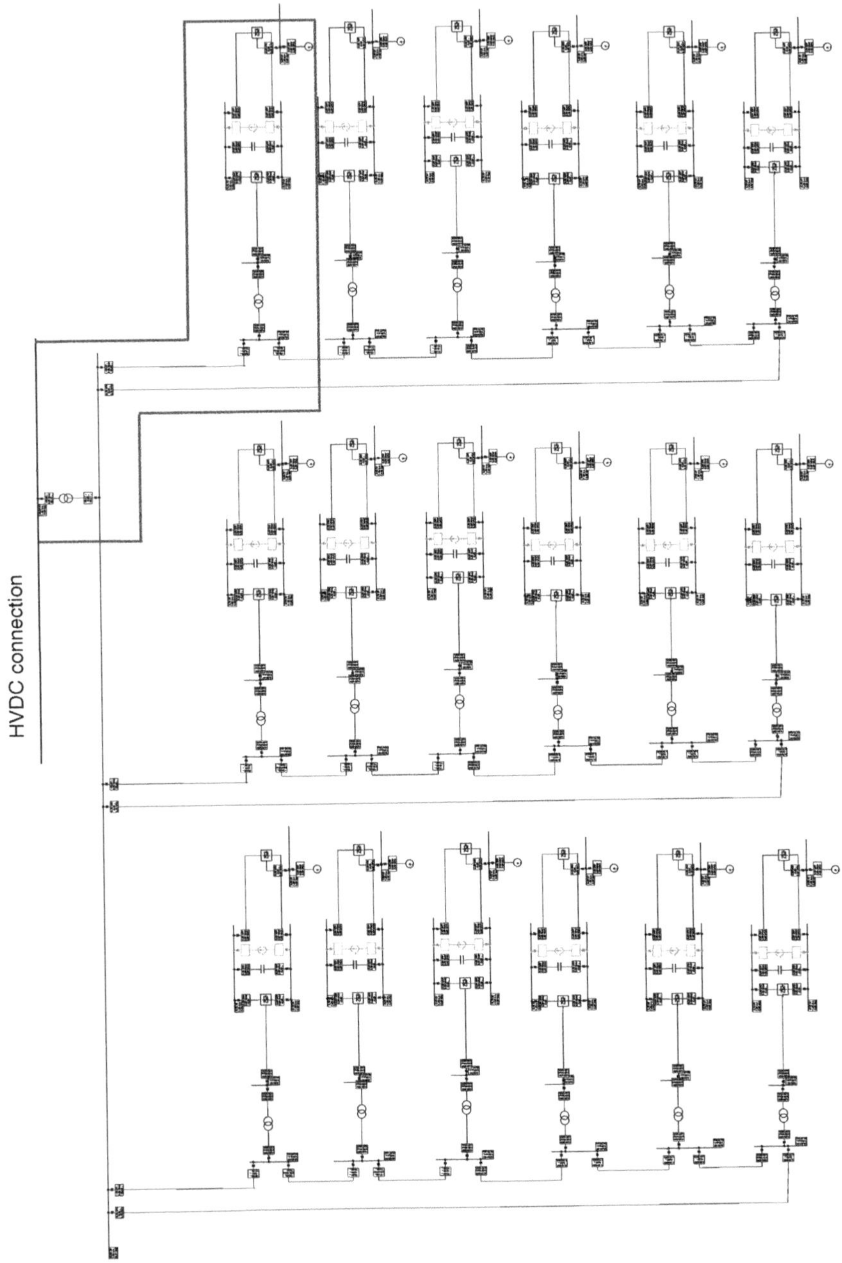

Figure 11.13 Layout of a 90-MW WPP represented DIgSILENT Power Factory®.

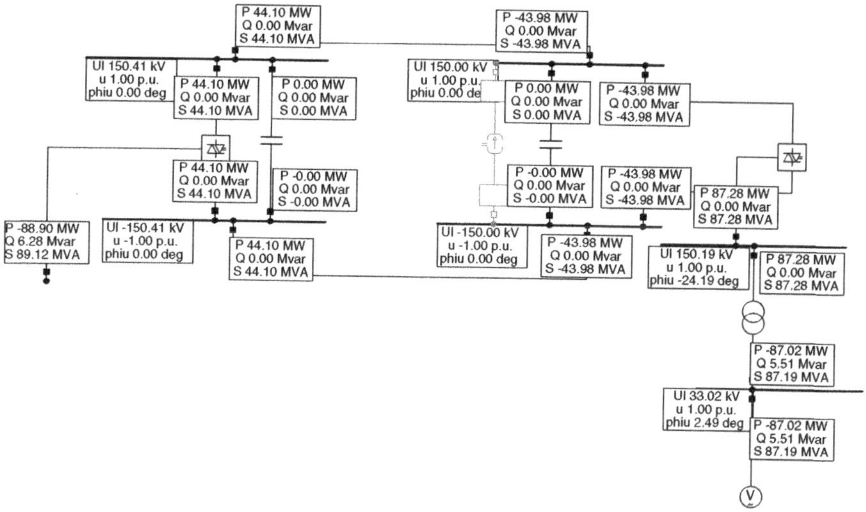

Figure 11.14 HVDC transmission load flow DIgSILENT Power Factory®.

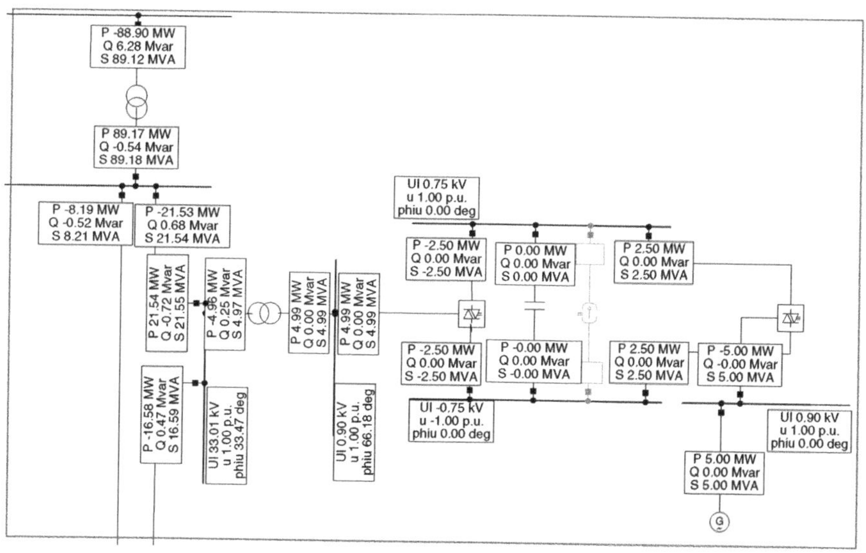

Figure 11.15 Wind turbine load flow in DIgSILENT Power Factory®.

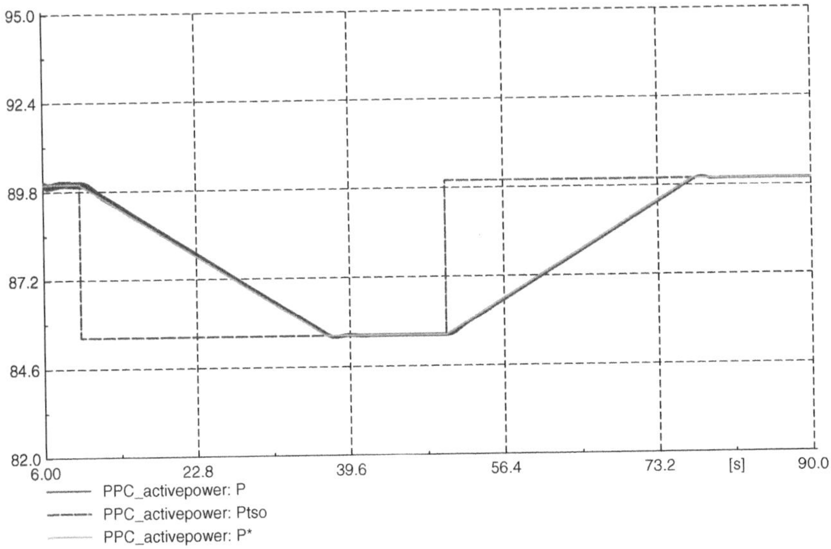

Figure 11.16 Active power reference sent by the TSO, ramp-limited, and active power measurement of the wind power plant output.

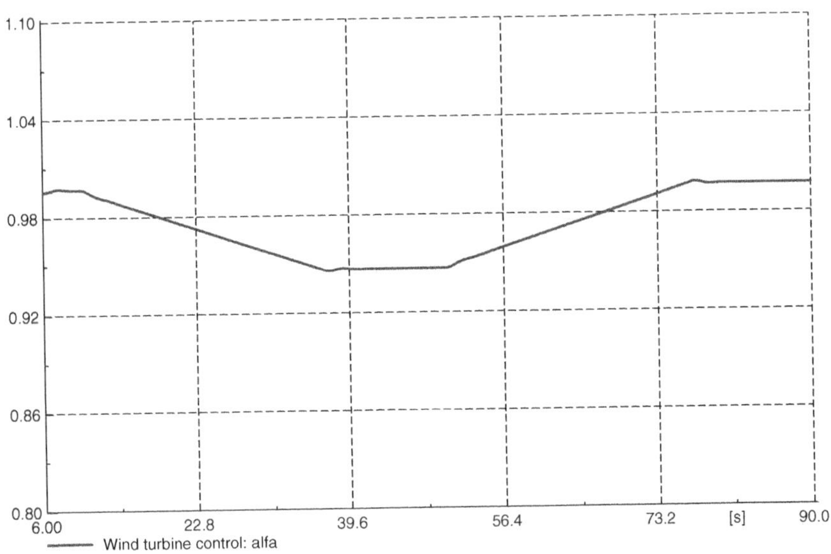

Figure 11.17 Signal sent from the power plant control to each wind turbine.

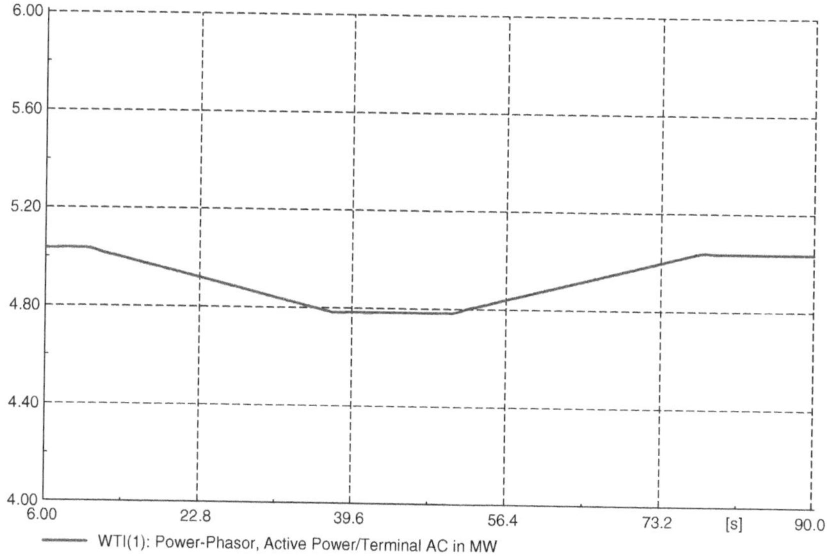

Figure 11.18 Active power output of a wind turbine inverter.

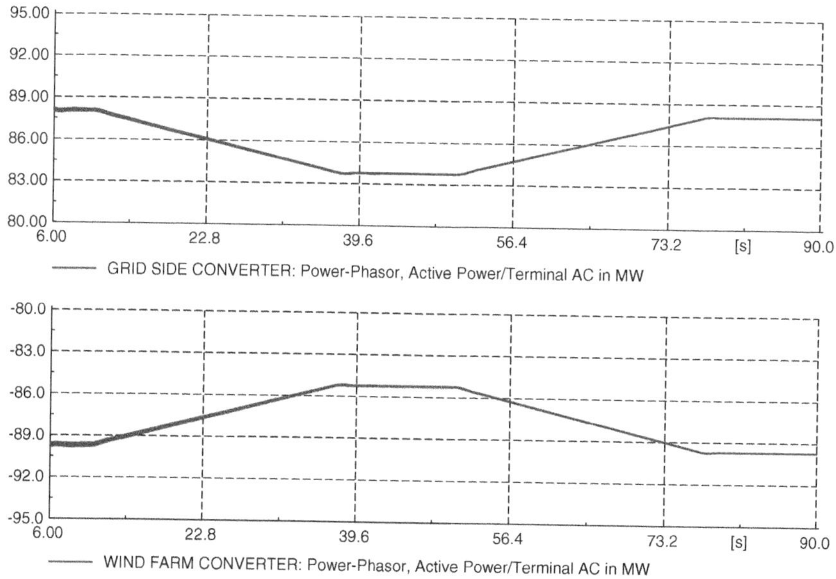

Figure 11.19 Active power measured at the offshore and onshore substations.

11.6 CONCLUSIONS

Offshore wind power plants connected to HVDC transmission systems are not synchronous with the main terrestrial grid. Therefore, such offshore wind power plants bring along novel and challenging requirements demanding strong coordination between the different power electronics converters of wind power plants: the VSC HVDC rectifier and the wind turbine converters. The offshore grid is maintained by means of the VSC HVDC rectifier, which is in charge of controlling voltage and frequency with appropiate coordination with the wind turbine converters. The wind power plants also provide support to the HVDC transmission system and the terrestrial grid or grids according to the relevant grid codes. The present chapter has introduced the most relevant requirements also highlighting different operation conditions in normal and fault condition. A simplified example is also introduced to show how the coordination between the different systems can be performed.

REFERENCES

1. C. Sourkounis and T. P., "Grid code requirements for wind power integration in Europe," *Conference Papers in Energy*, 2013.
2. M. Tsili and S. Papathanassiou, "A review of grid code technical requirements for wind farms," *IET Renewable Power Generation*, vol. 3, no. 3, pp. 308–332, 2009.
3. T. GmbH, "Requirements for offshore grid connections in the grid of Tennet TSO GmbH," 2010.
4. I. Erlich and U. Bachmann, "Grid code requirements concerning connection and operation of wind turbines in Germany," in *IEEE Power Engineering Society General Meeting*, vol. 2, 2005, pp. 1253–1257.
5. F. Díaz-González, M. Hau, A. Sumper, and O. Gomis-Bellmunt, "Participation of wind power plants in system frequency control: Review of grid code requirements and control methods," *Renewable and Sustainable Energy Reviews*, vol. 34, pp. 551–564, 2014.
6. Cigré, "Cigré B4-55 report HVDC connection of offshore wind power plants," Cigré, Technical Report, 2014.
7. E. Ela et al., "Active power controls from wind power: Bridging the gaps," in *NREL*, 2014.
8. B. Badrzadeh, "Wind power plant scada and controls," in *Power and Energy Society General Meeting, 2011 IEEE*, July 2011, pp. 1–7.
9. A. Egea-Alvarez, A. Junyent-Ferré, and O. Gomis-Bellmunt, "Active and reactive power control of grid connected distributed generation systems," in *Modeling and Control of Sustainable Power Systems*, ser. Green Energy and Technology, L. Wang, Ed. Berlin: Springer, 2012, pp. 47–81 [Online]. Available: http://dx.doi.org/10.1007/978-3-642-22904-6_3
10. Y. Jeong, K. Johnson, and P. Fleming, "Comparison and testing of power reserve control strategies for grid-connected wind turbines," *Wind Energy*, vol. 17, no. 3, pp. 343–358, 2014.
11. I. Erlich and M. Wilch, "Primary frequency control by wind turbines," in *Power and Energy Society General Meeting, 2010 IEEE*, July 2010, pp. 1–8.
12. H. Ma and B. Chowdhury, "Working towards frequency regulation with wind plants: Combined control approaches," *Renewable Power Generation, IET*, vol. 4, no. 4, pp. 308–316, July 2010.
13. J. R. Kristoffersen and P. Christiansen, "The Horns Rev wind farm and the operational experience with the wind farm main controller," *Wind Engineering*, vol. 27, no. 5, pp. 351–359.
14. M. Mata, J. Carulla, and O. Gomis, "Reactive power regulation of a windpark," *Alstom Wind, Patent application WO2012016585*, 2012.

PART IV

MODELING HVDC GRIDS

PART IV DEALS with the modeling of HVDC grids. Depending on the phenomena under study needed, different models are required. The relavant models for the different key system elements are introduced, and this for the different time frames.

The steady-state or power flow analysis of combined AC and DC systems is discussed first, considering different modeling approaches. Next, the optimal power flow for hybrid AC and DC systems is presented, including the application to relevant case studies. Security constraints are considered including preventive and curative measures.

The main control principles to ensure the voltage and power control of HVDC grids are described, including the introduction of a hierarchical control very similar to the AC system frequency control. The design of such controllers may require state-space (or equivalent) representation of the HVDC grids under consideration. A methodology to analyze such systems is developed and exemplified in a simple case study.

Faults in HVDC systems and the protection against them are one of the main technical challenges for the deployment of meshed HVDC grids and deserve particular attention. The analysis of the system in fault conditions requires specific detailed models of the cables and protections. Such models are analyzed, including simulation results on different fault conditions.

While computer-based simulation provides extremely useful information, some research may require experimental testing of part of the system under study in order to analyze some technical issues

HVDC Grids: For Offshore and Supergrid of the Future, First Edition. By Dirk Van Hertem, Oriol Gomis-Bellmunt, and Jun Liang.

associated with the real implementation of the proposed control schemes. Hardware-in-the-loop systems combine real devices with powerful real-time simulation allow the analysis of the system connected to new technologies and controls.

CHAPTER 12

MODELS FOR HVDC GRIDS

Jef Beerten and Dirk Van Hertem
Research Group Electa, Department of Electrical Engineering,
University of Leuven, Leuven, Belgium

12.1 INTRODUCTION

This chapter is an introductory chapter to Part IV. The objective is to highlight the need for different models within power system analysis and to indicate the specifics of HVDC grids with respect to modeling. The models themselves are introduced in greater detail in Chapters 13 to 18.

An electric power system consists of a number of different components (generators, lines, cables, transformers, capacitor banks, converters, etc.), each with their own operational limits and controllers. The overall power system can be modeled in terms of differential equations. In an electric power system, different time constants ranging from microseconds up to minutes and even days determine the dynamic system response dictated by these equations. This wide range introduces a number of challenges when it comes to modeling and controlling transmission systems. Including all time constants results in a very complex system to solve and is usually only feasible for a system relatively small in size. It is therefore common practice to focus on certain phenomena and develop a model which can be considered as a subset of the exact set of equations. In such a model, a number of simplifications are made: Interactions that are much slower than the event under study can be considered as constants, while the order of the model can be reduced further by modeling the interactions that are much faster as algebraic equations instead of differential equations, turning the mathematical description into a system of differential algebraic equations (DAEs). Next to a reduction in the temporal description of the mathematical model, it is also commonplace to reduce the system model by focusing on a particular area or subsystem and reduce the representation of the rest of the grid or remote areas.

12.2 POWER SYSTEM COMPUTATION PROGRAMS

As a result of these model reductions, different classes of power system programs and mathematical techniques have been developed to study power system interactions. A major classification be made based on the mathematical techniques used to get

HVDC Grids: For Offshore and Supergrid of the Future, First Edition. By Dirk Van Hertem, Oriol Gomis-Bellmunt, and Jun Liang.

insights in the system interactions. A general distinction can be made here between (a) time domain simulations and (b) routines to address the stability of the system by investigating the linearized state-space model. When time domain simulations are of interest, a distinction is traditionally made between electromagnetic transient-type programs (EMT), electromechanical stability programs and steady-state power flow programs. It has to be noted that other classes of program and methods have been developed as well, but within this chapter the focus will be on the three aforementioned classes.

As indicated by its name, EMT-type programs have historically mainly been used to simulate electromagnetic transients in the system—that is, changing currents and voltages that are a result of exchanges of the energy stored in the electric field in capacitances on the one hand and the energy stored in the magnetic field of inductances on the other hand. Typical examples of phenomena for which this type of programs has historically been used are switching transients, lightning impact, and transient overvoltages.

Electromechanical programs are designed to address the problems that can originate from the exchange of the kinetic energy stored in large rotating machines (primarily synchronous generators) and the power system. Typical phenomena studied with these types of programs are power oscillations and the associated transient stability of the system. A fundamental distinction between these two classes of programs, both intended to study dynamic interactions in the power system, is that in EMT-type programs the currents and voltages in the system are considered as state variables, whereas this is not the case for the electromechanical programs, where the periodically changing currents and voltages are typically modeled as quasi-stationary phasors. This simplification can be justified by the fact that the electromechanical energy exchange is largely dominated by relatively large time constants associated with the kinetic inertia of the generators. The timescales typically associated with these phenomena are several orders of magnitude higher than those associated with the energy exchanges between inductances and capacitances inside the AC system, and hence the faster dynamics of currents and voltages can typically be disregarded.

When the system dynamics are of no concern and one is only interested in the steady-state operation of the power system, the models can be reduced further by eliminating the temporal aspect from the analysis and by calculating the system using so-called power flow programs. The description of the system in this respect has been traditionally used for system contingency studies and for determining the optimal operation of the power system. A thorough discussion on the classification of different AC power system stability phenomena and the timescales involved is found in reference 1.

12.3 MODELING POWER ELECTRONIC CONVERTERS

What is clear from the previous discussion is that the AC system provides a relatively clear means to distinguish between slow and fast interactions and that two different classes of power system software originated from the corresponding model reduction: On the one hand, there are the EMT-type programs describing the currents and

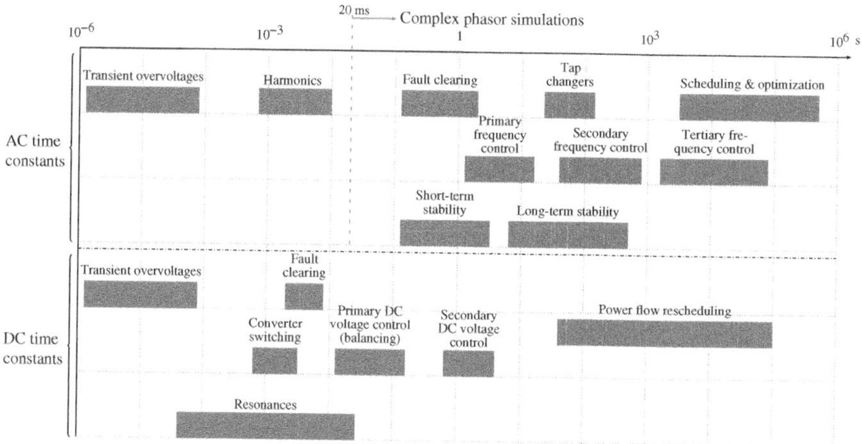

Figure 12.1 Overview of AC and DC grid dynamics [2].

voltages in the time domain, which causes a relatively high computational burden. On the other hand, the electromechanical programs start from the description of the system currents and voltages by means of quasi-stationary phasors. Using this quasi-steady-state representation, the focus of these programs inherently is on slower system dynamics of which the associated frequencies are lower than the fundamental frequency of the power system. For the sake of completeness, the concept of time-varying dynamic phasors, which is based on a Fourier approximation of the system quantities, should be mentioned here as a compromise to also include frequencies above the fundamental frequency while avoiding the high computational burden of a full time domain representation.

This rather distinct categorization for the power system phenomena based on whether their associated time constants are low or high compared to the fundamental period of the voltages and currents in the AC system has been graphically depicted in Figure 12.1.

The introduction of power electronic devices is challenging this approach to power system modeling. The switched nature of these devices and, consequently, the much higher bandwidth associated with the control loops make that the corresponding control actions can be very fast when compared to the timescales that traditionally have been associated with AC power system controller dynamics. This has been depicted graphically in the general classification of different dynamic phenomena and control dynamics in AC and DC systems in Figure 12.1.

12.3.1 Converter Models for Different Time Frames

Before addressing the details of the challenges raised by the introduction of HVDC grids to the picture, it is worth considering the different system models that were developed as a result of the introduction of power electronic devices in general and VSC HVDC specifically in the power system.

12.3.1.1 *Models for Electromagnetic Transient Studies* The detailed nature of the EMT approach, combined with the specific characteristics of the power electronic systems, make EMT-type programs an interesting choice for developing the models for these systems. Although the switched nature of the power electronic components can be challenging with respect to the representation of these components in EMT routines, the introduction of power electronics in the power system has given rise to new classes of models and software routines to simulate these components in detail. Using the most detailed models, the EMT approach thus allows us to accurately represent the switching dynamics and electromagnetic transients, but this modeling detail comes at a high computational cost. The main merit is that the VSC-HVDC system as such can be represented in great detail, allowing us to fine-tune control parameters and to investigate, among others, a detailed representation of new converter topologies [3, 4], as well as higher-frequency interactions, possible resonances with the local power system, DC fault analysis [5], and protection strategies [6]. The high computational burden makes these switching models less suitable for studies of large power systems. Furthermore, the models require a detailed knowledge of the converter topology, often down to the component level, which makes them all but generic. A major issue, however, is that such models and data are often not available and contain proprietary information since they can only be supplied by manufacturers of the equipment.

Alternatively, the converter model can also be simplified to an averaged model for use in EMT-type programs. Such averaged models are often encountered in power electronic simulations. One of the main aims when developing these models is to reduce the order of the system by eliminating the actual switching behavior of the components by only taking into account the overall, averaged result of the switched control action. As an example, a simple two-level converter can be simplified from a converter representation including switches in each converter arm to a voltage source in each phase, which produces a waveform that is mainly dictated by the DC voltage and modulation index. It is clear that, under steady-state operation, the voltage in the three phases will be sinusoidal and at grid frequency. Typical phenomena in HVDC grids that require EMT simulations include, for example, DC faults and converter switching. Chapter 17 discusses these models in greater detail.

12.3.1.2 *Models for Electromechanical Transient Studies* As discussed in the previous section, electromechanical stability programs allow us to simulate more complex power systems by representing the AC power system by phasors. When the modeling of VSC-HVDC systems is considered in this context, the description of the AC system in terms of fundamental frequency phasors justifies an approximation of the VSC that is similar to the way generators are modeled in the system. In other words, the observation made when discussing the technique of average modeling in power electronic software, namely that the converter will appear as a sinusoidal voltage source as regarded upon from the AC system, justifies the representation of the converter by an AC voltage source of which the amplitude and phase angle (or alternatively the d- and q-component) are controlled independently. Although switching models reveal more details, simplified and averaged models are well suited to study the VSC-HVDC system control response. Furthermore, these models are to a high degree independent of the actual converter topology. Typical applications of such models include the design of power oscillation damping (POD) control, or AC voltage dynamics. Chapters 15 and 16 discuss these models in greater detail.

12.3.1.3 Models for Power Flow Studies

Contrary to the previous classes of programs, power flow algorithms only represent the steady-state operation of the system after all transients have died out. The equations that are derived to study the power flows in the system can also be used in an optimal power flow (OPF) formulation, in either their most complete or most simplified form.

With respect to VSC-HVDC systems, power flow models allow us to study the overall effect of converter outages and, for example, the effect of a distributed voltage control on the AC and DC power flows. As dealt with later in this work in Chapters 13 and 14, in its most simple format, the differential equations describing the dynamic behavior of the converter can be simplified to power injections in the AC system.

12.3.2 Modeling of Other Grid Components for Different Time Frames

An aspect that has been left out of the discussion so far, but is equally important, is the appropriate modeling of the other components, such as cables, overhead lines, DC breakers, and dynamic breaking systems in the HVDC grid. It is worthwhile noticing that at present a lot of development is taking place in this area and a thorough discussion addressing all the developments would lead us too far for this introductory chapter.

As an example for different degrees of detail in DC grid models, we will rather take the DC cable as an example and shortly discuss the applicability of the models for the different classes of power flow studies. In general, a cable can be modeled in various ways:

1. Detailed frequency-dependent cable models.
2. (Cascaded) pi-sections including lumped cable resistance, inductance, and capacitance.
3. (Cascaded) pi-sections including lumped cable resistance and capacitance, but neglecting cable inductance, thereby eliminating line currents as state variables.
4. Lumped resistive element.

The modeling of the cable is important, since not using the appropriate model for a study might lead to either an oversimplification or an unnecessarily detailed description of the problem under consideration.

In EMT-type programs, the models that are typically encountered are frequency-dependent cable models. The model that is commonly used starts from a rational approximation by means of vector fitting of an analytic description of the cable geometry, yielding a similar frequency response [7]. This modeling approach results in a very accurate description of the cable frequency response. Such models are typically encountered in protection studies, but it is still unclear at the moment to what extent they are needed to model converter control interactions.

A commonly used alternative for studying control interactions is to use a lumped parameter representation, possibly cascading several line segments to increase the bandwidth of the model when using longer cables. Although a commonly used representation for control interactions, one has to beware of the fact that cascading several sections can create so-called "ringing" at the intersections [8] and might lead to false conclusions on, for example, the stability limits of the system.

The order of the model can be significantly reduced by also eliminating the cable inductances, since this removes the current as state variables from the equations. In this way, the HVDC grid thus simplifies to a network of capacitors interconnected by resistances, maintaining only the overall voltage dynamics of the system. Disregarding the capacitances as well also removes the voltages as state variables, which results in an all resistive network that can be used for steady-state power flow studies.

Although the applicability of this very simple representation is well-defined, the applicability of the other classes of models for the different classes of power system programs is not that well-defined yet. This specifically holds for models meant for inclusion in electromechanical transient software, since one of the fundamental assumptions of this class of programs is that the dynamics of interest can be described as phasors, thereby leaving out the AC voltages and AC currents as state variables. At the DC side, however, the overall system response and the control interactions between different converters is determined mainly by the DC voltages and currents. One has to consider to what extent the dynamics of the DC system need to be represented to obtain a valid overall system response, thereby taking into account the underlying simplifications of the steady-state phasor representation at the AC side.

12.4 HVDC GRIDS MODELING CHALLENGES

As the previous discussion already hints, the introduction of HVDC grids is challenging the way the power system and the different components have been modeled in the past. When compared to the traditional division of AC system dynamics, the phenomena in DC systems are a couple of orders of magnitude faster and pose a number of modeling challenges:

- The absence of a substantial level of energy storage makes DC voltage deviations much faster than AC frequency deviations, where the AC system inertia determines the rate of change of frequency. This results in quasi-instantaneous converter set-point changes when considered from a traditional AC system stability perspective.
- The absence of so-called "reactive" grid elements, caused by the presence of inductive elements in combination with a periodically changing system that limit the short-circuit current in AC systems, requires DC grid protection coordination to be a couple of orders of magnitude faster than traditional AC system protection.

- The introduction of new generations of VSCs using modular multilevel converters (MMC) with a large number of individually controlled modules poses challenges with respect to developing simplified, yet accurate converter models that can be used in system studies.
- An accurate representation of the dynamics of relevance for the DC voltage control is challenged by the introduction of inductors in present-day DC breaker designs, which, together with the DC cables, might cause interference with the control loops.
- The complex nature of the VSC converters and their control impede the development of generalized, yet detailed, converter models. Such models are needed when converters from different manufacturers, each with their own topology, control loops, and time constants, are interconnected in one single DC system.

The HVDC grid time constants are thus in general several orders of magnitude smaller than those in AC systems. Furthermore, the clear distinction between fast and slower system dynamics based on the AC system frequency no longer applies to HVDC grids and might need to be revised for hybrid AC/DC systems.

Since it would be out of the scope of this introductory chapter to discuss all modeling challenges, we focus on MMC converters and AC/DC system interactions.

12.4.1 Modular Multilevel Converter Modeling

Due to the introduction of the MMC concept, a number of new challenges arose with respect to converter modeling. Whereas two-level topologies are still straightforwardly simplified for EMT simulations by grouping all IGBTs in a converter arm in one equivalent switch, the high number of stacked modules makes full-detailed time domain simulations time-consuming. Recent approaches include a time-varying Thévenin equivalent approach [3], a continuous-variable dynamic model, considering the sum of the capacitor voltages instead of the individual capacitor voltages [9] and a continuous model including blocking and deblocking behavior of the converter [10]. An overview of different models available for EMT studies is given in reference 11.

12.4.2 Modeling AC/DC Dynamic System Interactions

Another challenge relates to the modeling of interactions of the HVDC grid with the existing AC grid. When both the detailed switching behavior of the converter and the AC transient responses are of interest, an interface can be provided between the switching model on the one hand and the transient stability program on the other hand [12]. The VSC HVDC system can also be simplified for electromechanical stability programs as described above. Alternatively, the electromechanical stability program can be combined with an EMT-type program as in [13], where an averaged converter model, implemented in an EMT-type program, was combined with an electromechanical stability software. One of the key issues is the interaction of the two types of software [14]. Challenges to this approach relate to the representation of the nearby AC grid surrounding the converter and the extent to which the AC and/or

DC grid can be simplified so that the models are still accurate enough to represent the dynamic AC/DC interactions of interest.

Other challenges arise from the introduction of voltage droop control since it invokes a system-wide response as the result of a DC system contingency. The introduction of HVDC grids therefore also challenges the secure operation of AC power systems, especially combined with the introduction of a massive amount of renewable energy sources. This urges for new assessments of system reliability that go beyond the existing $N-1$ criterion, which has to be replaced by more advanced statistical methods.

12.4.2.1 Interactions with the Nearby AC System A degraded AC system state in the vicinity of the converter can impact the performance of the DC grid. Future work in this direction could encompass the dynamic behavior of the DC system, thereby accounting for the AC system characteristics at the PCC (e.g., system strength, interconnections, etc.). Such a study could focus on the DC grid performance under normal and faulted conditions. An important aspect that needs to be addressed in this respect is the level of detail needed in the DC system representation when addressing such local interactions between AC and DC systems.

12.4.2.2 Meshed AC/DC System Interactions The initial fast system response after DC system contingencies has its repercussions on the AC system stability [15]. This possible thread for future work could encompass a further study of dynamic interactions between the AC system generators and the DC system converters. One of the questions to be assessed is the extent to which dynamics in the DC system are essential from the point of view of dynamic AC/DC system studies, especially when the interactions between slow-acting AC generators and fast-acting DC converters are investigated.

12.4.2.3 Interactions with Renewable Energy Sources Other control and modeling challenges arise when a massive integration of renewable energy sources is considered: With the connection of a multitude of offshore wind farms and possibly large-scale solar PV installations to a DC grid, it has to be investigated to what extent these sources can contribute to the control of the grid or to what extent a dynamic control of the DC grid converter invokes interactions with the wind farm or PV converters.

12.5 CONCLUSIONS

The expected advent of HVDC grids is challenging the way the power system has traditionally been modeled in the past. With ever-increasing computer performance, the traditional boundary between phasor models and EMT models are fading and it becomes possible to model networks in full detail, thereby not being hindered by the limitations traditionally imposed by phasor programs. However, accounting for very detailed DC systems (not to mention full-detailed MMC converter switching models) still poses a number of stringent limitations to the size of the network. One of the solutions can lie in the application and development of hybrid models, with a detailed EMT representation of the DC network and the AC network in the vicinity of the converter, combined with a standard electromechanical model for the rest of the AC network. Such combined approaches are the subject of ongoing research and could play an important role in future power system modeling for dynamic AC/DC system interactions.

REFERENCES

1. P. Kundur, J. Paserba, V. Ajjarapu, G. Andersson, A. Bose, C. Canizares, N. Hatziargyriou, D. Hill, A. Stankovic, C. Taylor, T. Van Cutsem, and V. Vittal, "Definition and classification of power system stability IEEE/CIGRE joint task force on stability terms and definitions," *IEEE Transactions on Power Systems*, vol. 19, no. 3, pp. 1387–1401, Aug. 2004.
2. J. Beerten, "Modeling and control of DC grids," Ph.D. dissertation, University of Leuven (KU Leuven), May 2013.
3. U. N. Gnanarathna, A. M. Gole, and R. P. Jayasinghe, "Efficient modeling of modular multilevel HVDC converters (MMC) on electromagnetic transient simulation programs," *IEEE Transactions on Power Delivery*, vol. 26, no. 1, pp. 316–324, Jan. 2011.
4. J. Peralta, H. Saad, S. Dennetiere, J. Mahseredjian, and S. Nguefeu, "Detailed and averaged models for a 401-level MMC HVDC system," *IEEE Transactions on Power Delivery*, no. 3, pp. 1501–1508, July 2012.
5. J. Yang, J. E. Fletcher, and J. O'Reilly, "Short-circuit and ground fault analyses and location in VSC-based DC network cables," *IEEE Transactions on Industrial Electronics*, vol. 59, no. 10, pp. 3827–3837, Oct. 2012.
6. K. De Kerf, K. Srivastava, M. Reza, D. Bekaert, S. Cole, D. Van Hertem, and R. Belmans, "Wavelet-based protection strategy for DC faults in multi-terminal VSC HVDC systems," *IET Generation, Transmission and Distribution*, vol. 5, no. 4, pp. 496–503, 2011.
7. B. Gustavsen and A. Semlyen, "Rational approximation of frequency domain responses by vector fitting," *IEEE Transactions on Power Delivery*, vol. 14, no. 3, pp. 1052–1061, July 1999.
8. J. Marti, L. Marti, and H. Dommel, "Transmission line models for steady-state and transients analysis," in *Athens Power Tech, 1993. APT 93. Proceedings. Joint International Power Conference*, vol. 2, Sept. 1993, pp. 744–750.
9. L. Harnefors, A. Antonopoulos, S. Norrga, L. Angquist, and H. Nee, "Dynamic analysis of modular multilevel converters," *IEEE Transactions Industrial Electronics*, vol. 60, no. 7, pp. 2526–2537, Jul. 2013.
10. N. Ahmed, L. Angquist, S. Norrga, and H.-P. Nee, "Validation of the continuous model of the modular multilevel converter with blocking/deblocking capability," in *Proceedings of IET ACDC 2012*, Birmingham, UK, Dec. 4–6, 2012, 6 pages.
11. H. Saad, J. Peralta, S. Dennetiere, J. Mahseredjian, J. Jatskevich, J. Martinez, A. Davoudi, M. Saeedifard, V. Sood, X. Wang, J. Cano, and A. Mehrizi-Sani, "Dynamic averaged and simplified models for MMC-based HVDC transmission systems," *IEEE Transactions on Power Delivery*, vol. 28, no. 3, pp. 1723–1730, July 2013.
12. G. Irwin, C. Amarasinghe, N. Kroeker, and D. Woodford, "Parallel processing and hybrid simulation for HVDC/VSC PSCAD studies," in *Proceedings of IET ACDC 2012*, Birmingham, UK, Dec. 4–6 , 2012, 6 pages.
13. A. van der Meer, R. Hendriks, and W. Kling, "Combined stability and electro-magnetic transients simulation of offshore wind power connected through multi-terminal VSC-HVDC," in *Proceedings of IEEE PES GM 2010*, Minneapolis, USA, July 25–29, 2010, 7 pages.
14. V. Jalili-Marandi, V. Dinavahi, K. Strunz, J. Martinez, and A. Ramirez, "Interfacing techniques for transient stability and electromagnetic transient programs—IEEE task force on interfacing techniques for simulation tools," *IEEE Transactions on Power Delivery*, vol. 24, no. 4, pp. 2385–2395, Oct. 2009.
15. R. Eriksson, J. Beerten, M. Ghandhari, and R. Belmans, "Optimizing DC voltage droop settings for AC/DC system interactions," *IEEE Transactions on Power Delivery*, vol. 29, no. 1, pp. 362–369, Feb 2014.

CHAPTER 13

POWER FLOW MODELING OF HYBRID AC/DC SYSTEMS

Jef Beerten
Research Group Electa, Department of Electrical Engineering, University of Leuven, Leuven, Belgium

13.1 INTRODUCTION

Power flow software is typically used to analyze steady-state interactions in power systems and to study the flow patterns in the system. When used to study the interactions in hybrid AC/DC power systems, it is possible to study the system-wide effect of converter outages and other contingencies in the HVDC grid by including the overall droop control law in the description of the problem. When solving the power flows in hybrid AC/DC power systems consisting of both AC and DC systems, different types of models can be used. In its simplest form, the DC system can be represented as injections to the AC system, thereby only including very limited or no information on the DC system quantities (e.g., voltages, currents, and line flows). For the detailed models, a general distinction can be made between unified and sequential power flow solution methods. In the unified methods, all AC and DC equations are solved simultaneously in one iterative run. In the sequential solution method, the AC and DC systems are solved sequentially.

A generalized picture of a hybrid AC/DC system is depicted in Figure 13.1. For convenience, the AC and DC bus numbers forming the interconnections to respectively the DC and the AC network have been sequentially numbered, starting from 1. Neither the AC nor DC systems have to be limited to these interconnected buses. In this example, AC systems 3 and 4 only share one converter with respectively DC grids 1 and 2. These systems can represent relatively small systems—for example, island systems or (offshore) wind farms connected to the offshore systems—but they can also represent larger AC systems that only have one interconnection with the other AC systems in the grid.

HVDC Grids: For Offshore and Supergrid of the Future, First Edition. By Dirk Van Hertem, Oriol Gomis-Bellmunt, and Jun Liang.

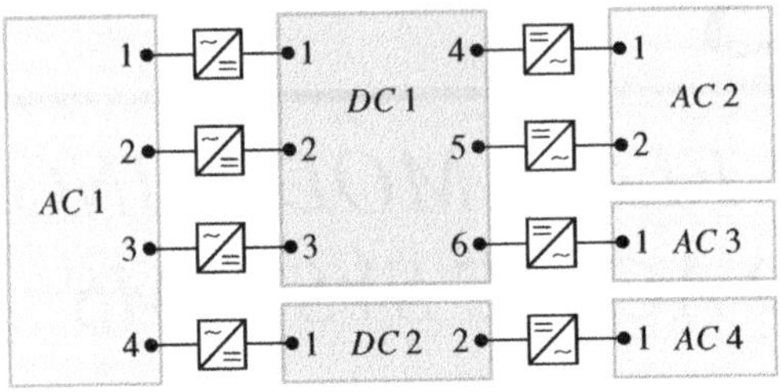

Figure 13.1 Interconnected AC/DC system.

13.2 SIMPLIFIED POWER FLOW MODELING

13.2.1 AC Power System Model

The power flow model of AC systems is well known from the literature [1, 2] and is only briefly introduced here for completeness.

The AC system can be represented by a number of nodes i where the generators and loads are connected, along with a connecting set of branches ij connecting the nodes. The problem is defined in terms of complex nodal voltage, $U_i\angle\theta_i$, and the apparent power injection, $S_{ij} = P_{ij} + jQ_{ij}$, in each node.

The nodes are typically subdivided into three types: the PV buses where the voltage and active power set-point are fixed or given, the PQ buses where active and reactive power are constant, and the slack bus which has a fixed reference angle and voltage.

The transmission lines are modeled as π-equivalent elements. The line between nodes i and j is represented by a series impedance $Z_{ij} = R_{ij} + jX_{ij}$ and a shunt

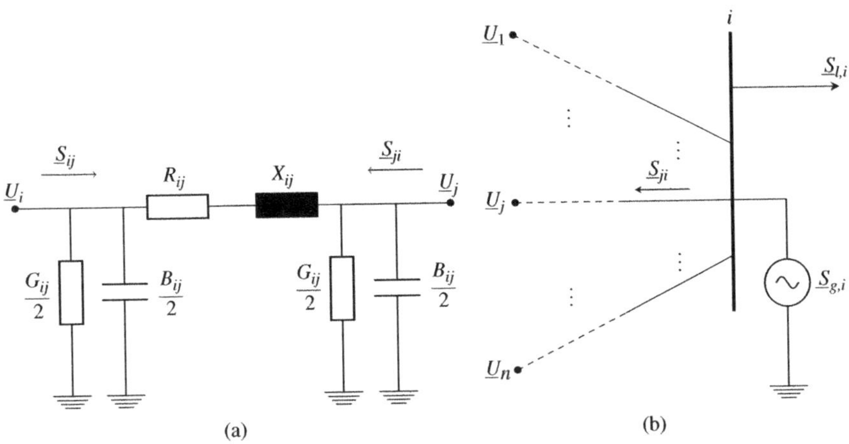

Figure 13.2 Representation of a transmission line and an AC node. (a) π-equivalent for a transmission line. (b) AC node i.

conductance G_{ij} and admittance B_{ij} The grid can than be represented by a nodal admittance matrix, Y_{bus}.

$$y_{bus,ij} = \begin{cases} \frac{1}{Z_{ij}}, & \text{when between nodes } i \text{ and } j \\ -\sum_{\substack{j=1 \\ j\neq i}}^{n} \frac{1}{Z_{ij}} \end{cases} \tag{13.1}$$

The power flow equations can then be written as a sum of the injections of generation $(P_{g,i}, Q_{g,i})$ and load $(P_{l,i}, Q_{l,i})$ and the sum of the branch flow towards that node $(P_{\Sigma br,i}, Q_{\Sigma br,i})$, which equal zero. For node i this gives

$$\begin{aligned} 0 &= P_{l,i} - P_{g,i} + P_{\Sigma br,i} \\ &= P_{l,i} - P_{g,i} + \sum_{j=1}^{n_b} Y_{bus,ij} U_i U_j \cos(\theta_i - \theta_j - \varphi_{ij}) \end{aligned} \tag{13.2a}$$

$$\begin{aligned} 0 &= Q_{l,i} - Q_{g,i} + Q_{\Sigma br,i} \\ &= Q_{l,i} - Q_{g,i} + \sum_{j=1}^{n_b} Y_{bus,ij} U_i U_j \sin(\theta_i - \theta_j - \varphi_{ij}) \end{aligned} \tag{13.2b}$$

Or in rectangular coordinates:

$$\begin{aligned} 0 &= P_{l,i} - P_{g,i} + P_{\Sigma br,i} \\ &= P_{l,i} - P_{g,i} + \sum_{j=1}^{n_b} U_i U_j (G_{bus,ij} \cos(\theta_i - \theta_j) + B_{bus,ij} \sin(\theta_i - \theta_j)) \end{aligned} \tag{13.3a}$$

$$\begin{aligned} 0 &= Q_{l,i} - Q_{g,i} + Q_{\Sigma br,i} \\ &= Q_{l,i} - Q_{g,i} + \sum_{j=1}^{n_b} U_i U_j (G_{bus,ij} \sin(\theta_i - \theta_j) - B_{bus,ij} \cos(\theta_i - \theta_j)) \end{aligned} \tag{13.3b}$$

When expressing the active and reactive power balance for each node as a function of the nodal voltages and angles as in (13.2a) (or (13.3a)) and (13.2b) (or (13.3b)), a system of nonlinear equations is obtained. This system is typically solved in an iterative manner, using the Newton–Raphson method:

$$\begin{bmatrix} \mathbf{\Delta P} \\ \mathbf{\Delta Q} \end{bmatrix}^{(i)} = \begin{bmatrix} \frac{\partial P}{\partial \theta} & U\frac{\partial P}{\partial U} \\ \frac{\partial Q}{\partial \theta} & U\frac{\partial Q}{\partial U} \end{bmatrix}^{(i)} \cdot \begin{bmatrix} \mathbf{\Delta \theta} \\ \frac{\mathbf{\Delta U}}{\mathbf{U}} \end{bmatrix}^{(i)} \tag{13.4}$$

13.2.2 DC Power System Model

In its most simple format, the DC system model can be limited to its overall effects on the AC systems to which it is connected. The DC system can then be represented as mere power injections into the AC system. This model is generally used to represent two-terminal systems in system studies, and it starts from the observation that for these studies the actual variables of the DC system are of lesser importance than the overall effects the HVDC system has on the AC system.

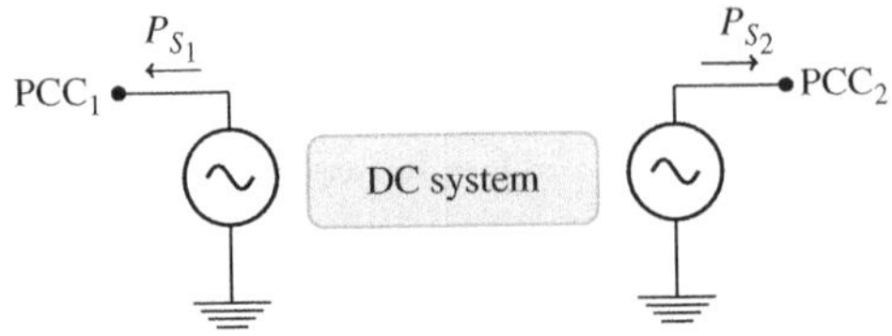

Figure 13.3 Simplified two-terminal HVDC model.

Figure 13.3 shows the model, which can be used to represent the two-terminal system 2 from Figure 13.1.

Neglecting the HVDC system losses, the HVDC system can conceptually be represented as two coupled generators with opposite power injections, where the AC injections at the point of common coupling (PCC), P_s, are equal in magnitude and opposite in sign

$$P_{s_2} = -P_{s_1} \tag{13.5}$$

Alternatively, DC system losses can be added to the model by having

$$P_{s_2} = -P_{s_1} - P_{DC,loss} \tag{13.6}$$

The losses can be assumed constant or can, for example, be a function of the converter power or the line current.

As discussed in greater detail in the next section, the model can be extended by accounting for the fact that reactive power injections (or alternatively the AC system voltages at the PCC) can be independently controlled by the two converters.

The simplified modeling approach can be used to represent more general DC systems, as shown for an n-terminal system in Figure 13.4. Neglecting the DC system losses, similarly to (13.5), the power injections at the PCC have to satisfy

$$\sum_{j=1}^{n} P_{s_j} = 0. \tag{13.7}$$

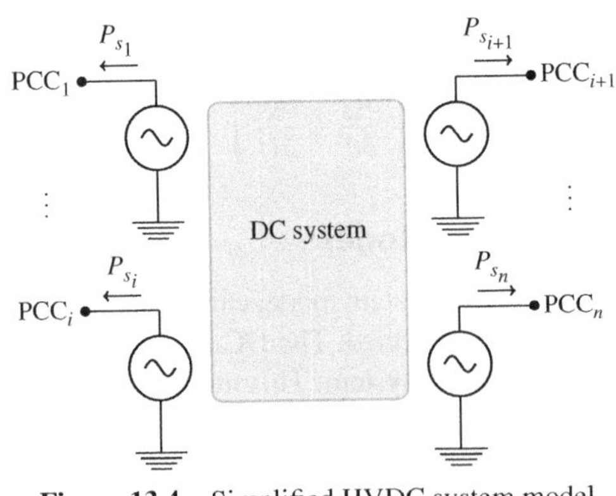

Figure 13.4 Simplified HVDC system model.

In analogy to (13.6), the model can be enhanced by including the DC system losses

$$\sum_{j=1}^{n} P_{s_j} = -P_{DC,loss} \tag{13.8}$$

The converter losses can be taken into account using a generalized loss formula with the converter losses quadratically depending on the converter current I_c [3]:

$$P_{loss} = a_i + b_i I_c + (c_i + r_i) I_c^2 \tag{13.9}$$

with

$$I_c = \frac{\sqrt{P_c^2 + Q_c^2}}{U_c} \tag{13.10}$$

the current of the generator on the AC side of the corresponding converter.

The factors a_i, b_i, and c_i stand for the individual loss factors of converter i. The content of the individual factors is:

- **No-Load Loss Factor a_i.** No-load losses of transformers, averaged auxiliary equipment losses (i.e., lightning, heating, cooling, control systems).
- **Linear Current Factor b_i.** Switching losses of the valves, more in particular the turn-off losses of the IGBTs and free-wheeling diodes.
- **Square Current Factor c_i.** Conduction losses of the valves. In the case where significant differences exist between the rectifier and inverter mode, c_i can depend on the operating mode.
- **Square Current Factor[1] r_i.** Load losses of transformers, series reactance, and other AC equipment.

Example values for the converter loss coefficients a, b and c of a 600-MW ±300-kV VSC converter are given in Table 13.1. The contribution of the different terms of (13.9) is discussed in greater detail in [3].

It is clear that the more complex the DC system becomes, the more challenging it gets to approximate the overall system losses as a function of the PCC power injections. Furthermore, by simplifying the DC system to its power injections into the AC system, all information on the DC system itself is lost, which makes this simple model insufficient for, for example, DC system contingency studies.

TABLE 13.1 Per Unit Converter Loss Coefficients

	a (×10³)	b (×10³)	c (×10³)
Rectifier	11.033	3.464	4.400
Inverter	11.033	3.464	6.667

[1] Please note that this additional term should only be included when this AC equipment is not explicitly represented in the power flow model.

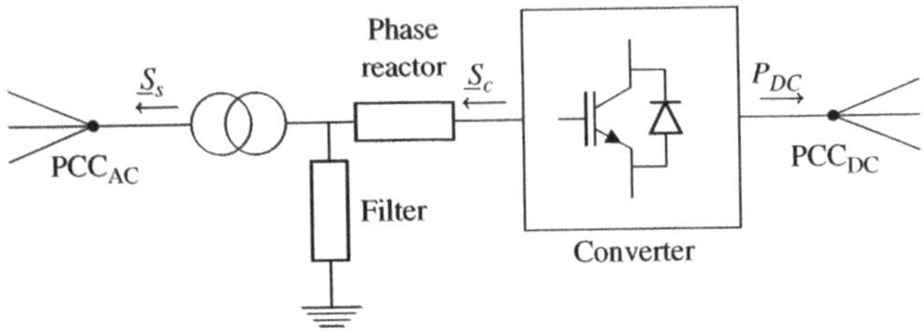

Figure 13.5 VSC-HVDC converter station.

13.3 DETAILED POWER FLOW MODELING

13.3.1 Converter Model

A general representation of a VSC-HVDC converter station, showing the different components, is depicted in Figure 13.5. As seen from the PCC in the AC system, the different components are:

- The converter transformer
- The AC filter
- The phase reactor
- The converter

The filtering requirements are dependent on the technology and on the switching method used to synthesize the converter voltage $\underline{U}_c$. Their effect at the fundamental grid frequency can typically be represented by a shunt capacitor. Recent developments in VSC HVDC technology based on multimodular converters (MMC) allow to strongly reduce or even eliminate the need of low-pass filtering.

As a result, in its most general format, a VSC-HVDC converter can be represented by a controllable voltage source $\underline{U}_c = U_c\angle\theta_c$ behind the phase reactor with a complex impedance $\underline{Z}_c = R_c + jX_c$. As shown in Figure 13.6, the phase reactor is connected to the susceptance B_f, which forms a part of the low-pass filter. The filter bus is connected to the AC grid through a transformer, represented by its impedance $\underline{Z}_{tf} = R_{tf} + jX_{tf}$. The equations for the power flowing to the AC network, following

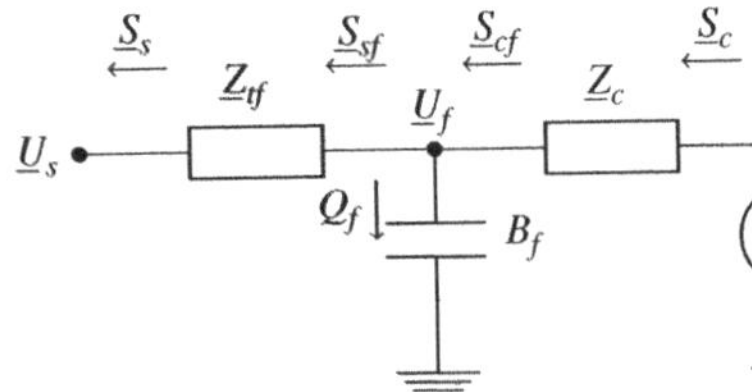

Figure 13.6 Equivalent single-phase power flow model of a converter station connected to the AC grid.

the standard power flow formulation, can thus be written as

$$P_s = -U_s^2 G_{tf} + U_s U_f [G_{tf} \cos(\theta_s - \theta_f) + B_{tf} \sin(\theta_s - \theta_f)] \tag{13.11}$$

$$Q_s = U_s^2 B_{tf} + U_s U_f [G_{tf} \sin(\theta_s - \theta_f) - B_{tf} \cos(\theta_s - \theta_f)] \tag{13.12}$$

with $\underline{U}_s = U_s \angle \theta_s$ and $\underline{U}_f = U_f \angle \theta_f$ respectively the complex grid side and filter bus voltage. The equations at the converter side can be written as

$$P_c = U_c^2 G_c - U_f U_c [G_c \cos(\theta_f - \theta_c) - B_c \sin(\theta_f - \theta_c)] \tag{13.13}$$

$$Q_c = -U_c^2 B_c + U_f U_c [G_c \sin(\theta_f - \theta_c) + B_c \cos(\theta_f - \theta_c)] \tag{13.14}$$

Assuming the AC filters to be lossless, the filter power reduces to its reactive part

$$Q_f = -U_f^2 B_f \tag{13.15}$$

while the expressions for the filter side complex power flowing through the transformer are written as

$$P_{sf} = U_f^2 G_{tf} - U_f U_s [G_{tf} \cos(\theta_s - \theta_f) - B_{tf} \sin(\theta_s - \theta_f)] \tag{13.16}$$

$$Q_{sf} = -U_f^2 B_{tf} + U_f U_s [G_{tf} \sin(\theta_s - \theta_f) + B_{tf} \cos(\theta_s - \theta_f)] \tag{13.17}$$

and those flowing through the phase reactor side are

$$P_{cf} = -U_f^2 G_c + U_f U_c [G_c \cos(\theta_f - \theta_c) + B_c \sin(\theta_f - \theta_c)] \tag{13.18}$$

$$Q_{cf} = U_f^2 B_c + U_f U_c [G_c \sin(\theta_f - \theta_c) - B_c \cos(\theta_f - \theta_c)] \tag{13.19}$$

In the absence of low-pass filters or with the filters omitted, the phase reactor and the converter transformer can be lumped together, eliminating the dependence on the complex filter bus voltage in (13.11)–(13.14). Similarly, if transformerless designs are considered, the filter bus and the AC grid bus coincide, hence simplifying the equations. These different representations are found in Figure 13.7. Figure 13.7c shows a representation that is generally used as a simplified model of a converter, namely a voltage source behind a complex impedance (or a pure inductive impedance when also omitting R_c). In that case, the power flow equations at the grid side simplify to

$$P_s = -U_s^2 G_c + U_s U_c [G_c \cos(\theta_s - \theta_c) + B_c \sin(\theta_s - \theta_c)] \tag{13.20}$$

$$Q_s = U_s^2 B_c + U_s U_c [G_c \sin(\theta_s - \theta_c) - B_c \cos(\theta_s - \theta_c)] \tag{13.21}$$

and the power flowing to the AC network at the converter end can then be written as

$$P_c = U_c^2 G_c - U_s U_c [G_c \cos(\theta_s - \theta_c) - B_c \sin(\theta_s - \theta_c)] \tag{13.22}$$

$$Q_c = -U_c^2 B_c + U_s U_c [G_c \sin(\theta_s - \theta_c) + B_c \cos(\theta_s - \theta_c)] \tag{13.23}$$

This representation is commonly used to represent a connection of a VSC, and for the majority of system studies, this representation is sufficient to represent the converters. This especially holds for MMC converters with small AC filters or even without AC filters. As discussed in Section 13.3.3, leaving out the filters for a two-level topology results in an underestimation of the converter reactive power capability at the PCC.

The converter losses can be taken into account by the loss function (13.9).

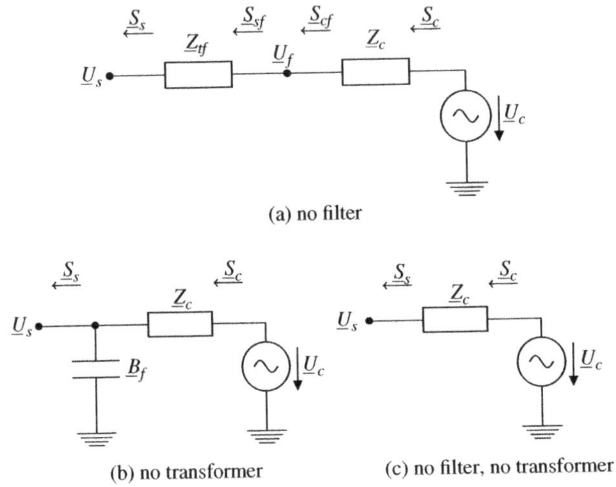

Figure 13.7 Single-phase model of a converter station. (a) Omitting the filter, (b) omitting the transformer, and (c) omitting both filter and transformer.

13.3.2 Control Mode Representation

A VSC HVDC converter can independently control the active and reactive power injection with respect to the AC system. The active power injection is modeled in different ways:

1. *Constant P*. The converter has a constant active power injection P_s into the AC grid.
2. *Constant U_{DC}*. The algorithm adapts the active power injection P_s to obtain a constant DC bus voltage U_{DC}.
3. *$P - U_{DC}$ Droop*. The actual active power injection depends on the DC bus voltage U_{DC} according to the droop characteristic

$$P_{DC} = P_{DC,0} - \frac{1}{k_{DC}}(U_{DC} - U_{DC,0}) \tag{13.24}$$

Figure 13.8 shows the *P–U* steady-state characteristics for these three control options.

Alternatively, the DC current can be controlled to a constant value and consequentially, the voltage droop characteristic can also be written as a function of the converter current. The characteristics from Figure 13.8 can also be combined to a so-called voltage-margin control or deadband droop control, or a piecewise-linear droop curve as discussed in greater detail in Chapter 15. This will not be treated further in this chapter and the description will be limited to a representation of the three control options shown in Figure 13.8.

The ability of the converter to independently control the reactive power results in two different representations in the power flow algorithm:

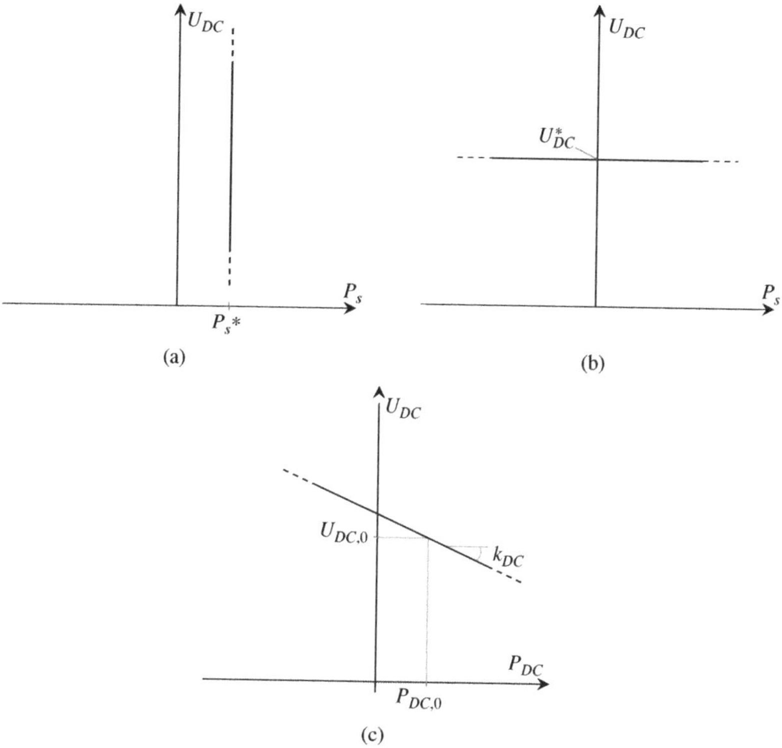

Figure 13.8 Converter steady-state operation characteristics. (a) P constant. (b) U_{DC} constant. (c) P–U_{DC} droop.

1. *Constant Q*: The converter has a constant reactive power injection Q_s into the AC grid.
2. *Constant U*: The converter adapts the reactive power injection to obtain a constant AC bus voltage magnitude U_s.

13.3.2.1 Two-Terminal Scheme The actual steady-state behaviour of a two-terminal VSC-HVDC link can be modeled in a power flow algorithm by one constant P converter and one constant U_{DC} converter. The active power injection of the latter is unknown prior to the power flow and depend on the losses in the VSC-HVDC system. This constant U_{DC} converter is referred to as the DC slack converter, due to its similarity to the slack bus in AC power flow algorithms.

13.3.2.2 Multi-Terminal Scheme The two-terminal representation can be extended to model x different MTDC systems with a total of y DC converters. With one DC slack bus per MTDC system, this results in y–x converters controlling their active power output (constant P). The x remaining DC slack converters, one per MTDC system, adjust their power to account for the DC system losses and to keep up the voltage of the DC grids.

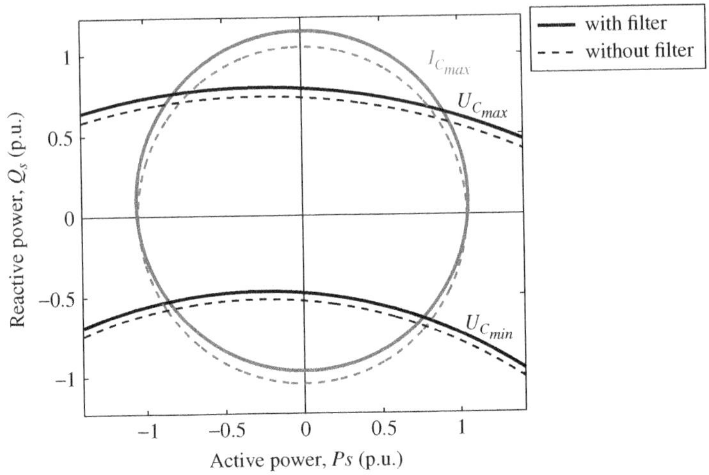

Figure 13.9 PQ-capability chart. Converter station current and voltage limits (with $I_{C_{max}} = 1.05$ p.u., $U_{C_{min}} = 0.85$ p.u., and $U_{C_{max}} = 1.2$ p.u.).

Alternatively, a voltage droop can be implemented to study the effect of distributing the DC voltage control function on the power flows. Doing so, the converters can be represented by (13.24), possibly combined with converters in constant power control representing, for example, offshore wind farms.

13.3.3 Converter Limits

To ensure an overall safe operation of the converter station, the steady-state working point must be situated within the PQ-capability chart of the converter, depicted in Figure 13.9.[2] The solid lines include the effect of the filters, and the dotted lines neglect the filters. The gray circle forms the converter current limit. The black arcs depict the upper and lower converter voltage limits. The PQ capability of the converter is given by the area inside these limits.

With the apparent power injection written as

$$\underline{S}_s = \underline{U}_s \underline{I}_s^* \tag{13.25}$$

the system bus voltage equation, which is equal to

$$\underline{U}_s = \underline{U}_f - \underline{Z}_{tf} \underline{I}_s \tag{13.26}$$

can be reformulated and substituted in the equation for the current at the filter bus, yielding

$$\underline{I}_s = \underline{I}_c - \frac{\underline{U}_s + \underline{Z}_{tf} \underline{I}_s}{\underline{Z}_f} \tag{13.27}$$

[2]See also Section 10.5.2: power capability curve.

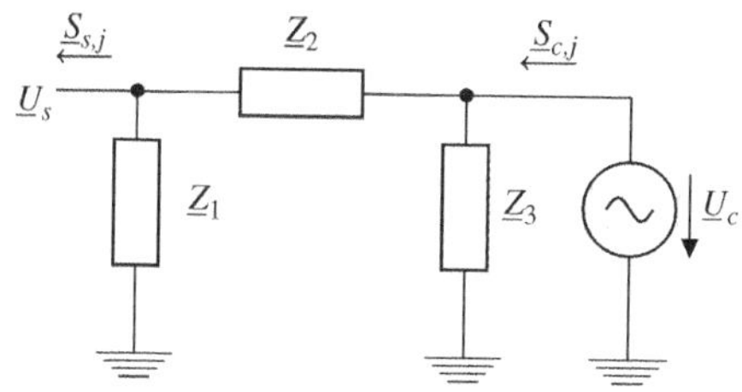

Figure 13.10 π-equivalent scheme of a VSC-HVDC converter.

Rewriting this equation in terms of the current $\underline{I}_s$ and substituting it in (13.25) results in a closed expression of the converter current limit

$$\underline{S}_s = -U_s^2\left(\frac{1}{\underline{Z}_f^* + \underline{Z}_{tf}^*}\right) + \underline{U}_s\underline{I}_{cm}^*\left(\frac{\underline{Z}_f^*}{\underline{Z}_f^* + \underline{Z}_{tf}^*}\right) \tag{13.28}$$

with $\underline{I}_{cm}$ the phasor of the maximum current

$$\underline{I}_{cm} = I_{c_{max}}\angle\theta_{I_c} \tag{13.29}$$

With the filters omitted, (13.28) simplifies to

$$\underline{S}_s = \underline{U}_s\underline{I}_{cm}^* \tag{13.30}$$

As shown in Figure 13.9, the circular current limit from (13.30) slightly scales and shifts to a new center point as a result of the filter inclusion. The filter capacitance thus leads to a small bias in Q_s-axis direction [4].

In VSC systems that use pulse-width modulation (PWM), the modulation factor has an upper and lower limit, respectively, to avoid overmodulation and the reappearance of harmonics. In [5], the lower reactive power limit is a constant at −0.5 p.u, which corresponds with the lower limit in this picture. With the modular multilevel converter (MMC) technology, the reactive power is limited by the current in the lagging region [6]. The lower voltage limit can thus be omitted in this case.

These limits are implemented as limits on the converter voltage. The voltage limits can be obtained by applying a Y–θ transformation to the equivalent model from Figure 13.6. This is shown in Figure 13.10, with the equivalent impedances $\underline{Z}_1$, $\underline{Z}_2$ and $\underline{Z}_3$, respectively:

$$\underline{Z}_1 = \frac{\underline{Z}_{tf}\underline{Z}_c + \underline{Z}_c\underline{Z}_f + \underline{Z}_f\underline{Z}_{tf}}{\underline{Z}_c} \tag{13.31}$$

$$\underline{Z}_2 = \frac{\underline{Z}_{tf}\underline{Z}_c + \underline{Z}_c\underline{Z}_f + \underline{Z}_f\underline{Z}_{tf}}{\underline{Z}_f} \tag{13.32}$$

$$\underline{Z}_3 = \frac{\underline{Z}_{tf}\underline{Z}_c + \underline{Z}_c\underline{Z}_f + \underline{Z}_f\underline{Z}_{tf}}{\underline{Z}_{tf}} \tag{13.33}$$

Rewriting $\underline{I}_s^*$ in (13.25) in terms of the currents $\underline{I}_1$ and $\underline{I}_2$, through $\underline{Z}_1$ and $\underline{Z}_2$, results in

$$\underline{S}_s = \underline{U}_s \cdot \left(\underline{I}_2^* - \underline{I}_1^*\right) \tag{13.34}$$

$$= \underline{U}_s \cdot \left(\frac{\underline{U}_c^* - \underline{U}_s^*}{\underline{Z}_2^*} - \frac{\underline{U}_s^*}{\underline{Z}_1^*}\right) \tag{13.35}$$

Rewriting this equation in terms of admittances results in the expression for $\underline{S}_s$ in terms of a converter voltage limit

$$\underline{S}_s = -U_s^2 \cdot \left(\underline{Y}_1^* + \underline{Y}_2^*\right) + \underline{U}_s \cdot \underline{U}_{cm}^* \cdot \underline{Y}_2^* \tag{13.36}$$

with $\underline{U}_{cm}$ the phasor of the minimum or maximum converter voltage $U_{c_{min}}$ or $U_{c_{max}}$. $U_{c_{max}}$ is the maximum voltage that can be generated by the converter while avoiding overmodulation. This approach implicitly assumes a constant DC voltage when calculating the limits. Alternatively, the modulation index can be included explicitly in the equations, as in [7]. The lower voltage limit $U_{c_{min}}$ in Figure 13.9 has been chosen to comply with the minimum reactive power limit from [5]. This limit can be omitted in an MMC or when no minimum reactive power limit is imposed on the converter.

Similar expressions can be found by applying l'Hôpital's rule to (13.31)–(13.33) for $\underline{Z}_f \rightarrow \infty$ or $\underline{Z}_{tf} \rightarrow 0$ when the filter or transformer are not included. The expression without the filter hence simplifies to

$$\underline{S}_s = -U_s^2\left(\frac{1}{\underline{Z}_{tf}^* + \underline{Z}_c^*}\right) + \frac{\underline{U}_s\underline{U}_{cm}^*}{\underline{Z}_{tf}^* + \underline{Z}_c^*} \tag{13.37}$$

Similar to the current limit, the inclusion of the filter shifts the voltage limit circles from (13.30) to a new center point, while their radii slightly change.

13.3.4 DC Grid Model

The power flows in the DC grid can be obtained in a way similar to a conventional AC power flow. The current injected at a DC node i can be written as the current flowing to the other $n-1$ nodes in the network

$$I_{DC_i} = \sum_{\substack{j=1 \\ j \neq i}}^{n} Y_{DC_{ij}} \cdot \left(U_{DC_i} - U_{DC_j}\right) \tag{13.38}$$

with $Y_{DC_{ij}}$ equal to $1/R_{DC_{ij}}$.

Combining all currents injected in an n bus DC network results in

$$I_{DC} = \mathrm{Y}_{\mathrm{DC}} U_{DC} \tag{13.39}$$

with $I_{DC} = [I_{DC_1}, I_{DC_2} \dots I_{DC_k}, 0 \dots 0]^T$ the DC current vector with $n-k$ zero elements due to DC buses without a power injection and converter outages. The vector $U_{DC} = [U_{DC_1}, U_{DC_2} \dots U_{DC_n}]^T$ denotes the DC voltage vector, and Y_{DC} denotes the DC bus matrix. When line outages are taken into account, the DC bus matrix needs to be updated accordingly. When x DC grids are considered, all DC bus matrices $\mathrm{Y}_{\mathrm{DC}j}$ of the different DC grids are combined in one single sparse band matrix Y_{DC}

$$\mathrm{Y}_{\mathrm{DC}} = \begin{bmatrix} \mathrm{Y}_{\mathrm{DC}1} & & \\ & \ddots & \\ & & \mathrm{Y}_{\mathrm{DC}x} \end{bmatrix} \tag{13.40}$$

The active power injected in node i can be written as

$$P_{DC_i} = U_{DC_i} I_{DC_i}, \qquad \forall i \leq k, \tag{13.41}$$

The current injections I_{DC} are not known prior to the power flow solution for the DC grid, whereas the active power injections P_{DC} are known for all buses as a results of the AC power flow, except for the DC slack and voltage droop buses. Combining (13.39) and (13.41), we obtain

$$P_{DC_i} = U_{DC_i} \sum_{\substack{j=1 \\ j \neq i}}^{n} Y_{DC_{ij}} \cdot \left(U_{DC_i} - U_{DC_j}\right) \tag{13.42}$$

In the case of a U_{DC}–P_{DC} voltage droop, the relationship between the active power and the DC voltage is written as

$$P_{DC_i} = P_{DC,0_i} - \frac{1}{k_{DC_i}} \left(U_{DC_i} - U_{DC,0_i}\right) \tag{13.43}$$

13.4 SEQUENTIAL AC/DC POWER FLOW

To solve the hybrid AC/DC system in steady state, a distinction can be made between unified models on the one hand and sequential models on the other hand. The distinction is based on the solution method: Unified methods solve the entire hybrid AC/DC system in one iterative run, whereas the sequential methods iterate between the AC and DC systems. Unified methods have the advantage that, in general, they need less calculation time, since they solve both systems at the same time. On the one hand, additional iterations between the AC and the DC systems in sequential methods may give rise to convergence problems due to oscillations between the AC and DC system solutions. On the other hand, sequential methods have the advantage that they can easily be combined with existing AC power flow packages. In the remainder of the chapter, focus will largely be on the sequential approach. However, the converter and DC system equations derived in the previous sections can equally well be used in unified methods. More information on unified solutions of hybrid AC/DC systems can be found in [8, 9].

Figure 13.11 shows the flow chart of a sequential power flow algorithm. Prior to solving the power flow, the algorithm starts with a pre-processing, including amongst others a per unit conversion and an internal bus renumbering. It is assumed in the remainder of the chapter that interconnected AC and DC buses have the same bus number. For convenience, the analysis will be confined to one AC grid and one DC

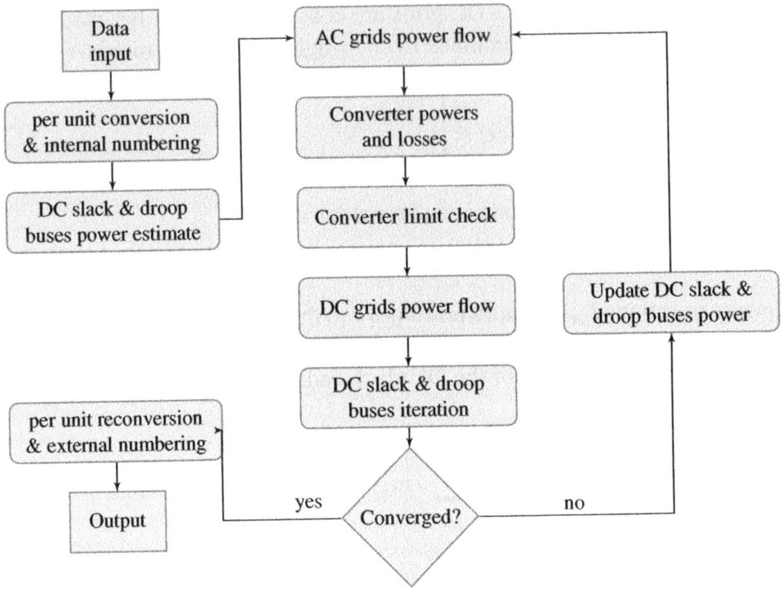

Figure 13.11 Flow chart of the sequential VSC-AC/DC power flow algorithm.

grid with n buses of which k have a connection to the AC grids. The method can easily be extended to include multiple AC and DC grids.

13.4.1 Initialization

To start the iteration, the system needs a first approximation of the converter active power injections, which can be done by assuming a lossless DC system and converters. Assuming the first converter to operate as a slack node in the DC system,[3] without lack of generality, we can write the vector of AC power injections as

$$P_s = [\ \underbrace{P_{s_1}}_{\text{slack}}, \underbrace{P_{s_2} \ldots P_{s_m}}_{\text{voltage droop}}, \underbrace{P_{s_{m+1}} \ldots P_{s_k}}_{P\text{–control}}, \underbrace{0 \ldots 0}_{\text{outage}}\ ]^T \tag{13.44}$$

The initialization of the active powers is given by

$$P_{s_i}^{(0)} = \begin{cases} -P_{dc,0_i} & \forall i:\ 2 \le i \le m \\ P_{s_i} & \forall i:\ m+1 \le i \le n \end{cases} \tag{13.45}$$

The active power delivered by the DC slack bus is given by

$$P_{s_1}^{(0)} = -\sum_{i=2}^{m} P_{s_i}^{(0)} - \sum_{j=m+1}^{n} P_{s_j} \tag{13.46}$$

[3]Note that when including a voltage droop at some converters, including a slack node is not strictly needed in order to find a solution. Mathematically, the voltage droop inclusion fulfills a function similar to a distributed slack bus.

or, in the case of multiple converters with a constant voltage,

$$P_{s_h}^{(0)} = -\frac{1}{l}\left(\sum_{i=l+1}^{m} P_{s_i}^{(0)} + \sum_{j=m+1}^{n} P_{s_j}\right) \qquad \forall h:\ 1 \le h \le l \tag{13.47}$$

Since multiple DC slack buses in a single DC grid might give rise to divergence in the DC power flow and are not likely to be implemented in real systems, the latter option is not preferred. In general, a DC system can also operate with all converters under droop control.

In the case where the system only has a constant voltage bus (slack node), but no voltage droop at other converters, the first estimate to initiate the iteration is given by

$$P_{s_1}^{(0)} = -\sum_{j=2}^{k} P_{s_j} \tag{13.48}$$

For subsequent iterations, the solution from the previous iteration is used for $P_{s_1}^{(k)}$.

13.4.2 AC Network Power Flow

The power flow equations for bus i in the AC grid can be written using (13.3a) and (13.3b).

The nonlinear set of power flow equations for all AC buses can be solved using a Newton–Raphson power flow, with the converter power injections P_s and Q_s included in the power mismatch vectors $\Delta P^{(j)}$ and $\Delta Q^{(j)}$ as negative loads. The power mismatch vectors can be rewritten as

$$\Delta P_i^{(j)} = P_i^{gen} - \left(P_i^{dem} - P_{s_i}\right) - P_i(U^{(j)}, \theta^{(j)}) \tag{13.49}$$

$$\Delta Q_i^{(j)} = Q_i^{gen} - \left(Q_i^{dem} - Q_{s_i}\right) - Q_i(U^{(j)}, \theta^{(j)}) \tag{13.50}$$

Converters in V-control are represented as dummy AC generators, and their AC buses are changed from PQ-nodes to PV-nodes. The active power injections of the slack buses are changed in order to control the DC grid voltages.

13.4.3 Converter Calculations

With the AC grid voltages $\underline{U}_s$ and power injection at the AC grid side $\underline{S}_s$ known, the converter side voltage and currents can be calculated. With (13.25)–(13.26), the filter bus voltage $\underline{U}_f$ can be written as

$$\underline{U}_{f_i} = \underline{U}_{s_i} + \underline{Z}_{tf_i} \cdot \left(\frac{\underline{S}_{s_i}^*}{\underline{U}_{s_i}^*}\right) \qquad \forall i < k \tag{13.51}$$

The converter current $\underline{I}_c$ can be extracted from (13.27), with $\underline{I}_s$ rewritten using (13.25)

$$\underline{I}_{c_i} = \frac{\underline{U}_{s_i}}{\underline{Z}_{f_i}} + \left(\frac{\underline{S}^*_{s_i}}{\underline{U}^*_{s_i}}\right) \cdot \frac{\underline{Z}_{f_i} + \underline{Z}_{tf_i}}{\underline{Z}_{f_i}} \qquad \forall i < k \tag{13.52}$$

which can be substituted in (13.9) to calculate the converter losses. The converter voltage $\underline{U}_c$ can be calculated as

$$\underline{U}_{c_i} = \underline{U}_{f_i} + \underline{Z}_{c_i}\underline{I}_{c_i}, \qquad \forall i < k \tag{13.53}$$

With all quantities on the AC side known, the DC grid's injected power becomes

$$P_{DC_i} = -P_{c_i} - P_{loss_i}, \qquad \forall i < k \tag{13.54}$$

13.4.4 Converter Limit Check

If the converter is operating outside its limits, the power injections are adapted to ensure a safe operation. In general, a weighting factor can be used to give a higher priority to active or reactive power. In this discussion, priority is given to active power. The grid-injected power P_s can remain unaltered if intersections with the capability chart can be found. If, on the contrary, P_s is too large, the active power is limited to its maximum value and the reactive power is adapted accordingly.

The current and voltage limits in (13.28) and (13.36) are circles in the PQ-plane. Hence, with P_s known, the two reactive power limits imposed by the current limit can be calculated as

$$Q_s = Q_o \pm \sqrt{r^2 - (P_s - P_o)^2} \tag{13.55}$$

with the circle's center $\underline{S}_o = P_o + jQ_o$ and a radius r. The center point and radius of the current limit are given by

$$\underline{S}_o = -U_s^2 \left(\frac{1}{\underline{Z}_f^* + \underline{Z}_{tf}^*}\right) \tag{13.56}$$

$$r = U_s I_{c_{max}} \left| \frac{\underline{Z}_f^*}{\underline{Z}_f^* + \underline{Z}_{tf}^*} \right| \tag{13.57}$$

The two solutions refer to the upper and lower reactive power limit imposed by the current limit.

Similarly, the voltage limits center point and radius are given by

$$\underline{S}_o = -U_s^2 \left(\underline{Y}_1^* + \underline{Y}_2^*\right) \tag{13.58}$$

$$r = U_s U_{cm} Y_2 \tag{13.59}$$

Replacing U_{cm} in (13.59) with $U_{c_{min}}$ or $U_{c_{max}}$ and substituting (13.58)–(13.59) in (13.55) gives an analytical expression for the lower and upper voltage limit reactive power. Only the solution with the positive sign is of interest, as the other solution refers to the lower crossing point with voltage limit circle.

The limit checks can be implemented similarly to those of generator limits: If a converter is operating outside its active power limits, the set-points must be reduced to their maximum values. Additionally, it must be ascertained that the DC slack buses are working within their bounds. Apart from the active power violations, the converter's reactive power must stay within limits. When a Q-limit is hit, the most stringent reactive power limit imposes the new converter operating point. If the converter is operating in V-control, the voltage controlled bus also has to be changed to a PQ-bus.

The Q-limits of the converters are updated every iteration due to changing grid conditions. As a consequence, it was found that simply enforcing the limits during each iteration proved unsatisfactory due to the fact that the limits differ from one iteration to another as a result of changing AC system voltages. Consequently, this can give rise to convergence problems. A solution to overcome this is to only update the limits when the incremental change in apparent power exceeds a minimum value. Alternatively, the limits can be checked after every iteration and only updated every three iterations. The algorithm should be properly tuned to achieve good convergence.

In summary, the following situations can occur:

- When an **active power controlling converter** has an active power setpoint outside of the PQ-capability chart, the active power order is reduced to comply with the most stringent limit.
- When a **DC voltage droop controlling converter** hits a limit, the converter is set to a constant active power injection equal to the maximum or minimum converter active power limit, depending on the direction of power flow.
- Similarly, when a **reactive power controlling converter** has a reactive power set-point outside of the PQ-capability chart, the reactive power order is reduced to comply with the most stringent limit, provided that no active power limit is hit. If this happens to be the case, the extent to which active power has to be prioritized over reactive power (or vice versa) needs to be defined beforehand.
- When an **AC voltage controlling converter** hits a reactive power limit, the converter is set to a constant reactive power injection determined by the most stringent limit. Similarly to the previous case, the priority of active or reactive power control has to be specified in case the active power limit is hit.

Whether a DC slack converter, if present, can be excluded from the analysis depends on whether changing the representation of the slack converter from constant DC voltage to constant active power still ensures finding a solution to the problem. As pointed out earlier, in the presence of voltage droop-controlled converters, a slack converter is not strictly needed from a mathematical point of view. Alternatively, the slack converter can be checked at the end of the power flow calculation and a redefinition of the power order set-points should be considered when it is observed that a limit is hit in this slack converter.

13.4.5 DC Network Power Flow

13.4.5.1 Constant Power Converters After calculating the AC power flow, all converter powers and losses are calculated to obtain the DC grid's injected powers P_{DC} for the k DC buses to which converters are connected. For converters 2 to k the DC power injection P_{DC} are known as a result of the AC power flow. Assuming that no droop-controlled converters are included in the system, expression (13.44) can be simplified into the following DC active power vector P_{DC}:

$$P_{DC} = [\ \underbrace{P_{DC_1}}_{\text{slack}}, \underbrace{P_{DC_2} \ldots P_{DC_k}}_{P\text{-control}}, \underbrace{0 \ldots 0}_{\text{outage}}\ ]^T \tag{13.60}$$

The nonlinear DC network equations from (13.42) can be solved with a Newton–Raphson method:

$$\left(U_{DC}\frac{\partial P_{DC}}{\partial U_{DC}}\right)^{(j)} \cdot \frac{\Delta U_{DC}}{U_{DC}}^{(j)} = \Delta P_{DC}^{(j)} \tag{13.61}$$

The power mismatch vector $\Delta P_{DC}^{(j)}$ is given by

$$\Delta P_{DC_i}^{(j)} = \begin{cases} P_{DC_i}^{(k)} - P_{DC_i}\left(U_{DC}^{(j)}\right) & \forall i \leq k \\ -P_{DC_i}\left(U_{DC}^{(j)}\right) & \forall k \leq i \leq n \end{cases} \tag{13.62}$$

with superscripts (j) and (k) respectively referring to the inner Newton–Raphson iteration and the outer AC/DC power flow iteration. The terms of the Jacobian are

$$\left(U_{DC_j}\frac{\partial P_{DC_i}}{\partial U_{DC_j}}\right)^{(j)} = -U_{DC_i}^{(j)} Y_{DC_{ij}} U_{DC_j}^{(j)} \tag{13.63}$$

$$\left(U_{DC_i}\frac{\partial P_{DC_i}}{\partial U_{DC_i}}\right)^{(j)} = P_{DC_i}^{(j)} + U_{DC_i}^{(j)\,2} \cdot \sum_{\substack{j=1 \\ j\neq i}}^{n} Y_{DC_{ij}} \tag{13.64}$$

The equations and terms corresponding to the slack bus are removed since its voltage is known prior to the DC network power flow. After convergence, the voltages on all DC buses are known, while the slack bus power injection can be found using (13.42).

13.4.5.2 Voltage Droop Converters In the case of voltage droop converters at buses 2 to m, only the DC power injection set-points $P_{DC,0}$ are known at these buses. The problem can now be rewritten by defining a modified active power vector P'_{DC} such that

$$P'_{DC} = [\ \underbrace{P_{DC_1}}_{\text{slack}}, \underbrace{P_{DC,0_2} \ldots P_{DC,0_m}}_{\text{voltage droop}}, \underbrace{P_{DC_{m+1}} \ldots P_{DC_k}}_{P\text{-control}}, \underbrace{0 \ldots 0}_{\text{outage}}\ ]^T \tag{13.65}$$

The DC bus voltages are calculated with a Newton method

$$\left(U_{DC} \frac{\partial P'_{DC}}{\partial U_{DC}} \right)^{(j)} \cdot \frac{\Delta U_{DC}}{U_{DC}}^{(j)} = \Delta P'^{(j)}_{DC} \tag{13.66}$$

with the modified power mismatch vector $\Delta P'^{(j)}_{DC}$ given by

$$\Delta P'^{(j)}_{DC_i} = \begin{cases} P_{DC,0_i} - P_{DC,0_i}\left(U^{(j)}_{DC}\right) & \forall i:\ 2 \le i \le m \quad \text{(voltage droop)} \\ P^{(k)}_{DC_i} - P_{DC_i}\left(U^{(j)}_{DC}\right) & \forall i:\ m < i \le k \quad (P\text{-control}) \\ -P_{DC_i}\left(U^{(j)}_{DC}\right) & \forall i:\ k < i \le n \quad \text{(outage)} \end{cases} \tag{13.67}$$

and $P_{DC,0_i}(U^{(j)}_{DC})$ as

$$P^{(j)}_{DC,0_i} = P_{DC_i}\left(U^{(j)}_{DC}\right) + \frac{1}{k_{DC_i}}\left(U^{(j)}_{DC_i} - U'_{DC,0_i}\right) \tag{13.68}$$

and, similarly to (13.62), superscripts (j) and (k) respectively refer to the inner Newton–Raphson iteration and the outer AC/DC power flow iteration. The Jacobian elements for the voltage droop buses are given by

$$\left(U_{DC_j} \frac{\partial P_{DC,0_i}}{\partial U_{DC_j}} \right)^{(j)} = \left(U_{DC_j} \frac{\partial P_{DC_i}}{\partial U_{DC_j}} \right)^{(j)} \tag{13.69}$$

$$\left(U_{DC_i} \frac{\partial P_{DC,0_i}}{\partial U_{DC_i}} \right)^{(j)} = \left(U_{DC_i} \frac{\partial P_{DC_i}}{\partial U_{DC_i}} \right)^{(j)} + \frac{1}{k_{DC_i}} U^{(j)}_{DC,i} \tag{13.70}$$

13.4.6 DC Slack Bus and Droop Buses Iteration

The AC bus active power injection P_s of the DC slack and droop buses is calculated from its DC power P_{DC} by accounting for the converter losses. As the converter losses from (13.9) depend on the yet unknown converter current, an additional iteration is needed to calculate the active power injection P_s. During this iteration, the grid side voltage $\underline{U}_{s_1}$ and reactive power injection Q_s are kept constant. The value of P_c can iteratively be updated according to

$$P^{(i)}_c = -P^{(k)}_{DC} - P^{(i)}_{loss} \tag{13.71}$$

with the superscripts (i) and (k) respectively referring to the DC slack bus iteration and the outer AC/DC power flow iteration. The converter losses P_{loss} are calculated from (13.9). The previous AC network power flow results are used to provide an initial estimate for the converter losses $P^{(0)}_{loss}$.

Figure 13.12 shows the flow chart of the DC slack bus iteration.

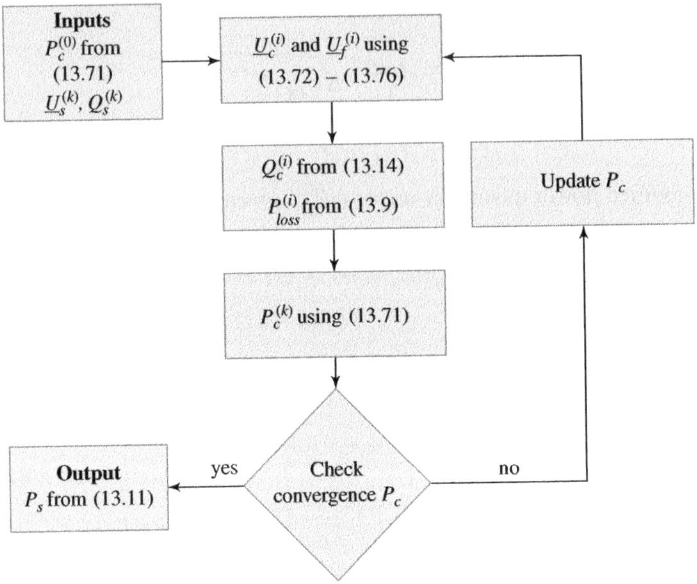

Figure 13.12 DC Slack and droop buses iteration flow chart.

13.4.6.1 Full Model A Newton–Raphson iteration based on $\underline{U}_c$ and $\underline{U}_f$ as variables is internally used to update the converter state in order to obtain a new value for $P_{loss}^{(i)}$,

$$\begin{bmatrix} \left(\frac{\partial P_c}{\partial \theta_c}\right)^{(j)} & \left(\frac{\partial P_c}{\partial \theta_f}\right)^{(j)} & \left(U_c\frac{\partial P_c}{\partial U_c}\right)^{(j)} & \left(U_f\frac{\partial P_c}{\partial U_f}\right)^{(j)} \\ 0 & \left(\frac{\partial Q_s}{\partial \theta_f}\right)^{(j)} & 0 & \left(U_f\frac{\partial Q_s}{\partial U_f}\right)^{(j)} \\ \left(\frac{\partial F_1}{\partial \theta_c}\right)^{(j)} & \left(\frac{\partial F_1}{\partial \theta_f}\right)^{(j)} & \left(U_c\frac{\partial F_1}{\partial U_c}\right)^{(j)} & \left(U_f\frac{\partial F_1}{\partial U_f}\right)^{(j)} \\ \left(\frac{\partial F_2}{\partial \theta_c}\right)^{(j)} & \left(\frac{\partial F_2}{\partial \theta_f}\right)^{(j)} & \left(U_c\frac{\partial F_2}{\partial U_c}\right)^{(j)} & \left(U_f\frac{\partial F_2}{\partial U_f}\right)^{(j)} \end{bmatrix} \cdot \begin{bmatrix} \Delta\theta_c^{(j)} \\ \Delta\theta_f^{(j)} \\ \frac{\Delta U_c}{U_c}^{(j)} \\ \frac{\Delta U_f}{U_f}^{(j)} \end{bmatrix} = \begin{bmatrix} \Delta P_c^{(j)} \\ \Delta Q_s^{(j)} \\ -F_1^{(j)} \\ -F_2^{(j)} \end{bmatrix} \tag{13.72}$$

with the functions F_1 and F_2 given by

$$F_1(\underline{U}_s, \underline{U}_c, \underline{U}_f) = P_{cf} - P_{sf} \tag{13.73}$$

$$F_2(\underline{U}_s, \underline{U}_c, \underline{U}_f) = Q_{cf} - Q_{sf} - Q_f \tag{13.74}$$

The power mismatches $\Delta P_c^{(j)}$ and $\Delta Q_s^{(j)}$ can be calculated by respectively using (13.13) and (13.12) and are given by

$$\Delta P_c^{(j)} = P_c^{(i)} - P_c\left(\underline{U}_f^{(j)}, \underline{U}_c^{(j)}\right) \tag{13.75}$$

$$\Delta Q_s^{(j)} = Q_s^{(k)} - Q_s\left(\underline{U}_s^{(k)}, \underline{U}_f^{(j)}\right) \tag{13.76}$$

The superscript (j) refers to the inner Newton–Raphson iteration. The elements of the Jacobian matrix can be analytically derived from (13.11)–(13.19) directly and by substitution in the expressions for F_1 and F_2:

$$\left(\frac{\partial P_c}{\partial \theta_c}\right)^{(j)} = -Q_c^{(j)} - U_c^{(j)\,2} B_c \tag{13.77}$$

$$\left(\frac{\partial P_c}{\partial \theta_f}\right)^{(j)} = Q_c^{(j)} + U_c^{(j)\,2} B_c \tag{13.78}$$

$$\left(U_c \frac{\partial P_c}{\partial U_c}\right)^{(j)} = P_c^{(j)} + U_c^{(j)\,2} G_c \tag{13.79}$$

$$\left(U_f \frac{\partial P_c}{\partial U_f}\right)^{(j)} = P_c^{(j)} - U_c^{(j)\,2} G_c \tag{13.80}$$

$$\left(\frac{\partial Q_s}{\partial \theta_f}\right)^{(j)} = -P_s^{(j)} - U_s^{(k)\,2} G_{tf} \tag{13.81}$$

$$\left(U_f \frac{\partial Q_s}{\partial U_f}\right)^{(j)} = Q_s^{(j)} - U_s^{(k)\,2} B_{tf} \tag{13.82}$$

$$\left(\frac{\partial F_1}{\partial \theta_c}\right)^{(j)} = Q_{cf}^{(j)} - U_f^{(j)\,2} B_c \tag{13.83}$$

$$\left(\frac{\partial F_1}{\partial \theta_f}\right)^{(j)} = -Q_{cf}^{(j)} + Q_{sf}^{(j)} + U_f^{(j)\,2} (B_c + B_{tf}) \tag{13.84}$$

$$\left(U_c \frac{\partial F_1}{\partial U_c}\right)^{(j)} = P_{cf}^{(j)} + U_f^{(j)\,2} G_c \tag{13.85}$$

$$\left(U_f \frac{\partial F_1}{\partial U_f}\right)^{(j)} = P_{cf}^{(j)} - P_{sf}^{(j)} - U_f^{(j)\,2} (G_c + G_{tf}) \tag{13.86}$$

$$\left(\frac{\partial F_2}{\partial \theta_c}\right)^{(j)} = -P_{cf}^{(j)} - U_f^{(j)} G_c \tag{13.87}$$

$$\left(\frac{\partial F_2}{\partial \theta_f}\right)^{(j)} = P_{cf}^{(j)} - P_{sf}^{(j)} + U_f^{(j)\,2} (G_c + G_{tf}) \tag{13.88}$$

$$\left(U_c \frac{\partial F_2}{\partial U_c}\right)^{(j)} = Q_{cf}^{(j)} - U_f^{(j)\,2} B_c \tag{13.89}$$

$$\left(U_f \frac{\partial F_2}{\partial U_f}\right)^{(j)} = Q_{cf}^{(j)} - Q_{sf}^{(j)} + U_f^{(j)\,2} (B_c + B_{tf} + 2B_f) \tag{13.90}$$

with the converter powers from (13.11)–(13.19).

13.4.6.2 Transformerless Model In the absence of the transformer, the active and reactive power injections can be rewritten in terms of $\underline{U}_s$ and $\underline{U}_c$ as

$$P_s = -U_s^2 G_c + U_s U_c [G_c \cos(\theta_s - \theta_c) + B_c \sin(\theta_s - \theta_c)] \tag{13.91}$$

$$Q_s = U_s^2 B_{cf} + U_s U_c [G_c \sin(\theta_s - \theta_c) - B_c \cos(\theta_s - \theta_c)] \tag{13.92}$$

with $B_c + B_f$ abbreviated as B_{cf}. By rewriting Q_s, the Jacobian matrix simplifies to a 2×2 structure retaining $\Delta\delta_c$ and $\frac{\Delta U_c}{U_c}$ as unknowns and retaining the derivates related to the P_c and Q_s equations as Jacobian elements

$$\begin{bmatrix} \left(\frac{\partial P_c}{\partial \theta_c}\right)^{(j)} & \left(U_c \frac{\partial P_c}{\partial U_c}\right)^{(j)} \\ \left(\frac{\partial Q_s}{\partial \theta_c}\right)^{(j)} & \left(U_c \frac{\partial Q_s}{\partial U_c}\right)^{(j)} \end{bmatrix} \begin{bmatrix} \Delta\theta_c^{(j)} \\ \frac{\Delta U_c}{U_c}^{(j)} \end{bmatrix} = \begin{bmatrix} \Delta P_c^{(j)} \\ \Delta Q_s^{(j)} \end{bmatrix} \tag{13.93}$$

with $\left(\frac{\partial P_c}{\partial \theta_c}\right)^{(j)}$ and $\left(U_c \frac{\partial P_c}{\partial U_c}\right)^{(j)}$ given by expressions (13.77) and (13.79). The other elements of the reduced Jacobian matrix are given by

$$\left(\frac{\partial Q_s}{\partial \theta_c}\right)^{(j)} = -P_s^{(j)} - U_s^{(k)\,2} G_c \tag{13.94}$$

$$\left(U_c \frac{\partial Q_s}{\partial U_c}\right)^{(j)} = Q_s^{(j)} - U_s^{(k)\,2} B_{cf} \tag{13.95}$$

with P_s, Q_s from (13.91)–(13.92) and P_c, Q_c from (13.13) – (13.14) with $\underline{U}_f$ replaced by $\underline{U}_s$.

13.4.6.3 Reduced Model (Without Filter) In the absence of the converter filters, the equations can be reduced by lumping together the transformer and converter impedance in $\underline{Z}_c$ (Figure 13.7c), resulting in the forms

$$\begin{bmatrix} \left(\frac{\partial P_c}{\partial \theta_c}\right)^{(j)} & \left(U_c \frac{\partial P_c}{\partial U_c}\right)^{(j)} \\ \left(\frac{\partial Q_s}{\partial \theta_c}\right)^{(j)} & \left(U_c \frac{\partial Q_s}{\partial U_c}\right)^{(j)} \end{bmatrix} \begin{bmatrix} \Delta\theta_c^{(j)} \\ \frac{\Delta U_c}{U_c}^{(j)} \end{bmatrix} = \begin{bmatrix} \Delta P_c^{(j)} \\ \Delta Q_s^{(j)} \end{bmatrix} \tag{13.96}$$

with the elements of the Jacobian matrix given by

$$\left(\frac{\partial P_c}{\partial \theta_c}\right)^{(j)} = -Q_c^{(j)} - U_c^{(j)\,2} B_c \tag{13.97}$$

$$\left(U_c \frac{\partial P_c}{\partial U_c}\right)^{(j)} = P_c^{(j)} + U_c^{(j)\,2} G_c \tag{13.98}$$

$$\left(\frac{\partial Q_s}{\partial \theta_c}\right)^{(j)} = -P_s^{(j)} - U_s{}^2 G_c \tag{13.99}$$

$$\left(U_c \frac{\partial Q_s}{\partial U_c}\right)^{(j)} = Q_s^{(j)} - U_s{}^2 B_c \tag{13.100}$$

with values for P_S, Q_S, P_C, and Q_C from (13.20)–(13.23). As pointed out in Section 13.3, this reduced model is sufficient in most practical cases. This especially holds when MMC converters with very low AC filter requirements are considered.

13.5 SOFTWARE IMPLEMENTATION

The advantage of using a sequential approach, as described in this chapter, is that the models can relatively easy be combined with existing AC power flow software. As an example, the program MATACDC is shortly discussed here. MATACDC allows the study of the steady-state interactions between multiple nonsynchronized AC grids and DC systems based on VSC HVDC technology. Besides regular power flow models, a variety of converter models and control representations, such as DC voltage droop control have been included, as well as a number of DC system contingencies (e.g., converter outages), which makes the program ideally suited for researchers and students working in the emerging field of HVDC grids. The program is freely available on the internet [10] and a fully documented user manual [11] and source code are provided. MATACDC is fully compatible with the power flow routines in MATPOWER [12], which is available online [13]. The program has a similar syntax, code structure and command line usage, which makes its use straightforward for persons who are already familiar with MATPOWER.

Figure 13.13 shows the program flow structure of MATACDC, as an example. The power flow algorithm sequentially solves the AC and DC system power flow, thereby respectively keeping all converter powers and voltages constant.

13.6 TEST CASE

As an example, results on the CIGRE DC Grid benchmark test system [14] are shown in Figure 13.14, for a converter outage of the offshore wind farm converter connected to bus Bb-D1. The HVDC networks consist of a symmetric monopolar link (±200 kV), a meshed 5-terminal bipolar HVDC grid (±400 kV), and a symmetric monopolar radial multi-terminal system with four converters (±200 kV). The networks consist of both overhead lines and cables. The three AC networks Ba-A0 to Ba-A1, Ba-B0 to Ba-B3, and Bo-C1 to Bo-C2 have operating voltages of respectively 380 kV, 380 kV, and 145 kV. Buses Ba-A0, Ba-B0, and Bo-C1 have been chosen as the AC slack nodes for the different AC grids. The converter connected to Bm-A1 is under constant DC voltage control (slack node), and the AC/DC converters connected to Bm-C1, Bb-C2, Bb-D1, Bm-E1, and Bm-F1 have a constant active power injection. All

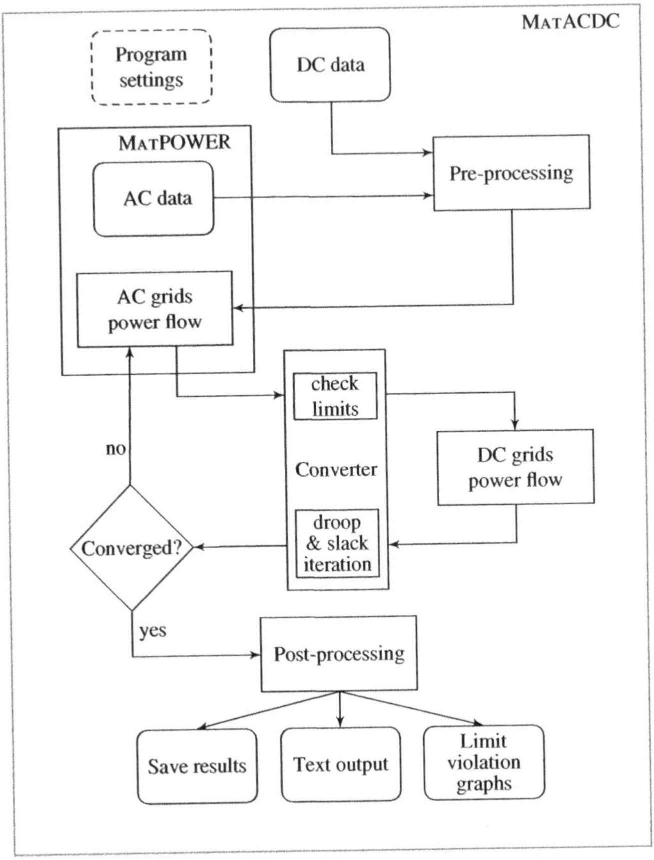

Figure 13.13 MATACDC program flow.

remaining converters have a power–voltage droop. The picture shows the overall control response of these droop-controlled converters to an outage of converter Bb-D1.

13.7 CONCLUSIONS

This chapter discussed the power flow of a combined AC/DC system. In its most simple form, the converters can be modeled as power injections to the AC system with coupled active power injections and independent reactive power setpoints. When the actual state of the HVDC system is of interest, a more accurate representation of the DC grid is needed. For steady-state purposes, the converter can generally be represented as a complex voltage source behind an impedance. In more elaborated models, the filters and transformers have to be modeled separately, which influences the converter's capability chart. The detailed models described in this chapter include the converter losses, and limits and allow for various representations of converters, either including transformers and/or filters. The power flow equations for the DC grid

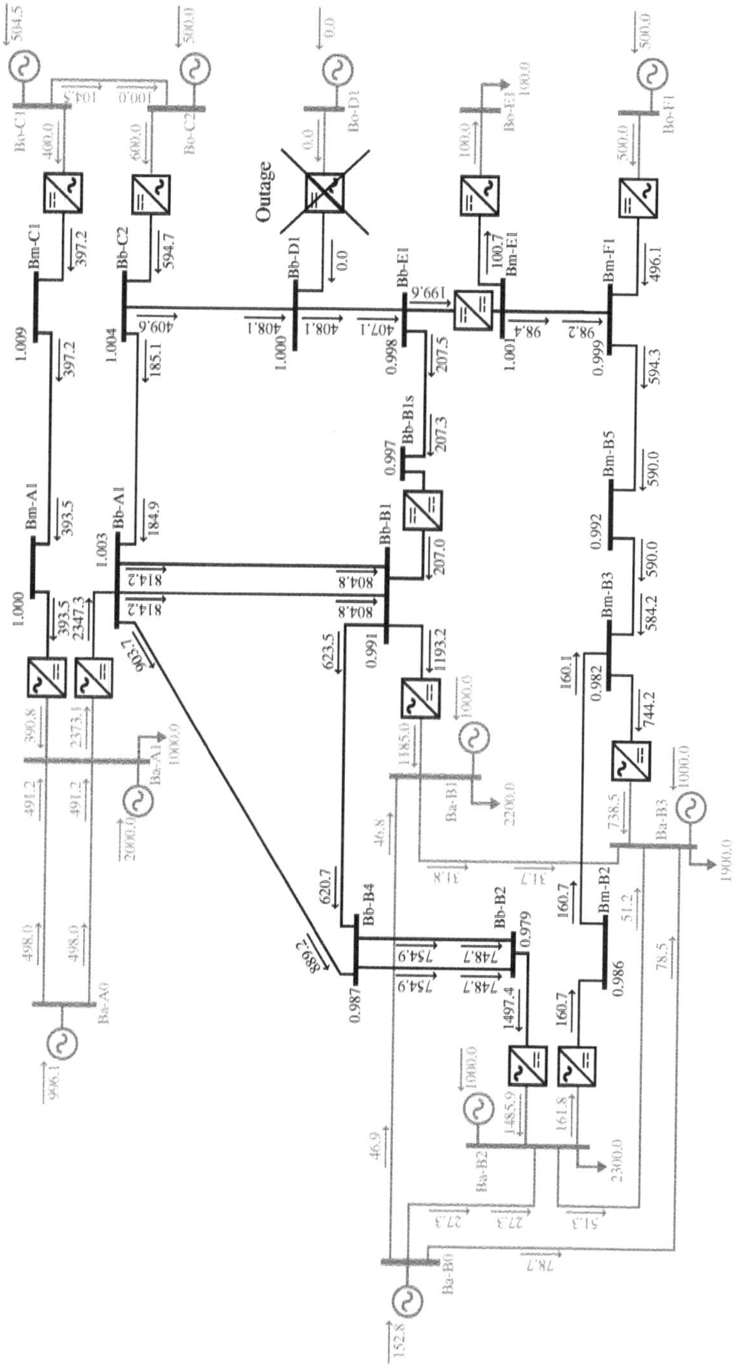

Figure 13.14 MatACDC power flow solution of a mixed AC/DC test system, the CIGRE DC grid benchmark test system with an outage of converter at bus Bb-D1.

were derived for various converter control representations, including constant active power, voltage droop, and constant DC voltage. Combining these different control representations in an HVDC grid allows us to study the overall system effect of these controls on the power flows in the AC/DC system.

REFERENCES

1. A. El-Abiad and G. Stagg, *Computer Methods in Power System Analysis. Stagg*. New York: McGraw-Hill, 1968.
2. J. Grainger and W. Stevenson, *Power System Analysis*, McGraw-Hill Series in Electrical and Computer Engineering: Power and Energy. New York: McGraw-Hill, 1994.
3. G. Daelemans, K. Srivastava, M. Reza, S. Cole, and R. Belmans, "Minimization of steady-state losses in meshed networks using VSC HVDC," in *Proceedings IEEE PES GM 2009*, Calgary, Canada, July 26–30, 2009.
4. S. Johansson, G. Asplund, E. Jansson, and R. Rudervall, "Power system stability benefits with VSC DC-transmission systems," in *Proceedings CIGRÉ 2004 Session*, Paris, France, 2004, 8 pages.
5. M. P. Bahrman and B. K. Johnson, "The ABCs of HVDC transmission technologies," *IEEE Power Energy Magazine*, vol. 5, no. 2, pp. 32–44, 2007.
6. G. P. Adam, O. Anaya-Lara, and G. Burt, "Steady-state and transient performance of DC transmission systems based on HVDC technology," in *Proceedings IET ACDC 2010*, London, UK, Oct. 20–21, 2010, 5 pages.
7. A. Pizano-Martinez, C. R. Fuerte-Esquivel, H. Ambriz-Perez, and E. Acha, "Modeling of VSC-based HVDC systems for a Newton-Raphson OPF algorithm," *IEEE Trans. Power Syst.*, vol. 22, no. 4, pp. 1794–1803, Nov. 2007.
8. X.-P. Zhang, "Multiterminal voltage-sourced converter-based HVDC models for power flow analysis," *IEEE Transactions on Power Systems*, vol. 19, no. 4, pp. 1877–1884, Nov. 2004.
9. M. Baradar and M. Ghandhari, "A multi-option unified power flow approach for hybrid AC/DC grids incorporating multi-terminal VSC-HVDC," *IEEE Transactions on Power Systems*, vol. 28, no. 3, pp. 2376–2383, Aug. 2013.
10. MatACDC website [Online]. Available: http://www.esat.kuleuven.be/electa/teaching/matacdc/
11. *MatACDC 1.0 User's Manual*, University of Leuven (KU Leuven) Department of Electrical Engineering (ESAT), Division ELECTA, Heverlee-Leuven, Belgium.
12. R. D. Zimmerman, C. E. Murillo-Sanchez, and R. J. Thomas, "MATPOWER: Steady-state operations, planning, and analysis tools for power systems research and education," *IEEE Transactions on Power Systems*, vol. 26, no. 1, pp. 12–19, Feb. 2011.
13. MATPOWER website. [Online]. Available: http://www.pserc.cornell.edu/matpower/
14. T. K. Vrana, S. Dennetière, Y. Yang, J. Jardini, D. Jovcic, and H. Saad, "The CIGRE B4 DC grid test system," *ELECTRA*, no. 270, pp. 10–19, 2013.

CHAPTER 14

OPTIMAL POWER FLOW MODELING OF HYBRID AC/DC SYSTEMS

Johan Rimez
Division TGX, Elia Engineering, Brussels, Belgium

14.1 INTRODUCTION

Next to the traditional power flow calculations as detailed in Chapter 13, the system operator also makes use of the optimal power flow (OPF) programs to deterime the optimal operating point of the system. This chapter extends the equations of the previous chapter into an OPF formulation. The theoretical and practical implementation of optimal power flow (OPF) for DC grids, both the standard and security constrained OPF are presented as an extension to the deterministic AC optimal power flow case. First, the AC optimal power flow case is briefly introduced, and the global convergence of optimization cases is discussed. A methodology is presented to extend a given optimization case with additional state variables and constraint equations in order to add new functionality to the original case.

The main part of the chapter consists of the extension of the AC OPF for hybrid AC/DC systems. This is done for a normal OPF and for the security constrained OPF. As the focus in this chapter lies on the OPF formulation for AC/DC hybrid systems, we make use of the simplified models of Section 13.2 to model the DC converter and the DC side. As such, in this chapter the droop, filters, and converter transformers are neglected and the system limits are simplified, although the formulation can easily be extended based on the equations of Chapter 13.

14.2 OPTIMAL POWER FLOW: STANDARD FORMULATION AND EXTENSION

14.2.1 Standard AC Formulation and Convergence

The deterministic optimization problem, together with its sufficiency conditions, is given below [1].

HVDC Grids: For Offshore and Supergrid of the Future, First Edition. By Dirk Van Hertem, Oriol Gomis-Bellmunt, and Jun Liang.

Assume the problem

$$\min_{x} \quad f(x) \tag{14.1}$$

$$x \in X \tag{14.2}$$

$$g(x) = 0 \tag{14.3}$$

$$h(x) \leq 0 \tag{14.4}$$

with $f : \Re^n \mapsto \Re$, $g : \Re^m \mapsto \Re$, $h : \Re^r \mapsto \Re$ and X a given subset of $\Re^n$. Let $x^\star$ be a vector which satisfies all constraints, but not necessarily the optimum of $f(x)$ in $\Re^n$, and which, together with the vectors $\mu^\star$ and $\lambda^\star$, minimizes the Lagrangian function:

$$\mathfrak{L}(x, \mu, \lambda) = f(x) + \mu^T g(x) + \lambda^T h(x) \tag{14.5}$$

over $x \in X$:

$$x^\star = \arg\min_{x \in X} \mathfrak{L}(x, \mu^\star, \lambda^\star) \tag{14.6}$$

and it satisfies the following conditions:

$$\lambda_i^\star = 0, \quad \text{for } \forall i \text{ with } h_i(x^\star) < 0 \tag{14.7}$$

$$\lambda_i^\star > 0, \quad \text{for } \forall i \text{ with } h_i(x^\star) = 0 \tag{14.8}$$

where $x^\star$ is the *global* minimum of the problem.

In the general case, this is a problem with a nonlinear objective function $f(x)$ to be minimized and taking into account the nonlinear equality constraints $g(x)$ and inequality constraints $h(x)$. The elements of the vector x are called the *state variables*.

The optimum can be found either by analytically solving (14.6), which proves unfeasible for large and/or complicated systems. Other methods use iterative algorithms that, starting from a initial point, try to find the optimum in an efficient way with respect to number of iterations and calculation speed. Such methods, and the solvers available, not only differ in convergence speed, but also in the ability to find the global optimum and in the types of problems that can be solved (e.g., nonlinearity or the management of integer decision variables).

For the remainder of this chapter, the barrier method is used. When the barrier method is used to augment the Lagrangian function [1], it can be proven that the found optimum is always the global optimum of the optimization problem.

In order to set up the optimization case for an AC system, assume a generic AC grid which contains the following network elements:

- n_b, the number of nodes in service.
- n_l, the number of network branches in service, typically overhead lines, underground cables, power transformers, and phase-shifting transformers.
- n_g, the number of active generators. Different generators connected to the same node can be aggregated to one single unit as a first simplification.
- Passive shunt elements, like capacitor banks and shunt reactors, which can only be switched on and off, are aggregated per node and treated in the same way as in power flow calculations.

- Active shunt elements, SVCs, STATCOM, … , are modeled as special-purpose network generators on a PV node.

The purpose of the OPF calculation is to find a set of state variables, defining a viable working point of the network, complying with the imposed constraints and which results in the minimum (or maximum) of the defined cost function. This cost function can be any function of the state variables: lowest losses, highest social welfare, operating point closest to the set-points, and so on.

The state variable vector used for the traditional AC optimal power flow formulation is

$$x = \begin{bmatrix} \Theta \\ U \\ P \\ Q \end{bmatrix} \tag{14.9}$$

where Θ are the n_b bus voltage angles, U the n_b bus voltage magnitudes, P the n_g generator active, and Q the n_g reactive power injections. The length of the vector x is thus $2n_b + 2n_g$ and contains only continuous values.

In order to build a more realistic model, the state vector x needs to be extended with a set of discrete value parameters: for example, tap position of power and phase-shifting transformers, state of capacitor banks, and so on. This extension of the optimization problem formulation can be found in literature, but has not been covered in detail here.

Linear and nonlinear constraints can be imposed on the optimization parameters in order not to exceed physical or other limitations of the network elements, and as such to find acceptable working points.

The upper and lower limits of the individual state variables are expressed as linear constraints:

- For the bus voltage angle vector Θ, in the normal case, only the voltage angle of the slack bus (SB) is set to 0°.
- Minimum and maximum voltage limitations are used for each of the bus voltage amplitudes U_i following operational experience, grid codes, and power quality requirement. For example, the maximum value is given by the equipment insulation level, and setting a minimum level lowers the risk of voltage collapse.
- Limitations on the vectors P_i and Q_i, the active and reactive power output of the network generators, give the power range for proper operation of that specific power plant.

The linear constraints thus create the following set of $2 + 2n_b + 4n_g$ linear inequality constraints:

$$\begin{aligned} &\theta_{SB} = 0 \\ &U_{i,min} \le U_i \le U_{i,max} \\ &P_{i,min} \le P_i \le P_{i,max} \\ &Q_{i,min} \le Q_i \le Q_{i,max} \end{aligned} \tag{14.10}$$

Linear equality constraints, such as the first equation of the slack bus reference in the set (14.10), are rewritten as (14.11) and count as a double linear inequality constraint.

$$x = x_0 \Leftrightarrow x_0 \leq x \leq x_0 \tag{14.11}$$

14.2.1.1 Nonlinear Equality Constraints The AC network configuration, the branch impedance, active and reactive power absorbed by the network loads, and the injections of the connected generators are mathematically described by nonlinear equality constraints. These expressions are the very same as the power flow equations, stating the active and reactive power balance in each node. Contrarily to the power flow calculation, only the reference bus angle, the network loads, and branch configuration are usually fixed. The notions slack bus, PV node, and PQ node are undefined in this context. As such, the expressions for all nodes, independent of their power flow context, are treated in the same uniform way. An additional advantage of splitting the equations into their active and reactive part is that direct calculation with complex numbers is avoided.

For a network with n_b buses, a total of $2n_b$ equality constraints is defined as

$$\mathbf{g}_P(\boldsymbol{\Theta}, \mathbf{U}, \mathbf{P}) = 0 \tag{14.12a}$$

$$\mathbf{g}_Q(\boldsymbol{\Theta}, \mathbf{U}, \mathbf{Q}) = 0 \tag{14.12b}$$

The expressions for $\mathbf{g}_P(\boldsymbol{\Theta}, \mathbf{U}, \mathbf{P})$ and $\mathbf{g}_Q(\boldsymbol{\Theta}, \mathbf{U}, \mathbf{Q})$ follow from the power flow equations as presented in the previous chapter: For an arbitrary node i, this results in Eqs. (14.13a) and (14.13b):

$$\begin{aligned} g_{P,i}(\boldsymbol{\Theta}, \mathbf{U}, \mathbf{P}) &= P_{l,i} - P_{g,i} + P_{\Sigma br,i} \\ &= P_{l,i} - P_{g,i} + \sum_{j=1}^{n_b} Y_{ij} U_i U_j \cos(\theta_i - \theta_j - \varphi_{ij}) \end{aligned} \tag{14.13a}$$

$$\begin{aligned} g_{Q,i}(\boldsymbol{\Theta}, \mathbf{U}, \mathbf{Q}) &= Q_{\Sigma br,i} + Q_{l,i} - Q_{g,i} \\ &= Q_{l,i} - Q_{g,i} + \sum_{j=1}^{n_b} Y_{ij} U_i U_j \sin(\theta_i - \theta_j - \varphi_{ij}) \end{aligned} \tag{14.13b}$$

with $P_{\Sigma br,i}$ and $Q_{\Sigma br,i}$ the active and reactive power flowing from node i into the branches. Y is the admittance matrix of the AC grid.

$P_{l,i}$ and $Q_{l,i}$ are fixed values (for the standard OPF case used here) and state the sum of active and reactive power extractions of the network loads connected to the AC bus i. And finally, $P_{g,i}$ and $Q_{g,i}$ are the total power *injections* (hence the minus sign) by the combined generators in that bus.

As these equations are forced to be zero, the optimization algorithm is forced to use the actual grid definition and power balance, hence the use of equality constraints. They are sufficient for the whole AC OPF case [2].

Other nonlinear equality constraints for the AC network case may apply, only (14.13a) and (14.13b) are used in this chapter.

14.2.1.2 Nonlinear Inequality Constraints Inequality constraints are used to limit computed function values using state variables to upper and lower bounds, instead of fixing them to predefined values. Direct limitations on the state variables are already introduced by the linear equation set (14.10).

Looking only at the passive network branches, current or power limits can be applied. The branch loading limits taken into account here are the long-term loading capabilities.

Because of the two connection points, each branch b adds two constraints to the OPF case: the current or power flowing into the link from its "from-" (i-) and the "to-" (j-) bus. These currents can be either positive or negative, depending on the current flow direction. One can set an upper and lower bound, multiplying the number of inequality constraints by two. The expression and the corresponding limits are squared, resulting for the branch b together with its current limit value $I_{max,b}$ in

$$\begin{aligned} h_{2b-1}(\mathbf{x}) &= |\underline{I}_{b,i}|^2 - I^2_{max,b} = I^2_{b,i} - I^2_{max,b} \leq 0 \\ h_{2b}(\mathbf{x}) &= |\underline{I}_{b,j}|^2 - I^2_{max,b} = I^2_{i,j} - I^2_{max,b} \leq 0 \end{aligned} \tag{14.14}$$

Here, $b = 1 \ldots n_l$ is the number of branches in the AC system, resulting in $2n_l$ constraints.

Alternatively, when it is desired not to limit the current, but the apparent power $S_{b,i}$ and $S_{b,j}$ (with the limitation $S_{max,b}$) flowing into or out of the branch b, one can express the inequality constraints by

$$\begin{aligned} h_{2b-1}(\mathbf{x}) &= S^2_{b,i} - S^2_{max,b} = (U_i I_{b,i})^2 - S^2_{max,b} \leq 0 \\ h_{2b}(\mathbf{x}) &= S^2_{b,j} - S^2_{max,b} = (U_l I_{b,j})^2 - S^2_{max,b} \leq 0 \end{aligned} \tag{14.15}$$

Note the use of the amplitude of the complex values of current, voltage, and power.

Putting all the above together, the base OPF case can be generically stated as an optimization problem:

$$\min_{x} \; f(x) \tag{14.16}$$

$$g(x) = 0 \tag{14.17}$$

$$h(x) \leq 0 \tag{14.18}$$

$$x_{min} \leq x \leq x_{max} \tag{14.19}$$

The full overview of all first and second derivatives needed to compile the Jacobian and Hessian[1] matrices of the above system to be implemented in the optimization algorithm is found in reference 3. The calculation of these matrices can be required by the solver algorithms in order to calculate the steepest descent gradient or the second-order (Newton–Raphson style) step updates [1,2].

[1] As the barrier uses the second-order derivatives, the Hessian matrix is needed as an input to the mathematical solver, increasing the convergence speed. For this, the constraints need to be twice differentiable.

14.2.2 Extending the Formulation

In order to provide additional functionality to an existing optimization case (e.g., the extension of an AC grid with a DC grid), often a definition of new state variables is needed. This procedure is quite straightforward in theory, but not many commercial software packages allow such an extension by user-defined parameter variables. Some (e.g., MATPOWER), allow the user to define additional state variables and the related linear equality and inequality constraints [4]. However, nothing has been foreseen for the user to implement his own nonlinear equations. That functionality is specifically required in order to implement the HVDC grid model and the security constraints.

Assume an optimization case with its state variable vector x (n_x elements) and the vector of the additional state variables z (n_z elements):

$$x_n = \begin{bmatrix} x \\ z \end{bmatrix} \tag{14.20}$$

The optimization case definition is affected as well, as the use of new state variables comes together with an additional set of equality and inequality constraints and a new cost function f_n, a function of the original and the new state variables. The subscript o is used for the original constraint functions and n for the added ones. The extended optimization case is

$$\min_{x,z} \quad f_n(x, z) \tag{14.21}$$

$$g(x_n) = \begin{bmatrix} g_o(x) \\ g_n(x, z) \end{bmatrix} = 0 \tag{14.22}$$

$$h(x_n) = \begin{bmatrix} h_o(x) \\ h_n(x, z) \end{bmatrix} \leq 0 \tag{14.23}$$

$$x_{min} \leq x \leq x_{max} \tag{14.24}$$

$$z_{min} \leq z \leq z_{max} \tag{14.25}$$

Whatever optimization algorithm is used, the nonlinear optimization techniques require us to extend the Jacobian and Hessian matrices for parameter functions g_n and h_n with each iteration step. A combination of submatrices depending on the original and the additional constraint vectors needs to be evaluated. Assume the original Jacobian ($n_o \times n_x$ matrix) matrix

$$J_o(x) = n_o\{ \overbrace{\begin{bmatrix} \dfrac{\partial g_o}{\partial x} \end{bmatrix}}^{n_x} \tag{14.26}$$

The number of functions in the original equality constraint vector g_o is assumed n_o and n_n for the extension constraint vector g_n. The extended Jacobian starts from previous expressions and adds all other necessary derivatives of the new constraints.

$$J(x,z) = \begin{matrix} n_o\{ \\ n_n\{ \end{matrix} \begin{bmatrix} \overbrace{J_o}^{n_x} & \overbrace{0}^{n_z} \\ \dfrac{\partial g_n}{\partial x} & \dfrac{\partial g_n}{\partial z} \end{bmatrix} \tag{14.27}$$

The Hessian matrix for a function vector is in theory three-dimensional, expressing the possible combinations of the second derivatives to its parameters (14.28). The Hessian consists of $n_0 + n_n$ "layers", with $n_0 + n_n$ the combined number of (nonlinear) functions in the function vector for which the calculation of its Hessian matrix is desired. Each $(n_x + n_z) \times (n_x + n_z)$-sized layer corresponds to all the second derivatives of one of these functions. Here, $n_x + n_z$ is the combined number of the original and new state variables of the optimization case.

$$H(x) = \left[\frac{\partial^2 g_i}{\partial x_{n,j} \partial x_{n,k}} \right] \tag{14.28}$$

The Hessian is calculated for each of the layers separately. Some optimization algorithms require the calculation of the dotproduct of the Hessian with a given column vector (for example, the vector of the corresponding Lagrange parameters μ). The result is a more conventional two-dimensional matrix

$$\mu \cdot H(x_n) = \sum_{i=1}^{n_o+n_n} \mu_i \left[\frac{\partial^2 g_i}{\partial x_{n,j} \partial x_{n,k}} \right] \tag{14.29}$$

The same matrix constructions apply for the Jacobian and the Hessian of the inequality function vector $h(x)$.

14.3 OPTIMAL POWER FLOW WITH DC GRIDS AND CONVERTERS

In this section, the general approach and the detailed development are given for the modeling of the meshed DC grid parallel to the AC grid through the AC/DC converter interface points [5]. It starts from the general approach, basic assumptions, and state-of-the-art. The VSC model and DC grid model are shown based on the simplified model of Chapter 13. A general OPF formulation for hybrid AC/DC systems is shown, including both AC and DC grid constraints.

14.3.1 General Approach and State-of-the-Art

The general approach to model the DC system and the AC/DC converters, whether it is a mere point-to-point link [6, 7], a radial grid, or a complex meshed grid with a multitude of AC/DC converters, remains the same. The implementation flexibility comes from the systematic approach of how the different components (converters and other grid elements) and DC grid structure are described and written out as (nonlinear) equality constraints.

In the following, the coupling between the AC and DC systems is exclusively made by VSC-HVDC converters as it is the technology which allows the construction of generic DC networks.

The number of VSC converters in the complete network is n_v, the number of DC branches n_c and the number of DC nodes n_N. A DC node in this context is defined as the location in the electrical DC network where two or more DC elements (converters and/or branches) are connected. A point-to-point link contains $n_N = 2$ DC nodes. Furthermore, it is assumed that all VSC converters in the same DC bus are modeled as one. Separate auxiliary DC buses can be introduced if this cannot be done (e.g., when the converters are not connected to the same AC bus).

The DC system uses the same power base as the AC grid. The reference voltage, expressed in *pu*, may differ. The model assumes that the rated voltage of the DC system is well-chosen so that all converters can operate in the given DC voltage ranges.

In references 8 and 6, the authors propose a unified optimal power flow algorithm as presented in this manuscript; however, only point-to-point HVDC systems have been developed and more emphasis is placed on the choice of control system parameters and market consequences. The method described in [9] follows the same outline as this research; it is able to do an optimization of the hybrid AC/DC meshed network in a unified way. The authors of reference 10 propose the optimization of control scheme parameters, in this case the droop control voltage-power on the DC side of the VSC converters. Their application to minimize the losses of the intermediate DC grid connect several large wind power plants to the main AC grid on several interface substations.

The optimization problem can be extended to the time domain, where a high number of controller gains and other parameters need to be chosen [11].

Not only HVDC systems, but also other flexible elements such as FACTS applications, can be incorporated into optimal power flow calculations [12].

14.3.2 Modeling VSC-HVDC Converters

The n_c converters are mathematically described as a set of nonlinear equality constraints. For each converter, the equation of the power balance between the AC and DC is used, differing by a loss term P_{loss}, i.e. the converter losses. On the AC side, the HVDC installation is emulated as a normal AC generator. On the DC side, an equivalent controlled DC source is chosen.

The notation and direction convention of active and reactive powers and the voltages are shown in Figure 14.1. Positive values of power on the AC side are injections into that grid (same convention as for the other generators), and positive values on the DC side are always directed towards the converter.

The loss equation for converter with index i, connected on the AC side to bus l modeled by the AC generator with index k, is

$$P_k + P_{loss,i} - P_{DC,i} = 0 \tag{14.30}$$

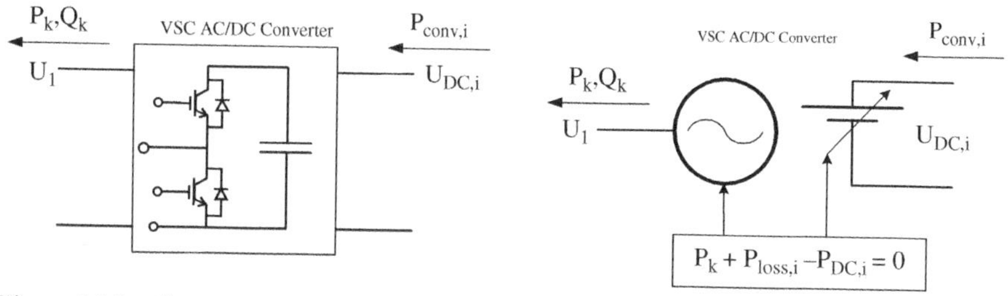

Figure 14.1 Converter model equivalents. Left: AC/DC VSC converter. Right: OPF equivalent.

Regardless of the operation mode, the converter losses are always positive, implying that more active power enters from one side than leaves to the other. In inverter mode, P_k and $P_{DC,i}$ are positive and $P_k < P_{DC,i}$. In rectifier mode, they are negative and again the expression $P_k < P_{DC,i}$ holds. From (14.30), the loss term $P_{loss,i}$ is always positive:

$$P_{loss,i} = P_{DC,i} - P_k > 0 \tag{14.31}$$

The general formulation of the converter i is (14.32) [6]

$$g_i(x) = P_k + P_{loss,i}(U_{m,l}, P_k, Q_k) - P_{DC,i} = 0 \tag{14.32}$$

with $i = 1 \ldots n_v$.

As discussed in Section 13.2.2, the loss equation can in general be represented by a quadratic function of the converter current (13.9).

Combining (14.32), (13.9), and (13.10) gives the complete expression (14.33) of one HVDC converter:

$$g_{conv,i}(x) = P_{k,i} - P_{conv,i} + a_i + b_i \frac{\sqrt{P_k^2 + Q_k^2}}{U_{m,l}} + (c_i + r_i)\frac{P_k^2 + Q_k^2}{U_{m,l}^2} = 0 \tag{14.33}$$

Obviously, in order to avoid impossible solutions, the AC voltage amplitude must be strictly higher than zero. In a realistic OPF case, minimum positive voltage limits are specified, solving this issue automatically. Note that (14.33) does not depend on the DC bus voltage.

Section 14.2.2 describes the methodology to extend the optimization case with additional state variables and constraint equations. For the n_v converters, the same number of extra variables, $\mathbf{P}_{conv} = [P_{conv,i}]$ are added and n_v constraint functions (14.33) are appended to the function vector $g(x)$.

$$\mathbf{J}_{conv,i}(\mathbf{x}) = \left[\frac{\partial g_i}{\partial x_j}\right] \tag{14.34a}$$

$$\mathbf{H}_{conv,i}(\mathbf{x}) = \left[\frac{\partial^2 g_i}{\partial x_j \partial x_k}\right] \tag{14.34b}$$

The commonly used way to define the initial point to start the iterative solving methods (i.e., when not starting from a previous calculation) is by taking the average of the respective upper and lower bounds of the state variables. For symmetric bidirectional grid devices (e.g., VSC converters), the initial power set-points are $P_{k,0} = Q_{k,0} = 0$. For this particular situation, some of the first and second derivatives in the Jacobian and Hessian are undefined. The implemented solution is altering the initial converter set-points to other values within their bounds. As all converters present no-load losses, it is practically excluded that the optimal set-point for a converter $P_k = Q_k = 0$.

14.3.3 Modeling DC Branches and Grid Topology

The same methodology as for the AC network branches and structure, using the admittance matrix, is applied. The equations turn out to be much simpler than in the AC case as only the DC bus voltage magnitudes need to be taken into account and phase angles are irrelevant. Shunt elements are omitted as well (i.e., there is no steady-state reactive charging of the system branches, nor are there iron losses in transformers).

A number of new state variables $\mathbf{U}_{DC}$, equal to the number of DC buses (n_N) in the system, is defined.

The admittance of the branch between the arbitrary nodes i and j is

$$y_{ij} = \begin{cases} 0\,, \text{ when no branch exists between nodes } i \text{ and } j \\ 0\,, \text{ when } i = j \\ \sum_k \frac{1}{r_{ij,k}} \end{cases} \tag{14.35}$$

Here, $r_{ij,k}$ is the series resistance of the kth branch ($k \geq 1$) between DC nodes i and j. The resistance is expressed in *pu*, using the voltage base of the DC network and the same power reference as in the AC network. For convenience, the individual admittances are already summed at the setup of the OPF DC case, should there be more than one branch in parallel between these nodes. Also for the DC case, the admittance matrix is symmetric ($y_{ij} = y_{ji}$).

For the (equivalent) branch between nodes i and j ($i \neq j$), with admittance y_{ij} from (14.35), the relation between currents (positive direction is inward towards the branch) and voltages becomes

$$I_{DC,i} = y_{ij}(U_{DC,i} - U_{DC,j}) \tag{14.36a}$$

$$I_{DC,j} = y_{ij}(-U_{DC,i} + U_{DC,j}) \tag{14.36b}$$

or in matrix notation:

$$\begin{bmatrix} I_{DC,i} \\ I_{DC,j} \end{bmatrix} = \begin{bmatrix} y_{ij} & -y_{ij} \\ -y_{ij} & y_{ij} \end{bmatrix} \cdot \begin{bmatrix} U_{DC,i} \\ U_{DC,j} \end{bmatrix} \tag{14.36c}$$

Using (14.36c), similarly to the summing method for the AC grid, the matrix of the current going out of the node i into the connected DC branches on that node is

$$[I_{br,i}] = [Y_{br,ij}] \cdot [U_{DC,j}] \tag{14.37}$$

with

$$
\begin{aligned}
Y_{br,ii} &= \textstyle\sum_{j=1}^{n_N} y_{ij} \\
Y_{br,ij} &= -y_{ij}\ , i \neq j
\end{aligned}
\tag{14.38}
$$

The active power flowing out of the node i towards the connected DC branches is

$$
\begin{aligned}
P_{br,i} &= U_{DC,i} I_{br,i} \\
&= U_{DC,i} \sum_j Y_{br,ij} U_{DC,j} \\
&= U_{DC,i}^2 Y_{br,ii} + U_{DC,i} \sum_{j \neq i} Y_{br,ij} U_{DC,j}
\end{aligned}
\tag{14.39}
$$

The power balance for the DC bus i is

$$
\begin{aligned}
g_{br,i}(\mathbf{x}) &= \sum_k P_{conv,k} + P_{br,i} \\
&= \sum_k P_{conv,k} + U_{DC,i}^2 Y_{br,ii} + U_{DC,i} \sum_{j \neq i} Y_{br,ij} U_{DC,j}
\end{aligned}
\tag{14.40}
$$

with $j = 1 \ldots n_N$ and $k \in$ {set of converters connected to bus i}. The previous set (14.40) defines n_N nonlinear equality constraints describing the DC branches and topology. Together with the n_v equations of the converters themselves, the DC grid is connected to the AC grid and optimized together as they define a single optimization case.

To calculate the Jacobian, all different nonzero derivatives need to be calculated to the state variables. Consider again the arbitrary bus i. The derivative of the DC power of the converter k connected to that node is

$$
\frac{\partial g_{br,i}(\mathbf{x})}{\partial P_{conv,k}} = 1 \tag{14.41}
$$

with $k \in$ {set of converters connected to bus i}. The derivatives of these constraint functions g_i to the bus voltages differ in nature: For the DC voltage on the same bus i the derivatives (14.42a) are obtained:

$$
\frac{\partial g_{br,i}(\mathbf{x})}{\partial U_{DC,i}} = 2Y_{br,ii} U_{DC,i} + \sum_{j \neq i} Y_{br,ij} U_{DC,j} \tag{14.42a}
$$

and (14.42b) for all other voltages

$$
\frac{\partial g_{br,i}(\mathbf{x})}{\partial U_{DC,j}} = Y_{br,ij} U_{DC,i} \quad \text{with } j \neq i \tag{14.42b}
$$

The nonzero elements of the Hessian are easily calculated by

$$\frac{\partial^2 g_{br,i}(\mathbf{x})}{\partial U_{DC,i}^2} = 2Y_{br,ii} \tag{14.43a}$$

$$\frac{\partial g_{br,i}(\mathbf{x})}{\partial U_{DC,i} \partial U_{DC,j}} = Y_{br,ij} \qquad \text{with } j \neq i \tag{14.43b}$$

Using this method to implement the DC system in the optimization case, $n_v + n_N$ new state variables are defined, using notation convention (14.20):

$$\mathbf{z} = \begin{bmatrix} \mathbf{P}_{conv} \\ \mathbf{U}_{DC} \end{bmatrix} \tag{14.44}$$

with

- $\mathbf{P}_{conv} = [P_{conv,i}]$ with $i = 1 \ldots n_v$, the DC power going into converter i.
- $\mathbf{U}_{DC} = [U_{DC,j}]$ with $j = 1 \ldots n_N$, the DC voltage of the node j.

The additional set of $n_v + n_N$ equality constraints, using the notation convention (14.22), for this method is

$$\mathbf{g}_n(x, z) = \begin{bmatrix} g_{conv,i} \\ g_{br,i}(z) \end{bmatrix} \tag{14.45}$$

with the n_v constraints $g_{conv,i}$ (14.33) for the converters and n_N constraints $\mathbf{g}_{br,i}$ (14.40) for the DC system topology.

14.3.4 Inequality Constraints for the DC Network

Converter and DC branch loading limitations are discussed in this section. The same reasoning as for the AC networks (Section 14.2.1) is followed.

The limitations for the active and reactive power exchange of the AC/DC converters with the AC network are specified by directly imposing the limits on the respective generator state variables (14.10). The limits of the DC power exchange and the DC bus voltages are set directly to the lower and upper limits of the corresponding state variables (14.44).

The branch loading limits need to be explicitly written, and are appended to the matrix of nonlinear inequality constraints (14.23).

For each DC branch, a set of two inequality constraints is defined, one for the maximum loading of the "from-" (subscript f) and the other for the "to-" (subscript t)

bus side of that branch. For DC branch i, they are

$$h_{i,f}(\mathbf{z}) = P_{i,f}^2 - P_{i,max}^2 \tag{14.46a}$$

$$h_{i,t}(\mathbf{z}) = P_{i,t}^2 - P_{i,max}^2 \tag{14.46b}$$

These equations cannot be integrated directly into the optimization case as they need to be written as functions of the defined state variables **z** (14.44). All expressions for the "to"-bus are the same, with the indices f and t interchanged. Only the state variables of the DC bus voltages are used:

$$\begin{aligned} h_{i,f}(\mathbf{x}) &= P_{i,f}^2 - P_{i,max}^2 \\ &= (U_{i,f} \cdot I_{i,f})^2 - P_{i,max}^2 \\ &= \left(U_{i,f}\left(\frac{U_{i,f} - U_{i,t}}{R_i}\right)\right)^2 - P_{i,max}^2 \\ &= \frac{1}{R_i^2}\left(U_{i,f}^4 - 2 \cdot U_{i,f}^3 \cdot U_{i,t} + U_{i,f}^2 \cdot U_{i,t}^2\right) - P_{i,max}^2 \end{aligned} \tag{14.47}$$

In these expressions the voltages $U_{i,f}$ and $U_{i,t}$ are the corresponding DC bus voltages of the *from*- and the *to*- bus of the DC branch i.

The nonzero equations of the first and second derivatives for the from-bus are

$$\begin{aligned} \frac{\partial h_{i,f}(\mathbf{x})}{\partial U_{i,f}} &= \frac{1}{R_i^2}\left(4 \cdot U_{i,f}^3 - 6 \cdot U_{i,f}^2 \cdot U_{i,t} + 2 \cdot U_{i,f} \cdot U_{i,t}^2\right) \\ \frac{\partial h_{i,f}(\mathbf{x})}{\partial U_{i,t}} &= \frac{1}{R_i^2}\left(2 \cdot U_{i,f}^2 \cdot U_{i,t} - 2 \cdot U_{i,f}^3\right) \end{aligned} \tag{14.48}$$

$$\begin{aligned} \frac{\partial^2 h_{i,f}(\mathbf{x})}{\partial U_{i,f}^2} &= \frac{1}{R_i^2}\left(12 \cdot U_{i,f}^2 - 12 \cdot U_{i,f} \cdot U_{i,t} + 2 \cdot U_{i,t}^2\right) \\ \frac{\partial^2 h_{i,f}(\mathbf{x})}{\partial U_{i,t}^2} &= \frac{2}{R_i^2} \cdot U_{i,f}^2 \\ \frac{\partial^2 h_{i,f}(\mathbf{x})}{\partial U_{i,f} \partial U_{i,t}} &= \frac{1}{R_i^2}\left(4 \cdot U_{i,f} \cdot U_{i,t} - 6 \cdot U_{i,f}^2\right) \end{aligned} \tag{14.49}$$

Together with an objective function $f(x)$, the equality constraints $g(x)$ and inequality constraints $h(x)$ form the optimization problem that is typically solved using a computational module or solver.

14.3.5 Simulation Results

Several scenarios using the IEEE 39 bus system have been simulated in order to test the features of the developed algorithms and methodologies. For each of the simulations, the chosen cost function is minimum system losses.

TABLE 14.1 Calculation Comparison of the IEEE 39-Bus Network Based on 10 Consecutive Runs

	AC	AC+DC	AC+DC (lim)
Elapsed time [s]	0.66	0.63	0.68
# iterations	17	15	16
Cost [MW]	6284.1	6294.8	6295.0

TABLE 14.2 Cost Function Differences of the IEEE 39-Bus Network

Cost Difference [MW]	AC	AC+DC	AC+DC (lim)
AC	0	10.63	10.84
AC+DC		0	0.21
AC+DC (lim)			0

Table 14.1 compares the cost function results, calculation time, and number of needed iterations for the reference AC case, for the case when the AC system is extended with a 10-bus meshed DC system (comprising 10 VSC converters and 12 DC branches). The calculation time and number of iterations remain about the same for this setup. The cost function, when applying the DC system, is significantly higher, due to the no-load loss part of the converter losses (Table 14.2). Applying restrictive limits on converter and DC branch power raises the overall grid losses even more.

The following convention has been used in the table headers to denote the calculation variants:

- AC: reference case, the AC network without the associated DC system;
- AC+DC: simulation of the AC together with the DC system;
- AC+DC (lim): simulation of the AC together with the DC system with limitations on converter (20 MW and 20 MVAr) and DC branch flows (10 MW).

14.4 ADDING SECURITY CONSTRAINTS

So far, only the optimization of the base system configuration has been treated. As one of the main tasks of the system operator is to operate the grid in a secure way taking into account possible future events, "*contingencies*," the above discussion and case definitions need to be extended with security constraints.

Mainly two different approaches to manage contingencies for operators and grid planners are possible depending on the nature of the contingency and the time available for corrective actions to pilot the grid back into a healthy state. Preventive security measures are taken in advance of a contingency which allow the system to remain secure after a possible outage. Curative security measures are only taken after the contingency.

14.4.1 Preventive Measures

Operators and planners can choose for a subset of security measures for preventive actions. For each of them, the conceptual structure of the grid together with an adequate choice of generator and converter set-points are such that, in case of the actual event, the operator can rely on the fact that the grid remains in an healthy state. No immediate action is required, and the operator can slowly alter the grid parameters and topology to go to the next calculated optimal state including new security measures. Typical measures during the planning stage build in redundancy in the system, for instance by installing parallel connections or meshing of the grid.

In order to rely only on preventive measures to restore the grid to a healthy state after a given credible contingency has taken place, the operator needs to take only these *first-stage* decisions, as grid security is the main objective. From the point of view of generation cost, the grid might be operated suboptimal.

14.4.2 Curative Measures

Curative or corrective measures are another subset of contingencies requiring immediate action of the operator right after an incident took place. In these cases, it is allowed to operate the grid in a state that cannot last indefinitely, due to overloading of grid elements for example. The operator has sufficient time and control possibilities to get the grid back into a healthy state. Before the contingency occurs, a screening of the possible contingencies and their required control actions is performed. Generally, outages of generators fall under this category.

These contingencies are not taken into account in the general cost function and are treated as *zero cost* decisions, which might not always be the case in practical cases. They are called *second-stage* decisions.

More and more, the network operators experience that an ever-growing part of the generator park is of a renewable energy nature. These depend on uncontrollable and fluctuating energy sources like wind or solar (biomass and hydro power plants do not fall under this category). It can put constraints on the feasibility of the predefined curative actions as these power sources could be offline at the moment needed or, on the contrary, are producing near-peak capacity in the contingency zone with possible danger that the generated power cannot be absorbed by the network. The complexity rises when external factors, like market prices or forecasted weather conditions, need to be taken into account to determine the possible curative measures. A more extensive overview of the philosophy of the operations of the power system with fully controllable injections can be found in Chapter 10.

14.4.3 Multistage Decisions

As pointed out before, due to the strong interconnection of the electrical grids, larger geographical regions need to be taken into consideration and single contingency surveys are no longer appropriate. The risk of cascading incidents needs to be taken into account, which could lead to a partial or complete blackout. Even if such an event is not plausible for single contingency situations, it is for an unfortunate set of strategic (not even necessarily close-by) outages and incidents, combined with unexpected generation and consumption patterns.

Furthermore, in such complicated grid situations, it is possible that the grid operator cannot solve the situation with one single action. Indeed, during, for example, cascading events (e.g., slow voltage collapse), multiple actions are required within a very short time. Combined with generation and load uncertainties, there is a strong need for quick decision tools to steer the operation in an online fashion, to a safe network state. The operator is able to work intuitively the first seconds of the unplanned outage, he is surrounded by tools that show him the severity of the situation and help him during the whole process.

14.4.4 Implemented Approach

In order to integrate the security constraints into the algorithms developed, some simplifications have been made. Only the curative and preventive actions are investigated for the simplified network. The control variables in the following scenarios are the active power generator output and the voltage amplitude on the generator buses, emulating the automatic voltage control of the generator units and converters. The free state variables are bus voltage angles and the reactive power output of the generators.

Branches outages may lead to situations for which the grid is split into several systems, isolated from each other. The optimization case becomes unsolvable if generators get isolated from the main grid or if a islanded system contains no slack bus. This leads to one of the main robustness issues of the security constrained OPF algorithms, leading to nonconvergence. As such, an algorithm is needed to detect the islanded systems for each contingency and has to redefine a generator node as slack bus for those islanded systems which have been separated from the islanded systems containing the original slack bus. The automatic islanded system detection and slack bus conversion is crucial for the robustness of the methodology. It is, however, not covered here. The discussion is applicable for both and AC and a DC system with isolated subsystems in contingency states.

As an example of, the bold elements in Figure 14.2 are those elements in the IEEE 39 bus grid (without DC grid) that produce a grid configuration that cannot be solved by the optimization algorithms using its original definition. Generator step-up transformer outages isolate their generator from the grid and taking branch L27 or transformer T32 out of service causes the creation of a islanded system without a slack bus (which is in this case node N32, associated with generator G02).

14.4.5 Mathematical Interpretation

Before launching the preventive security-constrained optimization case, first the results of the corresponding curative case are considered. If the curative optimization does not converge to a valid solution, the preventive actions (imposing even more severe constrains on the state variable set, including the constraints of the curative case) would not do either.

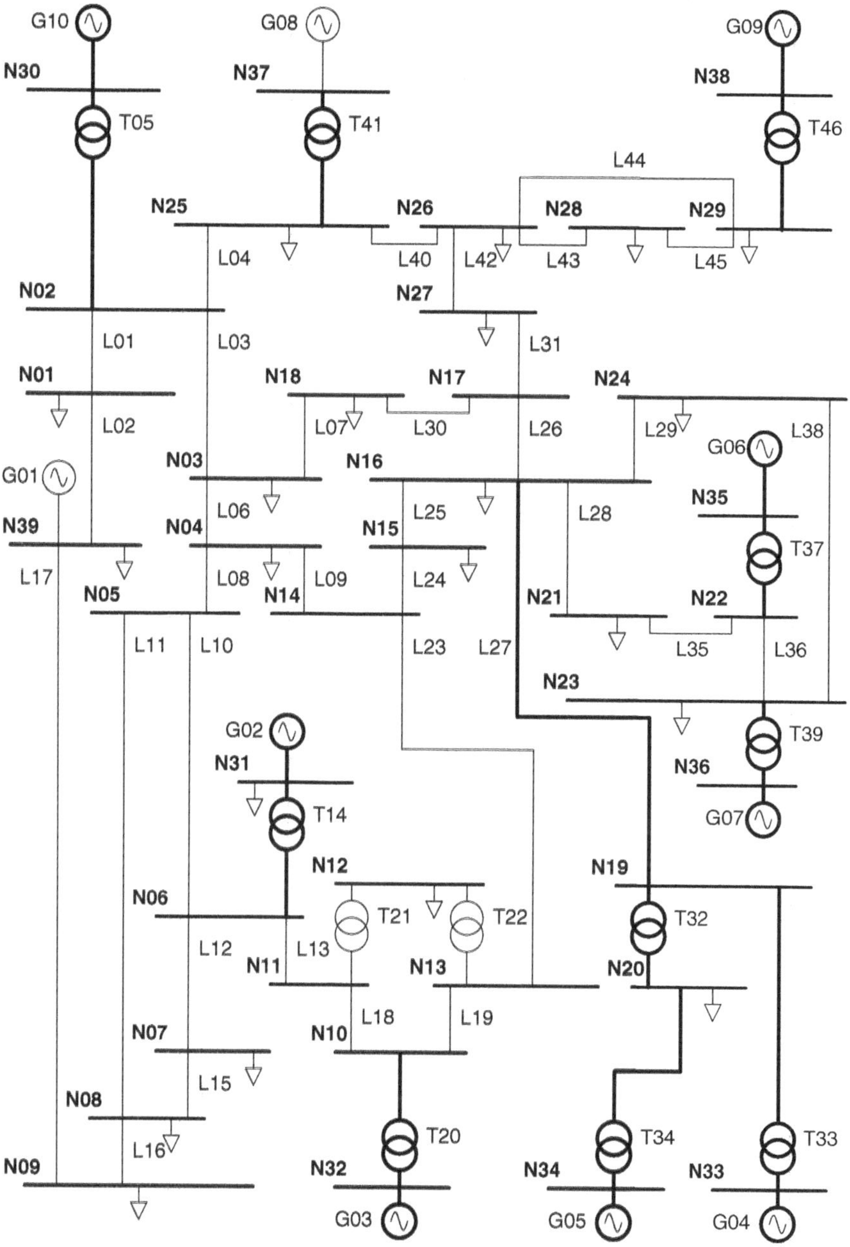

Figure 14.2 AC system of the IEEE 39-bus network with AC/DC converters. Outages of branches and generators in bold lead to a split in systems and are disregarded.

Furthermore, when assuming an almost unlimited time and flexibility for the operator to get the system back into a desired safe state, the curative case falls apart into k independent subproblems, which can be treated one after the other.

For the preventive security constraints, indeed, all different cases need to be optimized together. This is a challenging task for the optimization algorithm, as it is in general a nonconvex problem. Although, by reasoning, a solution should be feasible, the algorithm could fail to converge when bouncing on inequality constraints. The choice of the starting point near enough to the expected solution is essential.

The mathematical formulation of the preventive SCOPF becomes

$$\min_{x_0} \quad f_0(x_0) \tag{14.50a}$$

$$g_0(x_0) = 0 \tag{14.50b}$$

$$g_k(x_k) = 0 \tag{14.50c}$$

$$h_0(x_0) \leq 0 \tag{14.50d}$$

$$h_e(x_k) \leq 0 \tag{14.50e}$$

$$x_{min} \leq x \leq x_{max} \tag{14.50f}$$

$$x_k = x_0 \tag{14.50g}$$

with $k = 1 \dots c$ and c being the number of contingency cases.

For the inequality constraints, two different network limitation sets are defined: (14.50d) gives the normal state ratings (with the subscript 0), whereas (14.50e), gives the emergency state limitation set, denoted by the subscript e.

Referring to (14.20), some adaptations need to be made to translate the previous set of equations into the desired form. The vector of state variables becomes

$$x_n = \begin{bmatrix} x \\ z \end{bmatrix} = \begin{bmatrix} x_0 \\ x_1 \\ \vdots \\ x_c \end{bmatrix} \tag{14.51}$$

where the subscript 0 stands for the base case, and all others $(1 \dots c)$ denote the individual c preventive security constrained optimization cases.

The function vector $g(x_n)$ is reformulated following (14.22):

$$g(x_n) = \begin{bmatrix} g_0(x_0) \\ g_1(x_1) \\ \vdots \\ g_c(x_c) \end{bmatrix} \tag{14.52}$$

The function vectors $g_k(x_k)$ with $k = 1 \dots c$ describe the grid situation of the contingency k and differ from the base function vector $g_0(x)$ by having put the grid element of the corresponding contingency scenario out-of-service or nonexistent. This slightly modifies the grid topology locally.

Consider $J_{g,k}$ the Jacobian of the vector function $g_k(x_k)$; the Jacobian of the whole system can be composed of the individual elements:

$$J_g(x_n) = \begin{bmatrix} J_{g,0}(x_0) & 0 & 0 & \dots & 0 \\ 0 & J_{g,1}(x_1) & 0 & \dots & 0 \\ 0 & 0 & \ddots & & \\ \vdots & \vdots & & \ddots & \vdots \\ 0 & 0 & 0 & \dots & J_{g,c}(x_c) \end{bmatrix} \tag{14.53}$$

As these are preventive contingency cases, the control variables (active generator power output and generator bus voltage amplitudes) need to be forced equal to their initial state. An additional set of equality constraints has to be implemented.

$$g_a(x_n) = \begin{bmatrix} U_{i,k} - U_{i,0} \\ P_{j,k} - P_{j,0} \end{bmatrix} = 0 \tag{14.54}$$

with $k = 1 \dots c$, $i \in$ {set of generator buses} and $j \in$ {set of generators without the slack bus generators and the DC reference converter}.

In (14.54), the active power equality equations for the generators connected to the AC slack bus and the equivalent AC generator of the reference converter must be omitted to find a feasible solution. If they are still present, the optimization algorithm has to find a solution for which also the grid losses of the base and contingency cases are the same. It fails in most cases because the different topologies generate different transmission losses. Omitting the slack bus and DC reference generators still allows different system losses in both the AC and DC subsystems of each contingency case.

It is with this set of equations that the different contingency state sub-blocks are linked mathematically to the base case. The derivatives are incorporated in the Jacobian (14.53), and they are off-diagonal elements outside of the shown sub-blocks.

The same procedure needs to be applied for the inequality constraints $h(x)$:

$$h(x_n) = \begin{bmatrix} h_0(x_0) \\ h_e(x_1) \\ \vdots \\ h_e(x_c) \end{bmatrix} \tag{14.55}$$

together with its proper Jacobian

$$J_h(x_n) = \begin{bmatrix} J_{h,0}(x_0) & 0 & 0 & \dots & 0 \\ 0 & J_{h,e}(x_1) & 0 & \dots & 0 \\ 0 & 0 & \ddots & & 0 \\ \vdots & \vdots & & \ddots & \vdots \\ 0 & 0 & 0 & \dots & J_{h,e}(x_c) \end{bmatrix} \tag{14.56}$$

with the same $J_{h,e}$ repeated c times, the number of preventive contingency cases. The calculation of the Hessian of the function vectors $g(x_n)$ and $h(x_n)$ is given by (14.29).

14.4.6 Security Constraints Combined with DC Networks

All methods described in this chapter are entirely applicable when a DC grid and AC/DC converters are modeled. In (14.51), the vector x contains now the state variables of both AC and DC grid. Equations (14.52), (14.53), (14.55), and (14.56) assume that the DC grid equality and inequality constraints are incorporated.

For (14.54), linking the security cases to the pre-contingency state, different approaches apply. The fast automatic controls of the converters take immediate action after the grid incident and are able to find a new stable working state (e.g., using a droop control mechanism). In this case, it it not necessary to impose the same power and voltage constraints as in the initial state. Instead the converter control actions need to be modeled. For simplicity, the AC bus voltage and the converter active power are taken over from the initial state. Equation (14.51) is expanded in the same way with these converter parameters.

14.4.7 Simulation Results for the Security-Constrained Optimal Power Flow

In this section, security constrained optimal power flow calculations are presented for the IEEE 39 bus network. The results are presented without and with taking the associated DC system into account. Care should be taken when comparing the results: they cannot be treated as equivalent cases. The focus of the simulations lies on their robustness and correctness. The cost function for all the simulation in this section is the minimization of the active power generation.

Table 14.3 shows the results of the IEEE 39 bus network when all respective generators are disconnected one by one: the curative security contingency case for generators with and without DC grid. A dash (—) means that no convergence has been reached. The DC system clearly improves the overall grid security for this case. Only for the outage of generator G10, no feasible working point has been found. The bottom line in the tables indicates the highest deviation of the cost function.

Table 14.4 gives the optimization results and calculation times for the base case and with 1, 2, 3, and 4 preventive contingency constraints (without and with the DC network): the outage of branches L03, L06, L08, and L10. The calculation times are

TABLE 14.3 Cost Function Differences (%) of the Curative Security-Constrained OPF Disconnecting the Generators of the IEEE 39 Bus Network without and with the DC System

Generator Outage	Without DC system	With DC system
1	0.24	0.24
2	—	0.52
3	—	0.50
4	—	0.23
5	—	0.17
6	—	0.30
7	—	0.14
8	0.12	0.11
9	—	0.28
10	—	—
max	0.24	0.52

TABLE 14.4 Computation Characteristics of the Multiple Preventive security-Constrained OPF of the IEEE 39 Bus Network without and with the DC System

	Item	Base	1 SC	2 SC	3 SC	4 SC
w/o DC	Elapsed time (s)	0.52	1.26	1.77	2.41	3.35
	Iterations	12	15	15	14	16
	Cost function [MW]	6284.2	6284.4	6284.4	6284.9	6284.6
	Cost difference (%)		0.003	0.004	0.010	0.006
with DC	Elapsed time (s)	0.57	1.32	2.10	3.14	3.94
	Iterations	11	14	15	17	17
	Cost function [MW]	6295.0	6298.6	6298.2	6298.2	6298.7
	Cost difference (%)		0.056	0.050	0.051	0.058

the average of 10 consecutive runs of each scenario. The calculation time is with each case slightly higher, the number of iteration steps being almost constant. The change of the objective function value is negligible. The negative values are explained by the fact that the optimization algorithm immediately stops when its stopping criteria are reached. It is possible that the algorithm stops at a slightly better cost function value, when the optimization routine calculated a different state vector trajectory or when more constraints are added to the problem. In that case, the iterative algorithm needs more steps to find a solution complying to all constraints, giving the cost function also more steps to find even lower values than the less constrained problem. In practice, the cost function differences are so small that they can be treated as equal.

14.5 CONCLUSIONS

In this chapter, the extension of the AC optimal power flow case with a meshed DC grid is presented. The modeling of grid topology and definition of AC system components is presented first. Next a general deterministic OPF definition and a methodology to extend optimization cases with additional state variables and constraint equations to define additional functionality is provided. The nonlinear problem is formulated in a general manner, allowing any configuration of AC and DC power systems.

The developed optimal power flow algorithm for hybrid AC/DC power systems has been extended with security constraints. A methodology to construct optimal power flow cases taking into account several contingency cases is presented. Methodologies and algorithms for two distinct approaches have been developed. With the curative (or corrective) approach for a specific set of contingencies, the operator acts after the incident has occurred. The preventive security-constrained OPF optimizes the system already taking into account a parallel set of outages: The system remains secure after the occurrence of one of the contingencies in the set.

The discussed methodologies have been implemented and tested on a sample network. Specific attention is paid to the robustness of the algorithms when unsolvable grid topologies are present in the contingency set. The methodology is compatible with the inclusion of the meshed DC grids into any power system model. The simulations are a proof of concept, while not being exhaustive. Nevertheless, it is shown that the methodology is applicable for different power systems, even large ones, and this with reasonable convergence characteristics and computation time.

REFERENCES

1. D. P. Bertsekas, *Nonlinear Programming*. Nashua, NH: Athena Scientific, 1999.
2. J. A. Momoh, *Electric Power System Applications of Optimization*, vol. 11. Boca Raton, FL, CRC Press, 2001.
3. J. D. Weber, "Implementation of a Newton-based optimal power flow into a power system simulation environment," Ph.D. dissertation, University of Illinois, 1997.
4. R. D. Zimmerman, C. E. Murillo-Sánchez, and R. J. Thomas, "MATPOWER's extensible optimal power flow architecture," in *Power Engineering Society General Meeting*. New York: IEEE, 2009, pp. 1–7.
5. J. Rimez, "Optimal operation of hybrid AC/DC meshed grids," Ph.D. dissertation, Katholieke Universiteit Leuven, Belgium, 2014.
6. G. Daelemans, K. Srivastava, M. Reza, S. Cole, and R. Belmans, "Minimization of steady-state losses in meshed networks using VSC HVDC," in *Power Engineering Society General Meeting*. New York: IEEE, 2009, pp. 1–5.
7. X. Wei, J. H. Chow, B. Fardanesh, and A.-A. Edris, "A common modeling framework of voltage-sourced converters for load flow, sensitivity, and dispatch analysis," *Power Systems, IEEE Transactions on*, vol. 19, no. 2, pp. 934–941, 2004.
8. A. Pizano-Martinez, C. Fuerte-Esquivel, H. Ambriz-Perez, and E. Acha, "Modeling of VSC based HVDC systems for a Newton–Raphson OPF algorithm," *Power Systems, IEEE Transactions on*, vol. 22, no. 4, pp. 1794–1803, 2007.
9. J. Cao, W. Du, H. Wang, and S. Bu, "Minimization of Transmission Loss in Meshed AC/DC Grids With VSC-MTDC Networks," *Power Systems, IEEE Transactions on*, vol. 28, no. 3, pp. 3047–3055, Aug. 2013.
10. M. Aragüés-Peñalba, A. Egea-Alvarez, O. Gomis-Bellmunt, and A. Sumper, "Optimum voltage control for loss minimization in HVDC multi-terminal transmission systems for large offshore wind farms," *Electric Power Systems Research*, vol. 89, pp. 54–63, 2012.
11. K. Padiyar and N. Prabhu, "Modelling, control design and analysis of VSC based HVDC transmission systems," in *Power System Technology. International Conference on*, vol. 1. IEEE, 2004, pp. 774–779.
12. T. Chung and G. Shaoyun, "Optimal power flow incorporating FACTS devices and power flow control constraints," in *Power System Technology, International Conference on*, vol. 1. New York: IEEE, 1998, pp. 415–419.

CHAPTER 15

CONTROL PRINCIPLES OF HVDC GRIDS

Jef Beerten,[1] Agustí Egea,[2] and Til Kristian Vrana[3]

[1]Research Group Electa, Department of Electrical Engineering, University of Leuven, Leuven, Belgium

[2]Research Group CITCEA, Electrical Engineering Department, Technical University of Catalonia (UPC), Barcelona, Spain

[3]Norwegian University of Science and Technology, Trondheim, Norway

15.1 INTRODUCTION

One of the crucial aspects in HVDC grids is the control of the system voltage. Since the voltage differences throughout the system directly dictates the power flow through the lines, the nodal DC voltage plays a crucial role in the power flow scheduling. Furthermore, contingencies such as converter outages give rise to an imbalance that is reflected in the DC voltages in the grid. This imbalance has to be accounted for by changing the currents flowing in and out of the DC system. These DC node voltages can be directly influenced by controlling the DC current of the HVDC converter at that node. Different control strategies can be applied to balance the currents in a DC grid after a contingency. The actual implementation of this voltage control and of the scheduling of the power flows in a DC system is still subject to research and some of the solutions are subject to patent applications.

The aim of this chapter is twofold:

- First, it aims at classifying the different node voltage control methods, which have been proposed in the literature in recent years[1]. The control objectives and requirements and the steady-state working conditions of DC node voltage control are discussed, while the actual control implementation is not covered. The focus of this first part is purely on DC voltage control as a means of active power balancing, similar to primary frequency control in AC systems. The focus is on the concepts and general principles of the control.

[1]This classification has been adopted by the Cigré B4-58 working group on "Devices for load flow control and methodologies for direct voltage control in a meshed HVDC grid".

HVDC Grids: For Offshore and Supergrid of the Future, First Edition. By Dirk Van Hertem, Oriol Gomis-Bellmunt, and Jun Liang.

- Second, the basic principles of the converter control implementation as well as slower control actions, like power rescheduling, are discussed. Mathematically, these slower control actions can be regarded as set point changes. The general principles are introduced and an example of a possible implementation is given.

This chapter is partly based on shortened versions of two earlier contributions [1, 2]. The combination of DC voltage control with other control objectives (e.g., frequency control) is not discussed in this chapter.

15.2 BASIC CONTROL PRINCIPLES

All methods described in this chapter utilize local DC voltage measurements in order not to fully rely on external data for the control to work. However, the overall performance can be modified by using a common voltage signal for feedback, as proposed in reference 3.

In this chapter, positive current and power are defined as flowing out of the DC grid, which is coherent with the physical definitions of voltage, current, and power. In power systems, a reverse convention is often in use, which might be more intuitive because of production being positive and loads being negative. Given the definition on the DC side and the convention on the AC side, an HVDC converter's current and power are positive in inverter mode and negative in rectifier mode. The direction reference is therefore consistently defined for both the AC and the DC side. This leads to (a) regular AC frequency droop curves having a negative slope and (b) DC voltage droop curves (e.g., Figure 15.2, Section 15.3) having a positive slope.

The control of the DC voltage shows a lot of similarities with frequency control in an AC power system. Whenever a deficit occurs in the current that flows in/out the DC grid, the voltage at the different nodes will immediately react to this change as a result of the capacitors and cable capacitance discharging or charging. This current deficit can occur, for example, when a converter faces an outage. If the converter was exporting to the AC system, the outage will cause a current surplus leading to a voltage increase. On the contrary, if the converter was importing into the DC grid, the converter outage causes a current shortage leading to a voltage decrease. DC voltage can therefore be taken as a balance indicator, similar to AC frequency in AC power systems.

In AC systems, however, the frequency is the same throughout the entire system. The active power flow on a line is observed through the phase angle. In DC systems, DC voltage is influenced by the global balance as well as by the currents on the lines. Therefore DC voltage is not a true global measure, and it is therefore more challenging to use the DC voltage as a reference for balancing control. Despite these difficulties in terms of control, the DC voltage still appears to be the best indicator for stable DC grid operation.

Based on the described behavior, it is clear that the current balance has to be restored as soon as possible in order to keep the voltage from falling or rising. All control structures discussed further on in this chapter rely on the fact that the deviation of the controlled DC voltage (local measurement at each DC bus) is determined and corrected for by the control action, which comes down to restoring the current balance.

Assuming a lossless system, all node voltages would be identical and the current balance can be directly rewritten as a power balance. In a real system, DC line losses will add to the power balance equation and cause a mismatch between imported and exported power. Still, the general phenomena are the same: A change in the power balance will cause the DC voltage to deviate.

There are two basic principles regarding how DC voltage can be controlled: current-based control and power-based control [4, 5]. It is important to mention that both current-based control and power-based control behave identically at the operating point. The larger the DC voltage deviates from the reference value, the larger the differences between power-based and current-based control become.

15.2.1 Current-Based Control

In a current-based control scheme, the relationship to be used to control the DC voltage is an I–V characteristic [6, 7]. The main advantage of a current–voltage control characteristic is that it reflects linear control behavior in the sense that a voltage deviation will result in an equivalent current deviation. As the voltage is linked with charging or discharging the capacitive DC system, the control is linear and is the same for all voltage deviations from any reference value. The power–voltage relation is consequently nonlinear (parabolic).

Current-based control can be directly linked with the DC network dynamics: The charging of the capacitances in the DC network relies on a linear current–voltage relation, which has the physical dimension ohm. This makes current-based control somehow intuitive from a physical perspective as it has the same physical units as the DC line impedance.

15.2.2 Power-Based Control

As an alternative to the current-based control, the DC voltage control can be expressed in terms of active power, as discussed in references 8–10. Here the power-voltage relation is linear. Contrary to current-based control, power-based control shows nonlinear (hyperbolic) control behavior. One should take into account the fact that DC voltage control therefore is nonlinear.

Power-based control has the advantage that it is more intuitive from a power system perspective, where the focus often is on transmitted power. It is also somehow similar to power-frequency control in AC systems. The power–voltage relation has a physical unit of 1/A or Ω/V, which does not have a direct intuitive physical meaning. Power-based control can easily be integrated with the existing vector control schemes, which makes it well-suited to be used with the standard *dq*-current controllers used in VSC HVDC systems. As discussed in reference 4, it is also possible to use the current-based control to create an AC current reference.

15.2.3 Converter and System Limits

There are several limits for HVDC converter control. The three most important are briefly described here and shown in Figure 15.1. The values for the limits in the

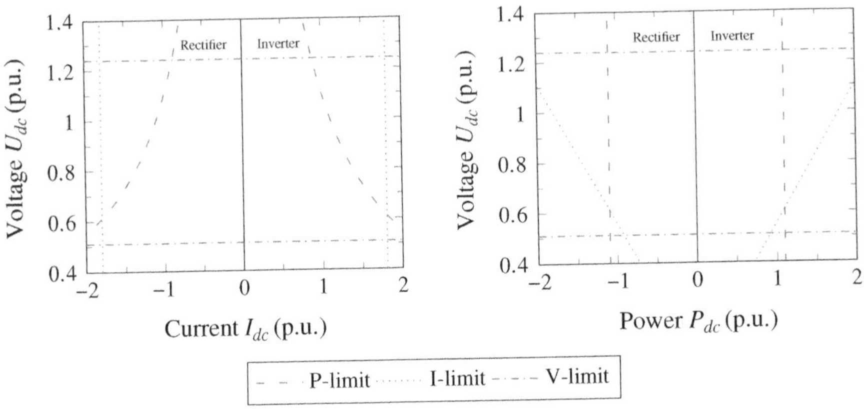

Figure 15.1 Converter and system limits.

figure are chosen for good graphic illustrations and do not represent realistic HVDC converter limitations. Figure 15.1 shows the same limits, respectively in the I–V and P–V plane. For example, the current limits are taken as constant and are hence given by vertical lines in the I–V plane and by a decreasing and increasing line in the P–V plane.

- **DC Voltage Limits.** The DC voltage has an upper and a lower limit. The upper limit is determined by the insulation of the switching components and the connected DC equipment (e.g., cables). The lower limit is more complex as it is based on a limitation of the modulation index, and it also depends on the converter topology and the converter control implementation.
- **Power Limits.** The active power has an upper limit. This power limit is caused by the semiconductor current limit, which limits the AC current and consequently the power (assuming a constant AC voltage). The power limit appears as a hyperbolic curve in the I–V plane and as a straight vertical line in the P–V plane.
- **DC Current Limits.** The DC current has an upper limit, based on the current rating of the connected DC components. This appears as a straight vertical line in the I–V plane and as a straight line towards the axes intersection in the P–V plane. As the limits depend on the connected DC equipment, there is no general answer whether these limits can be reached: The lower voltage limit might be reached before the current limit is hit.

15.3 BASIC CONVERTER CONTROL STRATEGIES

In the previous section, emphasis has been put on how a converter can control the DC voltage and the distinction that can be made based on the P–V or I–V relationship. In this section, it will be discussed how different control schemes can be approached as a generalized case of a voltage droop control. The discussion holds for both

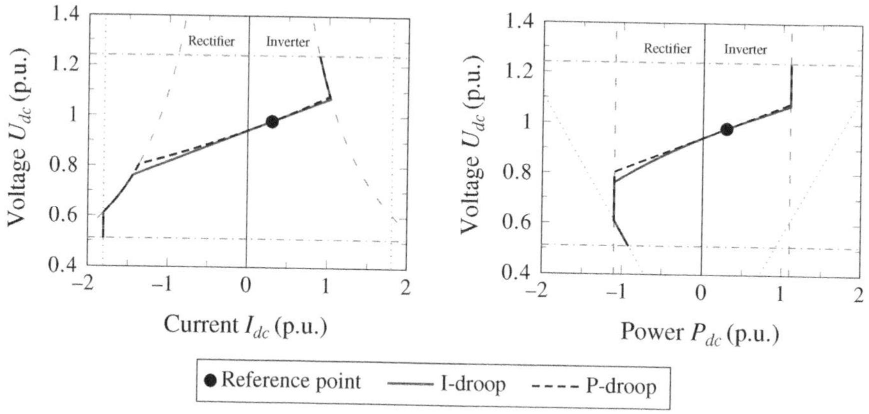

Figure 15.2 I–V curve and P–V curve for droop control.[2]

current- and power-based droop control. The control is complicated by the fact that the voltage in the DC grid varies as a result of the DC grid power flow. In order to obtain a certain power flow, the voltage set-points have to be different for all converters and should reflect the steady-state power flow solution of the DC network.

The explanation provided in this section reflects the mathematical representation of the control objective in terms of droop characteristics. The actual implementation, however, does not necessarily rely on the proportional droop controller, as discussed below. The term "basic strategies" refers to consistent control strategies, which do not depend on the operating point. The control strategies have only a single parameter, which is the droop value or control gain. The control is therefore linear. The advanced control schemes in Section 15.4 can be seen as a combination of the three control types discussed in this section.

15.3.1 Voltage Droop Control (Positive Droop Constant)

Voltage droop control creates a proportional relationship between the voltage and the control base (current or power) and is shown in Figure 15.2.

The droop constant is called k_{dc}. The control base is indicated by the subscript I or P. The subscript 0 indicates the voltage, current, and power set-points, with the natural relationship $P_{dc,0} = U_{dc,0} I_{dc,0}$. The voltage, current, and power deviations are defined as $\Delta U_{dc} = U_{dc} - U_{dc,0}$, $\Delta I_{dc} = I_{dc} - I_{dc,0}$ and $\Delta P_{dc} = P_{dc} - P_{dc,0}$. The I–V droop relation can be written as

$$\Delta I_{dc_I} = \frac{1}{k_{dc_I}} \Delta U_{dc} \tag{15.1}$$

[2]The term droop control only refers to the part of the curve between the converter limits from Figure 15.1. The same holds for the other figures in this chapter.

Rewriting (15.1) in terms of active power yields the corresponding power equation, which is parabolic:

$$\Delta P_{dc_I} = \left(\frac{U_{dc_0}}{k_{dc_I}} + \frac{P_{dc_0}}{U_{dc_0}} \right) \Delta U_{dc} + \frac{1}{k_{dc_I}} \Delta U_{dc}^2 \tag{15.2}$$

The expression for the P–V droop relation is similar to (15.1), namely

$$\Delta P_{dc_P} = \frac{1}{k_{dc_P}} \Delta U_{dc} \tag{15.3}$$

Rewriting (15.3) in terms of current yields the corresponding current equation which is hyperbolic:

$$\Delta I_{dc_P} = \left(\frac{1}{k_{dc_P}} - I_{dc,0} \right) \frac{\Delta U_{dc}}{\Delta U_{dc} + U_{dc,0}} \tag{15.4}$$

The slope of the I–V or P–V characteristic is determined by the droop value, which is the inverse of the proportional controller gain. The droop value k_{dc} in these equations is the slope of the characteristics in Figure 15.2 and is the inverse of the proportional controller gain used in the actual implementation. The stability of the droop controller depends on the droop value and the actual implementation. If well-designed, the voltage droop control leads to a stable operation. When a contingency occurs, the droop control is characterized by a steady-state deviation from the voltage reference as a result of the proportional control action.

15.3.2 Constant Current/Power Control (Infinite Droop Constant)

Constant current control and constant power control can respectively be represented as a vertical line segment in the I–V and P–V plane. Mathematically, the constant power/current control can be expressed as a limiting case of the aforementioned voltage droop control, with a droop constant equal to infinity, thereby not changing the current/power whenever the DC voltage changes. The converter will at all cost try to maintain the current/power injection constant, irrespective of the value of the DC voltage at its DC bus. Figure 15.3 shows the characteristic curves for both droop types in the P–V and I–V plane. Constant current control leads to a linear behavior in the P–V plane. Constant power control leads to hyperbolic behavior in the I–V plane.

In reality, constant current/power control is typically implemented using a PI controller to keep the current/power equal to its reference value. This is due to dynamic reasons, but theoretically, this comes down to the limiting case of an infinity droop value.

15.3.3 Constant Voltage Control (Zero Droop Constant)

Constant voltage control can be represented as a horizontal curve in the I–V or P–V plane (Figure 15.4). The DC voltage control can be regarded as the limiting case for which the droop constant goes to zero. From (15.1) and (15.3), it is clear that the

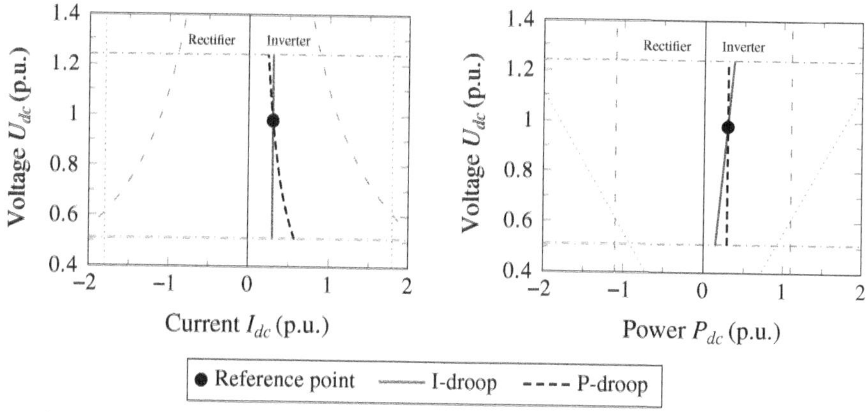

Figure 15.3 I–V curve and P–V curve for droop control with infinite droop constant.

converter in this case controls the voltage to its reference value. In Figure 15.4 no distinction between current- and power-based control is shown, since the curves are identical. A converter with this type of control is often referred to as a DC slack bus.

This control mode relates to a proportional control gain of infinity. This of course is unrealistic and will lead to instability. Constant voltage control is therefore in reality not realized with an infinite gain, but with a PI controller. In theoretic steady-state the outcome is the same, but due to dynamic reasons, the infinite control gain is not viable.

15.4 ADVANCED CONVERTER CONTROL STRATEGIES

In this section, control methods that cannot be described by a single control parameter are covered. They are therefore nonlinear—or, to be more precise, piecewise-linear.

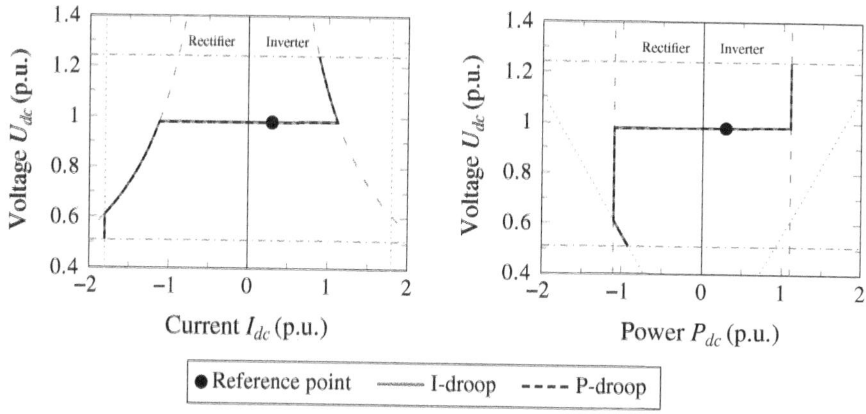

Figure 15.4 I–V curve and P–V curve for droop control with zero droop constant.

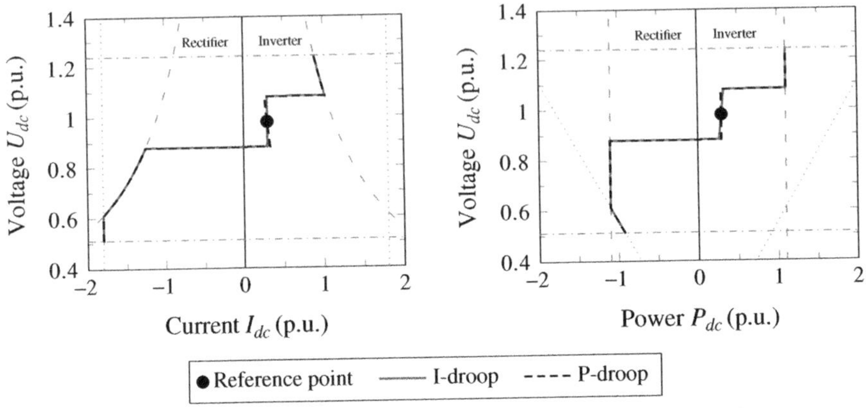

Figure 15.5 I–V curve and P–V curve for voltage margin control.

The actual control behavior depends on the operating point, thereby making a distinction between normal and disturbed operation. The advanced converter control strategies' behavior changes when large disturbances occur.

15.4.1 Voltage Margin Control

Voltage margin control was first proposed for DC networks in reference 11 and was later applied and developed further in references 12 and 13. It is a combination of constant current/power control and constant voltage control. A converter with voltage margin control normally operates in constant current/power control mode (as long as the voltage is within the normal operation voltage margin). If the voltage deviation reaches the limit of the voltage margin, the converter controller switches to constant voltage control, clamping the voltage at the margin limit to prevent further voltage deviation. This is shown in Figure 15.5.

15.4.2 Deadband Droop Control

Another advanced control strategy is droop control with an additional deadband [14]. Just like with voltage margin control, the converter operates at constant current/power control mode as long as the voltage is within the normal operation band or margin. If the voltage deviates to the limit of the deadband, the controller switches to droop control. Voltage margin control can be seen as the limiting case of deadband droop control with a zero droop constant. The control strategy is shown in Figure 15.6.

15.4.3 Undead-Band Droop Control

Another approach to DC voltage control, the so-called "undead-band" droop control, was proposed in reference 15 and developed further in reference 16. The control is fully based on voltage droop, but distinguishes between normal and disturbed operation by defining different droop constants for these two operation regimes. It

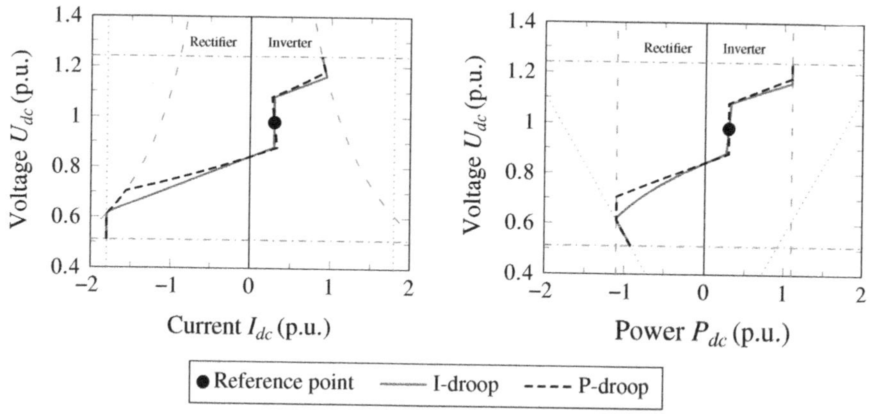

Figure 15.6 I–V curve and P–V curve for deadband droop control.

offers the possibility to optimize the droop constants and the overall control separately for normal and disturbed operation. This is shown in Figure 15.7. A similar concept has been proposed in reference 17.

The definition of undead-band droop control is the most general one, and all the other mentioned control methods can in fact be considered as specific examples of parameter sets of undead-band droop control. For example, deadband droop control can be seen as the limiting case of undead-band droop control with an infinite droop constant (constant current/power) under normal operation. Similarly, voltage margin control can be seen as the limiting case of undead-band droop control with an infinite droop constant under normal operation (constant current/power) and a zero droop constant under disturbed operation (constant voltage). Regular droop control can be seen as the limiting case of undead-band droop control with the same droop constant in both operation modes. Also other strategies, not mentioned in this chapter, can be perceived accordingly. As a complement to deadband droop control, a zero droop

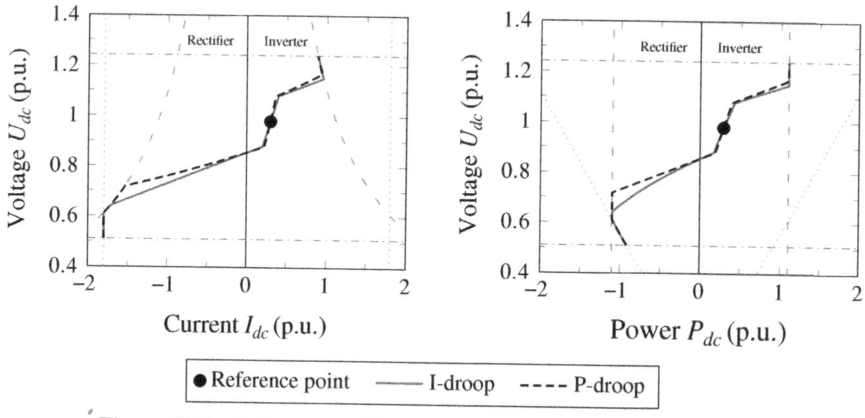

Figure 15.7 I–V curve and P–V curve for undead-band droop control.

constant under normal operation (constant voltage) can be combined with droop control under disturbed operation. Unlike a deadband, where the converter does not react on voltage changes, this scheme keeps the voltage constant under normal operation [18]. Another possible variation is a controller with droop control under normal operation and with zero droop (constant voltage) under disturbed operation [19].

15.5 BASIC GRID CONTROL STRATEGIES

In general, the DC voltage can be controlled to its reference value at one centralized voltage-controlling converter or by using a distributed control approach. The basic grid control strategies discussed in this section are based on the basic converter control strategies from Section 15.3.

15.5.1 Centralized Voltage Control

One converter has a droop value of zero, controlling the DC voltage to a constant value at its bus (Section 15.3.3), thereby acting as a DC slack bus. The other converters have a droop value of infinity (Section 15.3.2) and control their current/power to the set-point. The control concept is similar to the normal operation of a point-to-point VSC HVDC system.

Centralized voltage control gives rise to a well-defined operating point, since all but one of the converters operate at their current/power set-points. This strategy can hold for quite large networks under normal operating conditions. Since only one converter has to account for all disturbances within a grid, this method would only be applicable for relatively small fluctuations around the operating point in a larger system. With increasing network size, current/power fluctuations increase as well, thereby finally limiting the applicability of centralized voltage control. This type of operation would also have a significant influence on the AC system connected to that converter. The most critical problem, however, is how to handle severe disturbances, such as an imbalance exceeding the capabilities of the slack bus or the outage of the DC slack bus, for which no backup control is provided.

15.5.2 Distributed Voltage Control

Distributed balancing control is applied to all large AC power systems, and it is therefore a natural thought to consider this control method for DC power systems as well [20]. Distributed voltage control is implemented by applying droop control to several converters (Section 15.3.1). It is not necessary that all converters take part in the control of the DC voltage. Droop-controlled converters and current/power-controlled converters can easily be integrated into the same DC grid. This is similar to AC grids, where some power stations take part in primary frequency control while others do not. The integration of voltage controlled converters could cause problems: Similar to a frequency droop in AC systems, the power sharing amongst different droop-controlled converters is determined by the relative droop values of the different converters. The smaller the droop value, the higher the share of a particular converter in the power sharing. A voltage-controlling converter can be considered as the limiting case for the droop constant going to zero, thereby significantly reducing the power sharing of the other converters. In this way, adding a voltage-controlling converter would partly undermine the distributed control principle of droop control. In this case only the changes in the voltage drops on the lines determine the changes in the operating points of the droop-controlled converters, leading to smaller control contributions.

15.6 ADVANCED GRID CONTROL STRATEGIES

The advanced grid control strategies are based on the advanced converter control strategies from Section 15.4. Compared to the basic strategies, the main difference is the inclusion of backup mechanisms to handle large disturbances. For distributed voltage control, this backup is an improvement, while for centralized voltage control, it is a necessity as illustrated in Section 15.5.

15.6.1 Centralized Voltage Control with Centralized Backup

The backup system for the centralized control method can be implemented with a centralized approach as well. This means that a converter that normally controls current/power switches to voltage control to take over the task from the regular voltage controlling converter. The modification from centralized voltage control is to replace constant current/power converter controllers by a voltage margin control. If the voltage deviates significantly (i.e., the voltage controlling converter fails to do its task), a converter with voltage margin control hits the limit of the voltage margin, turns into the new DC slack bus, and starts to control the DC voltage at the limit of the voltage margin. In this way, the task of the slack can be passed back and forth between converters many times.

For smaller DC systems, this control can be a good choice, as long as the converter ratings and the system rating are in the same range. One slack bus converter can be selected, but for extraordinary conditions other converters help out. This is similar to the real implementation used in point-to-point connections. For large DC systems, this method might pose problems. With increasing system size, changes in operation mode will appear more often. This creates highly nonlinear system behavior, where local voltage measurements lose information on the general system state. Furthermore, when more converters participate in the voltage margin control, the number of voltage steps increases, which might give rise to voltage deviations that are too large. These two factors make it complicated to oversee and therefore control the network.

15.6.2 Centralized Voltage Control with Distributed Backup

A better way to provide a backup mechanism for centralized voltage control for large DC grids is to use deadband droop control instead of voltage margin control. If the voltage controlling converter fails, the voltage control task is not transferred to another converter (as with voltage margin control), but to all other converters with a distributed approach. At larger disturbances, the consequences are split among all converters, leading to a new stable operating point.

Compared to voltage margin control, this method can provide an improvement when it comes to handling large disturbances. However, it still faces similar problems under normal operation for very large power systems, where the converter ratings are very small compared to the system rating. The regular power fluctuations increase with system size, eventually exceeding the capability of one single converter, and therefore operation outside the deadband is likely to become more frequent.

15.6.3 Distributed Voltage Control with Distributed Backup

Generally, the distributed voltage control can safely operate a DC grid without a backup mechanism. But still, also with a distributed voltage control approach, it might be desirable to make a distinction between normal and disturbed operation—for example, stay close to the operating point under normal conditions, as discussed in [17], while having a backup control with different settings. In this way, converters with droop control that contribute to voltage control can increase their contribution (lower droop value) in the case of a severe disturbance. This can be implemented with undead-band droop control. Similarly, current/power controlled converters that do not contribute to voltage control could in some cases be implemented as deadband droop-controlled converters, thereby also contributing to the voltage control backup system.

15.7 CONVERTER INNER CURRENT CONTROL

The control concepts from the previous section can be realized in different ways, and several techniques have been proposed in the literature in order to control a VSC. However, most practical application use the principles of vector current control [21] for the inner control structure. In such a control scheme, the control of the current is realized using a mathematical transformation of all system quantities to a rotating reference frame locked to the periodically changing voltage at the point of common coupling. Doing so, all system variables appear as DC quantities and hence it becomes possible to independently control the d and q current components in a synchronous reference frame (SRF) while the synchronization is provided by a phase-locked loop (PLL) [22]. This is depicted schematically in Figure 15.8.

In the orientation used in this chapter the active current component is the q component, and the reactive current component is the d component. The PLL consists of a controller $G_{PLL}(s)$, typically a PI controller, which is connected in series to an integrator that evaluates the error of the system oriented component in order to obtain the system angle and velocity, which can then be used as an input for the mathematical transformation of the system variables (e.g., the converter currents). This transformation, known as the Park transformation, is depicted in Figure 15.8 as $T(\theta)$. The VSC is connected to the grid by means of a converter phase reactor, which, together with the converter transformer and a dedicated AC filter branch prevent higher-order harmonics to enter the AC system. In its simplest form, the AC system

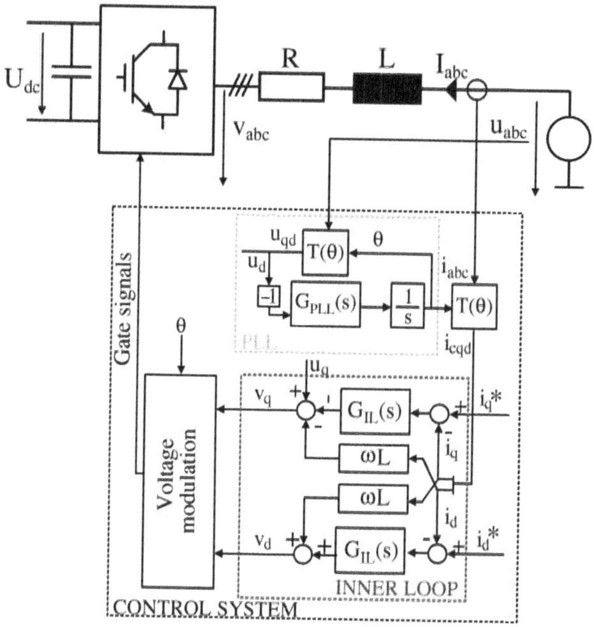

Figure 15.8 Physical power converter connection and PLL and current control implementation.

side can be simplified to a complex series impedance, as shown in Figure 15.8. This can be mathematically described as

$$\frac{d}{dt}\begin{bmatrix} i_q \\ i_d \end{bmatrix} = \begin{bmatrix} -\frac{R}{L} & -\omega \\ \omega & -\frac{R}{L} \end{bmatrix}\begin{bmatrix} i_q \\ i_d \end{bmatrix} + \begin{bmatrix} -\frac{1}{L} & 0 & \frac{1}{L} & 0 \\ 0 & -\frac{1}{L} & 0 & 0 \end{bmatrix}\begin{bmatrix} v_q \\ v_d \\ u_q \\ u_d \end{bmatrix} \tag{15.5}$$

where i is the current flowing through the filter, u is the voltage at the point of common coupling, v is the voltage applied by the power converter, R and L are the resistance and inductance of the coupling filter, and ω is the electrical system speed. According to (15.5), it is possible to design an independent control for each current component after adding decoupling terms for the d and q components (respectively $\omega L i_q$ and $-\omega L i_d$), together with a feed-forward term for the grid voltage, resulting in

$$\begin{bmatrix} v_q \\ v_d \end{bmatrix} = \begin{bmatrix} G_{IL}(s) & 0 & -G_{IL}(s) & -\omega l_c & 1 \\ 0 & G_{IL}(s) & \omega l_c & -G_{IL}(s) & 0 \end{bmatrix}\begin{bmatrix} i_q^* \\ i_d^* \\ i_q \\ i_d \\ u_q \end{bmatrix} \tag{15.6}$$

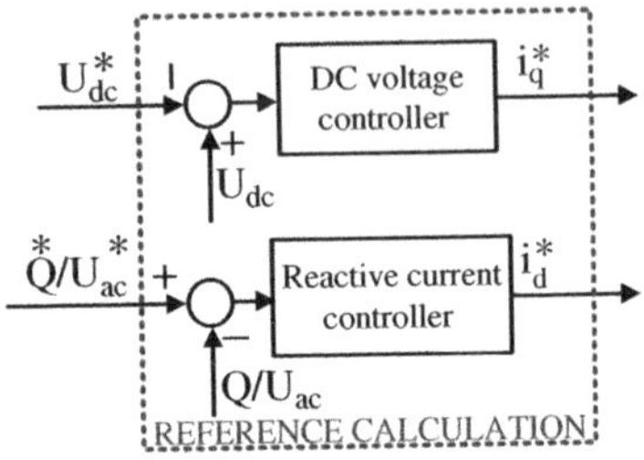

Figure 15.9 Current loop reference calculation controller scheme.

where i_q^* and i_d^* are the current component references and $G_{IL}(s)$ is the current controller, typically a PI controller. Further details on the design and tuning of the current control loop and PLL can be found in reference 23.

15.7.1 Current Control Reference Calculations

Since the PLL forces the system voltage to be along one of the axes in the SRF, the decoupled current control loop allows to independently control the current components associated with the active and reactive power. The references for these current components are given by an outer control loop. The active current reference corresponds to the so-called "active power converter control function" which can be either AC active power, DC active power, DC voltage control, DC current control, AC frequency control, or a combination thereof, as dealt with in the first parts of this chapter. Similarly, the reactive current component allows an independent control of the AC voltage or reactive power at each converter. An example of the current loop reference calculation can be seen in Figure 15.9.

As pointed out in Sections 15.3–15.4 , the DC voltage control can be conceived in different ways, each of which can be implemented in different ways. Although the fundamentals of these control concepts are currently well-known, their actual implementations are still under development and the subject of research. More information on different implementations available in literature can be found in references 1,5,15 and 16.

15.8 SYSTEM POWER FLOW CONTROL

It is clear from the previous discussion that it is common practice to build the actual control of the converters in a cascaded manner. Following this logic, a proper systemwide control of the converter power flows can be conceived as an outer control layer sending set-points to the different converters. Doing so, the power flows throughout the system can be changed and controlled. This section discusses the principles of such a systemwide power flow control. First, the analogy with the AC system control is discussed; afterwards, similar control principles for an HVDC grid are introduced.

15.8.1 AC System Power Flow Control

One of the fundamental controls in electric grids manages the balance between generation and load. Put differently, the balance between the injected and withdrawn power in a grid must remain equal. In AC systems, this balance is reflected in the value of the frequency. The power balance is maintained by the frequency control mechanism. This mechanism adjusts the power outputs of the generation (or possibly the load) to assure a constant frequency.

The first control action is an automatic reaction of all activated generators after a deviation of the frequency from the nominal frequency. In a multi-zonal system, this results in a change of output power throughout the system. The power deviation is proportional to the frequency deviation (droop control). The primary control is activated in a very short time period (15–30 seconds) and acts to limit the frequency deviation. The set-point for the active power injections in the different generating units during the primary control action remains at the initial value. The primary control actions are available for only a limited time period (e.g., 15 minutes).

The secondary control action restores the original exchanges between different zones, causing the ACE (Area Control Error) to get back to zero. The secondary control action changes the active power set-points in the area with a power deficit or excess to match the original schedule. Due to the change in set-point in the affected zone by the secondary control, the previously committed primary reserves are no longer required, and in fact overcompensate the initial fault. This effect is again automatically canceled by the primary control action which moves back to the original value in the entire system, thus releasing the previously activated primary reserves. The secondary control or power control can be described as a system that allows us to achieve the power set-points provided by a tertiary control. Furthermore, the secondary control corrects the power references after a contingency. In steady state, the main objectives of the secondary control are to ensure the power references and the power exchanges between the areas.

The tertiary control action shifts the power from the secondary reserves, which are available on a short time basis to more long-term power generation and is often manually activated by the system operator as the result of a system optimization.

15.8.2 Hybrid AC/DC System Power Flow Control

As an illustrative example of how the HVDC grid control can be conceived, this section introduces a mixed AC and DC grid control. The operation principles are based on the classic AC control structure from the previous section. More details on the actual implementation of this control structure is found in reference 2. A conceptual scheme of the secondary control can be seen in Figure 15.10.

In a DC grid, the injections to and from the system have to be balanced in a manner similar that of the AC system injections. Where the AC system power balance is reflected in a constant frequency, the DC voltage fulfills a similar role in the DC system. The power-frequency control in the AC system is therefore very similar to the power-DC voltage control scheme. The consensus seems to lead to a droop controlled DC voltage, much like the primary control actions which are used

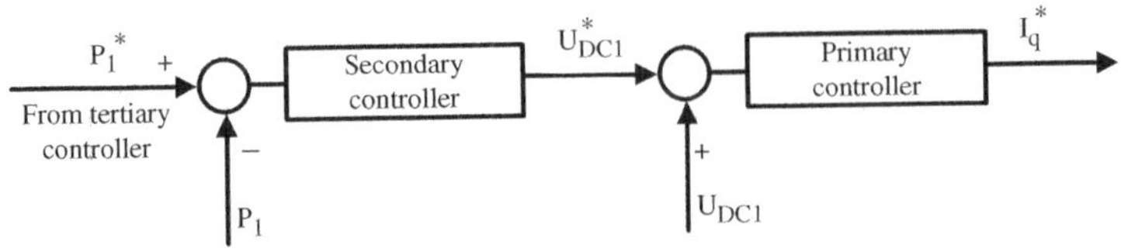

Figure 15.10 General scheme of the proposed control methodology.

in AC systems. Contrary to the AC system, which exhibits a significant inertia from the kinetic energy in the rotating machines, the stored energy in the DC system is very limited, with only the charge in the DC capacitors and cables. This makes that the voltages in the DC system change much faster than in the AC case (frequency power imbalance). The response of the controllers should therefore be equally fast. The time constant for the primary control loop is on the order of tens of milliseconds and is thus dictated by the DC grid characteristics.

Similarly to the AC system secondary control, a secondary control can be included for a DC grid to manage the power exchanges between the DC grid and the AC systems. Since the primary droop-controlled response already aims at stabilizing the system voltage, the settings for an outer secondary control action can be relaxed. Consequentially, this control can be set in the order of seconds or tens of seconds. Additionally, a tertiary controller can be included, which could, in analogy to the AC system, include an offline algorithms that react between 20 minutes and 1 hour.

Similarly to the AC system, when a fault occurs (e.g., when a converter station is disconnected), the control of the DC voltage causes the different converter systems to adjust their power injections immediately to cope with the system imbalance, thereby distributing the deficit according to the droop settings. As the voltage control at the DC side directly influences the power exchanges, there is a need to adjust the power injection set-points to meet the scheduled exchanges, especially if multiple zones or synchronous zones are connected to the DC grid. This requirement is very much in line with the secondary and tertiary control which is used in the AC system.

15.9 CONCLUSIONS

The currents in and out of a DC grid can be balanced by controlling the node voltages within defined limits. The DC node voltage can be influenced by controlling the current/power of the HVDC converter at that node. To ensure secure operation of the DC grid, the control should be robust and reliable and to some degree distributed and autonomous. It should also be flexible and easily expandable.

In this chapter, three overall converter control strategies are systematically introduced: constant voltage control, constant power/current control, and voltage droop control. It has been discussed to which extent these control strategies comply with or compromise the control requirements. It has been discussed that the constant current/power control and constant voltage control characteristics theoretically can be regarded as limiting cases of the droop control. The control concepts can be combined, leading to more advanced converter control strategies such as voltage margin control, deadband droop control, or undead-band droop control. These advanced converter control strategies distinguish between normal and disturbed operation.

Based on the introduced basic converter control strategies, two basic grid control strategies are introduced and classified: centralized and distributed voltage control. The application of the advanced converter control strategies results in advanced grid control strategies, which have a backup control mechanism to handle large disturbances.

There are different implementations possible for the DC voltage control. Most control concepts are based on vector control and use an inner current controller in addition to outer power or voltage control loops.

When the DC grid becomes an integral part of the transmission system, a grid control structure similar to the overall AC system control has to be conceived. Such a system controller involves changing converter set-point values, either to reestablish flow balances between different zones after an outage or to optimally control the flows inside the DC system.

REFERENCES

1. T. K. Vrana, J. Beerten, R. Belmans, and O. B. Fosso, "A classification of DC node voltage control methods for HVDC grids," *Electric Power Systems Research*, vol. 103, pp. 137–144, 2013.
2. A. Egea-Alvarez, J. Beerten, D. Van Hertem, and O. Gomis-Bellmunt, "Primary and secondary power control of multiterminal HVDC grids," in *AC and DC Power Transmission (ACDC 2012), 10th IET International Conference on*, Dec 2012, pp. 1–6.
3. B. Berggren, R. Majumder, C. Sao, and K. Lindén, "Method and control device for controlling power flow within a dc power transmission network," Patent WIPO International Publication Number WO 2012/000 549A1, International Filing Date: June 30, 2010.
4. J. Beerten and R. Belmans, "Modeling and control of multi-terminal VSC HVDC systems," *Energy Procedia*, vol. 24, pp. 123–130, Aug. 2012, selected papers from Deep Sea Offshore Wind R&D Conference, Trondheim, Norway, Jan. 19–20, 2012.
5. W. Wang, M. Barnes, and O. Marjanovic, "Droop control modelling and analysis of multi-terminal VSC-HVDC for offshore wind farms," in *Proceedings, IET ACDC 2012*, Birmingham, UK, Dec. 4–6, 2012, 6 pages.
6. E. Prieto-Araujo, F. D. Bianchi, A. Junyent-Ferre, and O. Gomis-Bellmunt, "Methodology for droop control dynamic analysis of multiterminal VSC-HVDC grids for offshore wind farms," *IEEE Transaction on Power Delivery*, vol. 26, no. 4, pp. 2476–2485, 2011.
7. O. Gomis-Bellmunt, J. Liang, J. Ekanayake, and N. Jenkins, "Voltage–current characteristics of multiterminal HVDC-VSC for offshore wind farms," *Electric Power Systems Research*, vol. 81, pp. 440–450, 2011.
8. R. Hendriks, G. Paap, and W. Kling, "Control of a multi-terminal VSC transmission scheme for connecting offshore wind farms," in *Proceedings, European Wind Energy Conference & Exhibition*, Milan, Italy, May 7–10, 2007, 8 pages.
9. T. Haileselassie, K. Uhlen, and T. Undeland, "Control of multiterminal HVDC transmission for offshore wind energy," in *Proceedings, Nordic Wind Power Conference Presentation*, Rønne, Denmark, Sept. 10–11, 2009, 7 pages.
10. J. Beerten, D. Van Hertem, and R. Belmans, "VSC MTDC systems with a distributed DC voltage control — a power flow approach," in *Proceedings, IEEE PowerTech '11*, Trondheim, Norway, June 19–23, 2011, 6 pages.
11. T. Nakajima and S. Irokawa, "A control system for HVDC transmission by voltage sourced converters," in *IEEE PES Summer Meeting, 1999*, vol. 2, 1999, pp. 1113–1119.

12. W. Lu and B.-T. Ooi, "DC overvoltage control during loss of converter in multiterminal voltage-source converter-based HVDC (M-VSC-HVDC)," *IEEE Transactions on Power Delivery*, vol. 18, no. 3, pp. 915–920, July 2003.
13. T. M. Haileselassie, M. Molinas, and T. Undeland, "Multi-terminal VSC-HVDC system for integration of offshore wind farms and green electrification of platforms in the North Sea," in *Proceedings, NORPIE 2008*, Espoo, Finland, June 19–21, 2008, 8 pages.
14. C. Dierckxsens, K. Srivastava, M. Reza, S. Cole, J. Beerten, and R. Belmans, "A distributed DC voltage control method for VSC MTDC systems," *Electric Power Systems Research*, vol. 82, no. 1, pp. 54–58, Oct. 2012.
15. T. K. Vrana, L. Zeni, and O. B. Fosso, "Active power control with undead-band voltage & frequency droop for HVDC converters in large meshed DC grids," in *EWEA 2012—European Wind Energy Conference & Exhibition*, Copenhagen, Denmark, Apr. 2012, pp. 612–616.
16. ———, "Dynamic active power control with improved undead-band droop for HVDC grids," in *Proceedings, IET ACDC 2012*, Birmingham, UK, Dec. 4–6, 2012, 6 pages.
17. C. Barker and R. Whitehouse, "Further developments in autonomous converter control in a multi-terminal HVDC system," in *Proceedings, IET ACDC 2012*, Birmingham, UK, Dec. 4–6, 2012, 6 pages.
18. CIGRÉ Working Group B4-52, "HVDC grid feasibility study," Final Report, Dec. 2012.
19. C. Barker and R. Whitehouse, "Autonomous converter control in a multi-terminal HVDC system," in *Proceedings, IET ACDC 2010*, London, UK, Oct. 20–21, 2010, 5 pages.
20. M. Aragüés-Peñalba, A. Egea-Àlvarez, O. Gomis-Bellmunt, and A. Sumper, "Optimum voltage control for loss minimization in HVDC multi-terminal transmission systems for large offshore wind farms," *Electric Power Systems Research*, vol. 89, pp. 54–63, 2012.
21. J. Svensson, "Synchronisation methods for grid connected voltage source converters," *IEE Proceedings in Generation, Transmission and Distribution*, vol. 148, no. 3, pp. 229–235, 2001.
22. S.-K. Chung, "A phase tracking system for three phase utility interface inverters," *IEEE Transactions on Power Electronics*, vol. 15, no. 3, pp. 431–438, 2000.
23. A. Egea-Alvarez, A. Junyent-Ferre, and O. Gomis-Bellmunt, "Active and reactive power control of grid connected distributed generation systems," in *Modeling and Control of Sustainable Power Systems*, series Green Energy and Technology, L. Wang, Ed. Berlin: Springer, 2012, pp. 47–81.

CHAPTER 16

STATE-SPACE REPRESENTATION OF HVDC GRIDS

Eduardo Prieto-Araujo[1] *and Fernando Bianchi*[2]

[1]Research Group CITCEA, Electrical Engineering Department, Technical University of Catalonia (UPC), Barcelona, Spain

[2]Research Group Wind Energy, Electrical Engineering Department, Catalonia Institute for Energy Research (IREC), Barcelona, Spain

16.1 INTRODUCTION

A methodology to obtain the state-space model of a generic multi-terminal HVDC grid (HVDC grid) is described. The procedure is based on combining the equations describing the different parts of the transmission grid. As a result, a simple model of a defined HVDC grid could be obtained to perform either steady-state or dynamic analysis.

16.2 MULTI-TERMINAL GRID MODELING

16.2.1 Grid Structure

A multi-terminal high-voltage direct current (HVDC grid) grid can be represented as an interconnection of nodes and branches [1]. The different AC systems connected to the grid through AC/DC VSC converters are considered either input or output power nodes. Transmission lines interconnecting the described nodes are defined as branches. Also, grid connections where only cables converge are defined as intermediate nodes. A simplified example of an offshore HVDC grid is shown in Figure 16.1. It is compounded by m AC systems injecting power to the grid, n AC systems extracting power from the grid, and p intermediate connection nodes and r branches. Notice that the number of intermediate connections and branches depend on the particular interconnection pattern.

HVDC Grids: For Offshore and Supergrid of the Future, First Edition. By Dirk Van Hertem, Oriol Gomis-Bellmunt, and Jun Liang.

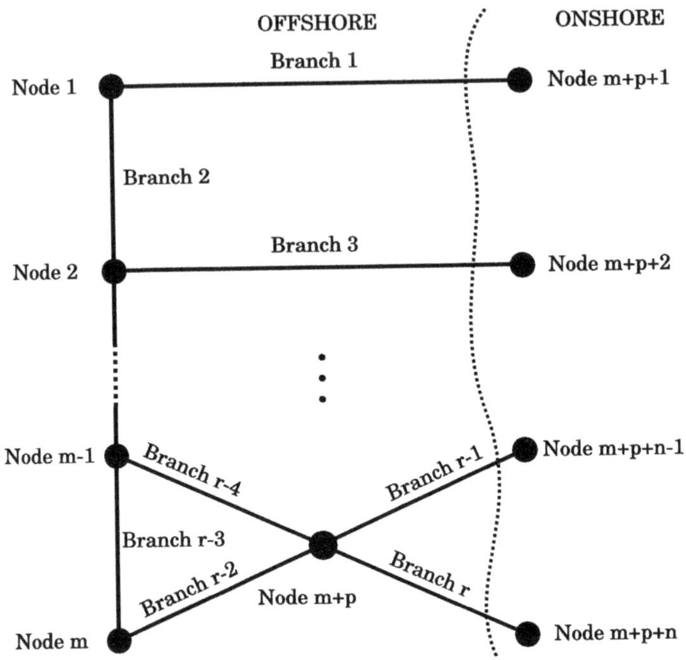

Figure 16.1 A node and branch scheme of an HVDC offshore grid.

16.2.2 AC Systems Connected to the HVDC Grid

AC systems are connected to the HVDC grid through VSC-HVDC power converters. An averaged model of the converter [2] is considered to represent its behavior inside the HVDC grid. This model approaches the AC side of the converters as three voltage sources and approaches the DC side as a current source together with a capacitor, as shown in Figure 16.2.

Using this simplified representation, each AC system is modeled from the DC grid side as a DC current source, as illustrated in Figure 16.3. At the converter DC side, the power flow in the node k is represented by a current i_k coming from a source of value

$$i_k = \frac{P_k}{E_k} \tag{16.1}$$

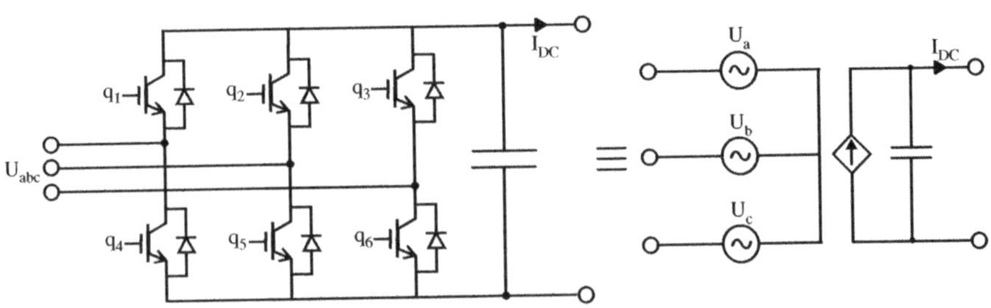

Figure 16.2 Average equivalent representation of a three-phase VSC converter.

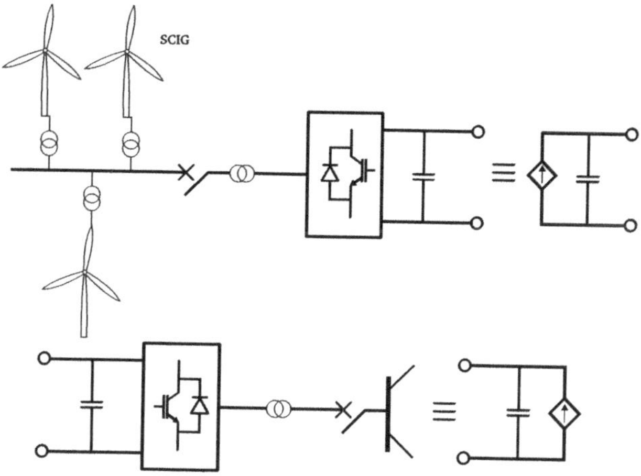

Figure 16.3 Equivalent representation of the wind farm and the AC grid converters for a DC grid analysis.

where P_k is the incoming power and E_k is the DC voltage at the node k. It will be assumed that the voltage E_k remains close to the nominal values E_0. Under this assumption, the current i_k can be assumed proportional to the power P_k.

16.2.3 Branches

Transmission lines interconnecting the different grid nodes are modeled by π-equivalent circuits [3]; see Figure 16.4. When several lines are connected to the same node, only one equivalent capacitor C_k is used to represent it, in order to reduce the number of the model variables.

$$C_k = \sum_{i=1}^{l} C_i \tag{16.2}$$

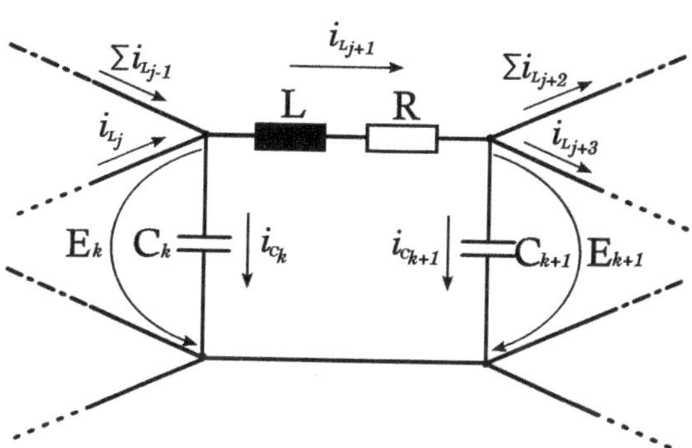

Figure 16.4 π-circuit modeling a branch element.

16.2.4 Intermediate Nodes

Several transmission lines in the HVDC grid may join two or more terminals at intermediate points. These types of nodes are considered intermediate nodes. Again, the number of capacitances placed in parallel can be reduced to an equivalent capacitance, considering the association of the π-equivalent corresponding circuits using (16.2).

16.2.5 Equivalent Circuit Model

An equivalent circuit can be obtained from the interconnection of the nodes and branches after the above-mentioned simplifications. An example of a HVDC offshore grid for wind power integration is shown in Figure 16.5.

Next, in Figure 16.5, equivalent scheme variables are described:

- $R_1 \ldots R_r$ are the branch equivalent resistances of the different lines of the grid.
- $L_1 \ldots L_r$ are the branch equivalent inductances of the different lines of the grid.
- $i_{L_1} \ldots i_{L_r}$ are the currents flowing through the different elements of the grids.

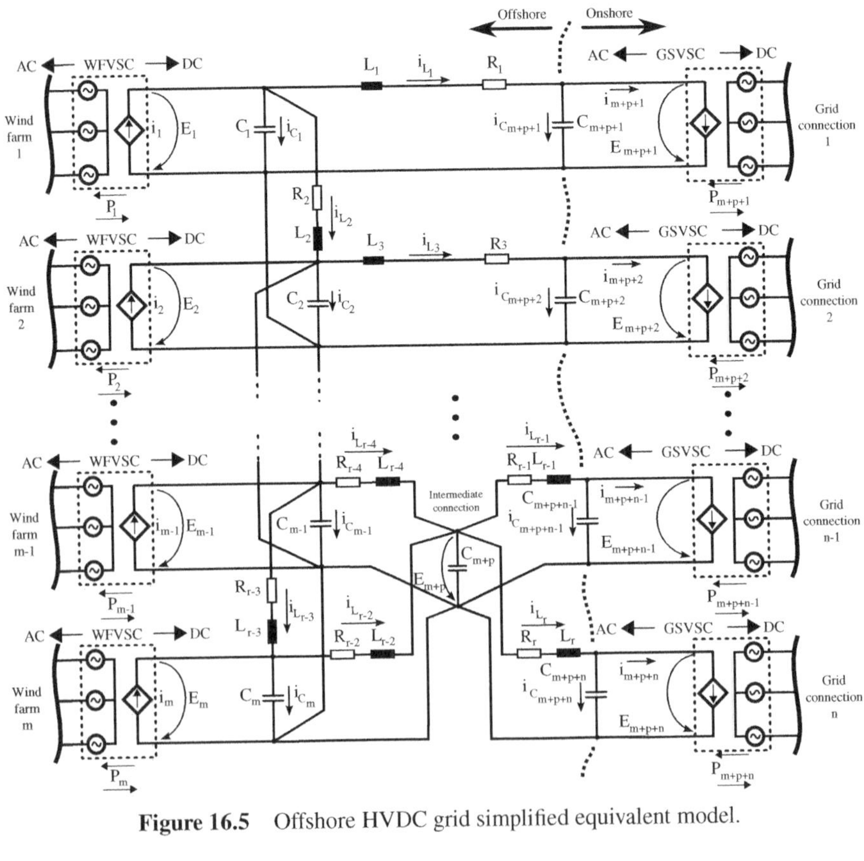

Figure 16.5 Offshore HVDC grid simplified equivalent model.

- $C_1 \ldots C_{m+p+n}$ are the equivalent capacitors resulting from the parallel association of the different capacitors of the models involved in the grid.
- $E_1 \ldots E_{m+p+n}$ are the $n+m$ voltages measured at the capacitors placed in parallel with the current sources of the equivalent scheme, in addition with the p voltages measured at the capacitors located in the intermediate connection points.
- $i_{C_1} \ldots i_{C_{m+p+n}}$ are the currents flowing each of the capacitors of the grid equivalent scheme.
- $P_1 \ldots P_m$ i $P_{m+p+1} \ldots P_{m+p+n}$ represent the power transmitted by each of the converters connected to the represented grid. The first m variables are related to the wind farm converters, and the following n variables are related to the grid side converters.
- $i_1 \ldots i_m$ i $i_{m+p+1} \ldots i_{m+p+n}$ represent the current flowing through the current sources presented at the equivalent scheme of the HVDC grid. The value of these sources is obtained from the relation between the power flowing the converter and the voltage present at its terminals:

$$i = \frac{P}{E} \tag{16.3}$$

16.2.6 State-Space Representation

Then, using circuits laws and after some variable manipulations, it is possible to find a set of first-order differential equations describing the dynamic behavior of the entire HVDC grid. These differential equations are known as the state-space representation and can be expressed in the following compact form [4, 5]:

$$\begin{aligned} \text{Dynamic equation: } \frac{dx}{dt} &= Ax + B_w w + B_u u \\ \text{Noncontrolled outputs equation: } z &= C_z x \\ \text{Controlled output equations: } y &= C_y x \end{aligned} \tag{16.4}$$

The equation variables are described as follows:

- x is the state vector.
- $\dot{x}$ is the derivative of the system state vector.
- u is the controlled input vector.
- w is the noncontrolled or perturbation input vector.
- y is the controlled output vector.
- z is the noncontrolled output vector.
- A is the state matrix.
- B_u is the controlled input matrix.
- B_w is the noncontrolled input matrix.
- C_y is the controlled output matrix.
- C_z is the noncontrolled output matrix.

A, B_w, B_u, C_z, and C_y are matrices of suitable dimensions. These matrices are obtained after arranging the variables and applying matrix computation laws.

The state vector x consists of internal variables that characterize the entire state of the system. In an electrical system, the currents in the inductors and the voltages in the capacitors are commonly selected as states. Therefore, in the case of the HVDC offshore grid in Figure 16.5, the state vector is given by

$$x = [E_1, \cdots, E_{m+p+n}, i_{L_1}, \cdots, i_{L_r}]^T$$

Each node has one capacitor and each branch has one inductor, therefore the total number of states is $n + p + m + r$.

The inputs are divided into two vectors: The vector u gathers the variables that can be used to control the system and w are disturbances, that is, external variables that are not possible to manipulate. In the case of the HVDC offshore grids, the inputs of the system are the current injected or extracted by the converters, therefore

$$w = [i_1, \ldots, i_j, \ldots, i_{n_{\text{nc}}}]^T, \qquad j \in \mathcal{J}_{\text{nc}}$$

$$u = [i_1, \ldots, i_j, \ldots, i_{n_{\text{c}}}]^T, \qquad j \in \mathcal{J}_{\text{c}}$$

where $\mathcal{J}_{\text{nc}}$ corresponds to the set of indexes of the nodes where the converters inject or extract power without voltage control and $\mathcal{J}_{\text{c}}$ denotes the set of indexes of the nodes where the voltage control is applied. Notice that the relation $n_{\text{c}} + n_{\text{nc}} = m + n$ must be held.

Similarly, the output is partitioned into two vectors. The vector y contains the variables that can be used in the control of the DC voltage. On the other hand, z stands for the vector of variables that are not available to be used by the controller. In the HVDC grid scheme, the controllers can only use the information provided by the voltage at the nodes where the voltage control is applied. The rest of the voltages must also be maintained close to the rated values but they cannot be fed back to the controllers. Hence,

$$z = [E_1, \ldots, E_j, \ldots, E_{n_{\text{nc}}}]^T \qquad j \in \mathcal{J}_{\text{nc}}$$

$$y = [E_1, \ldots, E_j, \ldots, E_{n_{\text{c}}}]^T \qquad j \in \mathcal{J}_{\text{c}}$$

The transfer matrix of the system

$$G(s) = \begin{bmatrix} G_{zw}(s) & G_{zu}(s) \\ G_{yw}(s) & G_{yu}(s) \end{bmatrix} \tag{16.5}$$

is obtained from the state-space equation (16.4), where

$$G_{zw}(s) = C_z(sI - A)^{-1}B_w$$
$$G_{zu}(s) = C_z(sI - A)^{-1}B_u$$
$$G_{yw}(s) = C_y(sI - A)^{-1}B_w$$
$$G_{yu}(s) = C_y(sI - A)^{-1}B_u$$

The transfers matrices G_{zu} and G_{yu} relate the currents imposed by the controller with the controlled and nondirectly controlled voltages, respectively, whereas the transfer matrices G_{zw} and G_{yw} connect the current not used in the control with the controlled and nondirectly controlled voltages, respectively. Figure 16.6 shows the DC multi-terminal matrix transfer function diagram representation, relating the controlled and non controlled inputs and outputs of the system. The existing relation between the system variables is again described schematically in Table 16.1.

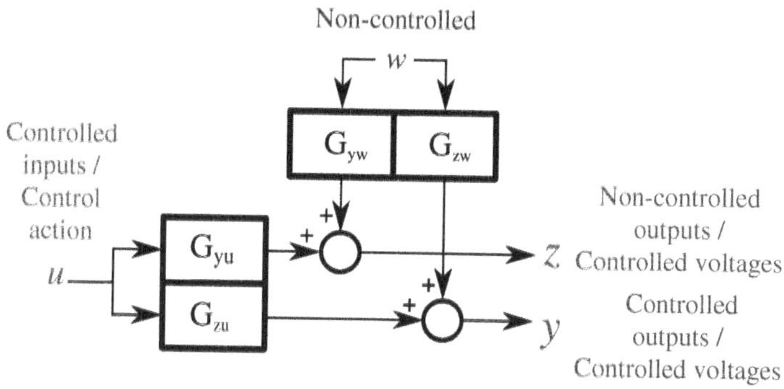

Figure 16.6 DC multi-terminal grid model matrix transfer function diagram representation.

TABLE 16.1 DC Multi-Terminal Grid Matrix Transfer Functions Classification

	Outputs	
Inputs	Noncontrolled (z)	Controlled (y)
Noncontrolled (w)	G_{zw}	G_{yw}
Controlled (u)	G_{zu}	G_{yu}

16.3 FOUR-TERMINAL GRID EXAMPLE

A simple four-terminal HVDC grid is used to illustrate the modeling methodology presented in previous Sections [6]. The four-terminal grid is depicted in Figure 16.7 and consists of two offshore wind farm converters WFC1 and WFC2 and two onshore grid side converters GSC1 and GSC2. The four-terminal HVDC grid has two power input nodes, two power output nodes, and three branches representing the cables linking the converters. The capacitors are the result of combining the capacitances of the nodes and the corresponding branch side, as explained in Section 16.2.

Two scenarios are analyzed. In the first case, it is assumed that droop voltage control is applied in both grid side converters whereas the wind farm converters inject all the wind power available. In the second scenario, due to a fault in the AC grid, both wind farm converters regulate the DC voltages and the grid side converters extract power from the HVDC grid at their maximum capacity. Thus, in these cases the state-space variables selection will change depending on which are the system

TABLE 16.2 Four-Terminal Example Variable Definition

Grid parameters	Nomenclature
Line resistance	R_1
Line resistance	R_2
Line resistance	R_3
Line inductance	L_1
Line inductance	L_2
Line inductance	L_3
Capacitances	C_k
Rated line current	i_L^{rtd}
Rated input current	i_k^{rtd}
Converter rated power	P_k
Rated DC voltage	E_k
Reference voltage	E_0

voltages that must be controlled. Table 16.2 summarizes the nomenclature used to represent the multi-terminal DC grid variables used in the model.

16.3.1 Case 1: Droop Voltage Control in the AC Grid Side

Applying circuit laws to the four-terminal grid in Figure 16.7, the following differential equations can be obtained:

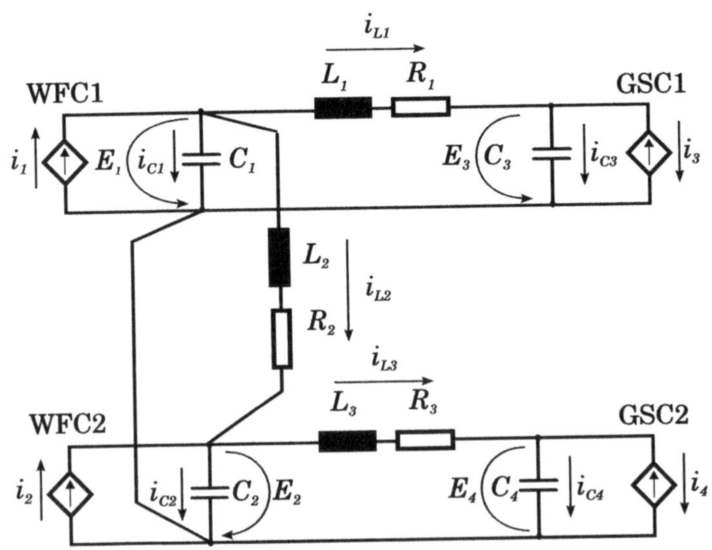

Figure 16.7 Four-terminal grid used to illustrate the modeling methodology.

$$\frac{dE_1}{dt} = \frac{1}{C_1} i_{C_1}, \quad \frac{dE_2}{dt} = \frac{1}{C_2} i_{C_2} \tag{16.6}$$

$$\frac{dE_3}{dt} = \frac{1}{C_3} i_{C_3}, \quad \frac{dE_4}{dt} = \frac{1}{C_4} i_{C_4} \tag{16.7}$$

$$\frac{di_{L_1}}{dt} = \frac{1}{L_1}(-R_1\, i_{L_1} + E_1 - E_3) \tag{16.8}$$

$$\frac{di_{L_2}}{dt} = \frac{1}{L_2}(-R_2\, i_{L_2} + E_1 - E_2) \tag{16.9}$$

$$\frac{di_{L_3}}{dt} = \frac{1}{L_3}(-R_3\, i_{L_3} + E_2 - E_4) \tag{16.10}$$

and the following algebraic equations can be obtained:

$$i_{C_1} = i_1 - i_{L_1} - i_{L_2} \tag{16.11}$$

$$i_{C_2} = i_2 - i_{L_3} + i_{L_2} \tag{16.12}$$

$$i_{C_3} = -i_3 + i_{L_1} \tag{16.13}$$

$$i_{C_4} = -i_4 + i_{L_3} \tag{16.14}$$

There are four capacitors and three inductors; therefore, the variables E_1, E_2, E_3, E_4, i_{L_1}, i_{L_2}, and i_{L_3} are sufficient to completely define the state of this system, that is,

$$x = [E_1\ E_2\ E_3\ E_4\ i_{L_1}\ i_{L_2}\ i_{L_3}]^T$$

As the WFCs are injecting all available power into the grid and in the GSCs are regulating the DC voltages, the input and output result divided into

$$w = [i_1\ i_2]^T, \qquad u = [i_3\ i_4]^T$$

$$z = [E_1\ E_2]^T, \qquad y = [E_3\ E_4]^T$$

The purpose of the droop voltage control applied on the right side of the four-terminal grid in Figure 16.7 is to maintain the DC voltage stable when the currents coming from the wind farm converters WFC1 and WFC2 change. Therefore, the vector of these currents is the disturbance w and the control input u is the vector of the currents of the GSCs i_3 and i_4. The voltages measured and fed back to the controller are the voltages E_3 and E_4 whereas the voltages E_1 and E_2 are not available for the controller, but it is desirable to maintain them close to the rated value.

After the previous definitions, substituting the currents in the capacitors in (16.6)–(16.7) by the relations (16.11)–(16.14) and reorganizing the differential equations, the matrices in the state-space representation (16.4) result:

$$A = \begin{bmatrix} 0 & 0 & 0 & 0 & -\frac{1}{C_1} & -\frac{1}{C_1} & 0 \\ 0 & 0 & 0 & 0 & 0 & \frac{1}{C_2} & -\frac{1}{C_2} \\ 0 & 0 & 0 & 0 & \frac{1}{C_3} & 0 & 0 \\ 0 & 0 & 0 & 0 & 0 & 0 & \frac{1}{C_4} \\ \frac{1}{L_1} & 0 & -\frac{1}{L_1} & 0 & -\frac{R_1}{L_1} & 0 & 0 \\ \frac{1}{L_2} & -\frac{1}{L_2} & 0 & 0 & 0 & -\frac{R_2}{L_2} & 0 \\ 0 & \frac{1}{L_3} & 0 & -\frac{1}{L_3} & 0 & 0 & -\frac{R_3}{L_3} \end{bmatrix}$$

$$B_w = \begin{bmatrix} \frac{1}{C_1} & 0 \\ 0 & \frac{1}{C_2} \\ 0 & 0 \\ 0 & 0 \\ 0 & 0 \\ 0 & 0 \\ 0 & 0 \end{bmatrix}, \qquad B_u = \begin{bmatrix} 0 & 0 \\ 0 & 0 \\ -\frac{1}{C_3} & 0 \\ 0 & -\frac{1}{C_4} \\ 0 & 0 \\ 0 & 0 \\ 0 & 0 \end{bmatrix}$$

$$C_z = \begin{bmatrix} 1 & 0 & 0 & 0 & 0 & 0 & 0 \\ 0 & 1 & 0 & 0 & 0 & 0 & 0 \end{bmatrix}$$

$$C_y = \begin{bmatrix} 0 & 0 & 1 & 0 & 0 & 0 & 0 \\ 0 & 0 & 0 & 1 & 0 & 0 & 0 \end{bmatrix}$$

16.3.2 Case 2: Droop Voltage Control in the Wind Farm Side

In the second case of study, it is assumed that a simultaneous fault in both AC grids forces the GSCs to enter in current limitation mode. In this circumstance, the WFCs are responsible for regulating the DC voltage. Hence, the control inputs are the currents injected by the WFCs and the disturbances are the currents extracted by the GSCs, that is,

$$w = [i_3\ i_4]^T, \qquad u = [i_1\ i_2]^T$$

On the other hand, the measured variables are the wind farm side voltages and the nondirectly controlled variables are the grid side voltages, that is,

$$z = [E_1\ E_2]^T, \quad y = [E_3\ E_4]^T$$

The state-space model have the same matrix A but the input and output matrices are now given by

$$B_w = \begin{bmatrix} 0 & 0 \\ 0 & 0 \\ -\frac{1}{C_3} & 0 \\ 0 & -\frac{1}{C_4} \\ 0 & 0 \\ 0 & 0 \\ 0 & 0 \end{bmatrix}, \quad B_u = \begin{bmatrix} \frac{1}{C_1} & 0 \\ 0 & \frac{1}{C_2} \\ 0 & 0 \\ 0 & 0 \\ 0 & 0 \\ 0 & 0 \\ 0 & 0 \end{bmatrix}$$

$$C_z = \begin{bmatrix} 0 & 0 & 1 & 0 & 0 & 0 & 0 \\ 0 & 0 & 0 & 1 & 0 & 0 & 0 \end{bmatrix}$$

$$C_y = \begin{bmatrix} 1 & 0 & 0 & 0 & 0 & 0 & 0 \\ 0 & 1 & 0 & 0 & 0 & 0 & 0 \end{bmatrix}$$

16.4 CONCLUSIONS

In this chapter the mathematical modeling of a HVDC grid is detailed. Using π-cable models and VSC converter equivalent representation, a complete grid model can be straightforwardly obtained. Different mathematical model representations are provided including state-space representation and transfer function matrices. Two case studies are presented to exemplify the procedure to develop HVDC grid model to analyze the grid during normal and AC fault conditions. The proposed model applies for both cases, considering that the controlled variables are different due to the fact that the voltage regulation is performed by GSC in normal operation and by WFC during voltage sags.

REFERENCES

1. J. Reeve, "Multiterminal HVDC power systems," *IEEE Transactions on Power Apparatus and Systems*, vol. PAS 99, no. 2, pp. 729–737, 1980.
2. G. Zhang, Z. Xu, and Y. Cai, "An equivalent model for simulating VSC based HVDC," in *Proceedings of the IEEE/PES Transmission and Distribution Conference and Exposition*, vol. 1, 2001, pp. 20–24.
3. P. Kundur, *Power System Stability and Control*. New York: McGraw-Hill Professional, 1994.
4. K. Ogata, *Modern Control Engineering*, ser. Instrumentation and controls series. Prentice Hall, 2010 [Online]. Available: http://books.google.es/books?id=Wu5GpNAelzkC
5. S. Skogestad and I. Postlethwaite, *Multivariable Feedback Control, Analysis and Design*, Hoboken, NJ: John Wiley & Sons, 2007.
6. E. Prieto-Araujo, F. Bianchi, A. Junyent-Ferre, and O. Gomis-Bellmunt, "Methodology for droop control dynamic analysis of multiterminal VSC–HVDC grids for offshore wind farms," *Power Delivery, IEEE Transactions on*, vol. 26, no. 4, pp. 2476–2485, Oct. 2011.

CHAPTER 17

DC FAULT PHENOMENA AND DC GRID PROTECTION

Willem Leterme and Dirk Van Hertem
Research Group Electa, Department of Electrical Engineering, University of Leuven, Leuven, Belgium

17.1 INTRODUCTION

This chapter gives an introduction to DC grid protection for short circuit faults at the DC side. Although DC grid protection must fulfill the same objectives as AC grid protection, its implementation will differ largely since DC fault currents are characterized by a high rate of rise, a large steady-state value, and the absence of natural zero crossings. The latter characteristic implies the need for specific DC breaker technology, whereas the former two impose limits on fault clearing time. This fault clearing time will lie in the order of only milliseconds due to sensitivity of power electronic components and limited interruption capability of DC breakers. Therefore, a new protection methodology is needed for the DC grid, comprising fault detection and clearing strategies.

In the first part of the chapter, introductory analysis of the phenomena occurring in the faulted DC grid is given. First, the impact of system grounding and configuration on steady-state fault currents and voltages in the DC grid is investigated. Thereafter, propagation of a fault transient in the DC grid is handled by means of traveling wave analysis. Next, fault current sources in the DC grid are analyzed. Finally, a description of modeling needs for electromagnetic transient analysis in DC grids is given.

The second part of the chapter deals with the protection system for the DC grid. First, the objectives of protection and protection philosophy for the DC grid are treated. Second, an overview of protection strategies based on DC fault clearance is given. Third, system relaying for selective protection in DC grids is explored. The last section provides a brief overview of DC breakers and measurement equipment. Within the chapter, the fault treatment of converter faults is not discussed as this is very technology (and manufacturer)-dependent.

HVDC Grids: For Offshore and Supergrid of the Future, First Edition. By Dirk Van Hertem, Oriol Gomis-Bellmunt, and Jun Liang.

17.2 SHORT-CIRCUIT FAULTS IN THE DC GRID

This section elaborates upon DC fault behavior by examining the factors influencing steady-state and transient fault phenomena. The analysis is based on converters without fault blocking capability and does not take into account surge arresters which protect transmission lines from overvoltages.

In Section 17.2.2, steady-state DC fault behavior is treated. Next, traveling wave phenomena and sources of DC fault current are studied. The transmission line and converter models for analyzing electromagnetic transients are subsequently treated. Finally, an overview of factors influencing the DC fault is given.

17.2.1 Sequence of Events of a DC Short-Circuit Fault

The time course of a DC fault is characterized by different phenomena. Consider a fault in the DC grid shown in Figure 17.1. An example time course of the DC fault current seen at line 12 for a fault occurring in the middle of that line is given in Figure 17.2. The overall time course of the fault current can be recognized as the response of switching a DC voltage source in an RL-circuit, in which the steady-state fault current is determined by $\frac{U_{DC}}{R_{line}}$ and the transient rise is determined by the time constant $\tau = \frac{R_{line}}{L_{line}}$. However, several other effects play a role as can be seen in Figure 17.3.

Figure 17.3 shows the sequence of events which occur during a DC fault and its interruption as also indicated in Figure 17.2. When a short circuit fault occurs at a transmission line at t_0, an electromagnetic wave is created which travels towards the terminations of the transmission line. After a certain time delay, the wave reaches the line termination at t_1, causing the locally measured current to increase and the voltage to decrease. The wave is partly reflected back to the fault and partly transmitted to the rest of the system. In this time period, the main current contribution from the

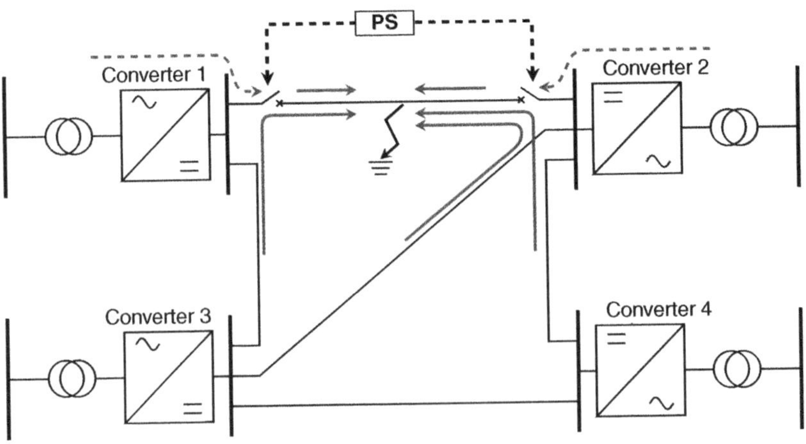

Figure 17.1 Fault propagation in DC grids.

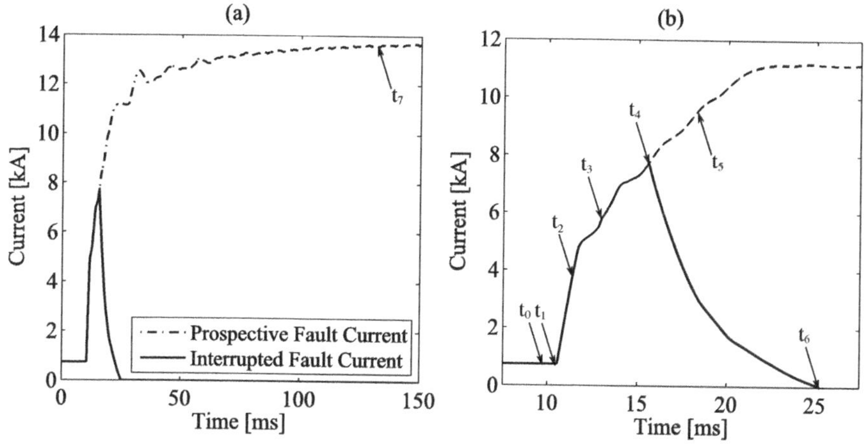

Figure 17.2 Example time course of a fault current in a DC grid. (a) DC fault current. (b) Detail of transient DC fault current.

converter is through discharge of its capacitors. After some more time, at t_2 the traveling wave created by the fault reaches the other terminals in the grid, which also start contributing fault current. At t_3, the converter blocks its IGBTs to avoid damage and current is fed through the AC side, until it is interrupted by DC breakers which trip at t_4. Without the breaker action, the fault current remains to be fed from the AC side up to either the steady-state short current is reached or a component is damaged (t_5). Due to the breaker action, the fault current diminishes and returns to zero at t_6. The total fault clearing time is hence from t_1 to t_6, whereas the time available for fault detection is only until t_4. If there is no breaker action, each converter keeps contributing to the fault through the AC side, converging towards a steady-state fault current at t_7, depicted in Figure 17.3.

17.2.2 Steady-State DC Fault Behavior

The prospective steady-state DC fault current is mainly determined by fault type and system grounding and configuration. The fault types can be split up in different categories (see Figure 17.4):

(a) pole-to-ground faults
(b) pole-to-neutral faults
(c) pole-to-pole faults.
(d) bus faults.
(e) neutral-to-ground faults.

The steady-state voltages and currents after the fault depend on the grounding and configuration of the DC grid.

For pole-to-ground faults (Figure 17.4a), one pole is connected to the ground through a fault impedance. In low-impedance grounded asymmetric or bipolar

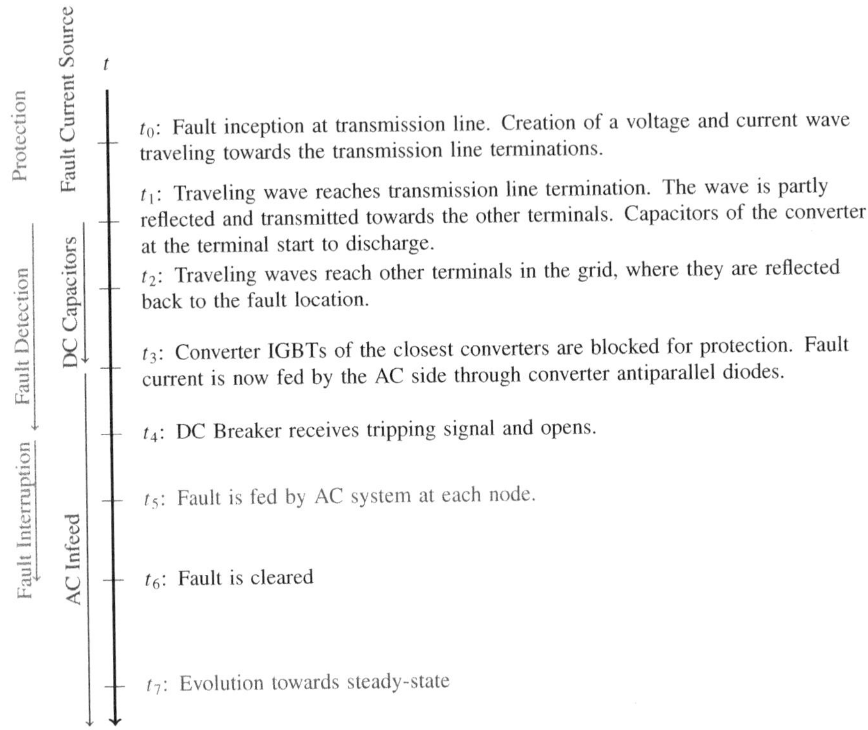

Figure 17.3 Time course of events during a DC fault.

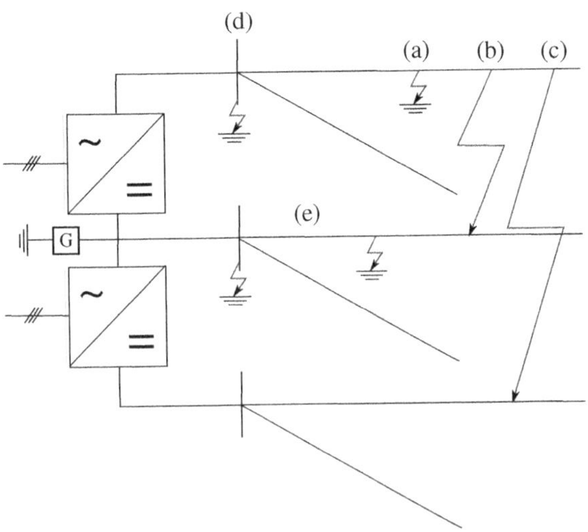

Figure 17.4 Types of DC faults.

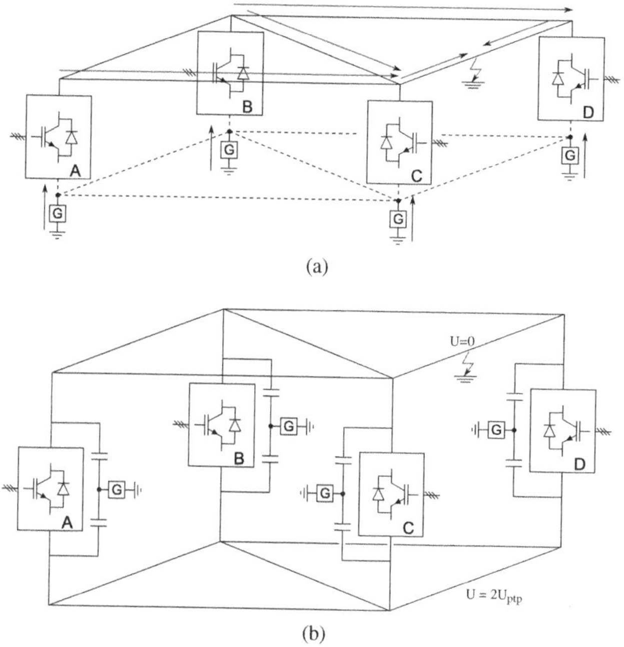

Figure 17.5 System response to pole-to-ground fault. (a) Asymmetric configuration. (b) Symmetric configuration.

configurations, this leads to large fault currents supplied through the grounding points, whereas the voltages are clamped to a low value (Figure 17.5a). For high-impedance grounded symmetric or bipolar configurations, the steady-state fault current is theoretically zero. Overvoltages on the healthy poles up to twice the nominal value can occur as the pole-to-ground fault defines a new ground reference (Figure 17.6b) [1]. If surge arresters with protective levels lower than twice the nominal voltage are used, the voltage is clamped to the level of the surge arresters and fault currents can flow in the DC grid.

Pole-to-pole faults (Figure 17.4c) are the connection of positive and negative pole. The fault response for pole-to-pole faults is similar for both low- and high-impedance grounded systems (Figure 17.6). For all systems, a direct path for the fault current exists, leading to large fault currents and low voltages. For cable systems, pole-to-pole faults are also connected to the earth. In overhead line systems, this is not necessarily the case.

Pole-to-neutral or neutral-to-ground faults (Figure 17.4b,e) are also possible for bipolar systems. Pole-to-neutral faults are similar in nature compared to pole-to-ground faults. In the case of neutral-to-ground faults, no large fault currents or undervoltages are noticed in the system.

Bus faults (Figure 17.4d) can also occur in the system. Depending on the type of fault, similar conclusions can be drawn for the different configurations. As opposed to a fault on a transmission line, the bus faults lead to loss of a complete bus, hence having a more severe impact in the grid.

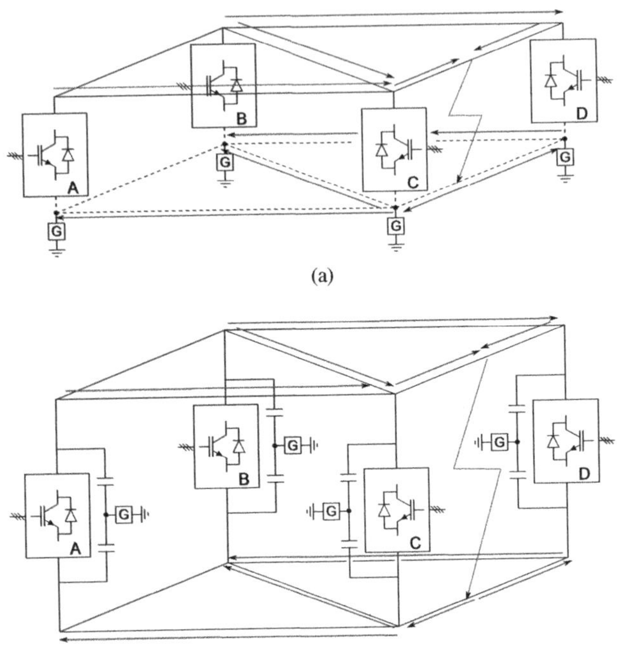

Figure 17.6 System response to pole-to-pole fault. (a) Asymmetric configuration. (b) Symmetric configuration.

17.2.3 Transient DC Fault Phenomena

This section provides an overview of the phenomena that occur in the DC grid during the fault transient. Due to the short timescale, electromagnetic transients such as traveling waves are prevalent. These effects are described in Section 17.2.3.1. Furthermore, the main source of fault current differs over time due to capacitive discharge of cables and protective actions of the converter, which is elaborated in the next section. The models for simulating electromagnetic fault transients in DC grids are treated in Section 17.2.3.3.

17.2.3.1 Fault Propagation in DC Grids As introduced in Section 17.2.1, traveling waves created by the fault propagate through the grid and are reflected and transmitted at each terminal. This section further elaborates on the behavior of traveling waves in a DC grid.

Fault Propagation in Transmission Lines If a solid pole-to-ground fault occurs at t_0, the voltage at the fault location abruptly decreases from the pre-fault voltage (U_{pre}) to zero voltage. This change in voltage is not immediately seen at the ends of the transmission line due to the finite speed of an electromagnetic wave that travels on the transmission line. To incorporate these effects, the transmission line must be modeled by a distributed parameters model [2].

For the following analysis, a lossless line with parameters independent from frequency is assumed. The differential equations for the voltage and current along the line and as function of time are [3]

$$\frac{\partial u(x,t)}{\partial x} = -L\frac{\partial i(x,t)}{\partial t} \tag{17.1}$$

$$\frac{\partial i(x,t)}{\partial x} = -C\frac{\partial u(x,t)}{\partial t} \tag{17.2}$$

leading to a solution in the form of

$$u(x,t) = f_1(x - v \cdot t) + f_2(x + v \cdot t) \tag{17.3}$$

$$i(x,t) = \frac{1}{Z_c}(f_1(x - v \cdot t) - f_2(x + v \cdot t)) \tag{17.4}$$

f_1 and f_2 are functions which represent the voltage distribution over the line, determined by the boundary conditions. v is the propagation velocity—that is, the velocity at which a wave travels over the transmission line and Z_c is the characteristic impedance of the transmission line. Expressions for v and Z_c for the lossless line are

$$v = \frac{1}{\sqrt{LC}} \tag{17.5}$$

$$Z_c = \sqrt{\frac{L}{C}} \tag{17.6}$$

The solution for the voltage along the line can hence be described using two functions f_1 and f_2, respectively representing waves traveling in the positive and negative x-direction. The voltage and current wave are linked by the characteristic impedance Z_c. Note that the current wave traveling in the positive direction has the same sign as the voltage wave, whereas an opposite sign is obtained for the current wave traveling in the negative direction.

The Bewley lattice diagram can be used to analyze traveling waves in the time domain. An example of such a lattice diagram is given in Figure 17.7. This figure shows that between t_0 and t_1, a wave traveling in the positive x-direction towards terminal 2 and in the negative x-direction towards terminal 1 is induced by the fault.

Typical order of magnitudes for the characteristic impedance for overhead lines are hundreds of ohms, whereas for cables this is on the order of tenths of ohms. The propagation velocity for overhead lines and cables is respectively around the speed of light and half the speed of light.

Reflection and Transmission at the Faulted Line Terminals The traveling wave emanating from the fault reaches the end of the transmission line with a certain delay. Consider the first wave, propagating from the fault location to bus 1. Take the amplitude of the voltage wave as u_+; then the amplitude of the associated current wave with is $i_+ = u_+/Z_c$. When the wave reaches the end of the transmission line, it perceives an impedance differing from the characteristic impedance Z_c. This impedance typically consists of the impedances of the converters and cables connected to bus 1. For now,

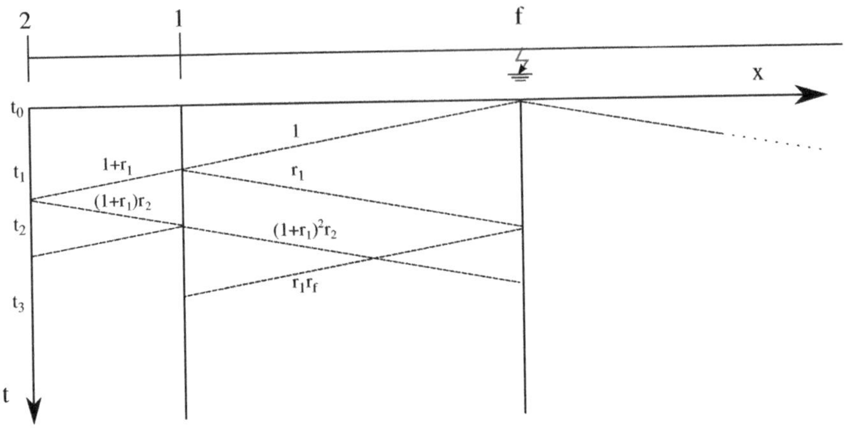

Figure 17.7 Bewley lattice diagram for analysis of traveling waves.

consider a constant impedance Z_l at the end of the transmission line. Then the voltage and the current at the terminal must satisfy

$$i = \frac{u}{Z_l} \tag{17.7}$$

and the solution for the wave, u_+ and i_+ related by the characteristic impedance Z_c, is no longer valid. Therefore, the wave is reflected at the end of the transmission line, creating a wave traveling in the opposite direction (i.e, from transmission line end to fault location). Naming the amplitudes of the reflected voltage and current wave u_- and i_-, an expression for the voltage at the terminal is accordingly obtained:

$$u = u_+ + u_- \tag{17.8}$$

$$i = i_+ + i_- = \frac{u_+ - u_-}{Z_C} \tag{17.9}$$

u_+ to u_- can be related by the voltage reflection coefficient r, obtained from (17.7), (17.8), and (17.9) as

$$r = \frac{u_-}{u_+} = \frac{Z_l - Z_c}{Z_l + Z_c}. \tag{17.10}$$

Hence, the voltage at the terminals can also be written as

$$u = (1 + r) \cdot u_+ \tag{17.11}$$

The right-hand side of (17.11) is also the amplitude of the wave which is transmitted to the line between converters 1 and 3.

Reflection and Transmission at Other Terminals Throughout the grid, waves propagate over the nonfaulted lines to the other buses, where they are reflected back to the faulted line. The voltage (or current) at each location in the grid and at each time instant can be obtained by summation of the waves that have passed before. This can be visualized using the Bewley lattice diagram—for example, for the voltage at terminal 1 (see Figure 17.7):

$$\Delta u_1 = 0 \, \forall t \in [t_0, t_1] \tag{17.12}$$
$$= u_+(1 + r_1) \forall t \in [t_1, t_2] \tag{17.13}$$
$$= u_+((1 + r_1) + (1 + r_1)^2 r_2)) \forall t \in [t_2, t_3] \tag{17.14}$$
$$= u_+((1 + r_1) + (1 + r_1)^2 r_2 + (1 + r_1) r_1 r_f + \cdots) \forall t \in [t_3, +\infty] \tag{17.15}$$

17.2.3.2 Sources Contributing to the Fault Current

The DC fault current is supplied by various sources during the period of the fault. The dominant fault current source changes from the beginning of the fault until the steady-state is reached.

Capacitive Discharge In cable networks, the energy stored in the capacitance of the cable is gradually discharged into the fault as the wave propagates through the cable. When the wave reaches the terminals, the current at the faulted cable increases depending on the number of connected cables. The reflection coefficient of a current wave from a faulted cable at a terminal where n cables are connected is

$$r_i = \frac{Z_c - \frac{Z_c}{n}}{\frac{Z_c}{n} + Z_c} = \frac{n - 1}{1 + n} \tag{17.16}$$

where r_i is the current reflection coefficient, relating the incoming wave to the reflected wave. In case of one cable, i.e. $n = 1$, both reflection coefficients are zero, indicating that the wave is fully transmitted to the other cable. Table 17.1 represents the current reflection coefficient and the current that is effectively measured at the terminal, taking the amplitude of the incoming current wave as 1. The current reflection coefficient increases with an increasing number of cables. The maximal reflection coefficient converges to 1 for a large number of cables, indicating that the amplitude of the incoming current wave is instantaneously doubled.

The capacitors of a converter also contribute to the fault current, depending on the converter topology. For some converter topologies (e.g, two-level), large DC capacitors are needed to maintain the DC voltage. The current reflection coefficient

TABLE 17.1 Reflection Coefficient for Different Number of Cables

n	r_i	$(1 + r_i)$
1	0	0
2	1/3	4/3
3	1/2	3/2
4	3/5	8/5
∞	1	2

for a capacitor with capacitance C terminating a transmission line with characteristic impedance Z_c is

$$r_i = \frac{Z_c - \frac{1}{j\omega C}}{Z_c + \frac{1}{j\omega C}} \tag{17.17}$$

Hence, the reflection coefficient reaches a value of 1 for large capacitors and large frequencies. The incoming current waves, with a steep wavefront are hence almost fully reflected by the capacitor. The decay rate of the current after the reflection depends on the size of the capacitor.

For modular multilevel converter topologies (e.g, half-bridge MMC), the DC capacitors are distributed in submodules within the converter arm. These topologies also include reactors in each arm which prevent the distributed capacitors from fast discharge into the fault. Moreover, the discharge of the capacitors can be interrupted by blocking the IGBTs of the converter [4].

Converter AC Contribution After the initial capacitive discharge phase, the fault current is supplied by the AC system, dependent on converter topology.

Converter IGBTs are blocked (t_{block}) as soon as the current through the converter surpasses the overcurrent threshold. IGBTs cannot withstand high currents, even for short time intervals, and they need to be protected as they contribute considerably to the total cost of the installation.

For most VSC-HVDC converters (two-level, half-bridge MMC), this does not mean that the current through the converter is blocked. The converter IGBTs also rely on antiparallel diodes, which are not controllable. When the IGBTs are blocked, the fault current is supplied by the AC side through these antiparallel diodes in converters without DC fault blocking capability. If the converter IGBTs are blocked, the AC side effectively perceives a three-phase short circuit. Hence, at the AC side a transient short-circuit current is supplied which eventually evolves into a steady-state fault current. The amplitude of the steady-state DC fault current depends on the strength of the AC grid at the converter terminals (including phase reactor and transformer), DC fault distance, and impedance [5]. The converter contribution is normally limited to 4–6 times the converter current as the phase reactor combined with the converter transformer have an impedance of approximately 0.15–0.25 pu.

Converters with DC fault blocking capability can prevent AC infeed and even reduce the DC fault current to zero. Therefore, a voltage opposing the AC voltage must be injected by the submodules. Examples of these converter topologies are the full-bridge modular multilevel converter, alternate arm converter, and derivatives [6–8]. Compared to the nonblocking converter types, the number of power electronic switches increases and hence the losses and costs are higher.

17.2.3.3 Electromagnetic Transient Analysis for Faults in DC Grids In the sections above, approximations regarding fault behavior in DC grids have been made. In real systems, transmission lines are not lossless and the line termination is not a constant impedance.

For detailed analysis of fault behavior in DC grids, offline simulations are typically performed using electromagnetic transient simulation tools (EMT-type tools). EMTP has been introduced in the late 1960 to analyze transients in power systems [9]. These programs provide a time domain solution by relying on numerical solution of differential equations of the electric circuit using very small timesteps [10].

Accurate evaluation of fault transients in DC grids require an accurate transmission line model and a detailed converter model.

Transmission Line Model The transmission line model must be able to accurately represent traveling wave phenomena during the fault transient. Lumped parameter approaches such as cascaded PI-sections are approximations because they are single-frequency parameter methods and do not accurately incorporate traveling wave propagation [11]. In most EMTP-programs, a transmission line model with distributed frequency-dependent parameters is available. This line model makes use of the traveling wave time between both sides to decouple both line ends.

The frequency-dependent line models depart from a linearized description of the line parameters in the frequency domain to obtain a time domain solution. The differential equations for the voltages and currents on the line in the frequency domain are

$$-\frac{dU}{dx} = (R + j\omega L)I = Z(\omega) \cdot I \tag{17.18}$$

$$-\frac{dI}{dx} = (G + j\omega C)V = Y(\omega) \cdot U \tag{17.19}$$

Consider a single phase line of length l, with terminal voltages and currents U_k, I_k, U_m, and I_m (shown in Figure 17.8). The voltage and current at location x can be related to the voltage and current t termination k as [11]

$$U_x + Z_c I_x = (U_k + Z_c I_k)e^{-\gamma x} \tag{17.20}$$

in which γ is the propagation function and Z_c is the characteristic impedance, both frequency-dependent:

$$\gamma = \sqrt{Y(\omega)Z(\omega)} \tag{17.21}$$

$$Z_c = \sqrt{\frac{Z(\omega)}{Y(\omega)}} \tag{17.22}$$

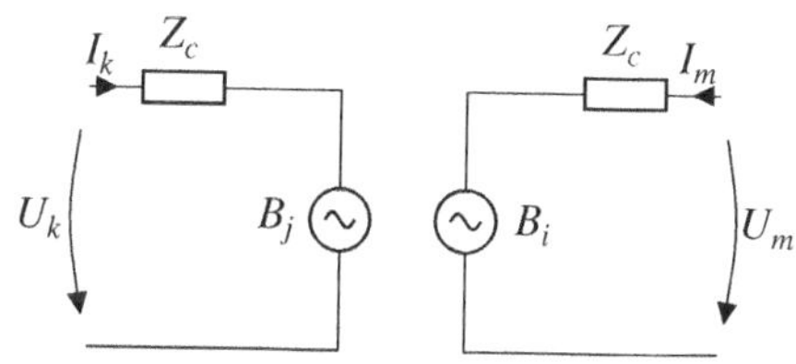

Figure 17.8 Frequency-dependent transmission line model [10, 11].

Define forward and backward traveling waves F_i and B_i at end i [10]:

$$F_i = U_i + Z_c I_i \tag{17.23}$$

$$B_i = U_i - Z_c I_i \tag{17.24}$$

The forward traveling wave at termination m can be related to the backward traveling wave at termination k by the propagation function:

$$B_k = F_m e^{-\gamma l} \tag{17.25}$$

With (17.20) and (17.25), one can come to a frequency domain representation of the line as shown in Figure 17.8. For this line model, only the voltages and currents at the line ends are available.

First the line parameters in (17.18)–(17.19) are obtained from the line geometry by means of a "line constants" or "cable constants" routine [12]. These are used to calculate the propagation function and characteristic impedance in the frequency domain. For an efficient conversion to the time domain, the parameters in the frequency domain are fitted using rational functions, eventually obtaining a linearized transmission line model [13, 10].

Converter Model The electromagnetic transients caused by DC faults are in the timescale of switching frequencies of the converter. Hence for an accurate representation of the fault transient, a detailed converter model is needed. It is possible to incorporate individual switches and even nonlinear switching events. However, from a power system perspective, not all detail regarding switching behavior is needed [14].

For modular multilevel converter topologies, difficulties arise due to the large number of nodes in the converter model. The MMC topology implies a large system matrix, which makes its simulation computational very heavy [15]. To overcome this issue, reduced converter models have been proposed in the literature. A brief description and overview of these models are given below and in Table 17.2.

In [16] a reduced half-bridge MMC converter model is obtained by using the implicit integration methods in the EMT-type software. Mostly, EMT-type software makes use of numerical integration schemes such as the trapezoid integration rule,

TABLE 17.2 Converter Models for EMT Simulation of MMCs

Model	Nonlinear Switching Model	Linear Switching Model	Continuous Model	Passive RLC
Detail	Each SM, each switch	Each SM, each switch	Each arm, no switches	Averaged DC side
Accuracy	Exact DC discharge	Approximate DC discharge	Approximate DC discharge	Approximate DC discharge
	Exact AC infeed	Approximate AC infeed	Approximate AC infeed	—
Simulation speed	Slowest	Faster than nonlinear	Faster than linear	Fastest

representing dynamic elements by current sources and resistances. Using the trapezoidal integration rule, each submodule can be represented by a Thévenin equivalent. By summation of each submodule in the converter arm, a converter arm can be represented by a Thévenin equivalent model using only two nodes. From the Thévenin model, a Norton equivalent can readily be obtained for integration in the system matrix. In this model, control is implemented up to firing pulses for each individual submodule. By adding a parallel path with a diode, representing the converter in blocked state, the model is also useful for DC fault simulations.

Further reduction of model detail is possible; a converter arm of a half-bridge MMC can be represented by a controlled voltage source [17]. The voltage source inserts a perfect sine wave, implying an infinite switching frequency and an infinite number of submodules. Instead of the voltage on each submodule, the total arm voltage, equaling the sum of the voltage of each submodule in the arm, is used. This total arm voltage is determined by the current through the arm. This model has also been extended with a blocking feature, making the model useful for DC fault simulations [18].

As a last step, the DC side transient can be approximated by a reduced model making use of only passive elements. The equivalent inductance and resistance can be modeled by $L_{eq} = (2/3)L_{arm}$ and $R_{eq} = (2/3)NR_{ON}$. The equivalent capacitance is given by $C_{eq} = 3C/N$ in reference 19, which approximates the transient response for the first milliseconds. In reference 20, this equivalent capacitance is given by $C_{eq} = 6C/N$, which is derived by using the energy conservation principle of all capacitors in the converter. This latter equivalent capacitance is better suited for dynamic simulations, where the energy conservation principle is more important than the transient response.

17.2.3.4 Example EMTP-Simulation for a Multi-Terminal DC Grid In this Section, an example of an EMTP-simulation for fault transients is given. The simulation is performed using PSCAD.

Multi-terminal DC Grid The multi-terminal DC grid considered is a radial three-terminal system, as shown in Figure 17.9. The multi-terminal system consists of three converters connected through two cables. The converters are considered to be solidly grounded, implying for the system either an asymmetric or the upper half of a bipole configuration. A solid pole-to-ground fault occurs at 25 km on the cable between converter A and C. The cable length between converters A and B is 150 km.

The cables are modeled using the frequency-dependent cable model available in PSCAD. The cables have two conducting layers, a copper core and screen, separated by XLPE insulation. Semiconducting screens are applied at the inner insulation, respectively of 1.7- and 1.9-mm thickness for inner and outer screen. XLPE is also used as outer insulator. The soil is modeled with a constant resistivity of 100 $\Omega \cdot$ m. Information about cable geometry and material parameters is given in Table 17.3. The characteristics of the cable at high frequencies are shown in Table 17.4.

The converters are modeled using the equivalent model with Thévenin representation of each arm [16]. Converter blocking is applied similarly as in reference 18. The converter rated power for each converter is 500 MVA, with a DC pole-to-ground

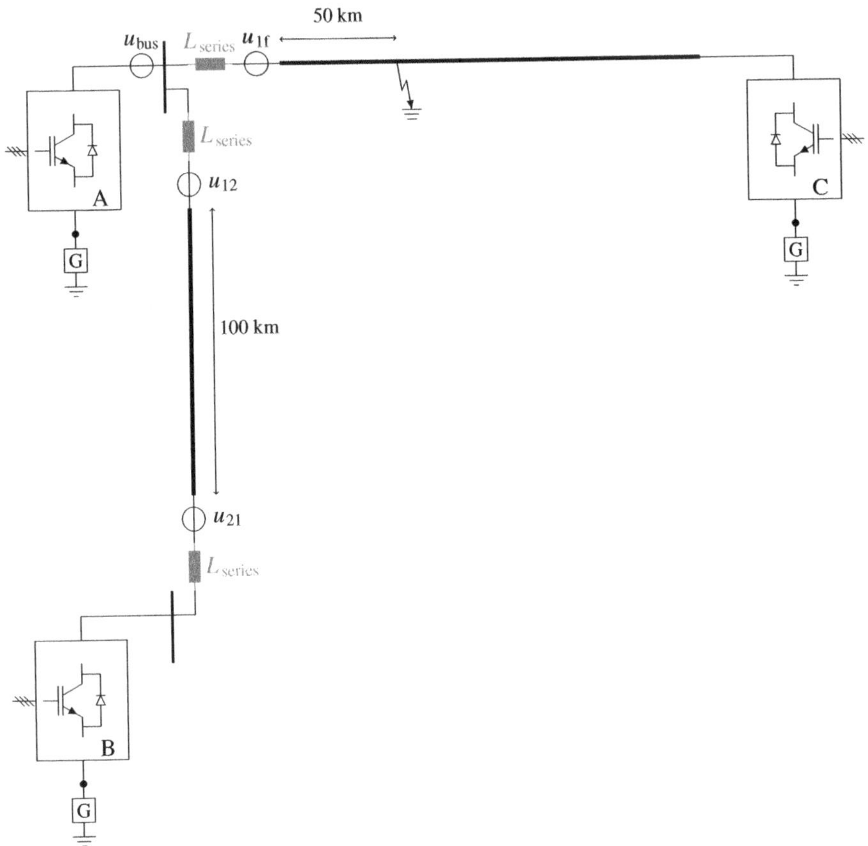

Figure 17.9 Multi-terminal DC grid with pole-to-ground fault.

voltage of 320 kV. The IGBTs of the converter are blocked when the current through one of the arms exceeds 2 kA. The AC grid representation is an equivalent representation with a short-circuit power of 4 GVA. A full overview of converter parameters is given in Table 17.5. Between the converters and the DC bus, a reactor of 10 mH is present for filtering the converter noise to the DC bus.

TABLE 17.3 Cable Parameters

	Outer Radius (mm)	ρ ($\Omega \cdot$ m)	ϵ_r (—)	μ_r (—)
Copper core	21.1	$1.72 \cdot 10^{-8}$	1	1
XLPE insulation	37.9		2.4	1
Copper screen	38.4	$1.72 \cdot 10^{-8}$	1	1
XLPE insulation	45		2.4	1

TABLE 17.4 Cable Parameters

Characteristic impedance	18.9 Ω
Wave propagation velocity	167 km/ms

TABLE 17.5 Grid and Converter Parameters

AC primary voltage U_{ac}	400 kV
AC inductance L_{ac}	0.1367 H
AC resistance R_{ac}	3.78 Ω
Transformer ratio	400/185 kV
Transformer leakage reactance u_k	0.1 p.u.
Converter rated power	500 MVA
DC voltage U_{dc}	320 kV
Converter reactor	32 mH
Number of submodules	100
Submodule capacitance	6.5 mF

Results In Figure 17.10, the contributions of the current to the fault current are shown for a pole-to-ground fault occuring at 50 ms. First concentrate on the first wave front. The current at the faulted cable at converter A, i_{1f} increases steeply to a value of about 12 kA. This increase can also be seen in the current in the healthy cable, i_{12}. This is due to the transmission of a current wave created by the fault, which is attenuated over the faulted cable and fully transmitted into the healthy cable. At the location of the DC bus at converter A, the current wave perceives an impedance consisting of the converter and the cable between terminals A and B. For the first current wave, the filter reactor acts as a large impedance, hence the current wave is transmitted mostly into the healthy cable. Within the first millisecond, the converter current increases steadily to about 5 kA and then abruptly diminishes, caused by blocking of the IGBTs of the converter. The current at converter terminal B increases only after 1 ms, due to

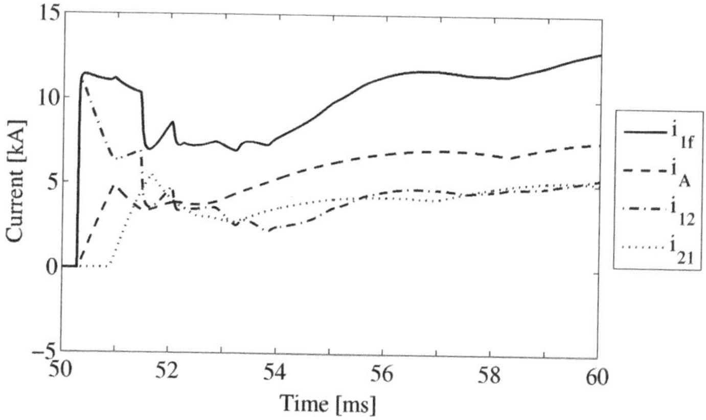

Figure 17.10 Fault currents in the multi-terminal system of Figure 17.9 for a fault at 50 ms.

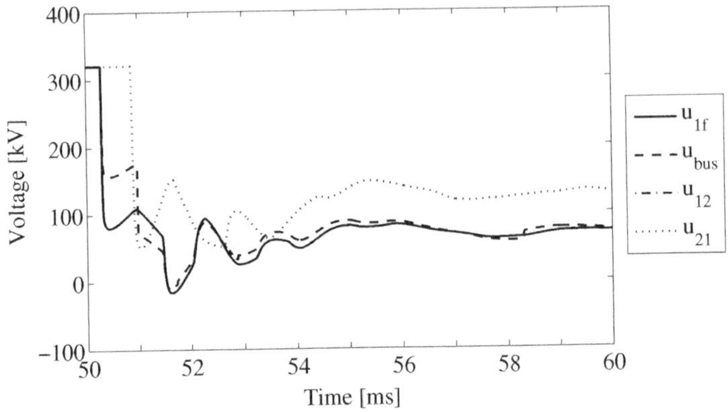

Figure 17.11 Voltages in the multi-terminal system of Figure 17.9 for a fault at 50 ms.

the traveling time of the wave over the healthy cable. In the following milliseconds, the behavior of the fault current i_{1f} is less clear due to superposition of reflected waves at converter B and at the fault location. After about 8 ms, the currents i_{21} and i_{12} start to converge, indicating that the solution moves towards a steady state.

The fault voltages are shown in Figure 17.11 for the same timeperiod. The voltage at the faulted cable, u_{1f}, and the voltage u_{12} are essentially the same because there is no impedance modeled in between. The voltage diminishes to a value of about 80 kV, indicating that the voltage wave of magnitude −320 kV is attenuated over the faulted cable. In this figure, the impact of the reflected voltage at node B can be clearly noticed half a milliscond later in the voltage at converter A by comparing u_{1f} and u_{21}.

The same exercise can be done when considering inductors in series with the cables. The inductors limit the rate of rise of the fault current, giving time for the protection system to detect and interrupt a fault. The fault current contributions are shown in Figure 17.12. Due to the series inductors, the rate of rise of the fault current

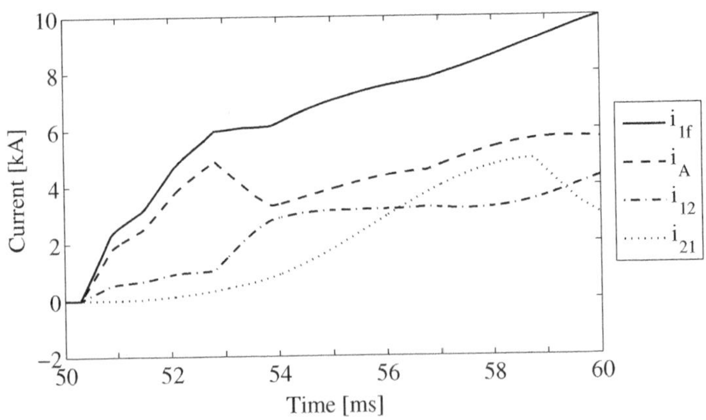

Figure 17.12 Fault currents in the multi-terminal system of Figure 17.9 for a fault at 50 ms. Inductors of 100 mH are in series with each cable.

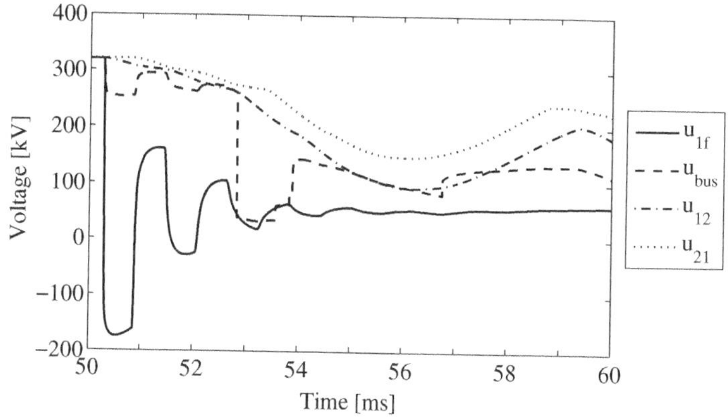

Figure 17.13 Voltages in the multi-terminal system of Figure 17.9 for a fault at 50 ms. Inductors of 100 mH are in series with each cable.

is largely diminished compared to the previous case. The initial fault current is now mainly supplied by the converter. After blocking the converter IGBTs, the healthy neighboring cable discharges more due to the sudden voltage drop. In Figure 17.13, the voltages are shown. The voltage at the faulted cable is inversed due to reflection of the voltage wave at the inductor. Roughly approximated, the inductor can be represented as an open circuit for the first incoming wave. Therefore, the initial current must be zero and the voltage is fully reflected. The reflections of the voltage at the fault, reaching the faulted cable end each half millisecond, are clearly noticeable in u_{1f}. The bus voltage and other voltages do not experience the transient as severe as in the previous case, as the voltage waves are largely reflected at the inductors. Only after converter blocking, the other voltages decrease more quickly.

17.3 DC GRID PROTECTION

This section provides an introduction to DC grid protection. First, the objectives of DC grid protection and its requirements are treated. In Section 17.3.3, fault clearing strategies and system relaying are treated. Finally, an overview of protection equipment, with a focus on DC breakers and measurement equipment, is given.

17.3.1 Objectives of Protection

As every power system, the DC grid is subjected to faults. Therefore, a protection system must be designed in the sense that the negative effects of these faults on the system are minimized. The main objectives of the DC protection system are similar to those of AC protection:

- Minimize the impact of faults on the grid functioning.
- Avoid damage of faults to components.
- Ensure safe operation of the system.

The objectives of protection can be translated into a DC grid protection philosophy, requiring the following [21, 22]:

Reliability. Reliability of the protection system includes dependability and security. Dependability implies the correct action of the protection against faults which need action. Security implies the non-action of the protection in cases when the protection must not act [23].

Speed. The protection must act fast to avoid damage to equipment, limit the fault current within the maximum interruptible current and limit the impact of the disturbance on the network. In the DC grid, time constraints are extremely stringent. Unlike in the AC grid, where tripping times are in the order of hundreds of milliseconds, only a few milliseconds are available before the fault must be cleared.

Sensitivity. Every fault must be detected; for example, protection of long transmission lines or detection of high impedance faults is harder than their counterparts.

Selectivity. Only the faulted zone of the system should be isolated. The protection system must define zones of protection which are separated by selectivity criteria in the fault detection. These zones can be single transmission lines but can also comprise DC subgrids.

Robustness. The protection system must be able to operate even in degraded situations. Duplication of protection systems to provide redundancy can aid to robustness.

Stability. After fault clearing, the system must reach stable operation within an acceptable time interval.

17.3.2 Differences with AC Protection

Although the DC grid protection philosophy generally coincides with the objectives of AC grid protection, there are some fundamental differences in constraints. In an AC grid, the fault clearing time is limited by the ability of the synchronous generators to keep synchronism during severe disturbances. This critical clearing time typically lies in the order of hundreds of milliseconds.

In a DC grid, the critical fault clearing time depends on other factors. First, damage to components must be avoided because of the large currents. Time constraints are hence imposed by the converter power electronics and the DC breakers. Second, due to converter blocking, controllability of the blocked converter within the DC grid is temporarily lost. Because a fault propagates rapidly throughout the grid, the number of nodes affected by the fault depend on the clearing time. The more nodes lost in the DC grid, the slower the system can be restored to normal operation. Finally, the maximum clearing time determines how much the AC system is affected by a disturbance at the DC grid.

Furthermore, the protection should be integrated in the HVDC grid without affecting normal operation. Due to more complex control systems in the HVDC grid, care must be taken regarding interference of the protection on the control system [24].

17.3.3 Protection Methodologies

The method of DC fault clearing determines the protection methodology for the DC grid. A protection methodology similar to current state-of-the art high-voltage AC protection requires fast-acting DC breakers at each transmission line. This is the ideal option regarding impact of a fault on the grid, but also the most expensive in terms of equipment. Alternatively, a DC fault can be cleared by switching off all terminals and disconnecting the faulted line in the de-energized system. A major advantage of this kind of protection is the absence of fast breakers in the DC grid. On the contrary, any fault in the DC grid leads to loss of the whole grid. While this can be feasible for small multi-terminal systems, losing a large DC overlay grid can be unacceptable. Below, DC grid protection strategies are described making use of different methods to interrupt the fault current.

Protection Using AC Breakers In this method, the AC side breakers of each converter are used to clear a DC side fault. During the opening of the AC breakers, the fault can be located in the DC grid [25]. After the DC grid is de-energized, disconnectors at the DC side must isolate the faulted section of the DC grid. These disconnectors do not need to be ultra-fast-acting or rated for high-fault currents.

As AC breakers are mechanical circuit breakers, requiring several cycles of the fundamental frequency to interrupt the current, this type of protection is relatively slow. Moreover, after the protection has acted, the DC grid must be re-energized and brought back to normal operation. For this purpose, multilevel topologies which retain charged capacitors by blocking IGBTs are an advantage. However, the loss of the full DC grid for several cycles of the fundamental AC frequency must be acceptable, limiting the applicability of this type of protection to small DC grids.

Protection by Interrupting the Converter Fault Current at the DC Side Similar to the previous protection method, the DC grid can be protected by using the fault-blocking capability of converters. Also with this type of protection, the whole DC grid is temporarily lost in case of DC faults. In contrast with the previous method, this method is faster, as the converters can almost instantaneously be blocked from the moment of detection. In the case of an isolated converter after DC fault clearing, the converter can still deliver reactive power to the AC grid.

This method requires that all of the DC converters have fault-blocking capability or require a fast DC breaker at converters without fault-blocking capability. Equal to the previous method, the whole DC grid is lost in case of a fault, whereas the interruption time can be smaller [26].

Protection Using DC Breakers By using DC breakers, the DC grid can be split into different protection zones. Each zone is covered by primary protection which clears a fault within the protected zone in the fastest time [27]. Backup protection must be provided should the primary protection in the faulted zone fail.

If each transmission line is covered by an individual zone, fast-acting DC breakers at each end of a protection zone are needed to isolate the fault. The size of the protection zones determines the number of DC breakers and the impact of a short-circuit fault within the DC grid. Each transmission line can be covered by an individual protection zone, requiring DC breakers at each transmission line. Alternatively, a protection zone can cover a larger portion of the DC grid, decreasing the number of fast DC breakers. Besides DC breakers, converters with fault-blocking capability [28] or DC–DC converters can be used to isolate the faulted section of the grid.

Due to low timescales for fault-clearing time in a DC grid, alternative protection sequences can be used to achieve selective protection. In an AC grid, the normal sequence for protection is detection of the faulted line, followed by isolation. In reference 29, the sequence of interrupting the fault current and locating the fault is reversed. Each DC breaker independently acts based on local signals, regardless of fault location. After interruption of the fault, each breaker of a nonfaulted protection zone recloses. An advantage of this method is sharing of energy absorption amongst different breakers, leading to smaller surge arrester ratings (see Section 17.4.1).

17.3.4 Relaying for DC Grids

Fault detection criteria are needed for correct relaying of DC breakers. In DC, the main quantities to consider for fault detection are the DC voltage and current magnitudes. Due to the small timescale and large influence of transient effects, AC relaying methods cannot be directly applied to DC grids [21]. The main criteria used in AC systems are given and assessed for relevancy to DC grid protection.

17.3.4.1 Overcurrent Protection Overcurrent relays trip the breakers whenever the current exceeds a threshold value. Coordination of overcurrent relays can be achieved by time or current grading or a combination of both [22]. A directional criterion provides extra selectivity. In radial AC systems, overcurrent protection has been succesfully applied, because fault currents are well known and are dominantly supplied from one side. In meshed AC transmission systems, overcurrent protection is generally not applied due to the lack of selectivity and large sensitivity of fault current levels to source variations [24].

The application of overcurrent protection in DC grids differs from AC grids. Because of the high rate of rise of the fault current and the time needed for fast DC breakers to interrupt the fault current, current thresholds need to be set low in order to limit the maximum fault current. Some degree of time grading is inherently present due to traveling time of the waves. However, coordination using overcurrent relays might be difficult as sensitivity towards grid layout and grid topology is large.

17.3.4.2 Distance Protection A distance relay calculates an impedance from the voltage and current measured at a point. In normal operation, the relay measures an apparent impedance depending on the power flow over the transmission line. In the case of a fault on the line, the impedance calculated is a measure for the distance to the fault. Typically, the relay characteristic, defining the reach of the relay, is provided in the complex impedance plane. By comparing the calculated impedance to the relay characteristic, faults can be discriminated as within or outside of the protection zone. As distance relays are rather insensitive to variations in source impedance, it is widely used in AC grid protection [22].

The application of distance protection to DC grids is not straightforward. During the first milliseconds of the transient, traveling wave reflections occur at the line terminations. Therefore, it is not straightforward to extract the line impedance from measurements at the converter terminals. Furthermore, calculating the impedance accurately can be hindered by other transients such as converter blocking or the fast changing nonlinear behavior of the voltage and current during the AC infeed phase. On a larger timescale, the fault does not result in a shift in a two-dimensional impedance plane. Only a shift in the resistance can be noted, making it hard to distinguish normal operation from fault situations. This requires steady-state fault currents, which is not acceptable in DC grids.

17.3.4.3 Unit Protection With unit protection, the zone of the protection is closed and its boundaries are clearly defined [24]. Both protection methods above are non-unit protection, meaning that only the start of the protection zone is clearly defined. The end of the protection zone varies with system conditions and measurement errors [22].

For unit protection of a transmission line, measurements need to be done at both ends. Essential to unit protection is the transmission of information between both ends. For DC grids, the delay of information transmission between both ends (protection signaling time) largely influences the feasibility of unit protection schemes. For a correct comparison, synchronization between measurements at both ends is needed. However, additional difficulty exists in the fact that transients in DC grids are not occuring at the same time. This is also true for AC systems, however, these effects are not dominating in the time frame in which the AC protection acts.

Differential Protection For differential protection, fault detection relies on difference between currents measured at both ends. Traditional differential protection uses current transformers and an interconnection between the two secondary sides to generate circulating current. With numerical relays, it is also possible to transmit information digitally to each end.

The application of differential protection in DC grids is not straightforward. Due to traveling wave phenomena, an increase in fault current due to a fault close to one line end is not immediately noticed at the other line end. Furthermore, it is not possible to wait for steady-state conditions. Therefore, adapted versions of differential protection must be used [30].

Differential protection might be very useful for systems with short line lenghts or for specific applications such as busbar protection.

17.4 DC PROTECTION COMPONENTS

In this section, DC breakers and measurement equipment as essential DC grid protection components are treated.

17.4.1 DC Breakers

Interruption of DC currents faces certain difficulties compared to interruption of AC currents. The functions that the DC breaker must fulfill can be split up as follows [31]:

- Create an opposing voltage to bring the fault current to zero.
- Dissipate the energy stored in the inductance of the system.
- Withstand the overvoltages due to the interruption.

For AC breakers, the first requirement is not needed as the current crosses zero at twice the fundamental frequency. With mechanical AC breakers, an arc which dissipates the energy in the circuit is drawn between moving contacts. When the arc is extinguished (either at a natural zero crossing or by forced extinguishing), the distance between the contacts provides isolation of the faulted area and the energized system. Speed of the breaker is typically expressed as the number of cycles of the fundamental frequency needed to interrupt the current.

For interruption of DC currents, this approach is not possible as there is no natural zero crossing of the current and an arc would remain between the contacts. Therefore, a voltage opposing the driving voltage must be inserted by the breaker to drive the current to zero.

Because of difficulties to integrate the three functions of the DC breaker into one element, current breaker designs split the functions over different devices by making use of parallel paths [31]. Depending on switch technology, DC breakers can be categorized into solid-state breakers (using power electronic switches), mechanical breakers, and hybrid breakers, making use of both switch technologies. Conceptual implementation of these breakers is shown in Figure 17.14. Typically, a surge arrester clamping the voltage to a value of 150% the nominal voltage is used [32]. This surge arrester also performs the energy absorbing function needed to dissipate the stored energy in the system inductance.

More information on HVDC breaker technology is available in Section 3.5.8.

17.4.2 Measurement Equipment

To capture the DC fault transient for fast fault detection, stern requirements are put to measurement equipment. As DC fault transients consists of voltages and currents with a high-frequency content, a high analog bandwidth and sampling frequency are needed. Moreover, the accuracy in a wide frequency range is required.

High-voltage measurement equipment typically consists of a converting device, transmission device, and recording device. Typically, the transfer behavior of the

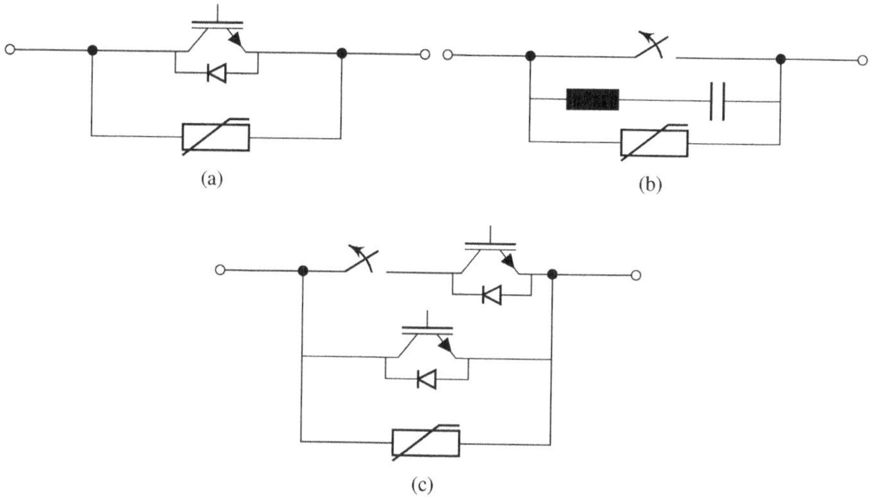

Figure 17.14 Circuit breaker concepts based on switch technology. (a) Solid-state breaker. (b) Mechanical breaker (passive-resonance type). (c) Hybrid breaker.

converting device largely determines the performance of the measurement equipment [33]. Below, a short overview of possible DC measurement equipment is given, with emphasis on the converting device.

Current measurement

Optical Current Transducer. Optical current transducers make use of the Faraday effect in an optical fiber. An optical fiber wound around the conductor experiences a magnetic field due to the current. This field influences optical signals induced in the fiber, which is used as a measure for the current. Advantages of optical current transducers are reduction in dimensions and reduction of electromagnetic interference [22, 34]. The bandwidth of an optical current sensor lies on the order of several to tens of kHz [35].

Hybrid Electro-optical Measurement. Hybrid electro-optical measurement combines an electrical sensor with optical transmission to provide isolation between the primary and secondary side of the measurement equipment e.g. a shunt measurement with optical transmission such as used in LCC HVDC [36].

Zero-Flux Current Transformers Zero-flux current transformers use a Hall sensor in a magnetic ring. The Hall effect voltage is a measure for the current in the conductor. The zero-flux current transformer can yield a high accuracy and a high bandwidth [22, 30]. This option is used for measuring currents in VSC HVDC [36].

Voltage Measurement

Hybrid Optical RC-Divider. The voltage measurement in VSC HVDC is performed using RC-voltage dividers [36]. Voltage measurements in HVDC make use of RC-dividers, separated from the secondary circuit by optics [34, 36].

17.5 CONCLUSIONS

This chapter highlighted the challenges for the protection of the future DC grid. The steady-state fault behavior is largely determined by the type of fault, system grounding, and configuration. The analysis of transient in the faulted system showed that the electromagnetic wave due to the fault, having a steep current and voltage wavefront, propagates quickly through the DC grid. The wave is reflected at every DC node, affecting every node within several milliseconds. To avoid damage to components and loss of the grid, the DC grid protection must locate and isolate the fault during this transient period. Therefore, AC protection means, relying mostly on distance protection and differential protection is not readily applicable.

In the last part of the chapter, DC protection equipment was treated. An overview of DC breakers and measurement equipment was given. The DC breaker technology evolves towards a hybrid breaker type, combining both mechanical and power electronic switches. For measurement equipment, stringent constraints are imposed regarding bandwidth and sampling frequency, as they must be capable to capture the fast fault transients.

REFERENCES

1. CIGRÉ WG B4-52, "HVDC grid feasibility study," CIGRÉ, Technical Report, 2013.
2. L. van der Sluis, *Transients in Power Systems*. Hoboken, NJ: John Wiley & Sons, Inc., 2001.
3. A. Gomez-Exposito, A. Conejo, and C. Canizares, *Electric Energy Systems, Analysis and Operation*. Boca Raton, FL: CRC Press, 2009.
4. S. Allebrod, R. Hamerski, and R. Marquardt, "New transformerless, scalable modular multilevel converters for HVDC-transmission," in *Proceedings of IEEE PESC 2008*, Rhodes, Greece, 15–19 June 2008, pp. 174–179.
5. J. Yang, J. Fletcher, and J. O'Reilly, "Multiterminal DC wind farm collection grid internal fault analysis and protection design," vol. 25, no. 4, pp. 2308–2318, 2010.
6. R. Marquardt, "Modular Multilevel Converter topologies with DC-short circuit current limitation," in *IEEE Proceedings of ICPE ECCE 2011*, Jeju, South Korea, 30 May–3 June 2011, pp. 1425–1431.
7. M. Merlin, T. Green, P. Mitcheson, D. Trainer, R. Critchley, W. Crookes, and F. Hassan, "The alternate arm converter: A new hybrid multilevel converter with DC-fault blocking capability," *IEEE Transactions on Power Delivery*, vol. 29, no. 1, pp. 310–317, Feb. 2014.
8. C. Davidson and D. Trainer, "Innovative concepts for hybrid multi-level converters for HVDC power transmission," in *Proceedings IET ACDC 2010*, London, UK, 19–21 Oct. 2010, 5 pages.
9. H. Dommel, *Electromagnetic Transients Program Reference Manual: (EMTP) Theory Book*. Bonneville Power Administration, 1986.

10. N. Watson, J. Arrillaga, and Institute of Electrical Engineers, *Power Systems Electromagnetic Transients Simulation*, series IEE power and energy series, Institute of Electrical Engineers, Ed.London: Institution of Engineering and Technology, 2003.
11. J. Marti, L. Marti, and H. Dommel, "Transmission line models for steady-state and transients analysis," in *Proceedings of IEEE APT '93*, Athens, Greece, 5–8 Sept. 1993, pp. 744–750.
12. J. Martinez-Velasco, *Power System Transients: Parameter Determination*. Boca Raton, FL: CRC Press, 2009.
13. B. Gustavsen and A. Semlyen, "Rational approximation of frequency domain responses by vector fitting," *IEEE Transactions on Power Delivery*, vol. 14, no. 3, pp. 1052–1061, 1999.
14. H. Saad, J. Peralta, S. Dennetiere, J. Mahseredjian, J. Jatskevich, J. Martinez, A. Davoudi, M. Saeedifard, V. Sood, X. Wang, J. Cano, and A. Mehrizi-Sani, "Dynamic averaged and simplified models for MMC-based HVDC transmission systems," *IEEE Transactions on Power Delivery*, vol. 28, no. 3, pp. 1723–1730, 2013.
15. J. Peralta, H. Saad, S. Dennetiere, J. Mahseredjian, and S. Nguefeu, "Detailed and averaged models for a 401-level MMC-HVDC system," vol. 27, no. 3, pp. 1501–1508, 2012.
16. U. Gnanarathna, A. Gole, and R. Jayasinghe, "Efficient modeling of modular multilevel HVDC converters (MMC) on electromagnetic transient simulation programs," vol. 26, no. 1, pp. 316–324, 2011.
17. A. Antonopoulos, L. Angquist, and H.-P. Nee, "On dynamics and voltage control of the modular multilevel converter," in *Proceedings of IEEE EPE 2009*, Barcelona, Spain, 8–10 Sept. 2009, 10 pages.
18. N. Ahmed, L. Ängquist, S. Norrga, and H.-P. Nee, "Validation of the continuous model of the modular multilevel converter with blocking/deblocking capability," in *Proceedings of IET ACDC 2012*, Birmingham, UK, 4–6 Dec. 2012, 6 pages.
19. W. Leterme and D. Van Hertem, "Reduced modular multilevel converter model to evaluate fault transients in dc grids," in *Proceedings of IET DPSP 2014*, Copenhagen, Denmark, 31 Mar. 3–Apr. 2014, 6 pages.
20. H. Saad, S. Dennetie?re, J. Mahseredjian, P. Delarue, X. Guillaud, J. Peralta, and S. Nguefeu, "Modular multilevel converter models for electromagnetic transients," *IEEE Transactions on Power Delivery*, vol. 29, no. 3, pp. 1481–1489, June 2014.
21. D. Van Hertem, M. Ghandhari, J. Curis, O. Despouys, and A. Marzin, "Protection requirements for a multi-terminal meshed DC grid," in *CIGRE 2011 Bologna Symposium*, Bologna, Italy, 13–15 Sept. 2011, p. 8 pages.
22. Alstom, *Network Protection and Automation Guide*. Stafford, UK: Alstom Grid, 2011.
23. L. Grigsby, *The Electric Power Engineering Handbook*, series The Electrical Engineering Handbook Series. London: Taylor & Francis Group, 2001.
24. P. Anderson, *Power System Protection*. Hoboken, NJ: J. Wiley & Sons, 1998.
25. L. Tang and B. Ooi, "Locating and isolating DC Faults in multi-terminal DC systems," vol. 22, no. 3, pp. 1877–1884, July 2007.
26. ———, "Protection of VSC-multi-terminal HVDC against DC faults," in *Proceedings IEEE PESC 2002*, vol. 2, Cairns, Australia, 24–27 June 2002, pp. 719–724.
27. S. Horowitz and A. Phadke, *Power System Relaying*. Hoboken, NJ: J. Wiley & Sons, 2008.
28. D. Schmitt, Y. Wang, T. Weyh, and R. Marquardt, "DC-side fault current management in extended multiterminal-HVDC-grids," in *Proceedings of IEEE SSD 2012*, Chemnitz, Germany, Mar. 2012, 5 pages.
29. C. Barker and R. Whitehouse, "An alternative approach to HVDC grid protection," in *AC and DC Power Transmission (ACDC 2012), 10th IET International Conference on*, Birmingham, UK, Dec. 2012, 6 pages.
30. J. Descloux, "Protection contre les courts-circuits des seaux courant continu de forte puissance," Ph.D. dissertation, Université de Grenoble, Grenoble, France, Sept. 2013.
31. C. Franck, "HVDC Circuit Breakers: A review identifying future research needs," *IEEE Transactions on Power Delivery*, vol. 26, no. 2, pp. 998–1007, 2011.
32. C. Meyer, M. Kowal, and R. De Doncker, "Circuit breaker concepts for future high-power DC-applications," in *Proceedings of IEEE Industrial. Applications Conference 2005*, vol. 2, Hong Kong, China, 2-5 Oct. 2005, pp. 860–866.

33. M. Ryan, *High-Voltage Engineering and Testing*, 3rd ed. London, UK: IET, 2013.
34. F. Jenau and G. Testin, "Modern Instrument Transformer Technologies for UHV AC and HVDC Networks," in *2nd IEC CIGRÉ 2009 Symposium*, New Delhi, India, 2009, 16 pages.
35. S. Kucuksari and G. Karady, "Experimental comparison of conventional and optical current transformers," *IEEE Transactions on Power Delivery*, vol. 25, no. 4, pp. 2455–2463, Oct. 2010.
36. Siemens Energy Sector, *Power Engineering Guide*, 7th ed., Berlin, Germany, 2012.

CHAPTER 18

REAL-TIME SIMULATION EXPERIMENTS OF DC GRIDS

Oluwole Daniel Adeuyi and Marc Cheah
Institute of Energy, School of Engineering, Cardiff University, Cardiff, Wales

18.1 INTRODUCTION

The study of DC grid technologies, as in many power system research, has to rely on simulation models especially in the first steps of the research. However, it is also important to implement the proposed solutions in real-time simulations or using experimental test rig, which allows a more accurate representation. The real-time simulation provides an improved model that considers the time limitations and makes possible the interconnection with real devices. On the other hand, the experimental test rigs are the most feasible representation of a real system and can be combined with real-time simulation schemes. This hybrid configuration allows the analysis of complex systems in more detail.

18.1.1 Benefits of Real-Time Simulation

A simulation can be defined as the representation of the behavior of a system using the behavior of another system. Advances in computing technologies have ushered in low-cost solutions with enhanced performance which improve the capability of simulation tools to solve complex problems in less time. With real-time simulator (RTS), it is possible to synchronize the clock of the real system and the simulation environment to achieve fast, flexible, and economic testing of new technologies without putting the live system to risk. Today, real-time simulation has wide range of industry applications in aerospace, power systems, automotive, electric motor designs, and so on, for rapid control prototyping (RCP), hardware-in-the-loop (HIL), and software in the loop (SIL) tests.

At the core, a RTS has three main components: a mechanism to accept the description of the system to be simulated, a digital hardware platform used for fast computing, and a software platform for managing the simulation [1]. In choosing a simulator, it is important to determine the frequency of the highest transients to be simulated, which in turn decides the minimum system frequency. Furthermore,

HVDC Grids: For Offshore and Supergrid of the Future, First Edition. By Dirk Van Hertem, Oriol Gomis-Bellmunt, and Jun Liang.

the complexity of the system to be simulated must be established together with the desired time-step duration to determine the computation power required. Moreover, the number of I/O channels that interface the simulator to external physical hardware devices and other controllers contributes to the overall cost and performance [2].

Over the years, systems and processes have become increasingly complex, and product development has come under severe time and cost constraints. Rapid control prototyping (RCP) enables engineers to quickly test proposed controls in a more flexible and iterative manner than implementing actual controller prototypes [3]. To achieve this, the plant controllers to be tested are simulated on an RTS and connected using real input/output interfaces to the physical plant. In RCP, internal controller states can be modified on the fly in a noninvasive manner, and this greatly reduces product development costs.

On the other hand, in hardware-in-the-loop (HIL) tests a virtual plant, which is simulated using the RTS, is connected via real input/output channels to physical controllers. This way, it is possible to verify the performance of controllers when virtual plants are subject to severe operating conditions, and at the same time eliminate any risk of plant failure. Furthermore, the complexity of the plant can be modified by adding mathematical models of related dynamic systems in the plant simulation. In this way, the repeatability of tests is assured at minimal cost.

Fully digital software-in-the-loop (SIL) tests combine the RCP and HIL techniques in a way that both the controller and plant are implemented on the same simulator platform. The advantage of SIL tests is that there is no requirement for external input/output signals, hence it is possible for timing to be out of synchronism with real-time without any impact on validity of result. Therefore, the simulation time step can be varied to reduce computation burden and extend capabilities of the RTS [2].

18.1.2 Benefits of Experimental Investigations

The last step to validate the feasibility of a study is to implement the solution in a real system. However, this option is usually difficult or expensive to apply except if the simulation model can provide an accurate representation that ensures the direct application to a real system. RTS solution has been presented in the previous section as a powerful tool that can emulate the analyzed system in real time, but it is still necessary to define a specific model able to consider all the necessary details that have to be studied.

A solution for this problem consists of using a scaled representation of the real system where it is possible to validate the results with more reliability. This experimental test rig as in the case of RTS, allows tests to be performed in a safe and controlled condition. The advantage of using a test rig is that it can represent most of the characteristics of the real system.

However, the accuracy of this scaled test rig will also depend on the elements that are used and the power and voltage levels. It is clear that a test rig with almost the same operational conditions will be more suitable to ensure the feasibility of the results, but at the same time it can be difficult to implement because of the requirement of more space, higher cost, or special electrical facilities. Therefore, during the design of a test rig there is a trade-off between accuracy and accessibility. For example, the power converters used in HVDC connections are expected to have modular multilevel configuration, but for an experimental platform it might be more suitable and cheaper to use two or three level converter configurations, except if the study is focused on the converter response or design. Another case is the modeling of the DC cables, where it is a good alternative to use PI sections based on real inductances, resistances, and capacitors to represent the cable, because the real cable could be impossible to set up in a standard-size laboratory.

The flexibility of the platform to model different systems is also an important point to consider. If the test rig is able to represent several configurations, it will be possible to analyze a larger number of case studies. Nevertheless, the accuracy could be reduced because the devices might not be able to model some specific features of the real system or the costs could be increased because more elements are requested to model different scenarios. For example, a multi-terminal HVDC grid for offshore wind power integration based on four nodes, two wind farms, and two connections to the main grid can be represented as a system with four connections to the main grid instead of two, where the dynamic response of the generator and the wind turbines is neglected and only their power flow is considered real.

On the other hand, it has to be taken into account that if the flexibility of the system is an essential point, then the use of RTS device might be a better solution, where almost any scenario is possible to create within the processing capabilities.

18.1.3 A Hardware-in-the-Loop (HIL) Test System

Figure 18.1 shows the general scheme of a HIL configuration. The main elements are:

- **Real-Time Simulator (RTS).** This is the platform where a virtual plant is going to be simulated.

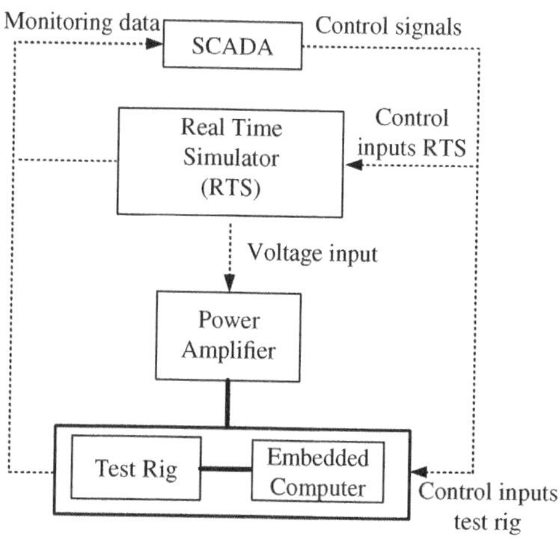

Figure 18.1 Scheme of a typical HIL simulator.

- **Test Rig.** This is the part of system represented using a scaled physical model.
- **Power Amplifier.** This element increases the power of the signal that comes from the RTS. In this case it is used as a voltage amplifier with a higher current rating.
- **Embedded Computer.** This acts as the controller of the test rig.
- **Supervisory Control and Data Acquisition (SCADA).** This element monitors the operation of the RTS and the test rig.

In this system the RTS (along with the electronic amplifier) and the test rig represent the plant that is analyzed, which in that case is a DC grid with all the necessary additional required elements (converters, cables, circuit breakers, generators, etc.). The SCADA and the embedded computer represent the physical elements to control and monitor the plant, which can be perfectly the same as in the real application.

There is a choice to represent the different elements to be implemented on either the RTS or the experimental platform. The RTS is more flexible and can represent a large-scale power system, but there is a precision compromise with the simulated models. On the other hand, the test rig has real elements that provide more accuracy, but at the same time the system tends to be more reduced.

This scheme is potentially useful for a wide range of systems studies, including the impact of emerging HVDC technologies on the operation and security of traditional ac grids. An example is illustrated in Figures 18.2 and 18.3, in which a physical three-terminal HVDC grid is interfaced to a main ac grid modelled using the RTS. All of this is carried out in real time and at no risk to the live network. This test rig

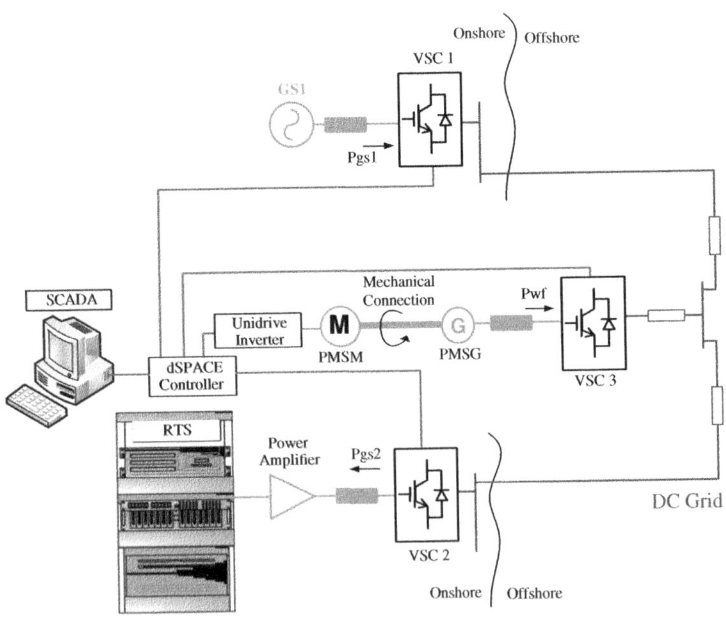

Figure 18.2 Configuration of an HIL test scheme using a three-terminal HVDC grid.

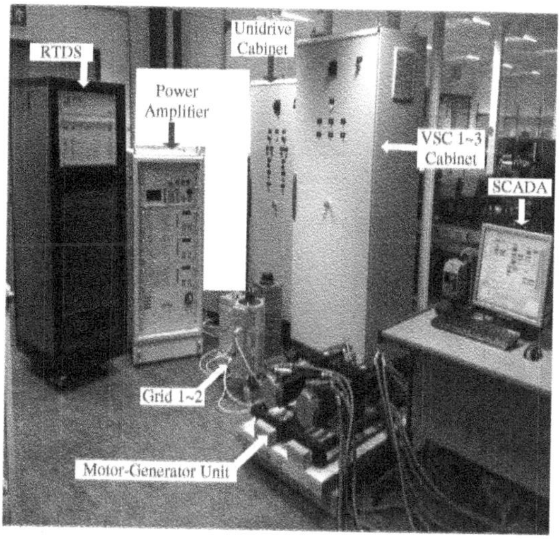

Figure 18.3 Example of an experimental setup of an HIL test scheme at the university of Cardiff.

provides a scaled-down platform suitable for laboratory investigations in educational institutions.

18.2 REAL-TIME SIMULATION IN POWER SYSTEMS

The application of real-time simulation to electromagnetic transients (EMT) phenomena of electrical power systems will be explored in this section. In most RTS, the overall network solution of the power system is by nodal analysis and the underlying solution algorithm are those introduced by H.W. Dommel in 1969 [4]. The focus here is on the modeling of power system components, basic architecture, operation and control of a RTS for HIL, and fully digital SIL tests.

18.2.1 Basic Architecture and Specifications

Digital simulators compute the state of the power system at discrete instants in time. The time between these discrete instants is the simulation time step (Δt) [2, 5]. It is common to use time steps in the range 40s–60 μs for frequency transients up to 2 kHz for acceptable results [2, 5]. Therefore, high-speed hardware processors connected in parallel are used to achieve the desired high computation rates. For example manufacturers of the RTDS Technologies use two types of processor cards [5]:

1. Triple processor card (3PC) which contains three analog devices ADSP21062 digital signal processors operating at clock speed of 40 MHz.
2. Reduced instruction set computer (RISC) processor card (RPC) with operating clock speed of up to 1.7 GHz.

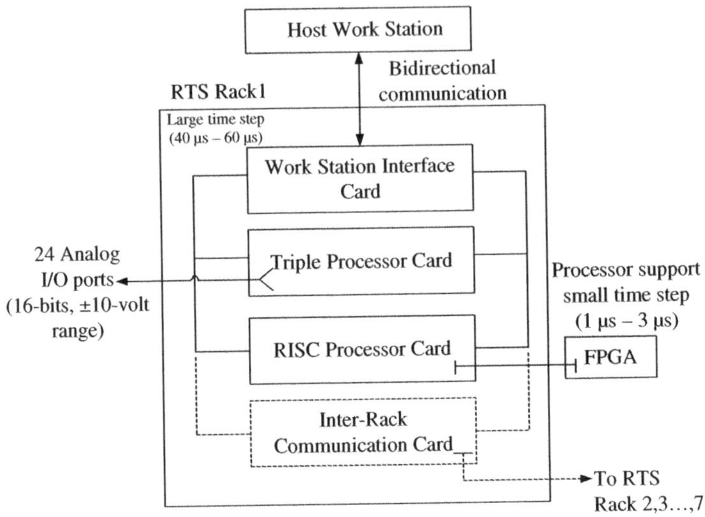

Figure 18.4 Basic architecture of an RTS manufactured by RTDS Technologies.

In Figure 18.4 the basic architecture of a RTDS is depicted [5]. The host work station is representative of part of the SCADA presented in Figure 18.1. A rack of this RTS can be configured to use either or both processor cards. In addition, a work station interface (WIF) card is provided for every rack and an inter-rack communication (IRC) card is provided for multi-rack systems.

The WIF handles rack diagnostics, RTS-to-computer workstation communication, multi-rack case synchronization, and backplane communication. The communication between the RTS and host workstation is via 10/100baseT Ethernet links which can be connected directly or via local area network (LAN). It is possible to connect an additional processor support, for small simulation time-step operation. Table 18.1 is an example comparison of RTS solutions provided by two manufacturers, RTDS Technologies [5, 6] and OPAL-RT Technologies [7].

18.2.2 Operation and Control of RTS

In order to achieve effective processor management, some power system components can be combined to run on a single processor card, while other components may require multiple processors. Components such as generators and transformers require the 3PC processors. The allocation of tasks to the processors is handled by the work station interface card and supervised by the operating system of the RTS [5].

Through the graphical user interface (GUI), it is possible to interact with the RTS before, during, and after the simulator operation. At the top level, predefined and user-defined power and control system models are managed and circuit diagrams are assembled. The schematic of the power system is completed by selecting and connecting individual components. This approach is user-friendly and very fast, as

TABLE 18.1 Comparison of Typical RTS Solutions

Specification	Technologies	OPAL-RT Technologies
Capability		
Large time step	50–60 μs	10–40 μs
Maximum number of three-phase buses simulated in 50 μs	50	150
Small time-step (with FPGA)	1.4–2.5 μs	100 ns–10 μs
Maximum number of one-phase nodes simulated in 1 μs	—	30
Real-Time Processor		
Name	PB5	eMEGASim E1-12
Type	Freescale MC7448 RISC	Fast INTEL XEON
Clock frequency	1.7 GHz	2.3 GHz
Number of cores	12	12
Number of processors per card	2	—
Operating system	VxWorks running on a PPC405	Linux Redhat
Software Toolboxes		
GUI	RSCAD	RT-LAB
Third-party applications	—	MATLAB, Simulink, SimPowerSystems, PLECS, LabView, etc.
Real-time solver for SimPowerSystems	—	ARTEMiS
Analog and Digital I/O		
Analog inputs (AI)	Differential ±10 V, 16 bits @6 μs (12 channels)	Differential ±16 V, 16 bits @2.5 μs (32 channels)
Analog outputs (AO)	Differential ±10 V, 16 bits @1 μs (12 channels)	Differential ±16 V, 16 bits @1 μs (32 channels)
Static digital input/output (DI/DO)	(+5- to +24-V) range (64 channels each)	(0 to +30 V) (16 channels each)

users may not have the knowledge of code models of power system components used in the simulation [5].

The input to the RTS compiler is the power system data entered by the user along with the hardware configuration file. At the mid-level, the compiler generates the code to be used in the parallel computing by the digital signal processors (DSPs) and sends this to the appropriate processor cards. At the lowest level, the library files contain the code modules of the individual power system and control components, which are based on machine language code. The compiled DSP code is transferred to the RTS when the start button is selected in the GUI. To achieve greater flexibility and ease of model sharing, it is possible to integrate off-the-shelf solutions of individual RTS manufacturers with third-party applications such as MATLAB, Simulink, SimPowerSystems, PLECS, LabView, etc. [7].

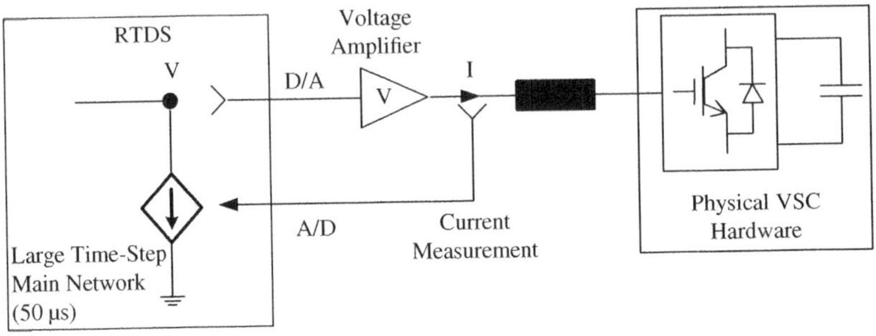

Figure 18.5 Simulation of power electronic converters for hardware-in-the-loop test.

18.2.3 Modeling of Power Electronic Converters in RTS

The real-time solution of a main network which has traditional power system components is achieved using large time steps in the range of 40–60 μs [2, 5]. Real-time simulation of power electronic converters, however, require high-frequency gate signals generated by digital controllers. These high frequencies introduce inaccuracies and numerical instability to the solution of the RTS due to limited computational bandwidth. Offline simulation is widely used as an alternative, but it is time consuming and required simplified average models which reduced precision [2]. To overcome this challenge, two common techniques to model power electronic converters in real time are [2, 5]: (i) analog hardware-in-the-loop test and (ii) fully digital software in the loop test, as illustrated in Figure 18.5 and Figure 18.6.

Users of first-generation RTS preferred to test performance of power electronic converters using the HIL technique [8]. Here, a physical model of the converter bridge is built and interfaced to the main network in RTS using D/A converter, power amplifier, current transducer, and A/D converter [5], as illustrated in Figure 18.5.

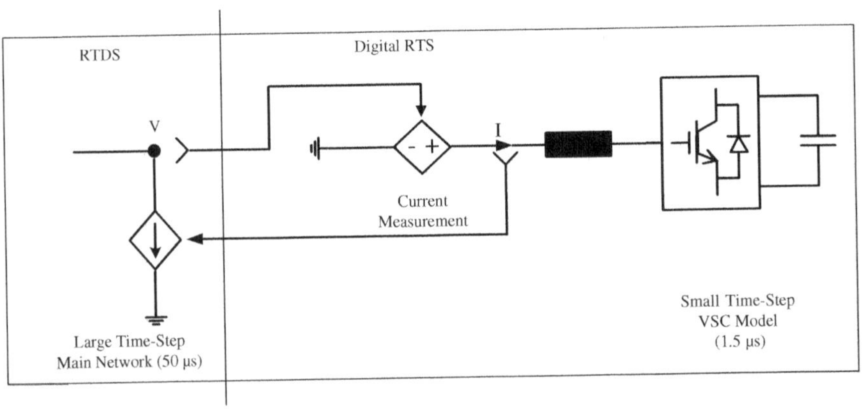

Figure 18.6 Simulation of power electronic converters for full digital software in the loop test.

This method is a good representation of the real performance of power converters and eliminates the need for precision compromise made on models in offline simulation. However, its major drawback is that analog input/output interface potentially introduces noise and delays to measurements. Moreover, the cost of building physical VSC can be prohibitive, and this does not assure ease of repeatability.

For the fully digital SIL technique, in Figure 18.6, state-of-the-art digital RTS use advanced processors which enable power electronic bridges to be modeled as subnetworks which is solved using small simulation time step in the range of 1.4–2.5 μs [2, 5]. This way, for example, a full digital switching model of converters can be implemented using the processors of the RTS. This method relies on emerging field-programmable gate array (FPGA) and other processor technologies which can enable submicrosecond time-step solutions to be executed and at low cost. The FPGAs support off-the-shelf processor cards of the RTS to provide the computational power required for power electronic models. With this method, there is a closer interface loop which results in shorter time delays, less noise, and better precision than the HIL technique. This SIL technique is very flexible and highly repeatable. Moreover, new algorithms and methods to provide the needed accuracy and efficiency for real-time network solution, embedded valve groups, and improved firing are an active research area [2, 5].

For example, in RTDS Technologies the concept of a subsystem is used for mathematical isolation of a portion of a system which is connected to another system only through traveling-wave transmission lines. This way, the network portions are decoupled and solved independently or by parallel computing to achieve faster simulation time.

On the other hand, OPAL-RT Technologies uses the ARTEMIS algorithm which is based on a suitable order approximation of matrix exponentials which are efficient for the fast simulation of power systems [9].

18.3 DESIGN OF EXPERIMENTAL TEST RIG

The design of the experimental test rig has to consider which elements are going to be used to represent the real system. The selection of these elements is usually dependent on the final application of the set-up. Therefore, if the test rig will have only an educational purpose or will be used for several studies, then it might be useful to have a more flexible and simple set-up. However, if the test rig has to be used to validate results of specific studies, then the set-up has to be more complex.

Another important aspect is the scaling definition of the platform to ensure that the results obtained using the test rig will be comparable to the real system. In AC systems the per-unit system is extensively used and in DC systems it is possible to use the same approach with some modifications and alternatives.

On the other hand, the test rig has to be monitored and controlled by another device, normally an embedded controller. The user will be able to implement their own control layout and check them in real-time through this device.

TABLE 18.2 Base Values for DC Grid Scaling Method

Quantity	Expression
Power	P_b
Voltage	U_b (voltage difference between the two poles)
Current	$I_b = \frac{P_b}{U_b}$
Impedance	$Z_b = \frac{U_b}{I_b}$

18.3.1 DC Grid Scaling Methods

The objective of scaling is to obtain an equivalent system that ensures the associated experimental results are valid for the real system. In the case of DC grids, it is possible to define two main different methods of scaling:

- Method based on per-unit system.
- Method based on time constants.

18.3.1.1 Method Based on Per-Unit System It is common to use the per-unit system in several studies of power systems such as power flow or electrical machine analysis. When DC grids have to be transformed in per-unit, two different options can be chosen:

- DC grid scaling method. This relies on a per-unit system only considering DC grids.
- Meshed AC/DC grid scaling method. This relies on a per-unit system considering AC/DC meshed grids [10].

The DC grid scaling method is directly the adaptation of the same method applied to AC grids. Table 18.2 shows the base values for that case.

This option does not provide a base value for the inductance and capacitance, because there is not a nominal frequency like in the AC side. It is necessary to define a scale factor using the impedance base values of the real system and the test rig:

$$k_{scale1} = \frac{Z_{b,rig}}{Z_{b,real}} \tag{18.1}$$

Then, the relation between the real system and the test rig for the inductance, resistance and capacitor is

$$k_{scale1} = \frac{R_{rig}}{R_{real}} = \frac{L_{rig}}{L_{real}} = \frac{C_{real}}{C_{rig}} \tag{18.2}$$

On the other hand, the meshed AC/DC grid scaling method considers the connection of power converters between the two grids, specifically VSCs. The main differences in relation to the conventional AC per-unit method are [10]:

- In the AC side, the peak value of voltage and current are used as base values instead of the rms.
- The base power considers the power transmission between the two sides of the converter.
- The DC side base voltage is considered two times the AC side base voltage. It is possible to define a unitary value for the index modulation when the two voltages are 1 pu.

This system has the advantage to use only one per-unit system to represent AC and DC grids, but the drawback is that the DC side base voltage will not be necessarily the same as the nominal DC voltage. However, this per-unit system defines a unitary modulation index and the real value is usually close to 1; therefore the two voltages will be approximately equal.

In Table 18.3 the AC and DC side base values are summarized.

In this case the inductance and capacitance base values can be obtained because there is a nominal frequency defined by the AC grid and a relationship between AC and DC side through the power.

TABLE 18.3 Base Values for Meshed AC/DC Grid Scaling Method

Quantity	Expression
Power (the same for AC and DC side)	$P_b = \frac{3}{2} U_{b,ac} I_{b,ac} = U_{b,dc} I_{b,dc}$
AC voltage	$U_{b,ac} = \hat{U}_s$ (amplitude of line-to-neutral nominal voltage)
AC frequency	$\omega_{b,ac} = \omega_0$ (power system nominal frequency)
AC current	$I_{b,ac} = \frac{2}{3} \frac{P_b}{U_{b,ac}}$
AC impedance	$Z_{b,ac} = \frac{U_{b,ac}}{I_{b,ac}}$
AC capacitance	$C_{b,ac} = \frac{1}{\omega_{b,ac} Z_{b,ac}}$
AC inductance	$L_{b,ac} = \frac{Z_{b,ac}}{\omega_{b,ac}}$
DC voltage	$U_{b,dc} = 2U_{b,ac}$
DC current	$I_{b,dc} = \frac{3}{4} I_{b,ac}$
DC impedance	$Z_{b,dc} = \frac{8}{3} Z_{b,ac}$
DC capacitance	$C_{b,dc} = \frac{3}{8} C_{b,ac}$
DC inductance	$L_{b,dc} = \frac{8}{3} L_{b,ac}$

Then, it is possible to define the following scale factor:

$$k_{scale2} = \frac{Z_{b,dc,rig}}{Z_{b,dc,real}} = \frac{R_{b,dc,rig}}{R_{b,dc,real}} = \frac{L_{b,dc,rig}}{L_{b,dc,real}} = \frac{C_{b,dc,real}}{C_{b,dc,rig}} \tag{18.3}$$

The scale factors obtained from the two per-unit systems will be different, but approximately the same if the two DC side voltage base values are close enough.

18.3.1.2 Method Based on Time Constants

The DC grid cable parameters have significant influence on the dynamic response of the system. Therefore, it is possible to find a relationship between these parameters and the time response to ensure that the real system and the test rig have a similar dynamic behavior. The advantage of this method is that it provides a basis for comparing the dynamic responses of the real system and the test rig.

Three time constants have been defined to accomplish this purpose [11]:

- **Power-Energy Time Constant.** It is related to the time that the system can transmit nominal power after a black out.

$$\tau_{WP} = \frac{W_{cap}}{P_{real}} = \frac{\frac{1}{2}C_{real}U_{dc,real}^2}{P_{real}} = \frac{\frac{1}{2}C_{rig}U_{dc,rig}^2}{P_{rig}} \tag{18.4}$$

- **Resonance Frequency Time Constant.** It is related to the resonance frequency in a specific line.

$$\frac{1}{\tau_{LC}} = \frac{1}{2\pi\sqrt{L_{real}C_{real}}} = \frac{1}{2\pi\sqrt{L_{rig}C_{rig}}} \tag{18.5}$$

- **LR Time Constant.** It is related to the ratio between the cable resistance and inductance.

$$\tau_{LR} = \frac{L_{real}}{R_{real}} = \frac{L_{rig}}{R_{rig}} \tag{18.6}$$

Therefore, it is possible to define a scale factor like in the previous methodology:

$$\frac{C_{rig}}{C_{real}} = \frac{P_{b,rig}}{P_{b,real}}\left(\frac{U_{dc,b,real}}{U_{dc,b,rig}}\right)^2 = \frac{Z_{real}}{Z_{rig}} \tag{18.7}$$

$$\frac{C_{rig}}{C_{real}} = \frac{L_{real}}{L_{rig}} \tag{18.8}$$

$$\frac{L_{real}}{L_{rig}} = \frac{R_{real}}{R_{rig}} \tag{18.9}$$

$$k_{scale3} = \frac{Z_{b,rig}}{Z_{b,real}} = \frac{R_{rig}}{R_{real}} = \frac{L_{rig}}{L_{real}} = \frac{C_{real}}{C_{rig}} \tag{18.10}$$

TABLE 18.4 Characteristics of Experimental Platform Example Case

Characteristics	Value
Nominal power transmission capacity	1 GW
Nominal AC side voltage	380 kV
Nominal DC voltage	±320 kV
Cable parameters (XLPE cable)	
Resistance	1.92 Ω
Inductance	24 mH
Capacitance	15.2 μF

Then, it is easy to check that this scale factor is the same that has been obtained through the DC grid scaling method in Section 18.3.1.1.

A drawback of this method is that it does not provide base values for power, voltage, and current. Hence, it is relies on one of the per-unit systems defined previously. Specifically, it is convenient select the system that only consider DC grids, because the scale factor is the same.

18.3.1.3 Example The previous methodologies are going to be presented through an example. In this case, the real system is based on a HVDC connection of 100 km with the characteristics of the Table 18.4, whereas the equivalent test rig has the characteristics presented in Table 18.5.

The objective is to find the equivalent cable parameters for the experimental platform considering the new power and voltage values.

18.3.1.4 Using DC Grid Scaling Method The base values in this case are in Table 18.6. Then, the associated scale factor is:

$$k_{scale1} = \frac{18.75}{410.26} = 0.076 \tag{18.11}$$

The scaled values of the cable parameters are

$$R_{rig} = 1.92 \cdot 0.076 = 0.146\ \Omega \tag{18.12}$$

$$L_{rig} = 24 \cdot 0.076 = 1.82\ \text{mH} \tag{18.13}$$

$$C_{rig} = 15.2/0.076 = 200\ \mu\text{F} \tag{18.14}$$

TABLE 18.5 Characteristics of Real System Example Case

Characteristics	Value
Nominal power transmission capacity	2 kW
Nominal AC side voltage	150 V
Nominal DC voltage	±125 V

TABLE 18.6 Base Values for Real System and Test Rig with DC Grid Scaling Method

Real System	Test Rig
$P_{b,real} = 1000$ MW	$P_{b,rig} = 2$ kW
$U_{b,real} = 640$ kV	$U_{b,rig} = 250$ V
$I_{b,real} = \frac{1000}{640} = 1.56$ kA	$I_{b,rig} = \frac{2000}{250} = 8$ A
$Z_{b,real} = \frac{640}{1.56} = 410.26\ \Omega$	$Z_{b,rig} = \frac{250}{8} = 31.25\ \Omega$

18.3.1.5 Using Meshed AC/DC Grid Scaling Method The base values in this case are in Table 18.7. Then, the associated scale factor is:

$$k_{scale2} = \frac{30}{384.85} = \frac{95.47}{1230} = \frac{8.27}{106.1} = 0.078 \tag{18.15}$$

The scaled values of the cable parameters are

$$R_{rig} = 1.92 \cdot 0.078 = 0.15\ \Omega \tag{18.16}$$

TABLE 18.7 Base Values for Real System and Test Rig with Meshed AC/DC Grid Scaling Method

Real System	Test Rig
$P_{b,real} = 1000$ MW	$P_{b,rig} = 2$ kW
$U_{b,ac,real} = 380\frac{\sqrt{2}}{\sqrt{3}} = 310.27$ kV	$U_{b,ac,rig} = 150\frac{\sqrt{2}}{\sqrt{3}} = 122.48$ V
$\omega_{b,ac,real} = 2\pi 50$ rad/s	$\omega_{b,ac,rig} = 2\pi 50$ rad/s
$I_{b,ac,real} = \frac{2}{3}\frac{1000}{310.27} = 2.15$ kA	$I_{b,ac,rig} = \frac{2}{3}\frac{2000}{122.48} = 10.89$ A
$Z_{b,ac,real} = R_{b,ac,real} = \frac{310.27}{2.15} = 144.32\ \Omega$	$Z_{b,ac,rig} = R_{b,ac,rig} = \frac{122.48}{10.89} = 11.25\ \Omega$
$C_{b,ac,real} = \frac{1}{2\pi 50 \cdot 144.32} = 22.06\ \mu$F	$C_{b,ac,rig} = \frac{1}{2\pi 50 \cdot 11.25} = 282.94\ \mu$F
$L_{b,ac,real} = \frac{48.42}{2\pi 50} = 459.4$ mH	$L_{b,ac,rig} = \frac{11.25}{2\pi 50} = 35.8$ mH
$U_{b,dc,real} = 2 \cdot 310.27 = 620.54$ kV	$U_{b,dc,rig} = 2 \cdot 122.48 = 244.96$ V
$I_{b,dc,real} = \frac{3}{4}2.15 = 1.61$ kA	$I_{b,dc,rig} = \frac{3}{4}10.89 = 8.17$ A
$Z_{b,dc,real} = R_{b,dc,real} = \frac{8}{3}144.32 = 384.85\ \Omega$	$Z_{b,dc,rig} = R_{b,dc,rig} = \frac{8}{3}11.25 = 30\ \Omega$
$C_{b,dc,real} = \frac{3}{8}22.06 = 8.27\ \mu F$	$C_{b,dc,rig} = \frac{3}{8}282.94 = 106.1\ \mu F$
$L_{b,dc,real} = \frac{8}{3}459.4 = 1.23$ H	$L_{b,dc,rig} = \frac{8}{3}35.8 = 95.47$ mH

$$L_{rig} = 24 \cdot 0.078 = 1.87 \text{ mH} \tag{18.17}$$

$$C_{rig} = 15.2 \cdot 12.83 = 195.02\ \mu\text{F} \tag{18.18}$$

18.3.1.6 Using Method Based on Time Constants The constant time values are

$$\tau_{WP} = \frac{\frac{1}{2} \cdot 15.2 \cdot 10^{-6} \cdot (640 \cdot 10^3)^2}{10^9} = 3.11 \text{ ms} \tag{18.19}$$

$$\tau_{LC} = 2\pi\sqrt{24 \cdot 10^{-3} \cdot 15.2 \cdot 10^{-6}} = 3.8 \text{ ms} \tag{18.20}$$

$$\tau_{LR} = \frac{24 \cdot 10^{-3}}{1.92} = 12.5 \text{ ms} \tag{18.21}$$

Then, the scaled values of the cables parameters are

$$C_{rig} = \frac{2 \cdot 2000 \cdot 3.11 \cdot 10^{-3}}{250^2} = 200\ \mu\text{F} \tag{18.22}$$

$$L_{rig} = \frac{0.0038^2}{4\pi^2 200 \cdot 10^{-6}} = 1.82 \text{ mH} \tag{18.23}$$

$$R_{rig} = \frac{1.82 \cdot 10^{-3}}{12.5 \cdot 10^{-3}} = 0.146\Omega \tag{18.24}$$

The base values will be the same used in the DC grid scaling method. Now, as a possible additional exercise the reader is encouraged to do a similar process, using the characteristics of the system presented in reference 12.

18.3.2 Instrumentation and Control Layout

The control layout is based on an embedded computer or embedded controller designed specifically for power electronic applications. Therefore, this element has to be able to provide:

- **Real-Time Processing.** The responses are usually on the order of milliseconds and microseconds. Concretely, this condition is an essential point for a DC grid experimental platform, where most of their elements are based on power electronic devices that require high-frequency switching.
- **Digital Signal Processing.** The main objective is to measure, filter, and even compress the analog signals. In DC grid test rigs these signals usually correspond to electrical or mechanical measurements (voltages, currents, speed, angle, etc.).

The embedded computer needs a central processing unit (CPU); in that case a microprocessor that is the responsible for providing these functions and performing mathematical operations quickly. This device requires peripheral elements to create the necessary hardware interface that ensures a correct operation. Typically, these peripherals include basically [13]:

Communication Systems. The embedded computer has to receive and send information with other devices during its operation or to upload the necessary programs during the implementation of the control system.

I/O Modules. The microcontroller has to receive analog, mainly measurements, and digital signals from sensors or other elements and at the same time send orders to the actuators. For DC grid applications the power electronic devices are the main type of actuators, therefore the PWM management is essential.

Memory. The microprocessors sometimes do not have enough memory and additional memory modules are required to process more amounts of data.

The management and supervision of the experimental platform is performed through a SCADA system, which can be implemented directly in the embedded computer or as an external element, for example a simple desktop PC or laptop.

The complexity of the embedded computer is associated with the final application and the cost. Therefore, the possible solutions range from simple controller boards, only having basic necessary functions, to processor boards, with more memory and processing capabilities. The latter option is usually combined with more powerful software tools that help the user to configure the microprocessor.

Another important feature of these devices is the possibility to be configured and programmed with a large level of flexibility, which is essential when different types of studies have to be performed with the same experimental platform. Therefore, the programming tools have to provide all the necessary capabilities to create almost any type of control system and the I/O modules has to be easy to reconfigure and use for as many applications as possible.

The embedded computer can be split in two or more microprocessors, where one group is responsible for the real-time and digital processing functions, whereas the other group can store or transmit the data and supervise the whole operation of the experimental platform. This option is especially important when there are several elements to control, for example in test rigs based on a multi-terminal DC grid configuration. In these cases, instead of only one embedded computer, it can be possible to find additional controller boards that are slaves of this main embedded system.

Two examples of commercial embedded computers are dSPACE and National Instruments products that offer several solutions in this field. Specifically, dSPACE devices are based on different types of microprocessor, such as TMS320 DSP, PowerPC or AMD Opteron [14], whereas National Instruments basically are based on FPGAs [15].

18.4 POTENTIAL APPLICATIONS OF HIL TESTS IN DC GRIDS

There are many applications in DC grids where HIL schemes can be used. In this section, four examples are discussed:

- Subsynchronous resonance (SSR) studies.
- DC grid protection.

- Integration of offshore wind power.
- Converter architecture and design.

18.4.1 Subsynchronous Resonance (SSR) Studies

HVDC connections can be used to solve SSR problems in the main grid by damping the torsional-mode frequencies. A possible HIL scheme can represent the HVDC connection through a test rig, whereas the main grid is modelled in the RTS. Therefore, it is possible to design and implement the necessary controls for the HVDC connection and analyze the effects in an emulated main grid. This option allows modeling different AC grid configurations using the same HVDC connection, which can be interesting to test the same solution for several scenarios.

18.4.2 DC Grid Protection

The detection of faults and the correct operation of the circuit breakers are an essential point in DC grids. Although there are no DC breakers in operation at present, HIL configuration can offer different possibilities in this case. One option could be to have a test rig based on scaled DC circuit breakers and model the DC grid and the associated converters with a RTS. Another option could be the opposite, to have a test rig based on a DC grid and model the circuit breakers in a RTS.

The first option has the advantage to use a real circuit breaker and consider all its dynamic response, but the model of the DC grid can have limitations during fault detection studies. For the second option the model of the circuit breaker can be insufficient to ensure its correct operation, but the fault detection procedures provide more reliable results. However, considering the difficulty to obtain a real circuit breaker and its cost, the second option can be more attractive and available for a larger number of studies.

Other options can be, for example, a test rig based on a DC grid with real circuit breakers and the representation of the AC grid connections in the RTS.

18.4.3 Integration of Offshore Wind Power

The integration of offshore wind power through HVDC connections is another important point of study. In this case a large number of possibilities exists.

There are three elements that can be integrated in a test rig or RTS:

- Wind farm.
- HVDC connection.
- AC grid.

If the study is focused on the wind farm response, then the test rig should include a real model of the wind turbine dynamic performance. However, if the study is related to the capabilities of the HVDC connection for wind power transmission, then the RTS can represent the response of the wind turbines. In all the cases, the AC grid is likely to be modeled using the RTS, because of its complexity.

18.4.4 Converter Architecture and Design

The RTS will be useful to model modular multilevel converter (MMCs) topologies, which have a high degree of complexity. To achieve this, the RTS can integrate additional processors to improve the computation capabilities required for the small time step simulation of power electronic switches.

Another alternative is to have a scaled model of a real MMC and a DC grid and represent the AC grid through the RTS. It is observed that in all these examples the main grid tends to be modeled using the RTS because of its complexity.

REFERENCES

1. V. Dixit, M. B. Patil, and M. C. Chandorkar, "Real time simulation of power electronic systems on multi-core processors," in *Power Electronics and Drive Systems, 2009. PEDS 2009. International Conference on*. New York: IEEE, 2009, pp. 1524–1529.
2. J. Bélanger, P. Venne, and J. Paquin, "The what, where and why of real-time simulation." Accessed: 31 Oct. 2015 [Online]. Available: http://www.opal-rt.com/sites/default/files/technical_papers/PES-GM-Tutorial_04%20-%20Real%20Time%20Simulation.pdf
3. OPAL-RT. (2013) About hardware in the loop and hardware in the loop simulation. Accessed: 03-Apr-2014.
4. H. Dommel, "Digital computer solution of electromagnetic transients in single- and multiphase networks," *Power Apparatus and Systems, IEEE Transactions on*, vol. PAS-88, no. 4, pp. 388–399, Apr. 1969.
5. RTDS Technologies, "Power system user's manual," RTDS Technologies, Technical Report, 2012.
6. RTDS Technologies. (2015) Pb5 card. Accessed: 31 Oct. 2015 [Online]. Available: www.rtds.com/the-simulator/our-hardware/pb5-card/
7. OPAL-RT Technologies. (2011) eMEGAsim-Power System and Power Electronic Real Time Simulator. Accessed: 29 Apr. 2014 [Online]. Available: www.opal-rt.com/new-product/emegasim-power-system-and-power-electronic-real-time-simulator
8. G. G. Parma and V. Dinavahi, "Real-time digital hardware simulation of power electronics and drives," *Power Delivery, IEEE Transactions on*, vol. 22, no. 2, pp. 1235–1246, 2007.
9. N. Léchevin, C. A. Rabbath, and P. Baracos, "Distributed real-time simulation of power systems using off-the-shelf software," *IEEE Canadian Review*, pp. 5–8, 2001.
10. A. Yazdani and R. Iravani, *Voltage-Sourced Converters in Power Systems*. Hoboken, NJ: John Wiley & Sons, 2010.
11. A. Egea-Alvarez, A. Junyent-Ferre, G. Gross, S. Galceran-Arellano, and O. Gomis-Bellmunt, "Multiterminal-hvdc scaled platform for offshore wind transmission systems emulation," in *Power Electronics and Applications (EPE), 2013 15th European Conference on*. New York: IEEE, 2013, pp. 1–10.
12. A. Egea-Alvarez, F. Bianchi, A. Junyent-Ferré, G. Gross, and O. Gomis-Bellmunt, "Voltage control of multiterminal vsc-hvdc transmission systems for offshore wind power plants: Design and implementation in a scaled platform," *Industrial Electronics, IEEE Transactions on*, vol. 60, no. 6, pp. 2381–2391, 2013.
13. M. Wolf, *Computers as Components: Principles of Embedded Computing System Design*. Amsterdum: Elsevier, 2012.
14. dSPACE. (2014) Hardware products. Accessed: 7 Apr. 2014 [Online]. Available: https://www.dspace.com/en/inc/home/products/hw.cfm
15. National Instruments. (2014) Embedded systems for monitoring and control. Accessed: 7 Apr. 2014 [Online]. Available: http://www.ni.com/embedded-systems/

PART V

APPLICATIONS

PART V COMPLEMENTS the other four parts of the book with selected applications which provide some advanced insights on how HVDC grids may be developed. This part does not intend to provide a comprehensive view on all the possible applications, but serves as an illustration of some examples on how HVDC grids can be implemented and what kind of services they can provide.

The applications studied include the utilization of HVDC systems to provide power system oscillation damping in the AC systems where they are connected, the design of an optimal droop for the control of the HVDC grid, the possible utilization of DC grid, power flow control devices in meshed HVDC grids, and finally the analysis of the offshore AC hub configuration.

HVDC Grids: For Offshore and Supergrid of the Future, First Edition. By Dirk Van Hertem, Oriol Gomis-Bellmunt, and Jun Liang.

CHAPTER 19

POWER SYSTEM OSCILLATION DAMPING BY MEANS OF VSC-HVDC SYSTEMS

José Luis Domínguez-García[1] *and Carlos E. Ugalde-Loo*[2]
[1]Research Group Wind Energy, Electrical Engineering Department, Catalonia Institute for Energy Research (IREC), Barcelona, Spain
[2]Institute of Energy, School of Engineering, Cardiff University, Cardiff, Wales

19.1 INTRODUCTION

Modernization of power systems, both through the renovation of its elements and by the introduction of new components, is modifying the way that electricity networks operate and thus the understanding we have of them. The inclusion of additional elements such as HVDC links, flexible AC transmission systems (FACTS) devices, or renewable energy generation, leads to concern about their capability to contribute to grid regulation.

Rapid wind power penetration into the generation share is affecting power systems stability and reliability. The characteristics of wind power plants (WPPs) are substantially different from those of conventional power plants [1]. This has led transmission system operators (TSOs) to establish Grid Codes for connecting wind farms to power systems [2]. In line with this, WPPs must be capable of complying with a number of requirements, including voltage sag ride through capability [3], frequency regulation [4], and active and reactive power regulation [5]. Further contributions by the system operators will be required in the future; among them, power system oscillations damping will play an important role. For example, the draft of the new Spanish Grid Code for wind power considers this capability [6]. Additionally, ENTSO-E has requested power system stabilizer capability as a requirement for any generation unit (including wind farms), as detailed in reference 7.

A substantial increase of HDVC installations has taken place in recent years, reinforcing the HVDC grid concept. As in the wind power case, the impact of the new technologies has contributed to the development of special Grid Codes. For example, ENTSO-E has presented a Grid Code draft for HVDC systems [8]. An

HVDC Grids: For Offshore and Supergrid of the Future, First Edition. By Dirk Van Hertem, Oriol Gomis-Bellmunt, and Jun Liang.

existing working group of CIGRÉ, B4-56 Guidelines for Preparation of Connection Agreements or Grid Codes for HVDC Grids is also playing an active role on this.

Power system stability mainly depends on the behavior of the generators connected to the grid. Therefore, the influence of wind power on power system dynamics is a relevant issue to be studied [9], which must be analyzed by considering different technologies and controls. As these have different dynamic characteristics, the effects caused on the grid will vary [10]. The contribution of a number of technologies such as HVDC links and their controllers, FACTS devices, and wind farms to enhance power system stability has been addressed in reference 11.

The transmission of power from distributed generation to consumption areas requires a substantial upgrade of existing power systems. Carrying out reinforcements of the transmission network requires a compromise between the difficulty in building new onshore overhead lines and the risks of using technologies new to the transmission system. Inland reinforcement will include fixed series capacitors, providing series compensation to increase the power transmission capability [12]. Alternatively, offshore reinforcement can be accomplished by the inclusion of HVDC links and HVDC grids [13], which provides more controllability to the power system [14].

This chapter shows the potential contribution that VSC-HVDC systems (both HVDC links and HVDC grids) can have to enhance power system stability—in particular, through damping of power system oscillations.

19.2 POWER SYSTEM STABILITY

Power system stability is understood as the capability of power systems to keep equilibrium during normal operation and to recover to an acceptable equilibrium point following a disturbance, maintaining variables within adequate boundaries [15].

Depending on the power system response to a fault, power system stability can be classified as follows [15]:

- **Rotor Angle Stability.** It is concerned with the ability of the synchronous generators of a power system to keep or restore the equilibrium between the electromagnetic and the mechanical torque of each machine and, also, to maintain the rotor angle synchronization between them.
- **Frequency Stability.** It is defined as the capability of a power system to recover the balance between generation and load, minimizing the loss of load.
- **Voltage Stability.** It is referred to as the capability of a power system to maintain steady (or within acceptable boundaries) all bus voltages both under normal operating conditions and after a disturbance.

Depending on the fault, rotor angle stability can be organized into two different groups: transient stability and small signal stability. Transient stability is related to the ability of the power system to maintain synchronism during a severe disturbance—for example, a three-phase fault on a transmission line. On the other hand, small signal stability is concerned with the effect of small disturbances—for instance, minor changes in load or generation [15]. (see Figure 19.1 [16].)

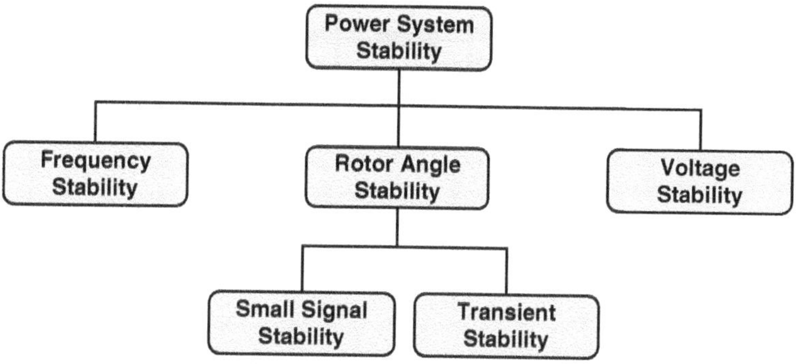

Figure 19.1 Classification of power system stability [16].

19.2.1 Low-Frequency Oscillations

Rotor angle stability analysis, within the context of small signal stability, studies the dynamic response of a power system—for instance, the nonoscillatory (or aperiodic) modes arising from the lack of synchronizing torque and the oscillatory modes appearing due to the absence of damping torque. The synchronizing torque is the component of torque in phase with the rotor angle of the synchronous generator; conversely, the component of torque in phase with the speed deviation is known as damping torque. The aperiodic problem has been largely solved through the introduction of an automatic voltage regulator (AVR), a form of excitation control, into the generators. Oscillation modes are usually damped by the inclusion of a power system stabilizer (PSS).

The small signal modes of oscillation, also called low-frequency oscillations (LFOs), can be classified as follows [16]:

- Power system oscillations
  - **Inter-area modes.** These are oscillation modes occurring between groups of generators far from each other. Generators belonging to one area oscillate against another group of generators from a different area. Inter-area modes usually appear when there is power transfer between areas connected through a tie-line. The frequency of these oscillation modes is approximately in the range of 0.1 to 0.7 Hz.
  - **Local modes or Intra-area modes.** These are related to the rotor angle oscillation of a single synchronous generator against the rest of the power system or of the local area where the generator belongs to. The frequencies of such modes are typically in the range of 0.7 to 2 Hz.
- Machine oscillations (> 2 Hz).
  - **Torsional modes.** These are related to the interaction between the mechanical system of synchronous generators and power system elements such as reactive power or series compensation.
  - **Control modes.** These oscillation modes consider the effect of controllers within the power system, such as excitation and power converters controllers.

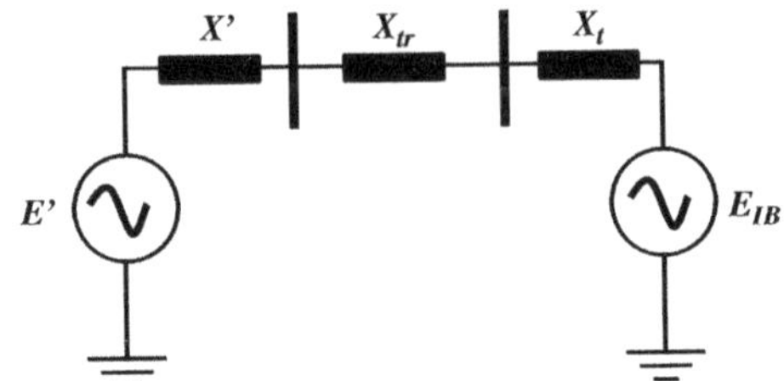

Figure 19.2 Synchronous machine connected to an infinite bus.

Numerous worldwide incidents traced to LFOs have been reported. Some of the most important occurred in: United Kingdom in 1980, with a frequency of oscillation of $\approx$ 0.5 Hz; West United States/Canada in 1996, where both power systems exhibited LFOs around 0.224 Hz; China on 6th March 2003, where a blackout occurred following oscillations of $\approx$ 0.4 Hz; and Italy on 28th September 2003, where a blackout occurred following oscillations of about 0.55 Hz.

Weak tie-lines and the nature of their longitudinal structure are causes for LFOs. Additionally, concentration of outputs to major power plants with insufficient reserve margins, heavy flow across transmission interfaces due to seriously imbalanced regional power, and pumped-storage units in pumping mode operation are other common causes for LFOs. Such issues may be avoided by suitably modifying the power flowing through the tie-line. This can be achieved through the use of power system elements with such capability, such as FACTS devices and HVDC links. Wind power can be also used to mitigate LFOs.

To give an insight into this phenomenon, consider the simple circuit shown in Figure 19.2, featuring a synchronous generator connected to an infinite bus via a tie-line reactance.

Let the synchronous generator be described by its classical model, based on the equation of motion. The linearized representation of the system in Figure 19.2 is given as

$$\begin{gathered} \frac{d\Delta\omega_r}{dt} = \frac{\omega_0}{2H}\left(P_m - P_e - K_D\Delta\omega_r\right) \\ \frac{d\Delta\delta_r}{dt} = \omega_0\Delta\omega_r \end{gathered} \tag{19.1}$$

where the generated electrical power (P_e) can be expressed as a function of the synchronous machine rotor angle (δ_r) as $P_e \approx K_S\Delta\delta_r$, following system linearization based on a Taylor decomposition. Therefore, Eq. (19.1) may be rewritten, in matrix form, as

$$\begin{bmatrix} \Delta\dot{\omega}_r \\ \Delta\dot{\delta}_r \end{bmatrix} = \omega_0 \begin{bmatrix} \frac{-K_D}{2H} & \frac{-K_S}{2H} \\ 1 & 0 \end{bmatrix} \begin{bmatrix} \Delta\omega_r \\ \Delta\delta_r \end{bmatrix} + \begin{bmatrix} \frac{\omega_0}{2H} \\ 0 \end{bmatrix} \Delta T_m \tag{19.2}$$

The eigenvalues of the linear system matrix in Eq. (19.2) can be be easily calculated as

$$\lambda_{1,2} = \frac{-\frac{K_D}{2H} \pm \sqrt{\left(\frac{K_D}{2H}\right)^2 - \frac{4K_S}{2H}}}{2} \tag{19.3}$$

These are directly related, as stated in the Lyapunov First Method, to the oscillatory behavior of the linear system. It can be easily seen from Eq. (19.3) that the eigenvalues are dependent on both the damping (K_D) and the synchronizing (K_S) torque components. It can be observed that due to the oscillatory behavior dependence on the synchronizing torque, LFOs may be damped by affecting the power flow of the power system.

19.2.2 Subsynchronous Resonance

Subsynchronous resonance (SSR) is an electric power system condition where the electric network exchanges energy with a turbine generator at one or more of the natural frequencies of the combined system below the synchronous frequency of the system [17, 18]. This dynamic phenomenon, upon occurrence, may have disastrous consequences if left unattended. SSR mainly occurs in series capacitor-compensated transmission systems [19]. When series compensation is in place, the natural frequency of the electrical system is effectively modified and may interact with mechanical natural frequencies of turbogenerators across the network. Under such conditions, undesirable subsynchronous oscillations may be exhibited, contributing to stress, fatigue, and potential damage to the turbine shafts [15, 18, 20].

The first appearance of SSR was reported in the 1970s at the Mohave power station in the United States, where two successive shaft failures occurred [20]. This station was part of a generation and transmission system in Arizona, Nevada, and Southern California—featuring levels of series compensation up to 70%. Several studies on SSR have been carried out since then and have been well documented in the literature [15, 17, 18]. Even though this is not a novel problem, the threat of SSR is still very real in modern transmission networks and is a topic of industrial interest [12, 13].

SSR is classified into the following categories [17]:

- **Induction Generator Effect (IGE).** This is caused by self-excitation of the rotor electrical dynamics. Since the rotating mmf produced by a subsynchronous current of the armature moves slower than the speed of the rotor, the rotor resistance (at the subsynchronous frequency, as seen from the armature terminals) is negative as the slip of the machine (which may be viewed as an induction generator) is negative. When the magnitude of this negative resistance exceeds the sum of the armature and network resistances at a resonant frequency, self-excitation occurs.
- **Torsional Interaction (TI).** This is caused by self-excitation involving electrical and mechanical dynamics and will be explained in more detail below.

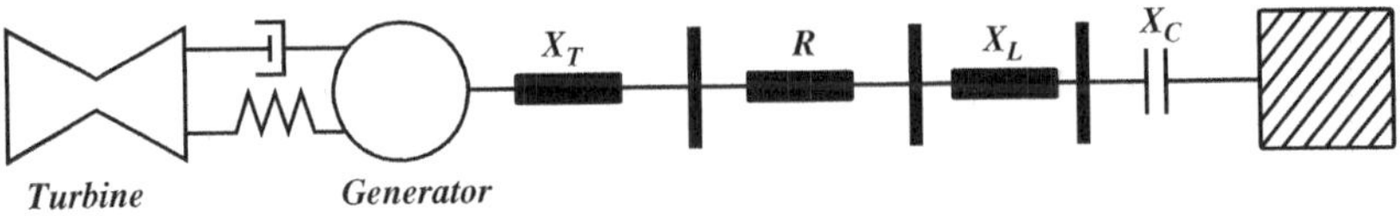

Figure 19.3 Series-compensated transmission line connected to a turbogenerator [18].

- **Transient Torques/Shaft Torque Amplification.** These result from system disturbances which can cause sudden changes in a series-compensated network. Current oscillations may occur, and if any subsynchronous network frequency coincides with a natural mode of the turbo-generator shaft, large peak torques may arise since they are proportional to the magnitude of the oscillating current.
- **Subsynchronous Control Interactions (SSCI).** These have been recently observed in wind farms connected to a series-compensated network. The phenomenon is due to the interaction among Type 3 wind farms and/or the power electronic converter control with the series-compensated line.

The example presented later in this chapter focuses on TIs. This type of SSR can be explained considering the series-compensated transmission line shown in Figure 19.3.

The natural resonant frequency f_{er} of this system can be calculated as [17–19]

$$f_{er} = f_0\sqrt{\frac{X_C}{X'' + X_T + X_L}} \tag{19.4}$$

where f_0 is the synchronous frequency, X the generator subtransient reactance, X_T the transformer leakage reactance, and X_C, X_L the inductive and capacitive reactances of the transmission line. It is assumed that the generator and turbine rotors in Figure 19.3 are mounted on elastic sections.

If rotor oscillations occur at a torsional mode frequency f_{TM}, induced voltages at the generator armature appear, presenting both subsynchronous ($f_{SUB} = f_0 - f_{TM}$) and supersynchronous ($f_{SUPER} = f_0 + f_{TM}$) components. These voltages, in turn, result in sub- and supersynchronous currents in the armature (and network) which produce electromagnetic torques with a frequency f_{TM}. If the subsynchronous component f_{SUB} is in the neighborhood of the electric resonant frequency f_{er} defined in Eq. (19.4), oscillations start building up as a consequence of this interplay between electrical and mechanical systems [17, 18]. If no corrective measures are taken, TIs can cause shaft damage and affect the system performance and stability.

If $X_C < X_L$ (which is the standard practice), it follows from Eq. (19.4) that $f_{er} < f_0$. Moreover, depending on the compensation level, $f_{er} \simeq f_0 - f_{TM}$, implying the presence of a TI. Therefore, the TI problem may be avoided by an adequate selection of series compensation levels. However, this can be difficult to achieve in practice since the values of X_C or X_L may vary with system operating conditions [18].

TIs are considered as small disturbance conditions. As such, they are analyzed through linear models and often assessed through eigenanalysis and time-domain simulations.

19.3 VSC-HVDC SYSTEMS DAMPING CONTRIBUTION: APPLICATION EXAMPLES

In this section, the potential enhancement of power systems stability by using VSC–HVDC systems including both HVDC grids and transmission links are analyzed. For the sake of simplicity, two application examples are presented to demonstrate the damping contribution capability of HVDC links, although the analysis can be extendable to HVDC grids of any size by considering the proper coordination. The first example considers the capability of offshore wind power generation connected onshore through an HVDC link to damp LFOs. A number of power system stabilizers that may be included in the power converter are compared. The focus of the second example is, through a simplified system representation, to use analytical and simulation tools to investigate the occurrence of SSR in an onshore transmission system. Damping of SSR using submarine VSC–HVDC links is demonstrated through numerical simulations, with experimental results validating the proposed control scheme.

19.3.1 Power System Stabilizers for Offshore Wind Power Plants Connected by Means of a VSC-HVDC System

The input of a PSS can be any signal affected by the oscillation that is desired to be damped. Thus, to avoid the use of wide-area communications in the control scheme, local signals are selected as inputs. The grid-side converter connection point is selected as a measurement point. However, it is important to consider that offshore wind power plants (OWPPs) are connected to the shore by HVDC links, which are usually far from conventional generation areas. Due to this fact, local modes could be unobservable. Therefore, PSSs must be designed to damp inter-area oscillation modes.

Within this context, the PSS control is based on the conventional scheme of a PSS for synchronous generators [15]. However, in this case, it is not necessary to match the phase compensation of the synchronous generator to produce damping torque; thus, the inclusion of lead/lag compensators can be avoided. According to reference 21, the controller can be composed of a proportional action, a limiter, and a band pass (or washout) filter to limit the frequency range where the controller is acting (Figure 19.4). Additionally, the inputs and outputs of the controller can be different from those of a conventional PSS. The input signal can be any variable affected by the power system oscillation. The output can be any variable capable of varying the power flow of the power system. In practice, the PSS introduces small variations to the reference values of the output signals. For a generic VSC-HVDC system (either links or a grid), the PSS controller is embedded in the grid side converter, generating references for the wind turbines, with the requirement of coordinated control if various converters are connecting the DC grid with AC main grid.

19.3.1.1 Case Study The power system under study is based on the well-known two area network described in reference 22. An OWPP is connected through an HVDC-link to the tie-line between the two areas, as shown in Figure 19.5. To simplify

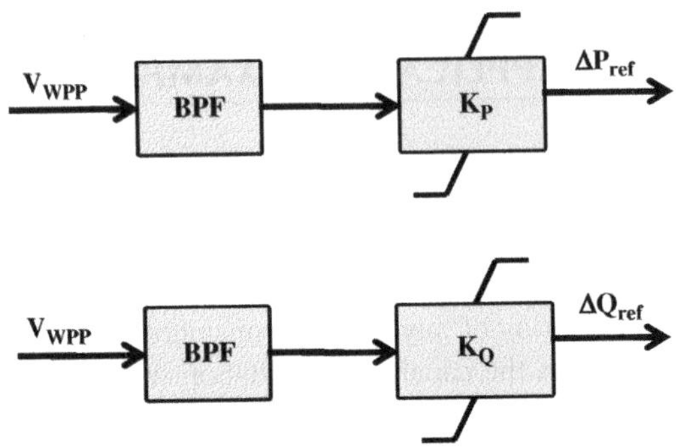

Figure 19.4 Block representation of the simple active and reactive power PSS controller.

the study, each area is represented as a large synchronous generator, with the governor and the excitation systems being included but without PSSs being considered. The OWPP is assumed as an aggregated model for simulation. In order to excite the power system oscillation, a three-phase fault of 0.133s is considered at the middle of the tie-line. A WPP of 50 MW is assumed, which is less than 5% of wind power penetration on the system. This value has been chosen to depict a conservative situation and to avoid a dominance of the OWPP on the power system dynamics.

PSS controllers are included to the grid side converter of the HVDC link. These consider the voltage magnitude of the connection point as an input and the active and/or reactive power reference as an output. The frequency of the band-pass filter lies between 0.08 Hz and 1.5 Hz, since the inter-area modes are in the range of 0.1 Hz to 0.7 Hz. The active (K_P) and reactive (K_Q) gains have been set as 10^4 and -10^5, respectively. These values are important as they can affect the stability of the

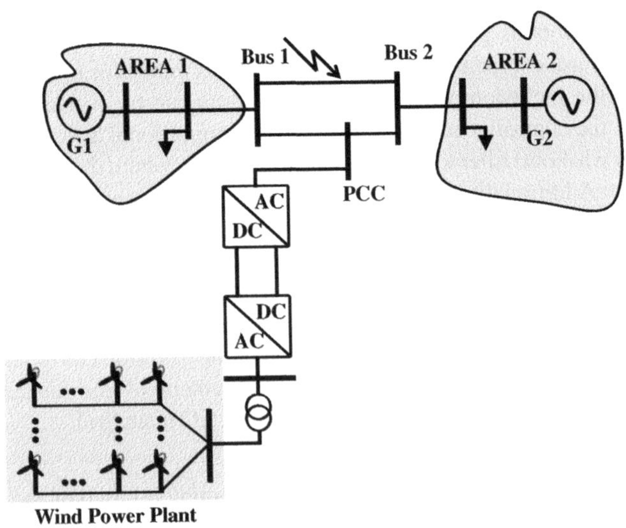

Figure 19.5 Representation of the power system used in this study.

system. Output saturations have been also included to limit the controller outputs between −0.15 and 0.15 p.u. This way, it is not necessary to overrate the HVDC converters [21].

The following analysis compares the capability of the OWPP to damp LFOs through the use of different PSS control schemes.

19.3.1.2 Comparison of Different PSS Schemes

A comparison between a number of control schemes is carried out to evaluate the PSS capability of the system to damp power oscillations [23]. The system under study is simulated under four scenarios:

- No PSS installed in the VSC-HVDC system (base case, No PSS).
- VSC–HVDC inverter with an extra active power loop to damp oscillations (P-PSS).
- VSC-HVDC inverter with an extra reactive power loop to damp oscillations (Q-PSS).
- VSC-HVDC inverter with both active and reactive power loops to damp oscillations (PQ-PSS).

The active power flowing through Bus 1 connecting Area 1 with the rest of the system is shown in Figure 19.6. Since Area 2 has larger loads than generation, active power flows from Area 1 to Area 2. It can be observed that the system without PSS compensation (No PSS) presents an almost critically stable dynamic behavior. The inclusion of an additional reactive power loop into the system shows a slight damping improvement (Q-PSS). On the other hand, the addition of an extra active power loop into the HVDC grid side converter presents an important improvement to damp the oscillation modes (P-PSS). Moreover, it can be observed that the simultaneous use

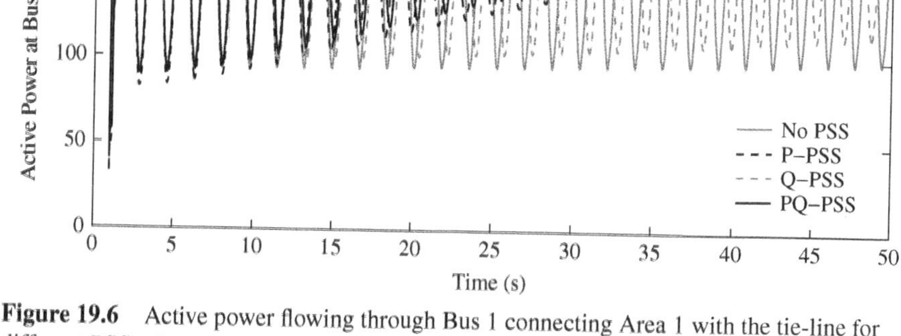

Figure 19.6 Active power flowing through Bus 1 connecting Area 1 with the tie-line for different PSS schemes.

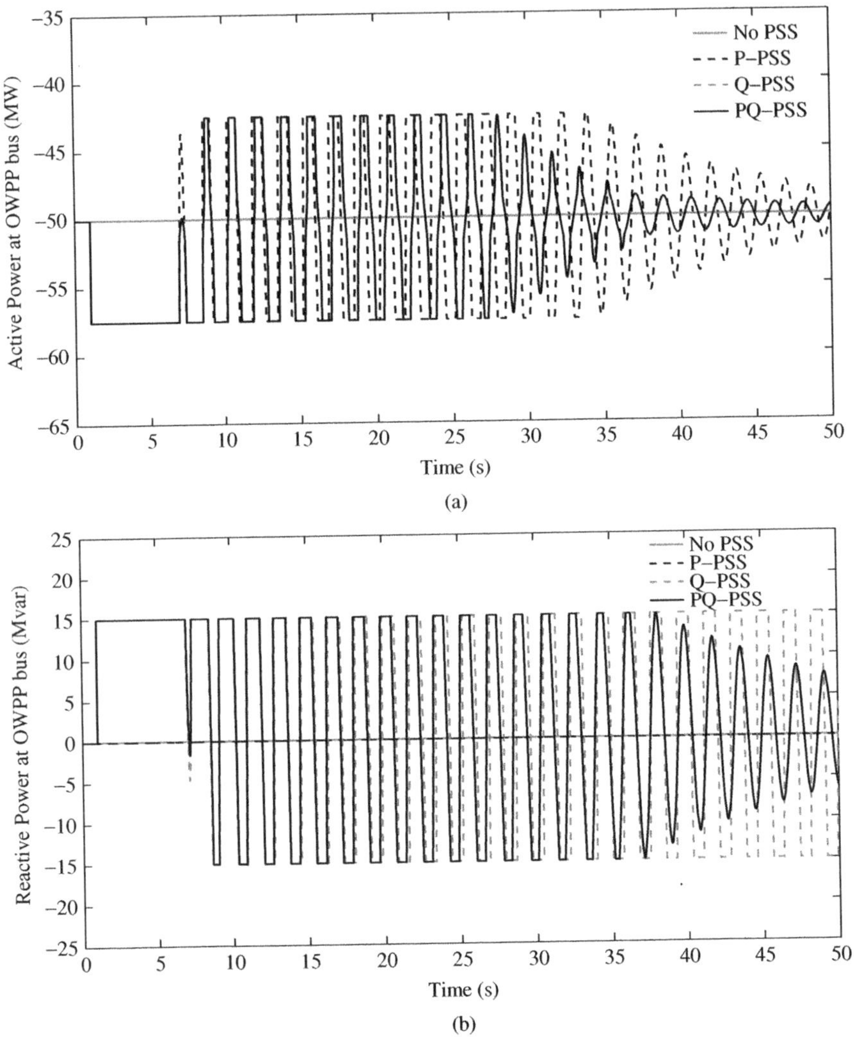

Figure 19.7 Active and reactive power delivered by the WPP through the HVDC link for different PSS schemes schemes. (a) Active power. (b) Reactive power.

of active and reactive power shows even a better damping capability (PQ-PSS). A better response obtained with P-PSS rather than with Q-PSS is an expected result since active power directly affects the mechanical dynamics, whereas reactive power indirectly affects the rotor dynamics.

Figure 19.7a presents the active power delivery response of the VSC-HVDC link. It can be seen that the converter delivers a constant value of active power (the reference value) for the No PSS and Q-PSS cases. On the other hand, the HVDC inverter delivers a compensating signal to damp oscillations when the P-PSS and PQ-PSS are included. Moreover, it can be observed that this response reaches the

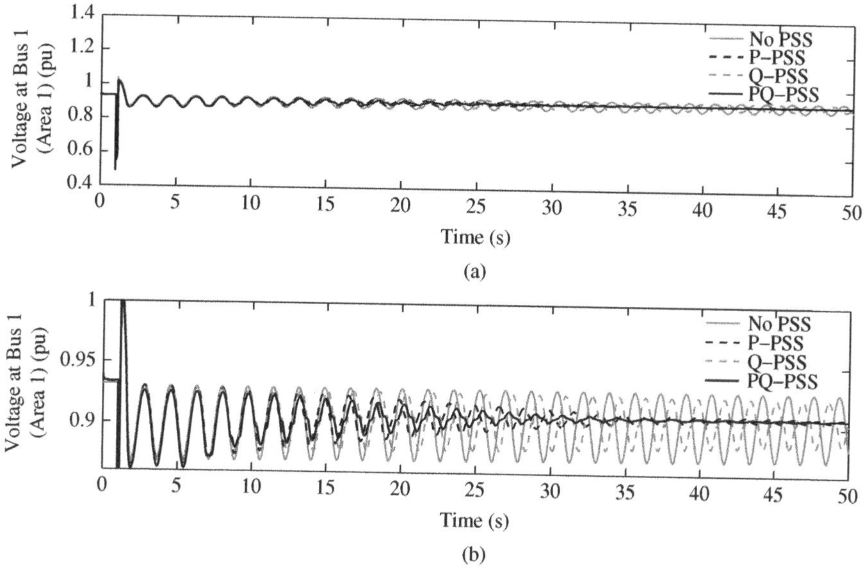

Figure 19.8 Voltage magnitude representation (including a closer view of it) at Bus 1 for different PSS schemes. (a) Voltage magnitude at Bus 1. (b) Zoom of voltage magnitude at Bus 1.

saturation limits for both cases. It can be noticed that the delivered active power enters into the linear zone in the PQ-PSS case sooner than when using a P-PSS. This is a consequence of the interaction with the reactive power loop. Figure 19.7b shows the reactive power provided by the VSC-HVDC system. As occurs with the active power, the reactive power of the VSC-HVDC link remains at its reference value (0 MVAr) for the cases No PSS and P-PSS. The Q-PSS and PQ-PSS react against the oscillation, but the reactive power reaches saturation limits. However, the Q-PSS remains saturated during all the simulation time because of its lower effect on the oscillatory mode, which requires more reactive power feeding from the converter.

Figures 19.8a and 19.8b show the voltage magnitude and a zoom of it at Bus 1, respectively. As happens with the active power, the PQ-PSS provides a better damping capability than the other controllers. The P-PSS also exhibits promising results.

The voltage magnitude and the corresponding zoom at point of connection of the VSC–HVDC link with the grid are presented in Figures 19.9a and 19.9b, respectively. It can be observed that the PQ-PSS and P-PSS rapidly stabilize the oscillation as in the voltage magnitude at Bus 1 shown previously. Since the voltage at the grid side converter bus is the input signal for the PSS controllers, a rapid stabilization of this signal implies less actuation time of the controllers.

19.3.2 Damping of SSR Using VSC-HVDC Links

SSR has been an extensive topic of research over the last decades. Countermeasures to this problem range from limiting series compensation to safe levels through system

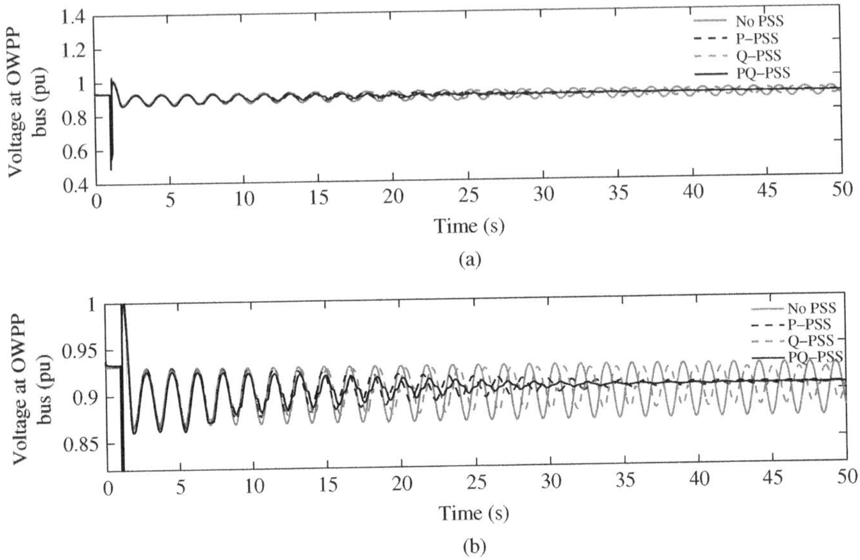

Figure 19.9 Voltage magnitude representation (including a closer view of it) at onshore converter connection Bus for different PSS schemes. (a) Voltage magnitude at onshore HVDC converter. (b) Zoom of voltage magnitude at onshore HVDC converter.

planning, use of filters, and damping schemes, as well as by employing FACTS devices [24–26]. The thyristor-controlled series compensator (TCSC), an established member of the FACTS technology and the electronically controlled counterpart of fixed series capacitors, has the capability to provide series compensation by reducing the electrical length of a transmission line [27]. Through an adequate control of the firing angle of its thyristors, the effective reactance of the TCSC can be adjusted—in turn, affecting the natural frequency of the electrical system. In addition to providing series compensation to maximize power transfer, an appropriate choice of the TCSC's boost factor allows inherent SSR damping: The equivalent impedance at specific subsynchronous frequencies may be designed to exhibit an inductive behavior with a strong resistive component [28, 29]. An active damping control loop specifically tailored to mitigate SSR can also be included [30].

The use of TCSC installations as an alternative to conventional series compensation comes at the expense of requiring additional primary plant. This involves a high capital cost and a large space (as compared to the use of fixed series capacitors). If system contingencies in a series-compensated transmission system exist and, as a consequence, SSR may result, an expensive TCSC solution will not be accepted as investment. Instead, fixed series capacitors could be used to provide the desired level of compensation, with SSR being damped through a different approach by making use of a power system plant already in place in the transmission network [14, 31].

Although line commutated HVDC is a well-established technology, the emergence of VSC-HVDC provides a strong alternative for bulk power transmission [32,33]. Even though the primary design objective of a VSC-HVDC link is to ensure that active power control is achieved along with reactive power support for the grid, it can provide SSR mitigation once an auxiliary control loop is implemented together with the primary control scheme [14].

19.3.2.1 Motivation Reduction of greenhouse gas emissions in Great Britain (GB) by means of grid integration of renewable sources comes at the expense of key operational challenges for the transmission network [13]. The prime candidate to supply renewable energy is wind power—particularly offshore, where wind is stronger and more consistent. Nevertheless, the transmission corridor between England and Scotland will not be adequate for power transmission with such an increased penetration of wind energy. Onshore and offshore infrastructure reinforcements have been proposed, as shown in Figure 19.10. These reinforcements of the transmission network have been split as a compromise between the difficulty in building new onshore overhead lines and the risks of using technologies new to the transmission system. Inland reinforcement will include fixed series capacitors, providing series compensation within the Anglo-Scottish inter-tie to increase the onshore power transmission capability [12], while offshore infrastructure support will be accomplished through submarine HVDC links at a rating of 2.1 GW [13]. The Western link is planned to connect Hunterston in Scotland with Deeside in Wales. The Eastern link will connect Peterhead in Scotland with Hawthorne Pit in England. It is reasonable to assume that at least one of these links would be a VSC-based link [14].

As described in Section 19.2.2, the use of series compensation demands caution as it can lead to SSR. In general, series compensation has been predominantly employed within regional and national grids featuring long transmission lines and relatively low fault levels [15]. However, the GB network is tightly meshed [12] and employs controllable devices (e.g., quadrature boosters), so the effect of series compensation should be evaluated within this context [14, 31].

Recognizing the possible risks of introducing series compensation in the GB transmission grid, two approaches have been discussed to prevent SSR [12]. The first alternative is a conservative one, restricting compensation levels below 35%. The second approach is based on TCSCs, where SSR damping is achieved through an active damping controller. An additional third possibility—not previously considered within the GB context—makes use of the VSC–HVDC link that has already been proposed [14]. This way, offshore infrastructure reinforcement is accomplished, with SSR damping provided by an auxiliary control loop.

19.3.2.2 System Under Study A simplified model of the GB network featuring planned infrastructure reinforcements is shown in Figure 19.11. Although a simplified case study of the national network, it represents a starting point for more detailed studies. A clear understanding of the SSR phenomenon is achieved by reducing system complexity while keeping an appropriate weighting of the size of generation [31]. The system splits the GB mainland system into three major generation areas: England/Wales, Southern Scotland, and Northern Scotland. The rating of the England/Wales generator is 21,000 MVA, with the Scottish generators having a considerably smaller capacity of 2800 and 2400 MVA respectively. A series capacitor has been included between Buses 4 and 5 to represent the series compensation of

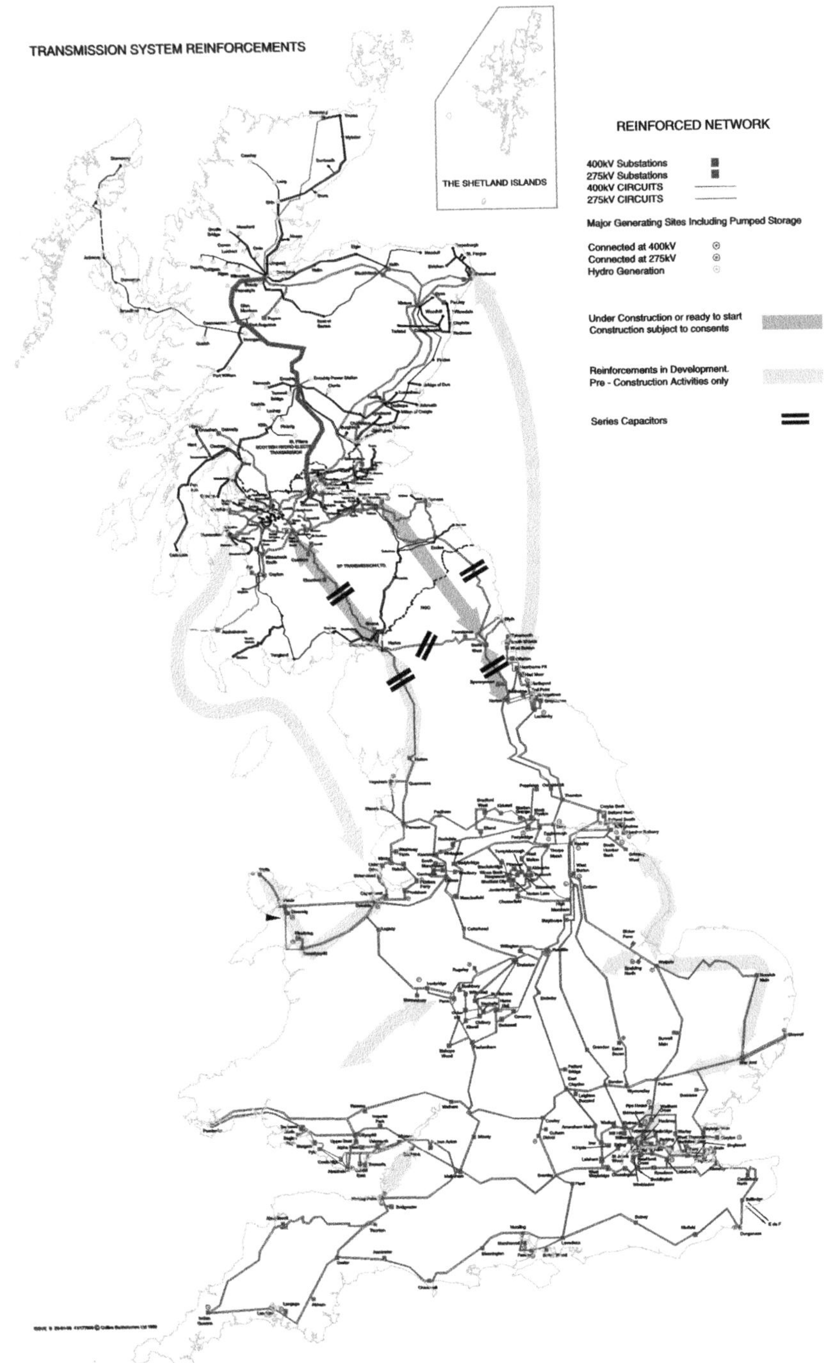

Figure 19.10 Map indicating approximate locations of a major project being progressed in relation to the 2012 updated ENSG "2020 Vision" reports, with the location of the series capacitors shown [13].

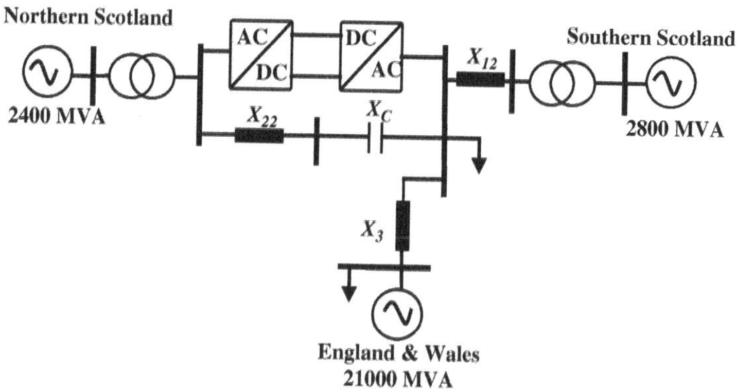

Figure 19.11 Three-machine network with onshore and offshore reinforcements.

transmission lines between England and Scotland. In order to assess the occurrence of SSR (in particular, TIs), the shaft of the Northern Scotland generator was modeled as a multi-mass system [14, 31], as shown in Figure 19.12. For the purpose of this example, a VSC-HVDC link is connected between Buses 4 and 5 to provide bulk power transfer. To accomplish this objective and to fulfill the additional goal of damping SSR, a hybrid control scheme has been designed.

The three-machine network was represented by a set of differential and algebraic equations, linearized around an operating point. The system operating point was obtained following a power flow solution, with initial conditions calculations as detailed in [15]. This way, the system may be represented in state-space form as described in [31]. As shown by Figure 19.13, the system is characterized by four unstable torsional modes (TM1–TM4) associated with the shaft of the Northern Scotland generator which may be excited at different series compensation levels. The ranges of series compensation where torsional instability arises are related to a network subsynchronous mode (termed SUB in the figure). The frequency of SUB decreases with an increase in series compensation, with instability occurring when such frequency is in the neighborhood of a torsional mode frequency, as illustrated by the real part of the eigenvalues of TM1–TM4.

19.3.2.3 Control Scheme Featuring Primary Control and an Auxiliary SSR Damping Loop

The overall control scheme has the following objectives: (i) to

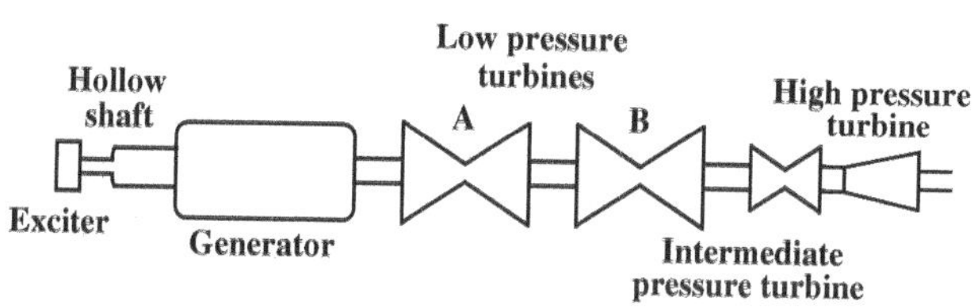

Figure 19.12 Shaft model of the Northern Scotland generator represented as a multi-mass system (Copyright ©The Institution of Engineering and Technology.)

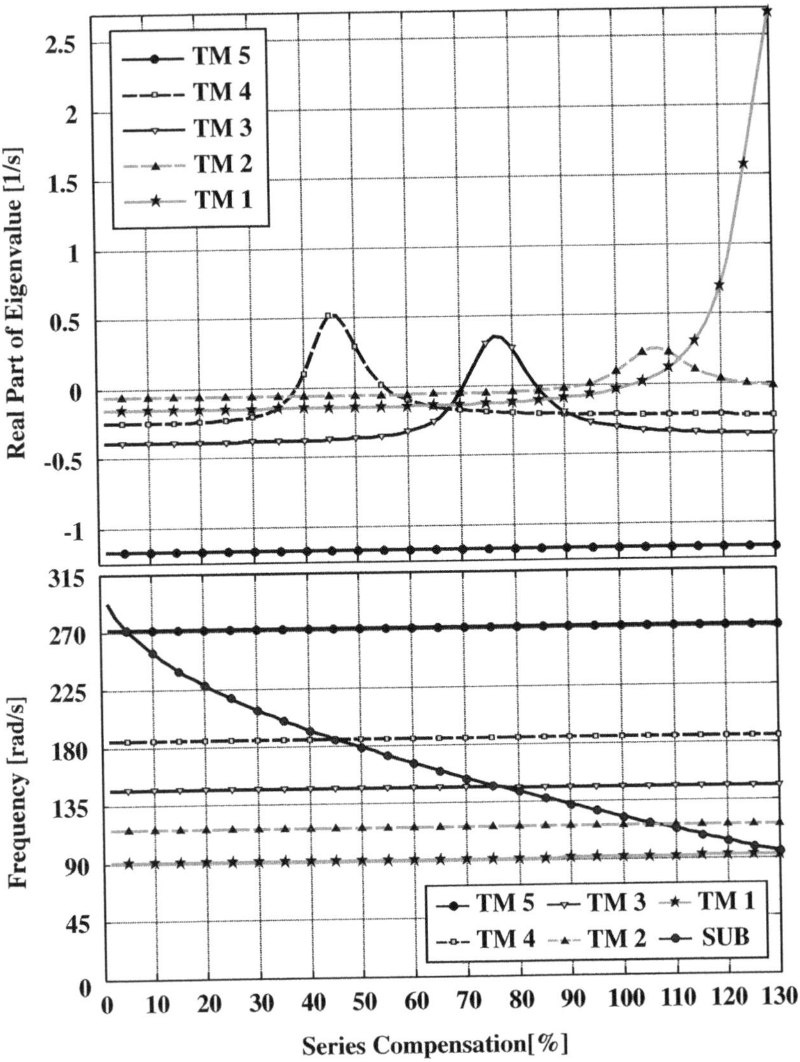

Figure 19.13 Stability of the torsional modes in terms of series compensation level and relationship with the relevant subsynchronous mode frequency [31]. (Copyright © The Institute of Engineering and Technology.)

control reactive power and to regulate the DC link voltage, while offering control over the magnitude and direction of active power flow, and (ii) to damp SSR in the series-compensated AC system after it has been affected by a fault or a system disturbance.

The approach behind control objective (i) is shown in Figure 19.14. Further details on its performance can be found in reference 14. In order to mitigate SSR and to comply with control objective (ii), an auxiliary damping loop has been designed and is also shown in the figure. In simple terms, this control structure produces a damping

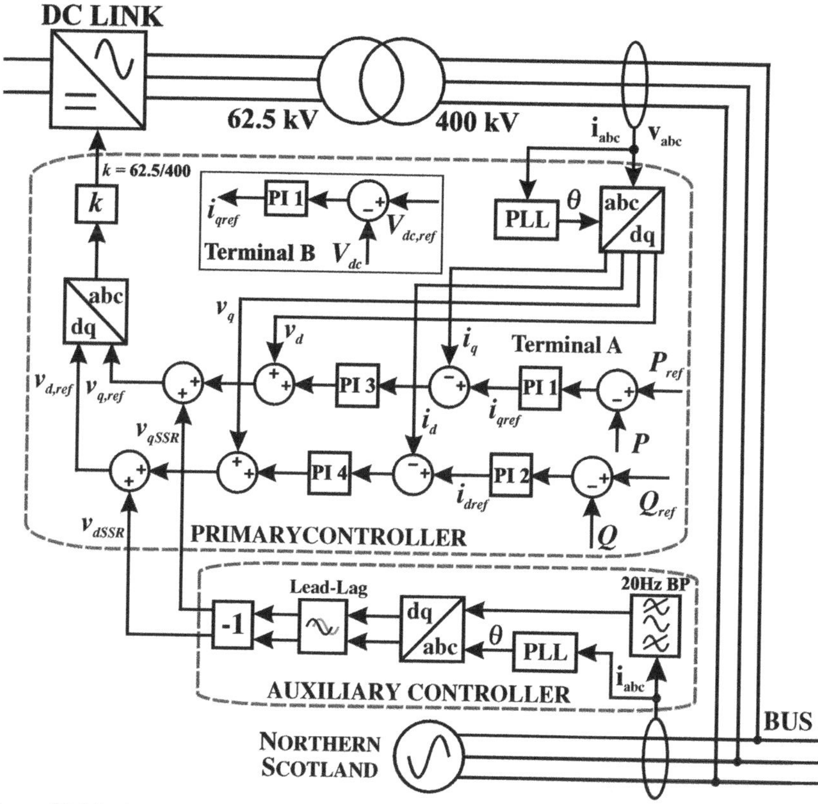

Figure 19.14 VSC-HVDC link control scheme to achieve power transmission and SSR damping [14].

signal by isolating the signal with the target frequency (i.e., the subsynchronous frequency of the electrical network) and injecting it back to the AC system with a 180-degree displacement to counteract SSR. In general, such a frequency can be identified using an estimation algorithm. More details about the auxiliary controller and all system parameters can be found in reference 14.

19.3.2.4 Simulation Results In order to evaluate the effectiveness of the proposed control scheme, a comparison between the system with and without the auxiliary SSR damping loop is carried out. Time domain transient simulations of the system were performed on PSCAD/EMTDC. The simulation starts in steady state with a 30% series compensation. At 2 s, the effective value of series compensation is increased to 40%, with results shown in Figure 19.15a. For the system featuring the primary control only, SSR is developed as no damping countermeasures have yet been included.

This is consistent with the eigenvalue analysis depicted in Figure 19.13. Moreover, the effect of the DC link dynamics does not influence the occurrence of SSR. However, if the damping controller embedded in the VSC-HVDC link is activated, it can be clearly seen that SSR is effectively mitigated (Figure 19.15b).

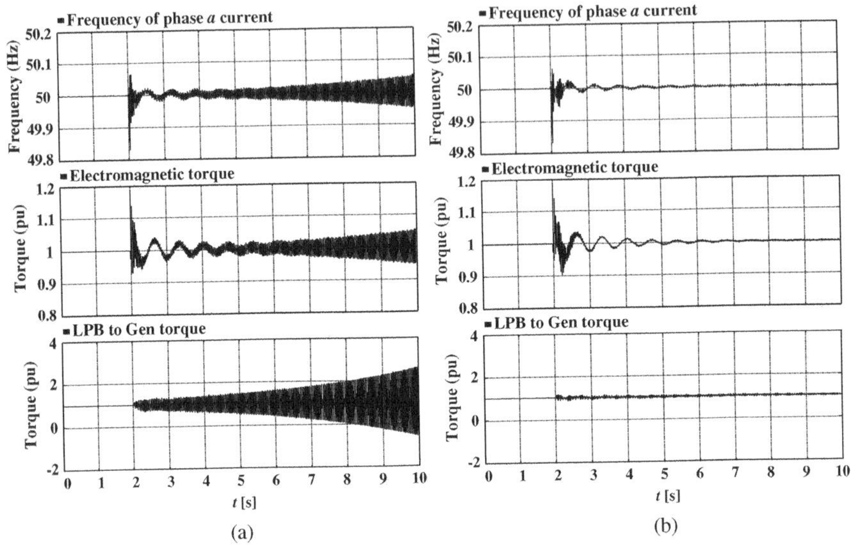

Figure 19.15 Simulation results featuring a VSC-HVCD link with (a) primary control only and (b) primary and auxiliary controllers: frequency of phase *a* current, electromagnetic torque, and torque (in the multi-mass shaft between low pressure turbine B and the generator) of the Norther Scotland generator.

19.3.2.5 Real-Time Experimental Results A real-time hardware-in-the-loop real-time digital simulator (RTDS)-HVDC test rig has been assembled to implement and validate the proposed damping scheme on an experimental platform. A three-terminal VSC-HVDC test rig is employed. It consists of three 2-kVA VSCs controlled through a dSPACE unit. The RTDS models the AC network of Figure 19.11 without the HVDC link. The three terminals of the VSC-HVDC test rig are set up independently as a DC voltage source, an amplifying converter, and a damping converter [14]. Unlike the simulation, there is no ordered power flow. The damping converter uses both active and reactive power to damp subsynchronous oscillations.

Time domain experiments are carried out in the multi-platform network. During the first experiment, the damping converter in the VSC-HVDC rig is not activated. As a consequence, the system develops SSR, which is exhibited through a rising AC system frequency and generator shaft torques, as shown in figure 19.16a. These show agreement with Figure 19.15a, the simulation undamped case. The experiment is repeated with the damping converter on the VSC-HVDC test rig activated, with results presented in Figure 19.16b. Following the change of series compensation from 30% to 40% at 2 s, the damping current injected into the AC system prevents SSR from developing. This is shown by the stability of the AC frequency and torques—consistent with the simulation results of Figure 19.15b.

Figure 19.17 shows a comparison between the simulation and experimental results. In Figure 19.17a, the undamped cases are compared, with both simulation and real-time results showing good agreement. Similarly, Figure 19.17b addresses the damped cases; both show a good agreement.

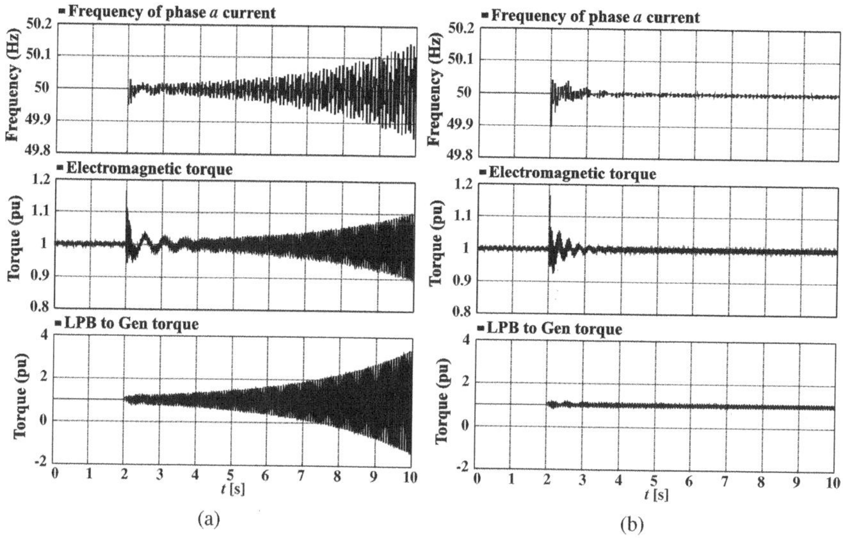

Figure 19.16 Real-time experimental results in an RTDS-HVDC test rig with (a) primary control only and (b) primary and auxiliary controllers: frequency of phase *a* current, electromagnetic torque, and torque (in the multi-mass shaft between low pressure turbine B and the generator) of the Norther Scotland generator.

The results presented in this application example show that the HVDC technology may contribute to SSR damping upon occurrence.

19.4 CONCLUSIONS

This chapter has demonstrated that VSC-HVDC systems (HVDC grids and transmission links) are capable to enhance power system stability, providing extra regulation. This fact has been evaluated for two different application cases focusing on the HVDC-link, but results obtained are fully generalizable to the rest of VSC-HVDC systems.

The LFOs damping capability of OWPP connected onshore by means of an HVDC link has been analyzed and compared under different PSS schemes. These

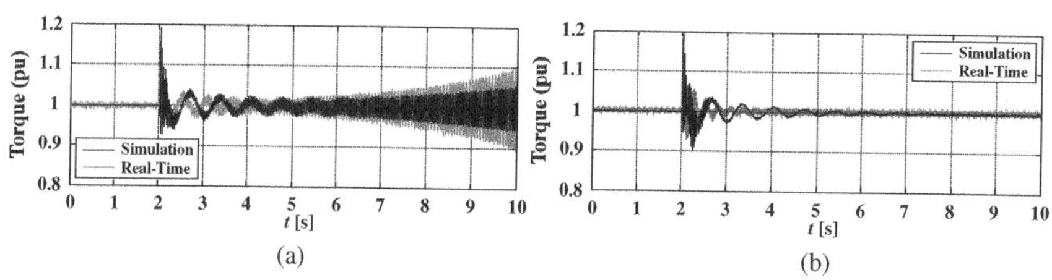

Figure 19.17 Comparison of simulation and real-time results. Northern Scotland electromagnetic torque with damping control: (a) not active, (b) active [31].

include the use of active power (P-PSS), reactive power (Q-PSS), and simultaneous active and reactive power (PQ-PSS) compensation signals. The best damping behavior has been observed when the PQ-PSS scheme is employed. A controller acting on the active power only has also exhibited a good performance. Results show that PSS controllers embedded in an HVDC link offer promising properties to damp inter-area oscillations.

The adverse effects in the form of SSR introduced by series compensation in a three-machine network, resembling the operation of the mainland GB system in the 2020, have been examined. Moreover, SSR mitigation has been achieved through an appropriate control strategy of a VSC-HVDC link. Such scheme combines the conventional active and reactive power control of an HVDC link and an auxiliary damping controller to damp SSR. This has been done both by simulation and through a real-time experiment. Simulation and experimental results showed good agreement.

As HVDC converters are utilized more in the future hybrid AC/DC power system, their flexibility can be used to damp power system oscillations in adjacent AC transmission systems. This may be cost effective by eliminating the requirement for additional power electronics equipment (such as TCSCs).

REFERENCES

1. T. Ackermann, *Wind Power in Power Systems*. Hoboken, NJ: Wiley, 2005.
2. M. Tsili, C. Patsiouras, and S. Papathanassiou, "Grid code requirements for large wind farms: A review of technical regulations and available wind turbine technologies," in *Proceedings of the European Wind Energy Conference & Exhibition (EWEC'08)*, 2008.
3. O. Gomis-Bellmunt, A. Junyent-Ferre, A. Sumper, and J. Bergas-Jane, "Ride-through control of a doubly fed induction generator under unbalanced voltage sags," *IEEE Transactions on Energy Conversion*, vol. 23, pp. 1036–1045, 2008.
4. M. Stiebler, *Wind Energy Systems for Electric Power Generation*. Berlin: Springer-Verlag, 2008.
5. O. Anaya-Lara, N. Jenkins, J. Ekanayake, P. Cartwright, and M. Hughes, *Wind Energy Generation: Modelling and Control*. Hoboken, NJ: Wiley, 2009.
6. M. de Industria Turismo y Comercio: Secretaria de Estado de Energía, "Instalaciones conectadas a la Red de Transporte y equipo generador: requisitos minimos de diseño, equipamiento, funcionamiento, puesta en servicio y seguridad," Red Eléctrica, Technical Report, Oct. 2008.
7. ENTSO-E, "Entso-e network code for requirements for grid connection applicable to all generators," March 2013.
8. ———, "ENTSO-E draft network code on high voltage direct current connection and DC–connected power park modules," Nov. 2013.
9. J. L. Domínguez-García, O. Gomis-Bellmunt, F. Bianchi, and A. Sumper, "Power oscillation damping supported by wind power: A review," *Renewable and Sustainable Energy Reviews*, vol. 16, no. 7, pp. 4994–5006, 2012.
10. A. Jha, *Wind Turbine Technology*. Boca Raton, FL: CRC Press, 2011.
11. R. Eriksson and L. Söder, "Wide area measurement system-based subspace identification for obtaining linear models to centrally coordinate controllable devices," *IEEE Transactions on Power Delivery*, vol. 26, pp. 988–997, 2011.
12. C. Hor, J. Finn, G. Thumm, and S. Mortimer, "Introducing series compensation in the uk transmission network," in *Proceedings of IET ACDC*, 2010, pp. 1–5.
13. Electricity Networks Strategy Group (ENSG), Our electricity transmission network: A vision for 2020.ENSG,2012.

14. L. Livermore, C. E. Ugalde-Loo, Q. Mu, J. Liang, J. B. Ekanayake, and N. Jenkins, "Damping of subsynchronous resonance using a voltage source converter-based high-voltage direct-current link in a series-compensated Great Britain transmission network," *IET Generation, Transmission & Distribution*, vol. 8, no. 3, pp. 542–551, 2014.
15. P. Kundur, *Power System Stability and Control*. New York: McGraw-Hill, 1994.
16. IEEE/CIGRE Join Task Force on Stability Terms And Definitions, "Definition and classification of power system stability," *IEEE Transactions on Power Systems*, vol. Vol 19, pp. 1387–1388, 2004.
17. Subsynchronous Resonance Working Group, "Reader's guide to subsynchronous resonance," *IEEE Transactions on Power Systems*, vol. 7, no. 1, pp. 150–157, 1992.
18. P. M. Anderson, B. L. Agrawal, and J. E. Van Ness, *Subsynchronous Resonance in Power Systems*. New York: IEEE Press, 1990.
19. J. W. Ballance and S. Goldberg, "Subsynchronous resonance in series compensated transmission lines," *IEEE Transactions on Power Apparatus and Systems*, vol. 92, no. 5, pp. 1649–1658, 1973.
20. R. G. Farmer, A. L. Schwalb, and E. Katz, "Navajo project report on subsynchronous resonance analysis and solutions," *IEEE Transactions on Power Apparatus and Systems*, vol. 96, no. 4, pp. 1226–1232, 1977.
21. G. Tsourakis, B. Nomikos, and C. Vournas, "Contribution of doubly fed wind generators to oscillation damping," *IEEE Transactions on Energy Conversion*, vol. 24, pp. 783–791, 2009.
22. M. Klein, G. Rogers, and P. Kundur, "A fundamental study of inter-area oscillations in power systems," *IEEE Transactions on Power Systems*, vol. 6, no. 3, pp. 914–921, 1991.
23. J. L. Domínguez-García, F. Bianchi, and O. Gomis-Bellmunt, "Analysis of the damping contribution of power system stabilisers driving wind power plants," *Wind Energy*, vol. 17, no. 2, pp. 267–278, 2014.
24. IEEE Subsynchronous Resonance Working Group, "Countermeasures to subsynchronous resonance problems," *IEEE Transactions on Power Apparatus and Systems*, vol. 99, no. 5, pp. 1810–1818, 1980.
25. K. V. Patil, J. Senthil, J. Jiang, and R. M. Mathur, "Application of STATCOM for damping of torsional oscillations in series compensated AC systems," *IEEE Transactions on Energy Conversion*, vol. 13, no. 3, pp. 237–243, 1998.
26. Hammad, A. E. and El-Sadek, M., "Application of a thyristor controlled var compensator for damping subsynchronous oscillations in power systems," *IEEE Transactions on Power Apparatus and Systems*, vol. 103, no. 1, pp. 198–212, 1984.
27. E. Acha, C. Fuerte-Esquivel, H. Ambriz-Perez, and C. Angeles-Camacho, *FACTS. Modelling and Simulation in Power Networks*. Hoboken, NJ: John Wiley & Sons, 2004.
28. W. Zhu, R. Spee, R. R. Mohler, W. A. Mittelstadt, and D. Maratukulam, "An EMTP study of SSR mitigation using the thyristor controlled series capacitor," *IEEE Transactions on Power Delivery*, vol. 10, no. 3, pp. 1479–1485, 1995.
29. L. A. S. Pilotto, A. Bianco, W. F. Long, and A. A. Edris, "Impact of TCSC control control methodologies on subsynchronous oscillations," *IEEE Transactions on Power Delivery*, vol. 18, no. 1, pp. 243–252, 2003.
30. R. A. Hedin, S. Weiss, D. Torgerson, and L. E. Eilts, "SSR characteristics of alternative types of series compensation schemes," *IEEE Transactions on Power Systems*, vol. 10, no. 2, pp. 845–852, 1995.
31. C. E. Ugalde-Loo, J. B. Ekanayake, and N. Jenkins, "Subsynchronous resonance in a series-compensated Great Britain transmission network," *IET Generation, Transmission & Distribution*, vol. 7, no. 3, pp. 209–217, 2013.
32. P. Bresesti, W. L. Kling, R. L. Hendriks, and R. Vailati, "HVDC connection of offshore wind farms to the transmission system," *IEEE Transactions on Energy Conversion*, vol. 22, no. 1, pp. 37–43, 2007.
33. N. Flourentzou, V. G. Agelidis, and G. D. Demetriades, "VSC-based HVDC power transmission systems: An overview," *IEEE Transactions on Power Electronics*, vol. 24, no. 3, pp. 592–602, 2009.

CHAPTER 20

OPTIMAL DROOP CONTROL OF MULTI-TERMINAL VSC-HVDC GRIDS

Fernando D. Bianchi[1] *and Eduardo Prieto-Araujo*[2]

[1]Research Group Wind Energy, Electrical Engineering Department, Catalonia Institute for Energy Research (IREC), Barcelona, Spain

[2]Research Group CITCEA, Electrical Engineering Department, Technical University of Catalonia (UPC), Barcelona, Spain

20.1 INTRODUCTION

The stability of electrical networks including HVDC grids has been analyzed in the literature, see for example, references 1 and 2. These studies are mainly focused on HVDC links and their effects on the stability of the AC networks where they are inserted. They also analyze their possible contribution to improve AC system stability. However, these results are not directly applicable to HVDC grids in the context of offshore wind farms. HVDC networks connecting offshore wind power plants to land AC grids exhibit different characteristics, and in particular they offer new control alternatives [3]. In wind power applications, an HVDC transmission may result in a MTDC network in which its terminals are connected to voltage source converters (VSCs) [4, 5]. These converters permit the transfer of power from wind farms to AC grids where the loads are connected.

The aim of control in these MTDC networks is to ensure a proper power transmission, which implies to maintain the DC voltage almost constant and also to attenuate the effects of changes in the incoming power and faults in the AC grids [3, 6]. Some DC grid management strategies based on coordinated closed-loop DC voltage control and DC droop characteristics were proposed and simulated in reference 7. In reference 3, the authors address the modeling and simulation of multi-terminal VSC-HVDC transmissions for offshore wind power. However, there are just a few results on stability analyses and systematic control design procedures for multi-terminal VSC-HVDC grids connecting offshore wind farms to AC systems.

Generally, in these MTDC grids, a decentralized proportional control, the so-called droop control, is employed to fulfill the specifications on the DC voltage

HVDC Grids: For Offshore and Supergrid of the Future, First Edition. By Dirk Van Hertem, Oriol Gomis-Bellmunt, and Jun Liang.

behavior. In reference 8, a stability analysis based on multivariable frequency response is presented for different operating conditions. In reference 9, the control design is expressed as an optimization procedure using Lyapunov stability theory to provide droop gains ensuring stability in several operating conditions. This chapter presents more extensive discussion of these results.

20.2 CONTROL OF MULTI-TERMINAL VSC-HVDC GRIDS

A generic configuration of N_{in} offshore wind farms and N_{out} terrestrial AC grids connected by a multi-terminal VSC-HVDC is depicted in Figure 20.1. The aim of the system is to transfer the energy captured by the offshore wind farms to the onshore loads connected to land AC grids. The voltage sources converters (VSC) on both sides of the MTDC network regulate the injected or extracted power. The balance between the injected power and the extracted power is reflected in the DC voltage. Therefore, an affective way to regulate the power transmission through the multi-terminal VSC-HVDC network is by maintaining the DC voltages within a given range.

The complete system in Figure 20.1, including the AC grids corresponding to the wind farms and land installations, may be extremely complex. However, useful conclusions can be drawn from rather simple mathematical descriptions. This kind of model is also more adequate to design DC voltage controllers, thus ensuring a proper power transmission. In simplified models, the submarine cables are represented by circuits of concentrated parameters, and the VSC converters are represented by currents sources obeying a given static characteristic [8] (see also Chapter 16).

The control of the VSCs consists of two feedback loops, an inner current loop and an outer loop regulating the DC voltage. Therefore, the latter is the control that governs the power transmission between wind farms and terrestrial grids. This

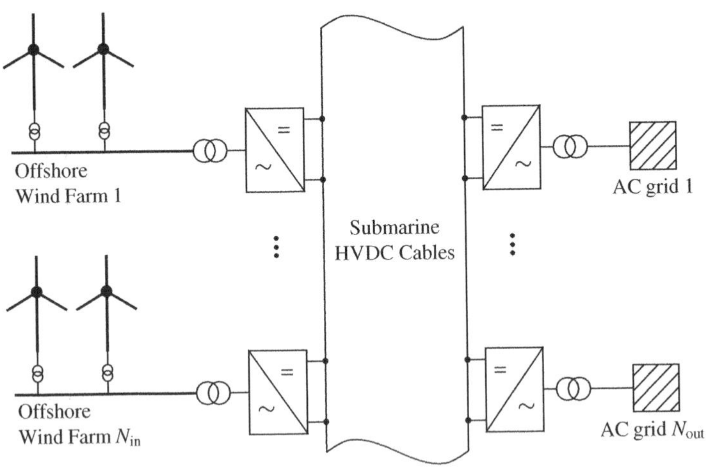

Figure 20.1 Schematic representation of a generic HVDC multi-terminal network.

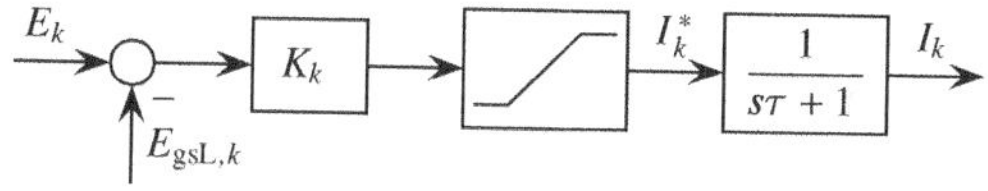

Figure 20.2 Simplified model of the VSC at each terminal of the MTDC grid.

control should be decentralized; that is, the outer loop of each VSC should only include variables measured at the terminal of the same converter. In MTDC grids for offshore wind power applications, the converters are located at great distances from each other. Any centralized scheme requires relying on communications to send voltages measured at distant converters, which are not sufficiently reliable to be used in the DC voltage control.

Commonly, the DC voltage control of MTDC for wind power integration is a proportional law, known as droop control. This control law at a terminal k is given by the expression

$$I_k^* = K_k(E_k - E_{\mathrm{gsL},k})$$

where E_k is the DC voltage and E_{gsL} is a voltage to achieve a desire power sharing among terminals. The current I_k^* is the reference for the inner loop of the VSC. The simplified model for a GSC can be seen in Figure 20.2. The first-order system describes the current loop. The saturation limit is given by

$$I_{\mathrm{gsH}} = \frac{P_{\max}}{E} = \frac{\sqrt{3}V_{AC}\cos\theta}{E} I_{AC}^{\max} \tag{20.1}$$

where V_{AC} and $I_{AC}^{\max}$ are the AC voltage and the AC current, respectively. The VSCs on the grid side and the wind farm side have different operation modes depending on the voltage E_k and the current I_k.

As a consequence of the converter limitations, the GSCs present two operating modes. The GSCs operates in normal mode when the DC voltage is below E_{gsH}. The objective in this operating mode is to regulate the DC voltage at the converter terminals. The current limitation mode corresponds to the situation when the current reaches the limit I_{gsH} and the DC voltage exceeds the limit E_{gsH}. The static current–voltage curve for a generic node k is illustrated in Figure 20.3 and obeys the following expression

$$I_k^* = \begin{cases} K_k(E_k - E_{\mathrm{gsL},k}), & E_k < E_{\mathrm{gsH},k} \\ P_{\max,k}/E_k, & \text{otherwise} \end{cases} \tag{20.2}$$

with $k = 1, \dots, N_{\mathrm{out}}$. The shaded area indicates the range $[\underline{E}_{\mathrm{gs}}, \overline{E}_{\mathrm{gs}}]$ of the acceptable DC voltage values. Commonly, converters on the grid side work in normal mode. The current limitation mode is only reached when the power transfer capability of the converter falls as a consequence of severe voltage faults in the AC grid. In this circumstance, the AC voltage (V_{AC}) drops and thus the current limit I_{gsH} also decreases. From Eq. (20.1), it is clear that a voltage sag of 50% causes a reduction of the current limit I_{gsH} by 50%, which in turn reduces the normal operation area. This situation is also illustrated in Figure 20.3 (thinner line).

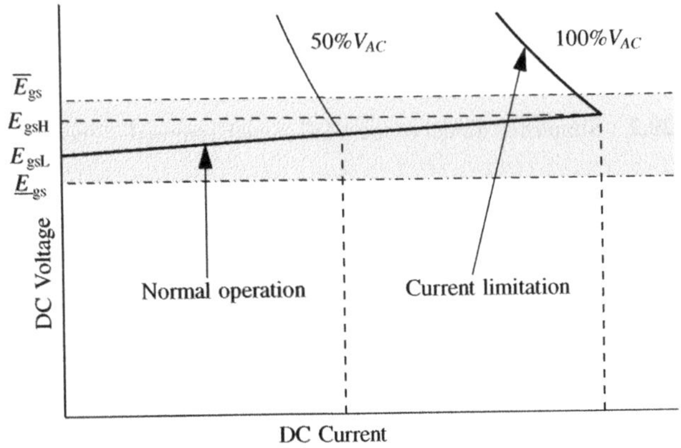

Figure 20.3 Static current–voltage characteristic of a GSC. The dashed line shows the characteristic under a voltage sag of 50%.

The WFCs present three operating modes as can be seen in Figure 20.4 [4]. In normal operation mode, the objective is to inject the power coming from the wind farm into the HVDC grid. In this circumstance, the control of the AC side of the WFC maintains the voltage and frequency constant and it can be assumed that the VSC on the DC behaves as a current source. The hyperbolic sector of the curve corresponds to normal operation. When the DC voltage exceeds the limit E_{wfL}, WFC enters droop mode with the aim of keeping the DC voltage within limits. Commonly, WFCs enter droop mode when the power transmission capability of the GSCs have been reduced due to a voltage fault in the AC grids. The power reduction in this case is performed by changing the power reference in the converter controlling the AC voltage in the wind farm grid. WFCs have also a current limitation mode for cases when the current

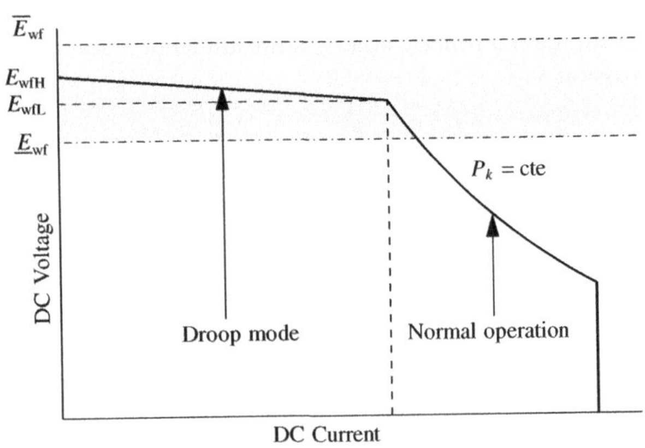

Figure 20.4 Static current–voltage characteristic of a WFC.

reaches the maximum value I_{wfH}. The current–voltage characteristic for a generic node k is summarized as follows:

$$I_k^* = \begin{cases} K_k(E_{\mathrm{wfH},k} - E_k), & E_{\mathrm{wfL},k} < E_k < E_{\mathrm{wfH},k} \\ P_k/E_k, & E_k \le E_{\mathrm{wfL},k} \;\&\; I_k < I_{\mathrm{wfH},k} \\ I_{\mathrm{wfH},k}, & I_k \ge I_{\mathrm{wfH},k} \end{cases} \tag{20.3}$$

with $k = 1, \ldots, N_{\mathrm{in}}$. The limit of the droop mode ($E_{\mathrm{wfL},k}$) depends on the power P_k delivered by the wind farm. The shaded area indicates the range where the DC voltages must remains, that is, $E_k \in [\underline{E}_{\mathrm{wf}}, \overline{E}_{\mathrm{wf}}]$.

The operating mode of every VSC is determined by the load state and the DC voltage level at each terminal of the MTDC. In normal operation, the WFCs are controlled to establish the AC voltage of the offshore wind farms. The AC voltage and frequency must be maintained nearly constant in order to ensure the maximum power injection into the MTDC grid. The GSCs are responsible for maintaining the DC voltages within limits using the droop control laws. This implies that only the controllers of the GSCs are active. However, this control configuration changes under different operating conditions. For example, under a voltage fault on the AC grids, the roles of the converters may be reversed. In this circumstance, the current in the GSCs may reach the converter limits causing an increment of the DC voltage. As a consequence, the WFCs must now keep the DC voltage within limits whereas the droop controls in some or all GSCs remain inactive.

Owing to the continuous changes of operating mode depending on the particular values of voltage and current, the decentralized control law must also change in order to adapt itself to different operating conditions. If the converters obey the static curves in Figures 20.3 and 20.4, each VSC will find the operating mode that maintains the DC voltage within the acceptable limits. That is, the controller adaptation is done automatically if the converters obey the static characteristics (20.2) or (20.3). Nevertheless, the droop gains K_k must be designed in order to ensure stability and performance under any possible operating condition and also during the transitions among them. That means that the decentralized control law must ensure stability and performance for a time-varying system.

Notice that instead of defining the droop control as

$$K = \mathrm{diag}(K_1, \ldots, K_{(N_{\mathrm{in}}+N_{\mathrm{out}})}) \tag{20.4}$$

they can also be defined as

$$K = K_G \cdot \mathrm{diag}(q_1, \ldots, q_{(N_{\mathrm{in}}+N_{\mathrm{out}})})$$

where the parameters q_k are constants obtained from a steady-state study [7]. These constants are associated with the resistance values of the line and the amount of power incoming or outgoing from each terminal. These constants are positive in the case of power output nodes and negative in the case of power input nodes. This approach reduces the degrees of freedom in the droop design (only the gain K_G needs to be determined), which has to ensure stability besides an adequate power sharing. For droop control defined as (20.4), the power sharing can be achieved with a proper selection of the references $E_{\mathrm{gsL},k}$ and E_{wfH}. That will be the approach adopted here.

20.3 TIME-VARYING DESCRIPTION FOR DROOP CONTROL DESIGN

As explained in Chapter 16, following a systematic procedure, it is possible to find a state-space description as

$$\mathcal{M}: \begin{cases} \dot{x} = Ax + B_q q \\ p = C_p x \end{cases} \tag{20.5}$$

where the state vector is

$$x = [E_1, \ldots, E_{(N_{\text{in}}+N_{\text{out}})}, E_{(N_{\text{in}}+N_{\text{out}}+1)}, \ldots, E_{(N_{\text{in}}+N_{\text{out}}+N_{\text{int}})}, I_{L_1}, \ldots, I_{L_{N_{\text{br}}}}]^T \tag{20.6}$$

the input

$$q = [I_1, \ldots, I_{N_{\text{in}}}, I_{(N_{\text{in}}+1)}, \ldots, I_{(N_{\text{in}}+N_{\text{out}})}]^T$$

are the currents from the VSCs, the output

$$p = [E_1, \ldots, E_{N_{\text{in}}}, E_{(N_{\text{in}}+1)}, \ldots, E_{(N_{\text{in}}+N_{\text{out}})}]^T$$

are the DC voltages at each power node, and

$$A = \begin{bmatrix} 0 & 0 & A_{13} \\ 0 & 0 & A_{23} \\ A_{31} & A_{32} & A_{33} \end{bmatrix}, \qquad B_q = \begin{bmatrix} B_1 \\ 0 \\ 0 \end{bmatrix}, \qquad C_p = \begin{bmatrix} I \\ 0 \\ 0 \end{bmatrix}^T$$

with $A_{13}, A_{23}, A_{31}, A_{32}, A_{33}$, and B_1 matrices obtained from the electrical equations corresponding to the HVDC network.

In order to find a control design procedure that fulfills the previously discussed requirements, a time-varying representation of the multi-terminal VSC-HVDC grid is proposed. First, the nonlinear relation between the current and the voltage at each power node is represented as a sector bounded uncertainty. Hence, the incoming/outgoing current at each power node is given by

$$I_k = \alpha_k(w - \delta_k E_k)$$

where δ_k is a time-varying parameter taking values in the sector $[\underline{\delta}_k, K_k]$,

$$w = [E_{\text{wfH},1}, \ldots, E_{\text{wfH},N_{\text{in}}}, E_{\text{gsL},1}, \ldots, E_{\text{gsL},N_{\text{out}}}]^T$$

and α_k is 1 in the case of a power input node and -1 in the case of a power output node. The lower limits $\underline{\delta}_k$ are given by $P_{\text{min},k}/\underline{E}_{\text{wf},k}$ in the case of WFCs and $\sigma P_{\text{max},k}/\overline{E}_{\text{gs},k}$ in the case of GSCs, with σ the maximum accepted voltage sag. Figure 20.5 illustrates the sector-bounded uncertainty representation for both static voltage–current curves.

Therefore, the MTDC network can be modeled as the linear system which affinely depends on parameter $\delta = [\delta_1, \ldots, \delta_{(N_{\text{in}}+N_{\text{out}})}]^T$, that is,

$$\mathcal{M}(\delta): \begin{cases} \dot{x} = (A + B_q \Delta C_p)x + B_q q \\ p = C_p x \end{cases} \tag{20.7}$$

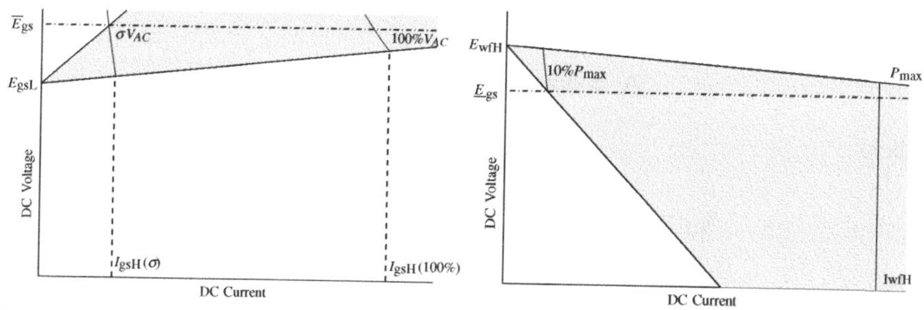

Figure 20.5 Illustration of the sector-bounded uncertainty representation for the static voltage–current curves.

where $\Delta = \text{diag}(\delta_1, \ldots, \delta_{(N_{\text{in}}+N_{\text{out}})})$ taking values in the hyperrectangle $\boldsymbol{\Delta} = [\underline{\delta}_1, K_1] \times \cdots \times [\underline{\delta}_{N_{\text{in}}+N_{\text{out}}}, K_{N_{\text{in}}+N_{\text{out}}}]$.[1]

Therefore, for control design purposes, the system (20.7) can be described as in Figure 20.6, where

$$G_c(s) = C_c(s\mathbf{I} - A_c)^{-1}B_c = \frac{1}{s\tau + 1} \cdot \mathbf{I}_{(N_{\text{out}}+N_{\text{in}})} \tag{20.8}$$

represents the dynamics of the current loops of the GSCs, where $\mathbf{I}$ is the identity matrix.

Then, the input to the plant $\mathcal{M}$ is

$$q = G_c(s)\Lambda_j K(w + p) + (\mathbf{I} - \Lambda_j)Hp \tag{20.9}$$

where

$$K = \text{diag}(K_1, \ldots, K_{(N_{\text{in}}+N_{\text{out}})})$$

is the decentralized control law to be determined and

$$H = \text{diag}(\underline{\delta}_1, \ldots, \underline{\delta}_{(N_{\text{in}}+N_{\text{out}})})$$

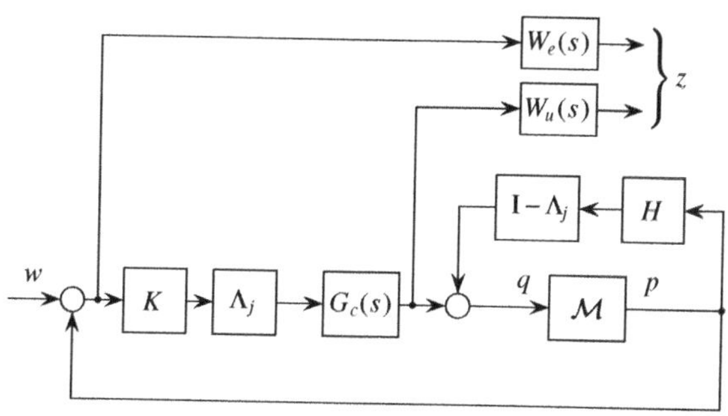

Figure 20.6 Polytopic closed-loop representation of the multi-terminal VSC-HVDC grid.

[1] This kind of system is known as a linear parameter varying (LPV) system.

The matrix Λ_j is defined as

$$\Lambda_j = \mathrm{diag}(\Lambda_{1,j}, \dots, \Lambda_{(N_{\text{in}}+N_{\text{out}}),j})$$

where

$$\Lambda_{k,j} = \begin{cases} 1, & \text{if droop at the node } k \text{ is active} \\ 0, & \text{otherwise} \end{cases} \tag{20.10}$$

This matrix determines the control configuration—that is, which nodes are in droop control and which are just injecting or extracting power without voltage control. For example, in normal operation, where the WFCs inject all the wind power available into the MTDC network and the GSCs are in droop control to regulate the DC voltage, the matrix results

$$\Lambda = \mathrm{diag}(\underbrace{0, \dots, 0}_{N_{\text{in}}}, \underbrace{1, \dots, 1}_{N_{\text{out}}})$$

This configuration changes if, for instance, a voltage fault occurs in the AC grid. This causes some of the converters on the AC grid side to enter current limitation and some of the WFCs to enter droop control. If all WFCs are in droop control and the GSCs in current limitation, the matrix Λ takes the form

$$\Lambda = \mathrm{diag}(\underbrace{1, \dots, 1}_{N_{\text{in}}}, \underbrace{0, \dots, 0}_{N_{\text{out}}})$$

In fact, there are $N = 2^{(N_{\text{in}}+N_{\text{out}})}$ possible configurations.

In the scheme in Figure 20.6, all outputs are fed back into the plant $\mathcal{M}$ but the only feedback gains to be designed are those associated with the droop control (the matrix K). The other feedback paths are caused by the relationship between the current and the voltage when the converter does not control the DC voltage.

Therefore, the closed-loop relationship between the disturbance w and the performance signal z is given by

$$\begin{aligned} \dot{x} &= A_{cl,j}x + B_{w,j}w \\ z &= C_z x + D_{zw}w \end{aligned} \tag{20.11}$$

where

$$A_{cl,j} = \begin{bmatrix} B_1(\Lambda_j K + (\mathbf{I} - \Lambda_j)H) & 0 & A_{13} & B_1 C_c \\ 0 & 0 & A_{23} & 0 \\ A_{31} & A_{32} & A_{33} & 0 \\ 0 & 0 & 0 & A_c \end{bmatrix}$$

$$B_{w,j} = \begin{bmatrix} (B_1 C_c)^T & 0 & 0 & B_c \end{bmatrix}^T \Lambda_j$$

$$C_z = \begin{bmatrix} \mathbf{I} & 0 & 0 & 0 \end{bmatrix}$$

$D_{zw} = \mathbf{I}$, with $j = 1, \ldots, N$. Due to the changes in the matrix Λ_j, the MTDC grid represented by (20.11) is a time-varying representation.[2]

20.4 DESIGN OF OPTIMAL CONTROL DROOPS

The purpose of the droop control is to maintain the DC voltages almost constant in spite of disturbances caused by changes in the wind power and the faults in the AC grid. This objective should be fulfilled even when some converters change from one operating mode to other. This objective will be stated in the form of the $\mathcal{L}_2$ criterion

$$\max_{w \neq 0} \frac{\|z\|_2}{\|w\|_2} < \gamma \tag{20.12}$$

The Lyapunov function

$$\mathcal{V}(x, \Delta) = x^T X x + z^T z - \gamma^2 w^T w$$

can be used to establish stability and the performance condition (20.12). Hence, the closed loop system (20.11) must satisfy the condition

$$\frac{d\mathcal{V}(x, \Delta)}{dt} \leq 0, \qquad \forall \Delta \in \mathbf{\Delta} \tag{20.13}$$

It can be proved that the condition (20.13) can be expressed as

$$\begin{bmatrix} XA_{cl,j} + (\star) & XB_{w,j} & C_z^T \\ B_{w,j}X & -\gamma\mathbf{I} & D_{zw}^T \\ C_z & D_{zw} & -\gamma\mathbf{I} \end{bmatrix} \prec 0, \qquad \forall j = 1, \ldots, N \tag{20.14}$$

where $\prec 0$ denotes negative definite [10].

By substituting the closed loop matrices $A_{cl,j}$ in (20.14), the design of the decentralized control K results in a robust decentralized static output feedback problem. This is a difficult problem to be solved in general. In order to simplify the control design, a relaxation can be introduced using the particular structure of the matrix B_w. The relaxation is based on the idea proposed in reference 11 of using a particular Lyapunov matrix

$$X = \rho X_o + Q X_q Q^T$$

where $\rho > 0$ is a scalar and Q is matrix such that

$$Q^T [\,(B_1 C_c)^T \quad 0 \quad 0 \quad B_c\,] = 0 \tag{20.15}$$

Hence, the gain computation can be divided into two optimization procedure with linear matrix inequality (LMI) constraints, which can be effectively solved with the available software [12, 13].

[2]More precisely, it is a polytopic LPV system.

Procedure 20.4.1 *The decentralized proportional control K that stabilizes the system (20.5) and ensures the performance criterium (20.12) can be computed as*

(i) Find a positive definite matrix X_o such that

$$X_o(A + B_1 H(\mathbf{I} - \Lambda_j)) + (\star) \prec 0 \quad (20.16)$$

for all $j = 1, \ldots, N$.

(ii) Find a matrices X_q, L and scalar ρ such that

minimize: γ,

subject to:

$$\begin{bmatrix} \rho X_o \bar{A}_j + (\star) & 0 & C_z^T \\ \star & -\gamma \mathbf{I} & D_{zw}^T \\ \star & \star & -\gamma \mathbf{I} \end{bmatrix} + \begin{bmatrix} Q X_q Q^T \bar{A}_j + U_j L + (\star) & U_j L & 0 \\ \star & 0 & 0 \\ \star & \star & 0 \end{bmatrix} \prec 0$$

$$\rho X_o + Q X_q Q^T \succ 0$$

where $\bar{A}_j = A + B_1 H(\mathbf{I} - \Lambda_j)$, $U_j = X_o B_{w,j}$, $K = L/\rho$ and $\star$ is deduced by symmetry.

Notice that in order to guarantee stability and performance against changes in the operating modes, the existence of a common Lyapunov matrix X must be proved.

20.5 FOUR-TERMINAL VSC-HVDC NETWORK EXAMPLE

The application of the procedure proposed in the previous section is illustrated with the four-terminal VSC-HVDC grid shown in Figure 20.7. The system consists of two wind farms (WF1 and WF2) and two AC grids (GRID1 and GRID2). The DC sides of the four converters are connected by submarine cables. The power generated by Wind Farm 1 and Wind Farm 2 is injected into the HVDC network through WFC1 and WFC2 and then delivered to the main grids (AC GRID1 and AC GRID2) through the GSC1 and GSC2 converters. The three branches (Cable 1, 2, 3) represent the DC link interconnecting the converters (see Figure 20.7). The parameter values can be found in Table 20.1.

The four-terminal grid in Figure 20.7 has two power input nodes, two power output nodes, and three branches representing the cables linking the converters. There are four capacitors and three inductors; therefore, the variables E_1, E_2, E_3, E_4, I_{L_1}, I_{L_2}, and I_{L_3} are sufficient to completely define the state of the system, that is,

$$x = [E_1 \; E_2 \; E_3 \; E_4 \; I_{L_1} \; I_{L_2} \; I_{L_3}]^T$$

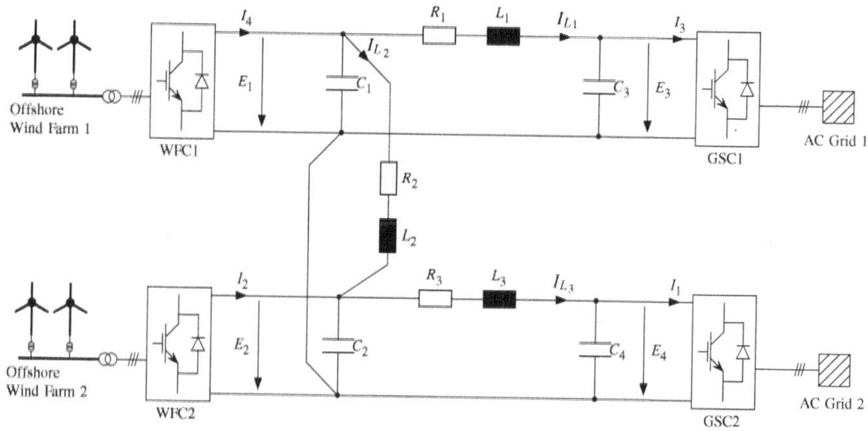

Figure 20.7 Four-terminal VSC-HVDC network example.

As explained in Chapter 16, the open-loop matrix in the state-space representation (20.5) is

$$A = \begin{bmatrix} 0 & 0 & 0 & 0 & -\frac{1}{C_1} & -\frac{1}{C_1} & 0 \\ 0 & 0 & 0 & 0 & 0 & -\frac{1}{C_2} & -\frac{1}{C_2} \\ 0 & 0 & 0 & 0 & -\frac{1}{C_3} & 0 & 0 \\ 0 & 0 & 0 & 0 & 0 & 0 & -\frac{1}{C_4} \\ \frac{1}{L_1} & 0 & -\frac{1}{L_1} & 0 & -\frac{R_1}{L_1} & 0 & 0 \\ \frac{1}{L_2} & -\frac{1}{L_2} & 0 & 0 & 0 & -\frac{R_2}{L_2} & 0 \\ 0 & \frac{1}{L_3} & 0 & -\frac{1}{L_3} & 0 & 0 & -\frac{R_3}{L_3} \end{bmatrix}$$

TABLE 20.1 Parameters of the Four-Terminal Example

Grid Parameters	Value
Line resistance R_1 and inductance L_1	0.50 Ω, 5.0 mH
Line resistance R_2 and inductance L_2	0.25 Ω, 2.5 mH
Line resistance R_3 and inductance L_3	0.40 Ω, 4.0 mH
Capacitances C_k $(k = 1, \ldots, 4)$	150 μF
Rated line current i_L^{rtd}	667 A
Rated input current i_k^{rtd}	667 A
Converter rated power P_k	100 MW
Rated DC voltage E_k	150 kV
Reference voltage E_0	145 kV

$$
B = \begin{bmatrix}
\frac{1}{C_1} & 0 & 0 & 0 \\
0 & \frac{1}{C_2} & 0 & 0 \\
0 & 0 & -\frac{1}{C_3} & 0 \\
0 & 0 & 0 & -\frac{1}{C_4} \\
0 & 0 & 0 & 0 \\
0 & 0 & 0 & 0 \\
0 & 0 & 0 & 0
\end{bmatrix}
$$

There are $N_{\text{in}} + N_{\text{out}} = 2 + 2 = 4$ converters therefore there are $2^{(N_{\text{in}}+N_{\text{out}})} = 16$ possible control configurations. The application of Procedure 20.4 for the 16 possible Λ_j produces the following droop gain

$$K = \text{diag}(-0.0778, -0.0782, 0.0795, 0.0807)$$

with a performance level of $\gamma = 0.57$, where the output z was weighted with the function

$$W_{e,j}(s) = \frac{0.25}{s/0.5 + 1} \cdot (0.25(\mathbf{I} - \Lambda_j) + 0.75\Lambda_j)$$
$$W_{u,j}(s) = 0.1(0.25(\mathbf{I} - \Lambda_j) + 0.75\Lambda_j)$$

for $j = 1, \ldots, 16$. The optimization problem was solved with the free available software Sedumi [12] and Yalmip [13].

The droop gains computed with the proposed procedure were evaluated by simulation. The first scenario analyzed consists in two simultaneous changes in the power injected into the multi-terminal VSC-HVDC grid by the WFCs. Initially, the wind power coming from WF1 and WF2 is 20 MW. At $t = 0.2$ s, the power injected by both wind farms increases from 20 MW to 90 MW. Then, the power drops to 20 ms at time 0.7 s. The evolution of the power at each converter can be seen in Figure 20.8a. During the increment in the power injected by both wind farms, the power extracted by GSC1 and GSC2 increases according to the power balance, that is, since $P_4/P_3 \approx I_4/I_3 = K_4/K_3 \approx 1$. The power extracted by GSC1 and GSC2 is approximately the same. Figure 20.8b shows the voltage at each terminal. Initially, the voltages on the wind farm side are 143 kV, after the change in the power these values increase about 6 kV. The grid side voltages are slightly lower due to the voltage drop across cable resistances. The temporal evolutions at all terminals are similar. Figure 20.8c shows the currents at each converter. The responses are similar to the power flow (except for a scale factor) because of the change in voltages is small.

The second scenario corresponds to a three-phase voltage sag 80% deep at both AC grids at 0.1 s. Previous to the fault, each wind farm is delivering 60 MW. In this circumstance, the GSCs reach the maximum current limits ($I_{\text{gsH},3} = I_{\text{gsH},4} \approx 667$ A) causing that both WFCs enter droop mode. Figure 20.9 shows the responses corresponding to this scenario. The power is initially at the values given by the power

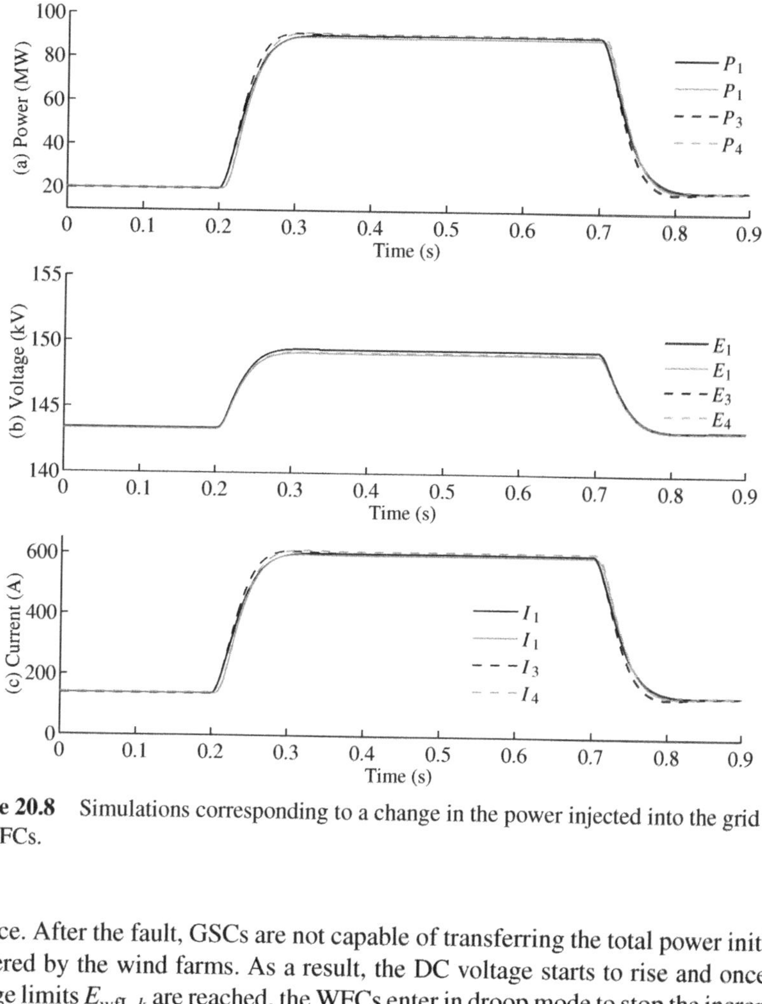

Figure 20.8 Simulations corresponding to a change in the power injected into the grid by the WFCs.

balance. After the fault, GSCs are not capable of transferring the total power initially delivered by the wind farms. As a result, the DC voltage starts to rise and once the voltage limits $E_{\text{wfL},k}$ are reached, the WFCs enter in droop mode to stop the increment of the voltage. The fault provokes a substantial reduction of the power delivered to the AC grids. The power extracted by each GSCs falls 30 MW, which causes an increment of DC voltage. Due to the operation of the GSCs in current limitation mode, the voltages are higher than in normal operation, this can be seen in Figure 20.9b, where all voltages are around 152 kV. It can also be observed that during the fault, the voltages at GSCs are slightly lower than the values at the wind farm terminals. This is a consequence of the increment of the current through the cables causing higher voltage drop across cable resistances. The effect of the droop control is clear. The voltages are maintained within the allowed limits and the power injected by each WFC is reduced, from 60 MW to 30 MW. Once the fault is cleared at 0.6 s, all the variables return to the initial levels. The simulations in Figure 20.9 show that the computed droop gains are able to achieve a satisfactory voltage control during a severe fault avoiding the disconnection of the converters.

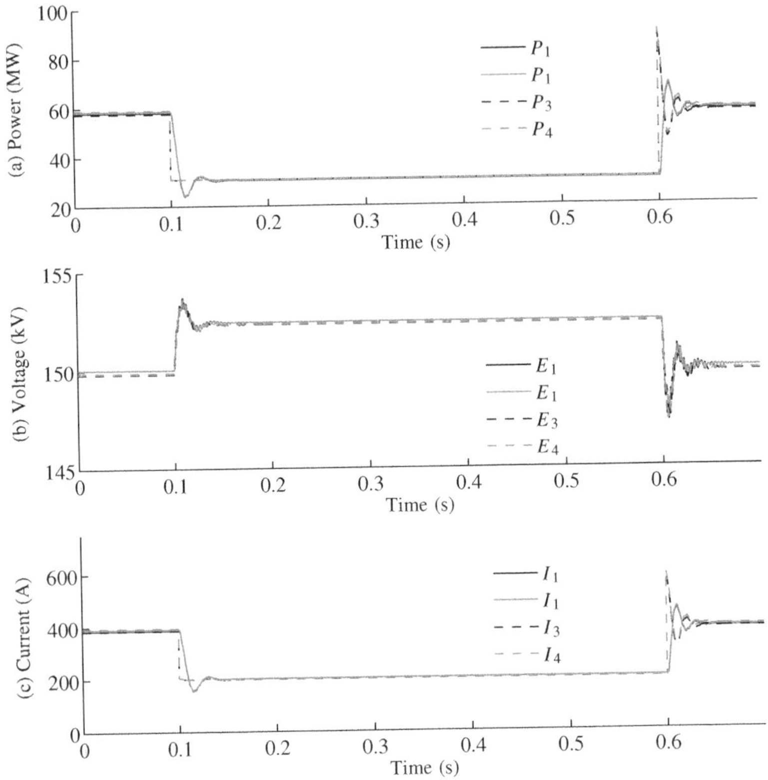

Figure 20.9 Simulations corresponding to voltage sags in the AC grids.

20.6 CONCLUSIONS

The integration of an increasing number of offshore wind farms demands the construction and control of multi-terminal VSC-HVDC networks. Voltage regulation of these networks is crucial to achieve an efficient and safe power transmission of the energy captured by the offshore wind farms. Multi-terminal VSC-HVDC networks in the context of wind farms integration exhibits a time-varying behavior as a consequence of changes in the operation conditions. In order to describe this behaviour, a general time-varying description of multi-terminal VSC-HVDC networks has been introduced. The control of multi-terminal VSC-HVDC networks requires the design of a robust decentralized static controller that ensures stability and performance under changes in the operating conditions. The synthesis of these decentralized controllers has been expressed as nonconvex optimization problem. The particular structures of the system and the control have been exploited to propose a relaxation in order to convexify the synthesis procedure and thus simplify the computation of the static feedback. As a result, the decentralized static controller can be computed by solving a convex optimization problem with LMI constraints. With this formulation, the designed static control stabilizes and ensures a given performance level under any operating condition of the multi-terminal VSC-HVDC grid.

REFERENCES

1. P. Kundur, *Power System Stability and Control*. New York: McGraw-Hill Professional, 1994.
2. J. Machowski, J. Bialek, and J. Bumby, *Power System Dynamics and Stability*. New York: Wiley, 1997.
3. J. Liang, O. Gomis-Bellmunt, J. Ekanayake, and N. Jenkins, "Control of multi-terminal VSC-HVDC transmission for offshore wind power," in *Proceedings of the 13th European Conference Power Electronics and Applications (EPE '09)*, 2009, pp. 1–10.
4. O. Gomis-Bellmunt, J. Liang, J. Ekanayake, and N. Jenkins, "Voltage–current characteristics of multiterminal HVDC-VSC for offshore wind farms," *Electrical Power Systems Research*, vol. 81, no. 2, pp. 440–450, 2011.
5. O. Gomis-Bellmunt, J. Liang, J. Ekanayake, R. King, and N. Jenkins, "Topologies of multiterminal HVDC-VSC transmission for large offshore wind farms," *Electrical Power Systems Research*, vol. 81, no. 2, pp. 271–281, 2011.
6. J. Liang, T. Jing, O. Gomis-Bellmunt, J. Ekanayake, and N. Jenkins, "Operation and control of multi-terminal HVDC transmission for offshore wind farms," *IEEE Transactions on Power Delivery*, vol. 26, pp. 2596–2604, 2011.
7. L. Xu, L. Yao, and M. Bazargan, "DC grid management of a multi-terminal HVDC transmission system for large offshore wind farms," in *Proceedings of the International Conference on Sustainable Power Generation and Supply (SUPERGEN '09)*, 2009, pp. 1–7.
8. E. Prieto-Araujo, F. Bianchi, A. Junyent-Ferr, and O. Gomis-Bellmunt, "Methodology for droop control dynamic analysis of multi-terminal VSC–HVDC grids for offshore wind farms," *IEEE Transactions on Power Delivery*, 2011.
9. F. D. Bianchi and O. Gomis-Bellmunt, "Droop control design for multi-terminal VSC–HVDC grids based on LMI optimization," in *50th IEEE Conference on Decision and Control and European Control Conference (CDC-ECC)*, Dec. 2011, pp. 4823–4828.
10. P. Apkarian, P. Gahinet, and G. Becker, "Self-scheduled h_∞ control of linear parameter-varying systems: a design example," *Automatica*, vol. 31, no. 9, pp. 1251–1261, 1995.
11. A. Zecevic and D. D. Siljak, "Control design with arbitrary information structure constraints," *Automatica*, vol. 44, no. 10, pp. 2642–2647, 2008.
12. J. Sturm, "Primal-dual interior point approach to semidefinite programming," PhD Thesis, Tilburg University, Dept. of Econometrics, 1997.
13. J. Lofberg, "YALMIP: A toolbox for modeling and optimization in MATLAB," in *Proceedings of the CACSD Conference*, Taipei, Taiwan, 2004 [Online]. Available: http://users.isy.liu.se/johanl/yalmip

CHAPTER 21

DC GRID POWER FLOW CONTROL DEVICES

Chunmei Feng,[1] Sheng Wang,[2] and Qing Mu[3]
[1]Engineering Department, Power Conversion, GE Energy Management, Stoke-on-Trent, United Kindom
[1]Institute of Energy, School of Engineering, Cardiff University, Cardiff, Wales
[1]Power System Department, China Electric Power Research Institute (CEPRI), Beijing, China

IN ELECTRICITY transmission networks, electric power flows are the result of the interaction between power generation, consumption, and available transmission paths. Without intervention on the natural power flow around the transmission networks (AC or DC systems), some paths may be lightly loaded and some overloaded. To avoid this problem, controlling the power flow is necessary.

Power flow control (PFC) is an important feature and one technical challenge of interconnected power systems. This is because power flows in these paths, as per Kirchhoff laws, are determined by the line parameters, topology of the network, generation, and load location. The power flows are not directly dependent on transmission line ownerships, contracts, thermal limits, or losses.

AC grids have existed for over a century and have dealt with power flow control with many different devices and methods [1]. For example, system operators could connect or disconnect lines, change the tap settings of transformers, connect capacitors for reactive power injection, and give active or reactive control signals to power plants to change the power flow on transmission paths. The principle and some methods for the power flow control used in AC grids can apply to DC grids such as variable series voltage injection, variable series impedance insertion, and topology control (line switching), but techniques and devices employed to control the power flow in AC grids are not directly applicable to DC grids due to different characteristics of AC systems and DC systems. In AC systems the dominant impedance is the transmission line inductance; by inserting capacitor banks or inductors, the line reactance can be changed, thus regulating the power flow around the grid. But in DC systems, the current flow is governed entirely by the DC resistance of the conductors and the transmission path inductance plays no part in the steady-state power flow. DC resistance needs to be inserted in DC systems if variable series impedance insertion method is used. If the topology control method is employed, AC breakers are needed

HVDC Grids: For Offshore and Supergrid of the Future, First Edition. By Dirk Van Hertem, Oriol Gomis-Bellmunt, and Jun Liang.

for AC systems while DC breakers are installed for DC systems. Due to different characteristics between AC systems and DC systems, devices for power flow control implementation in DC systems are different from those in AC systems.

Among different DC grid configurations, meshed multi-terminal HVDC systems have superiority over radial and point-to-point topologies, such as less converter stations and lower conversion losses. Meshed DC grids also provide multiple DC lines for the power delivery within the system to enhance the system redundancy and reliability. However, although the power injected into a DC grid is controlled by converters, the power flow within a meshed grid is mainly determined by resistances according to Ohm's Law. In such meshed grid the DC current will flow from one node to another via the path of least resistance, which may lead to one line that becomes overloaded. In order to achieve flexible and economic operation of DC grids, further flexible control devices and methodologies will be required to achieve the full controllability of the power flow. In addition, the flexible power flow control can be also used to prevent the overload of a transmission line/cable in post-fault conditions, by transferring the excessive current through an alternative DC transmission line.

The aim of the chapter is to present DC power flow control devices (DCPFCs) [2] and the effects of DCPFCs on DC grids. The chapter is organized as follows: Topologies of power flow control devices are described in Section 21.1. Sections 21.2 and 21.3 provide generic modeling of DCPFCs for power flow calculation of DC grids and the sensitivity analysis. Case studies using a five-terminal DC grid with these facilities are presented in Section 21.4, and case tests to investigate effects of DCPFCs on a DC grid based on the CIGRE test system are carried out in Section 21.5.1. Section 21.6 provides a comparison of DCPFCs, and Section 21.7 concludes the chapter.

21.1 DC POWER FLOW CONTROL DEVICES (DCPFC)

There are mainly three types of DCPFCs to realize the flexible power flow control: DC transformers, variable series resistors (VSR), and series voltage sources (SVS). DC transformers are shunt connected devices, while VSR and SVS are series connected devices. Their topologies, modeling, and control schemes are discussed in this section.

21.1.1 DC Transformer

Like an AC grid, a DC grid will consist of different DC voltage levels. The multiple DC voltage levels must be connected to increase the flexibility of power transfer and enhance the overall economic operation. However, DC currents and voltages cannot step up or down through an AC transformer, because DC current could not produce alternating flux to induce the voltage. DC/DC converters in DC transmission systems, called DC transformers, have been developed to realize the similar function of an AC transform in the AC transmission system.

A typical DC transformer, as shown in Figure 21.1, is mainly composed of a DC/AC converter, an AC transformer, and an AC/DC converter. The DC/AC and

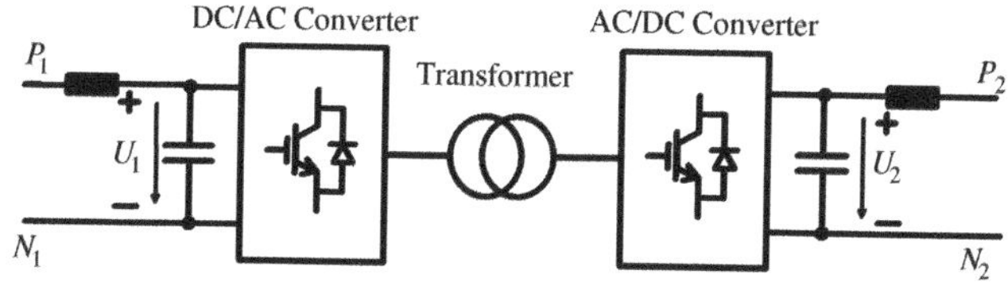

Figure 21.1 Topology of DC transformer.

the AC/DC converters can use different topologies such as two-level or multilevel topology. The converter in the prime side provides the conversion from the DC voltage to AC. Various AC voltage waveforms can be generated depending on the topology of converters and the application. For example, a flyback converter can produce the square waveform with low frequency and low loss but high current harmonics. A PWM converter can produce an AC voltage waveform using a high switching frequency which leads to a nearly sinusoid current waveform. A multilevel converter produces the step voltage waveform which results in nearly sinusoid current waveform with low harmonics. The AC transformer provides electrical isolation between the two converters. The turns ratio of the transform is also used to change the voltage level. A high-frequency AC waveform can decrease the size of the AC transformer and magnetic elements to get a higher performance at the output (bandwidth, ripple, etc.). At the secondary side, the AC/DC converter controls the DC output voltage by adjusting the modulation index of the converter. When needed, inductors and capacitors are used for smoothing DC currents and voltages.

There are various topologies of DC transformers. In reference 3, the bidirectional resonant converter is to employ a capacitor to achieve the high-step-up conversion ratio. Salient feature of this converter is to achieve soft switching for all switches and diodes. Moreover, the three-phase series resonant converter in reference 4 is developed to achieve longer turn-off time and fewer harmonics than the single-phase one.

An equivalent model of a DC transformer is shown in Figure 21.2. In this model, the primary side is a current source I_1, with a parallel equivalent capacitor C_E, while the secondary side is a voltage source U_2, with the series equivalent inductor L_E. The relationship of the voltage and the current of the two sides of the transformer

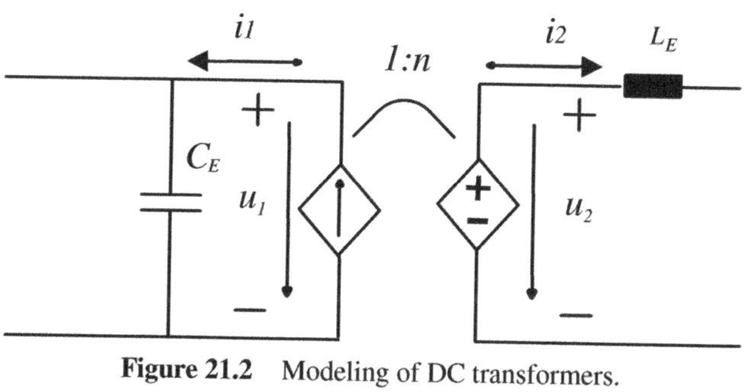

Figure 21.2 Modeling of DC transformers.

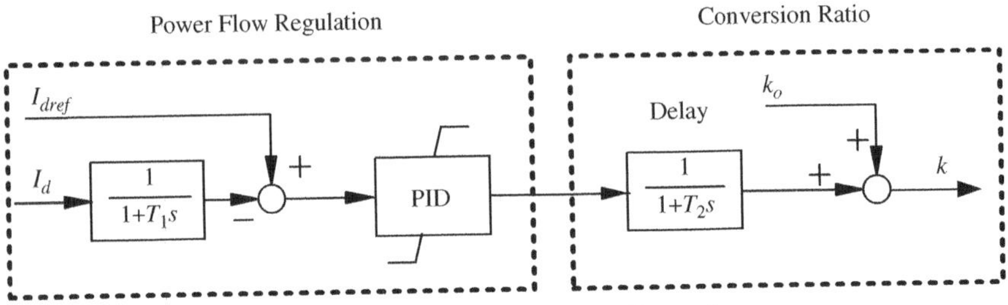

Figure 21.3 Control scheme of DC transformer.

can be presented as in (21.1a).

$$U_2 = k \cdot U_1 \tag{21.1a}$$

$$I_2 = \frac{-1}{k} \cdot I_1 \tag{21.1b}$$

$$k = \frac{m_1}{m_2} \cdot n \tag{21.1c}$$

n represents the turns ratio of the AC transformer, m_1 and m_2 represent the modulation index of AC/DC converters, respectively. By regulating the modulation index, m_1 or m_2, the voltage U_2 and the current I_1 can be controlled to a desired value. This simple model can achieve sufficient accuracy in the steady-state analysis of a DC transformer [5]. DC transformers can have the following functions: (1) voltage stepping, (2) voltage and power regulation, and (3) dc fault current interruption [3]. Obviously, a DC transformer can play an important role in the power flow control in DC grids. However, DC transformers have high construction costs, large power losses, and the low reliability for high-power applications, compared to AC transformers.

When a DC transform is applied in power flow control, the DC transformer will adjust the modulation index to control its DC current. A scheme of the DC transformer for the power flow control is given in Figure 21.3. The first part is the power flow regulation. I_{dref} is the reference current of the DC transmission line to be controlled, while I_d is the measured current of the line (which can be I_1 or I_2 in Figure 21.2). The first-order filter represents the measurement process. The error between the two current signals is fed to a PID regulation. A voltage signal is produced and fed to the second part for generating the conversion ratio k. k_0 is the rated value to ensure the normal operation of the DC transformer. The modulation index m_1 and m_2 are then determined using (21.1c) as k is known.

21.1.2 Variable Series Resistor (VSR)

A series resistor can be inserted in a DC grid to change the equivalent resistance of the DC line. Figure 21.4 shows the topology of variable series resistors. This topology is composed of a fixed resistor R, a pair of bidirectional IGBTs, IGBT 1 and IGBT 2, and diodes connected with the IGBT in series for withstanding the reverse voltage. The power dissipation on the resistor may require a large cooling system. In this

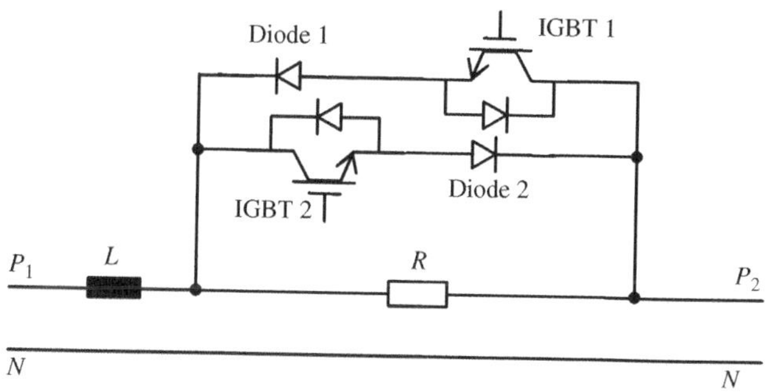

Figure 21.4 Series variable resistor diagram.

figure, only one unit of VSR is connected to the positive pole of the DC line in series, although another unit can be also connected to the negative pole.

The regulation of the average series resistor value R_{ave} is determined by the duty cycle, D, of the IGBT control (21.2).

$$R_{ave} = \frac{T_{off} \cdot R}{T} = (1 - D) \cdot R \tag{21.2}$$

where R_{ave} is the average resistance, and T_{off} represents the off state of the IGBT in each switching period T. Duty cycle D is the percentage of one period T in which IGBT is on.

Within a DC grid, a VSR regulates the active power through a DC line (not necessary the line where VSR is located) by adjusting the inserted resistance. The control system of a VSR has also two different parts: the power flow regulation and the resistance reference, as shown in Figure 21.5. In the power flow regulation, the error between the reference current, I_{dref}, and measured current of the DC line to be controlled, I_d, is fed to a PID regulation. The second part receives the resistance signal generated from the power flow regulation and produces the reference signal of the variable resistor. The delay link is to mimic the delay in practical equipment, including the data process and the signal transmitting. R_o represents the resistance value for normal operation of the DC network, which might be zero.

The proposed IGBT pair in the variable resistor meets the requirement of VSC-HVDC in bidirectional applications. The average value of the resistor can vary

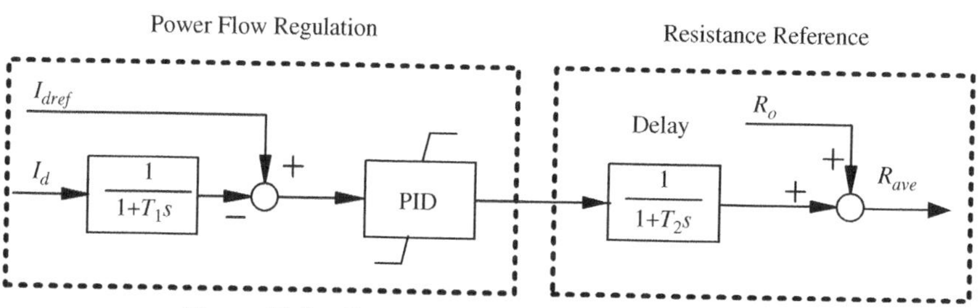

Figure 21.5 Control scheme of variable series resistor.

smoothly according to the reference. The main advantage of using a VSR would be its low investment cost. This is because a VSR is connected in series with a DC line and is therefore entirely at the line potential. The voltage rating of such a device can be very small. However, the resistance inserted into the DC circuit will cause additional power losses which in turn may require a large cooling system. Moreover, if the VSR is located on a DC line where the DC current is very small, its power controllability will be reduced. Consideration of the device location has to be drawn before the use of VSR.

21.1.3 Series Voltage Source (SVS)

Series voltage sources have been used in AC systems [6]. The same principle can be applied to DC grids but with DC voltage sources inserted into DC lines in series to change power flows. This series device acts as a controllable voltage source with a small voltage value. Normally it has a voltage magnitude on the order of 1–5% of the DC grid voltage. An SVS can be implemented by different devices and topologies.

One of SVS topologies used in DC grids is proposed in reference 6, as shown in Figure 21.6, in which the SVS is connected to the positive pole only. An AC transformer, an AC/DC converter, and a DC/DC converter are used to connect the AC source and the DC voltage. The AC/DC converter can be either a two-level three-phase PWM-controlled converter, as illustrated in Figure 21.6, or a multilevel 3-phase converter, which is controlled to maintain the voltage of the DC capacitor, U_d. The DC/DC converter is typically an H bridge, and an inductor L is connected in series for smoothing the current. This DC/DC converter is also called a four quadrant chopper which regulates the DC line currents by controlling the DC output of this device, ΔU, in order to regulate the power flow of DC grids.

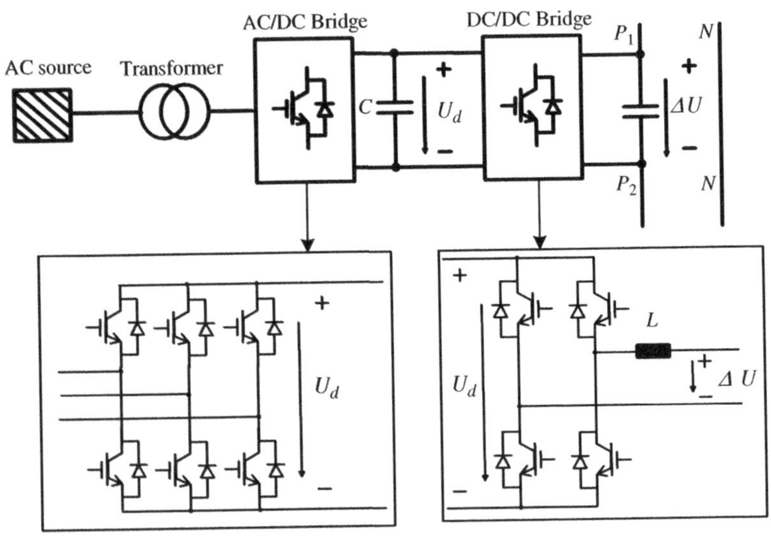

Figure 21.6 Series variable voltage source: IGBT-based PFC.

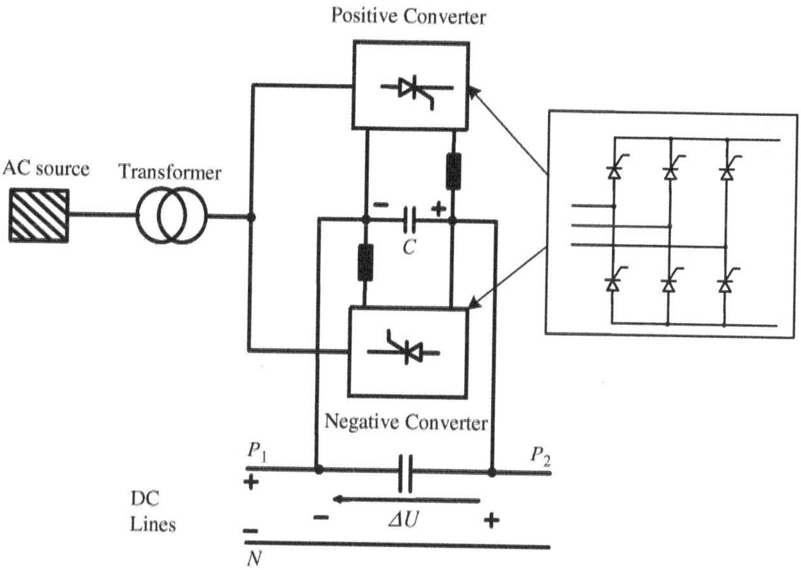

Figure 21.7 Series variable voltage source: thyristor-based PFC.

Figure 21.7 illustrates a thyristor-based PFC which has two three-phase thyristor based converters, and they are connected in anti-parallel to ensure four-quadrant operation of the PFC. The power flow controller is powered from AC side where the AC side voltage is stepped down by an AC transformer. The firing angles of converters are related in such way that both converters produce the same terminal voltage, one converter operating as a rectifier and other operating as an inverter. However, only one converter receives the firing pulses at any time instant and the other converter is blocked by the current polarity. When the DC line current falls below a threshold both converters receive the firing pulses. Therefore, there is a possibility of a discontinuous DC line current. During this period of operation, circulating current flows between the converters to allow a smooth change in the direction of DC line current. A large inductor is required to limit the circulating current between the converters. However, introduction of a large inductor will increase the power loss due to a large DC line current flow through the inductor.

The SVS exchanges real power with the AC system through a three-phase transformer. By inserting positive or negative DC voltage, the SVS can import/export power from/to the AC system. Therefore, the power loss of SVS is relatively smaller compared to that of VSR. The main drawback of these topologies of SVS (as shown in either Figure 21.6 or 21.7) is the need of a three-phase transformer which has to be rated for the DC system voltage.

Figure 21.8 shows dual H-bridge PFC which consists of two H-bridges (DC/DC converters) connecting to two different DC lines but sharing a common DC voltage U_C. Average DC voltages will be inserted into each DC line. This will lead to the power exchange between these two DC/DC converters and thus controls the power flow through DC lines. One of two H-bridges controls the capacitor voltage and

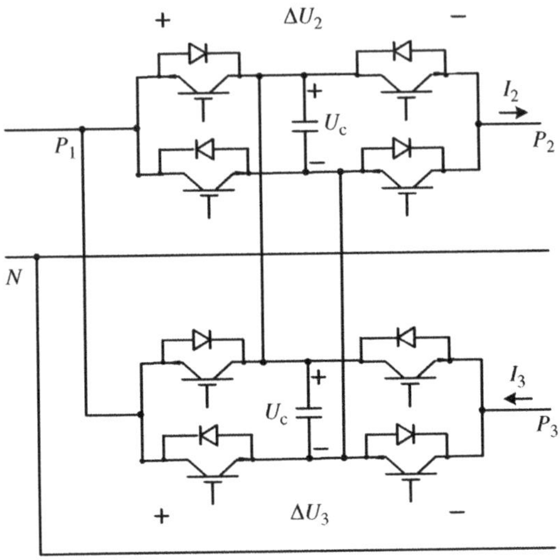

Figure 21.8 Series variable voltage source: dual H-bridge PFC.

another H-bridge controls the DC line current by chopping the voltage U_c. In this case, power taken from one line is equal to the power added to the other line. This structure allows the power exchange within the DC grid and therefore avoids the use of a three-phase transformer.

The control scheme of the device is given in Figure 21.9. The control system is composed of the power flow regulation and the voltage reference generator. For the power flow regulation, the error between the reference current, I_{dref}, and measured current, I_d, of the DC line to be controlled is fed to a PID regulator to generate a voltage variation signal. In the voltage reference block, a first-order delay link represents the delay in the signal processing and transmitting. ΔU_{rated} is the rated output voltage of the device in the normal condition. The final output signal ΔU_{ref} is the desired output voltage ΔU of the converter [7], which is achieved through the DC/DC converter control. In the case of dual H-bridge configuration, the controller in Figure 21.9 controls the inserted voltage of one H-bridge converter, such as ΔU_2 in Figure 21.8. Another H-bridge converter controls the common DC voltage U_c and generates another series voltage ΔU_3. The relationship between the voltages and currents is $\Delta U_2 \cdot I_2 = \Delta U_3 \cdot I_3$.

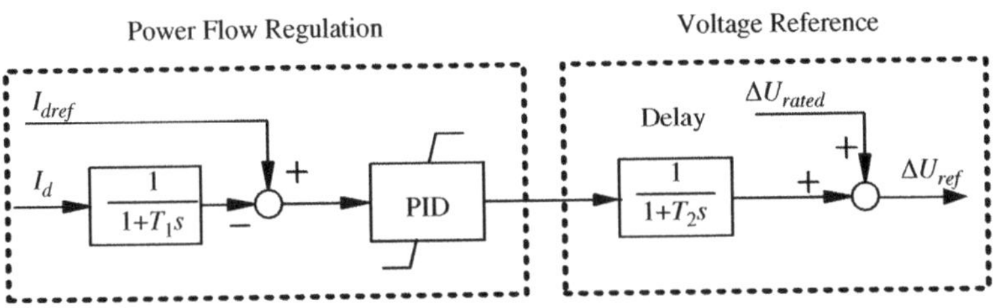

Figure 21.9 Control scheme of series variable voltage source.

This topology allows the use of various types of sources, such as a diesel generator, tapped from DC lines, or a source supplied from the AC bus in the converter station, either single phase or three phases. Comparing to the variable resistor, the benefit of an SVS is that there is no power consumption except converter losses.

Notice that all types of DCPFCs can be located either on both poles of one DC branch or only on one pole. If a DCPFC is installed on both poles, the DC voltage of positive pole and negative pole is balanced. Installation of a DCPFC on only one pole will lead a small DC voltage unbalance between two poles (i.e., equal to the voltage across the DCPFC). However, this reduces the number of switching devices (such as IGBTs) by half.

21.2 GENERIC MODELING OF DC POWER FLOW CONTROL DEVICES

In a DC transmission system equipped with DCPFCs, the DCPFCs and converters are all controllable devices which can regulate the power flow within the system. A steady-state power flow in a simple DC system is shown in Figure 21.10.

In Figure 21.10a, the DC power along the branch, from the converter VSC1 to another converter VSC2, P_{12} is

$$P_{12} = U_2 \cdot \left(\frac{U_1 - U_2}{R} \right) \tag{21.3}$$

The power flow can be controlled by changing the difference between converter voltages U_1 and U_2. If a DCPFC is installed between VSC1 and VSC2, the power flow equation is given by

$$P_{12} = U_2 \cdot \left(\frac{U_1 - U_2 - U_x}{R} \right) \tag{21.4}$$

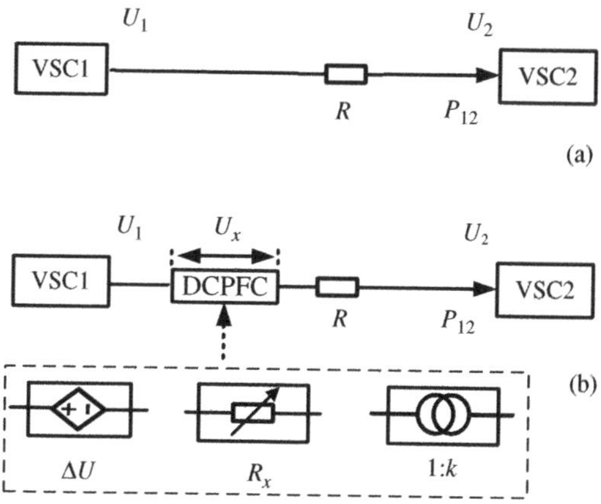

Figure 21.10 DC power transmission without DCPFC (a) and with DCPFC (b).

where U_x is the voltage across the DCPFC (21.5). Due to the small resistance in DC grids, a small U_x can result in a large variation of the DC power.

The SVS controls the power flow by inserting a series voltage ΔU. This inserted voltage U_x can be positive or negative. The VSR inserts a variable R_x to regulate the power flow. The resultant voltage V_S controls the power flow. This inserted voltage U_x is also affected by the DC current. The current effect is reduced when the DC current is close to zero. The DC transformer adjusts the voltage ratio of its two sides to regulate the power flow. A DC transformer is shunt connected and can change the power flow significantly because the resultant equivalent voltage $U_x = (k-1) \cdot U_1$ which is very sensitive to k since U_1 is near constant around the system rated voltage.

$$U_x = \begin{cases} \Delta U & \text{for SVS} \\ I \cdot R_x & \text{for VSR} \\ (k-1) \cdot U_1 & \text{for DC transformer} \end{cases} \tag{21.5}$$

However, this also indicates the DC transformer must be sized at the system voltage rate. The investment cost of the DC transformer will be much higher than that of the SVS and the VSR because the SVS and the VSR are rated only at a portion of system voltage rate. Therefore, the DC transformer is most suitable for connecting two HVDC systems with different voltage levels. The power flow control is an accessory function of such a device. A DC transformer is not an economical option for the power flow control purpose only. The SVS and VSR are expected to be more economical if only power flow control is required.

21.3 SENSITIVITY ANALYSIS OF DCPFC IN DC GRID

In an n-terminal MTDC system without DCPFCs, the power flow through VSCs is given by

$$P = U \otimes (G \cdot U) \tag{21.6}$$

where $\otimes$ is the entry-wise matrix multiplication operator, G is the conductance matrix without DCPFCs and $P = [P_1 \ldots P_i \ldots P_n]^T$, $U = [U_1 \ldots U_i \ldots U_n]^T$ are vectors representing converter power and converter DC voltages.

Switching-station DC buses in a MTDC which are not connected to any converter can mathematically be considered as connected to converters with zero power flow.

When a DCPFC is inserted into a DC branch between two DC buses; N_i and N_j , as shown in Figure 21.11, the power of the DCPFC, P_x, is a function of the DC voltages, U_i and U_j, and its own DC voltage, U_x:

$$P_x = U_x \cdot \left(\frac{U_i - U_x - U_j}{R_i + R_j} \right) \tag{21.7}$$

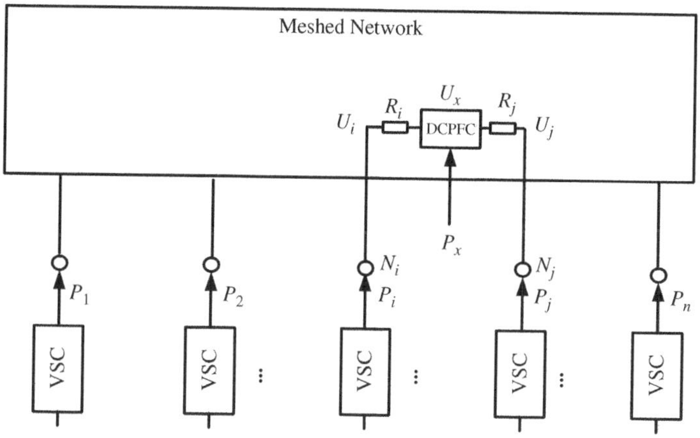

Figure 21.11 MTDC system with a DCPFC inserted in one branch.

This power can be the power losses of a VSR or the power exchange of a SVS. By combining (21.6)–(21.7), an expression for power flows in an n-bus MTDC system with one DCPFC inserted between node N_i and N_j (Figure 21.11) can be written as

$$\begin{bmatrix} P_1 \\ \vdots \\ P_i \\ \vdots \\ P_j \\ \vdots \\ P_n \\ \vdots \\ P_x \end{bmatrix} = \begin{bmatrix} U_1 \\ \vdots \\ U_i \\ \vdots \\ U_j \\ \vdots \\ U_n \\ \vdots \\ U_x \end{bmatrix} \otimes \left[\begin{array}{ccccccc|c} G_{11} & \cdots & G_{1i} & \cdots & G_{1j} & \cdots & G_{1n} & 0 \\ \vdots & \ddots & & & & & & \vdots \\ G_{i1} & \cdots & G_{ii} & \cdots & G_{ij} & \cdots & G_{in} & G_{ij} \\ \vdots & & & \ddots & & & & \vdots \\ G_{j1} & \cdots & G_{ji} & \cdots & G_{jj} & \cdots & G_{jn} & -G_{ji} \\ \vdots & & & & & \ddots & & \vdots \\ G_{n1} & \cdots & G_{ni} & \cdots & G_{nj} & \cdots & G_{nn} & 0 \\ \hline 0 & \cdots & G_{ij} & \cdots & -G_{ji} & \cdots & 0 & G_{ij} \end{array}\right] \cdot \begin{bmatrix} U_1 \\ \vdots \\ U_i \\ \vdots \\ U_j \\ \vdots \\ U_n \\ \vdots \\ U_x \end{bmatrix} \tag{21.8}$$

where G_{ij} and G_{ji} are the conductance between node N_i and N_j both equal to $\frac{1}{R_i+R_j}$.

If m DCPFCs are installed in an n-bus MTDC system, Eq. (21.8) can be written into a general expression representing the power flow:

$$\left[\frac{P}{P_x}\right] = \left[\frac{U_1}{U_x}\right] \otimes \left[\begin{array}{c|c} G & G_x^T \\ \hline G_x & diag(G_m) \end{array}\right] \cdot \left[\frac{U}{U_x}\right] \tag{21.9}$$

where P_x and U_x are vectors representing the local power across DCPFCs and their measured voltage. G_x is an additional $m \times n$ admittance matrix for solving the device power of DCPFCs. Nonzero elements in G_x are the conductance of branches where DCPFCs are located such as those shown in (21.8). $diag(G_m)$ is an $m \times m$ diagonal matrix where its nonzero elements are also the admittance of branches where DCPFCs are located.

The small signal equation of (21.9) is then expressed as

$$\begin{bmatrix} \Delta P \\ \Delta P_x \end{bmatrix} = \begin{bmatrix} J_{n\times n} & J_{n\times m} \\ J_{m\times n} & J_{m\times m} \end{bmatrix} \cdot \begin{bmatrix} \Delta U \\ \Delta U_x \end{bmatrix} \tag{21.10}$$

where $J_{n\times n}$, $J_{m\times n}$, $J_{n\times m}$ and $J_{m\times m}$ form the Jacobian matrix of the MTDC system integrated with DCPFCs. Equation (21.10) reflects the power deviation respecting to the change of converter DC voltages and the voltages across DCPFCs. A ratio σ is introduced to analyze the relationship between the power of a DCPFC and the line power to be controlled, P_c:

$$\sigma = \frac{P_x}{P_c} \tag{21.11}$$

The droop control of each VSC is

$$diag(k) \cdot (U_0 - U) = P_0 - P \tag{21.12}$$

Then the power flow within a MTDC system can be represented as

$$\begin{bmatrix} P \\ P_x \end{bmatrix} = \begin{bmatrix} diag(1) & 0 \\ 0 & diag(\sigma) \end{bmatrix} \cdot \begin{bmatrix} P_0 \\ P_c \end{bmatrix} + \begin{bmatrix} diag(k) & 0 \\ 0 & 0 \end{bmatrix} \cdot \left(\begin{bmatrix} U_0 \\ 0 \end{bmatrix} - \begin{bmatrix} U \\ 0 \end{bmatrix} \right) \tag{21.13}$$

where $diag(\sigma)$ is an $m \times m$ diagonal matrix that refers to the ratio between P and P_0. $diag(k)$ is an $n \times n$ diagonal matrix containing the droop gains of each converter controllers and $diag(1)$ is an $n \times n$ identity matrix. U_0 is the vector representing the DC voltage reference of converter control and P_0 is vector representing the power reference for both converters and DCPFCs.

Equation (21.13) can be written as small-signal matrix as

$$\begin{bmatrix} \Delta P \\ \Delta P_x \end{bmatrix} = \begin{bmatrix} diag(1) & 0 \\ 0 & diag(\sigma)^{(0)} \end{bmatrix} \cdot \begin{bmatrix} \Delta P_0 \\ \Delta P_c \end{bmatrix} + \begin{bmatrix} diag(k) & 0 \\ 0 & 0 \end{bmatrix} \cdot \left(\begin{bmatrix} \Delta U_0 \\ 0 \end{bmatrix} - \begin{bmatrix} \Delta U \\ 0 \end{bmatrix} \right) \tag{21.14}$$

Notice that changes in power reference or local power across DCPFDC will change the ratio between P_x and P_0. The elements in a ratio matrix $diag(\sigma)^{(t)}$ can be updated using the equation below:

$$\sigma^{(t+1)} = \sigma^{(t)} + \frac{d\sigma}{dP_x} \cdot \Delta P_x + \frac{d\sigma}{dP_c} \cdot \Delta P_c \tag{21.15}$$

If there is no set-point change of converters (i.e., $\Delta P_0 = 0$ and $\Delta U_0 = 0$), combining (21.10) and (21.14) results in

$$\begin{bmatrix} \Delta P \\ \Delta P_x \end{bmatrix} = \left\{ \begin{bmatrix} diag(1) & 0 \\ 0 & diag(1) \end{bmatrix} - \begin{bmatrix} diag(k) & 0 \\ 0 & 0 \end{bmatrix} \cdot \begin{bmatrix} J_{n\times n} & J_{n\times m} \\ J_{m\times n} & J_{m\times m} \end{bmatrix} \right\}^{-1} \cdot \begin{bmatrix} diag(1) & 0 \\ 0 & diag(\sigma)^{(0)} \end{bmatrix} \cdot \begin{bmatrix} 0 \\ \Delta P_c \end{bmatrix} \tag{21.16}$$

Equation (21.16) shows the impact of changing controlled power ΔP_c of the DCPFCS on the power flow ΔP and ΔP_x.

It can be seen that the impact of ΔP_c on the power flow through each converter is a function of the droop of converter ($diag(1)$), system topology and line resistance (Jacobian matrix), and the ratio index at a certain state ($diag(\sigma)^{(0)}$).

The impact of ΔP_c on system DC voltage change (ΔU) can also be obtained from (21.10) and (21.14):

$$\begin{bmatrix} \Delta U \\ \Delta U_x \end{bmatrix} = \left\{ \begin{bmatrix} diag(k) & 0 \\ 0 & 0 \end{bmatrix} + \begin{bmatrix} J_{n\times n} & J_{n\times m} \\ J_{m\times n} & J_{m\times m} \end{bmatrix} \right\}^{-1} \cdot \begin{bmatrix} diag(1) & 0 \\ 0 & diag(\sigma)^{(0)} \end{bmatrix} \cdot \begin{bmatrix} 0 \\ \Delta P_c \end{bmatrix} \tag{21.17}$$

Expression (21.17) gives the voltage change across each DCPFC ΔU_x when the power reference changes. If the DCPFCs are SVS, the change of inserted voltages of SVS will simply be equal to ΔU_x. If VSR are used, the inserted resistances can be estimated by calculating the conductance of the DC lines where VSR are located.

In general, Eqs. (21.16) and (21.17) show the sensitivity of converter power and voltage to the power reference of DCPFCs. These equations can be used to further decide the voltage and power rating of a DCPFC and the best location of a DCPFC for controlling a certain line power.

21.4 CASE STUDY OF POWER FLOW CONTROL DEVICES IN DC GRIDS

A five-terminal DC grid is used to verify the feasibility and effectiveness of the proposed power flow control devices.

21.4.1 System Topology

The example five-terminal system has 7 lines and is shown in Figure 21.12. Lines 1 to 5 connect the Nodes 1 to 4 together to form a complex meshed topology. A wind farm is connected to Node 5 through a converter. Table 21.1 gives the relevant parameters of the DC grid.

21.4.2 Test Conditions

Table 21.2 shows the operation conditions of converters. The converter on Node 3 controls the DC voltage to ±200 kV, and other terminals control the power. The power and voltage references are listed in Table 21.2.

21.4.3 Simulation Results

Each DCPFC device is installed in Line 1 between Node 1 and Node 3. This case evaluates their effectiveness to control the power of Line 6 between Node 5 to Node 1.

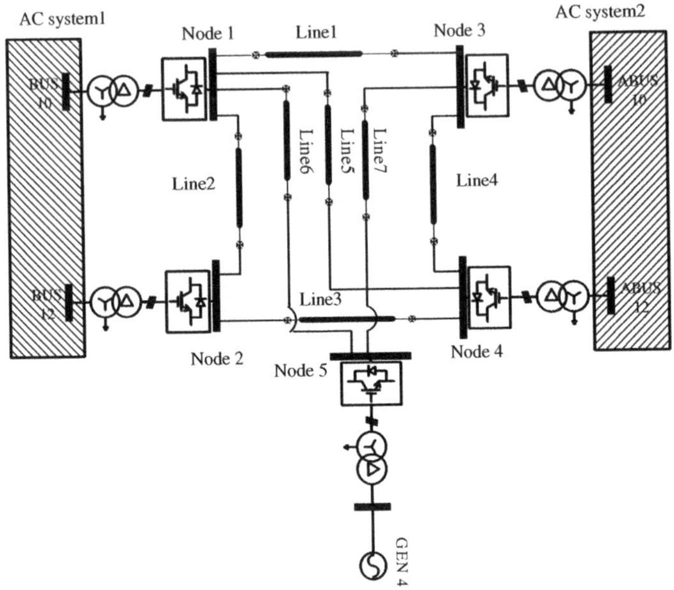

Figure 21.12 A five-terminal DC system.

TABLE 21.1 Line Parameters of the Five-Terminal DC Grid

Line	From	To	Length [km]	R [Ω/km]	L [mH/km]	C [μF/km]
1	1	3	300	0.0121	0.1056	0.2961
2	1	2	100	0.0121	0.1056	0.2961
3	2	4	150	0.0121	0.1056	0.2961
4	3	4	150	0.0121	0.1056	0.2961
5	1	4	350	0.0121	0.1056	0.2961
6	1	5	180	0.0121	0.1056	0.2961
7	5	3	180	0.0121	0.1056	0.2961

TABLE 21.2 Preconditions of Converters in the Five-Terminal DC Grid

Node	Mode	P [MW]	U [kV]	Q [MVAr]
1	PQ	320	n/a	40
2	PQ	240	n/a	40
3	VQ	n/a	±200 kV	40
4	PQ	−280	n/a	40
5	PQ	100	n/a	40

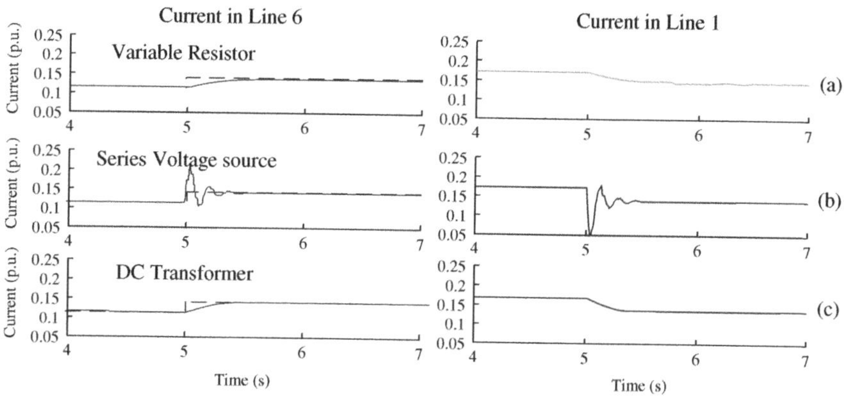

Figure 21.13 Control responses of line currents.

21.4.3.1 Step Change of Current Reference In this case, the current reference of Line 6 changes from 0.115 p.u. to 0.14 p.u. at $t = 5$ s, and the results are shown in Figure 21.13.

The first row shows the responses using the series variable resistor. The current of Line 6 is only 0.136 p.u., which cannot reach the reference value because of the limitation of resistance of the VSR. The current of Line 1, where the device is installed, decreases from 0.175 p.u. to 0.144 p.u..

The second row shows the performance of the SVS. The current of Line 6 can quickly reach the reference value of 0.14 p.u. at 5.3 s. But oscillations in the transient procedure can be observed. The third row shows the responses using the DC transformer. The DC transformer can achieve very fast current tracking, with less oscillation.

21.4.3.2 Disturbance Test In this test, the Line 5, from Node 1 to Node 4, is disconnected at $t = 5$ s. Without any flexible control, the current of Line 6 will increase from 0.114 p.u. to 0.155 p.u., which is above its current limit, assumed to be 0.13 p.u. The flexible devices are used to reschedule the power flow and avoid the overload of Line 6. Simulation results are given in Figures 21.14 to 21.16, in which curves "A" stand for results with the flexible power flow control and curves "B" are for results without the control.

Figure 21.14 shows the results using the series variable resistor. At 5.4 s, the system reaches a new steady state and the current of Line 6 has increased by only 10%,

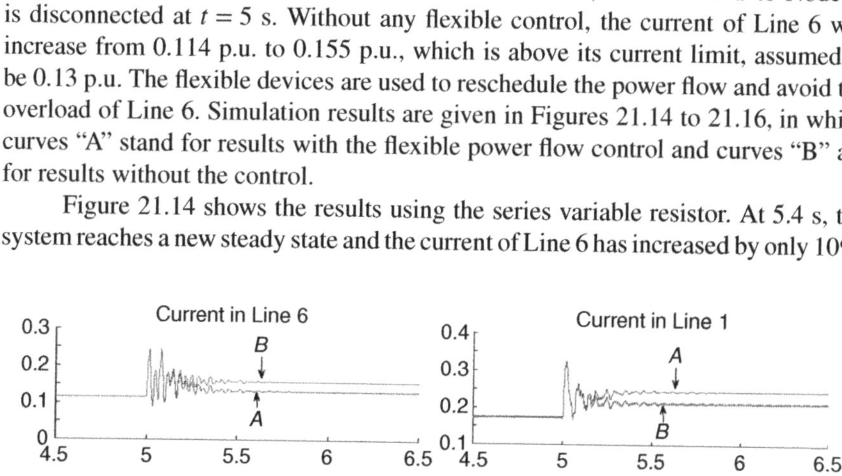

Figure 21.14 Power flow control using the series variable resistor.

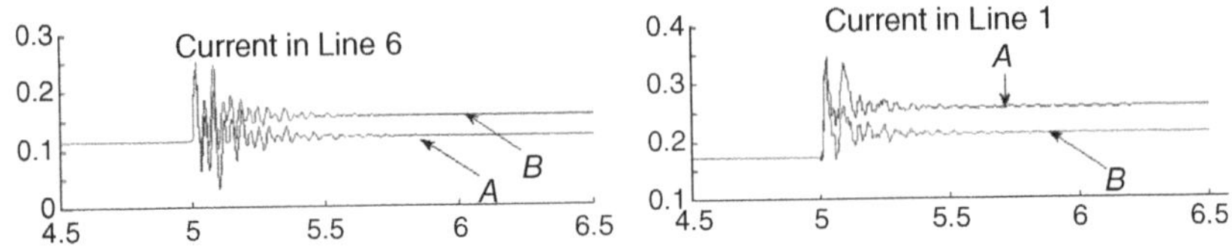

Figure 21.15 Power flow control using the series voltage source.

from 0.114 p.u. to 0.123 p.u. Although the current does not reach the pre-disturbance value, the Line 6 current is within the security range. The power has been rescheduled with 0.04 p.u. more current flowing through Line 1. However, the current on other lines, Line 2 to Line 4, keep nearly unchanged with or without the control of the variable resistor. This is due to the fact that the flow in Line 1 is more sensitive to the power control device in Line 6.

The results using the series voltage source (SVS) are given in Figure 21.15. The current of Line 6 can be maintained to the original value after the disconnection of Line 5. Compared to the situation without the SVS, an increase in the current through Line 1 of 0.06 p.u. can be seen. Similarly, the currents of other lines are nearly unaffected by the SVS.

The results using the DC transformer are given in Figure 21.16. The DC transformer can avoid the overload in Line 6, and it has the smaller impact on other lines. It is observed that the post-fault conditions when the power controller is not activated for the DC transformer are different from that with the VSR or the SVS. The current of Line 6 has only increased to 0.143 p.u., less than the ones before. This is because the DC transformer has a constant output voltage when the power flow control is disabled.

21.5 CONTROL SENSITIVITY OF DCPFC IN DC GRIDS

This section analyses power flow control sensitivities of various DCPFCs in a DC grid and validates the analysis method developed in Section 21.3.

21.5.1 Test System

The CIGRE DC grid test model, given in Figure 21.17, is used to test the DCPFC.

This model consists of 9 AC/DC converters rated at 2400 MW. The DC Grid is rated at ±400 kV. The DC network has 12 buses and is highly meshed by both DC overhead lines and DC cables. More detailed parameters of the system can be found

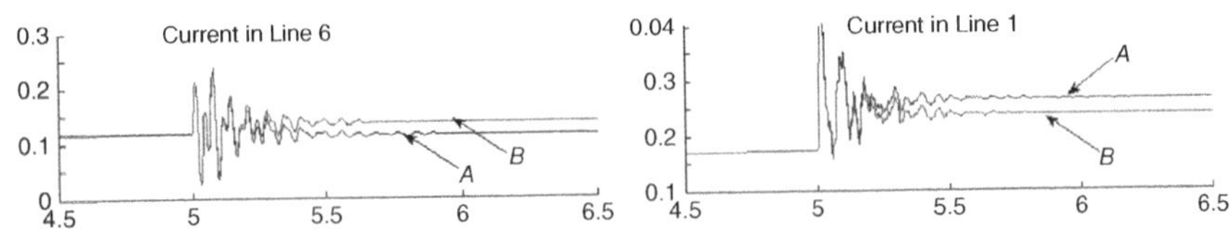

Figure 21.16 Power flow control using DC transformer.

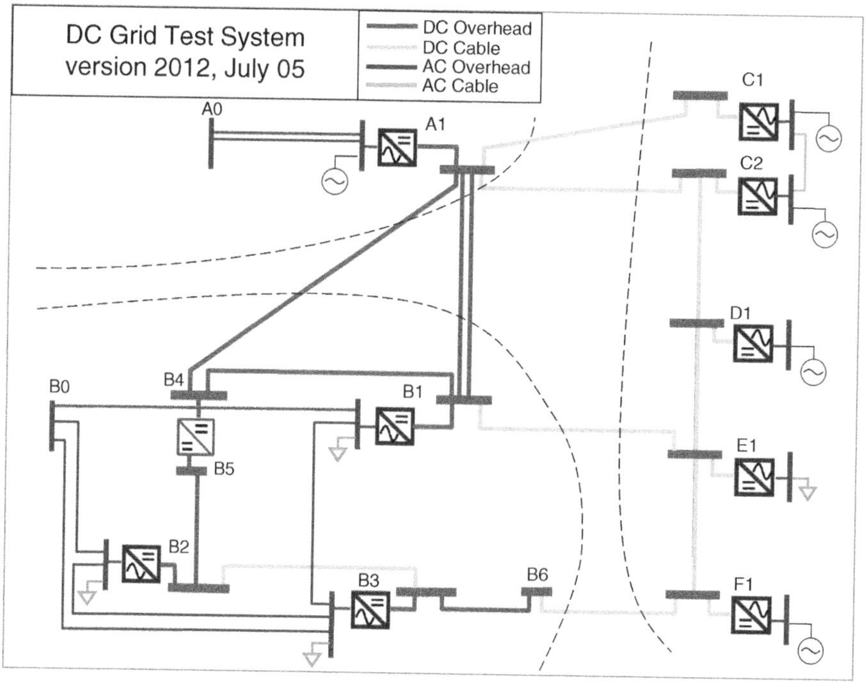

Figure 21.17 CIGRE WG B4-58 DC grid test model [8].

in [8]. The original model contains a DC transformer between buses B4 and B5. This DC transformer is also replaced by other DCPFCs in this study. Control modes of converters are given in Table 21.3.

The offshore buses are in power control mode and thus have no droop. The buses B4 and B6 are ride-through buses which are considered as power controlled buses regulating their power to zero (i.e. no power exchange with AC systems). B2 operates

TABLE 21.3 Control Modes of Converters

		Control Parameters		
Bus	Control Mode	k [MW/kV]	U_0 [kV]	P_0 [MW]
A1	Droop	−60	818.17	1500
B1	Droop	−100	809.54	−1900
B2	Voltage control	∞	800.00	—
B3	Droop	−50	802.05	800
B4	"Power control"	0	—	0
B6	"Power control"	0	—	0
C1	Power control	0	—	500
C2	Power control	0	—	500
D1	Power control	0	—	1000
E1	Power control	0	—	−100
F1	Power control	0	—	500

in voltage control mode and thus has an infinite droop. In simulation, the infinite droop can be represented by a very large droop of B2, for example, 10^6 MW/kV.

The power flow control of this test system was coded using MATLAB script in which the generated results were all validated by PSCAD simulation. Two cases are studied.

Case 1: Power control of line B1–E1 using a DCPFC

Case 2: Location change of a DCPFC, control target same to Case 1

21.5.2 Test Results

21.5.2.1 Case 1: Power Control of Line B1–E1 Using a DCPFC When the DCPFC inserts zero voltage or resistance, the power of line B1–E1 is computed to be −463.4 MW. The DCPFC then receives a power order to control the power of line B1–E1 to be −415 MW. Thus ΔP_c of the DCPFC is −48.3 MW. The power and voltage change (ΔP and ΔU) of each bus are shown in Table 21.4.

The results show that DCPFC operation has different impacts on buses controlled by different modes. A1, B1, and B3 are droop-controlled buses where both their DC voltage and power are changed (ΔU, ΔP). Amongst these, B1 has the biggest droop gain; this leads to the largest change of its power. ΔP is proportional to ΔU defined by the droop ($\Delta P = k \cdot \Delta U$).

B2 is the voltage-controlled bus, and thus its DC voltage is unchanged. However, its power is changed to balance the power change of other buses.

The other buses are power-controlled buses. Power of these buses is not influenced while their DC voltages are changed. B4, which is the bus with the DCPFC, has the largest DC voltage deviation while B3 has least DC voltage deviation. This is because B2 controls the DC voltage, B3 can then be considered as having the longest

TABLE 21.4 Results of Case 1

	SVS		VSR	
Bus	ΔU [kV]	ΔP [MW]	ΔU [kV]	ΔP [MW]
A1	1.038	−62.282	1.044	−62.662
B1	1.082	−108.237	1.088	−108.908
B2	0	181.289	0	182.742
B3	0.235	−11.766	0.236	−12.223
B4	1.983	0	1.971	0
B6	0.399	0	0.391	0
C1	1.035	0	1.018	0
C2	0.978	0	0.961	0
D1	0.892	0	0.876	0
E1	0.840	0	0.823	0
F1	0.546	0	0.534	0
DCPFC	3.9010	3.324	3.9012	3.324

TABLE 21.5 Results of Case 2

	Location of DCPFC				
	B1–B4	B1–E1	B4–B5	B3–B6	C2–D1
σ	1/207500	1/1081.9	1/124.85	1/193.47	1/1278.4
SVS power exchange [MW]	0.0022	0.3836	3.324	−2.149	0.322
SVS voltage inserted [kV]	2.864	0.747	3.901	−1.42	1.604
VSR power losses [MW]	0.0022	0.3836	3.324	−2.149	0.322
VSR resistance inserted [Ω]	3728.4	1.456	4.579	−0.94	8.001

electrical distance to the location of the DCPFC. The sum of the power changes of all buses indicates the power losses changes in the DC grid.

The control effects of the SVS and the VSR are similar. The SVS inserts a voltage of 3.9010 kV to adjust the power of the line B1–E1 to −415 MW. It acts as an inverter exporting 3.324 MW power to the AC system. The VSR inserts an average resistance of 4.579 Ω which leads its voltage to be 3.9012 kV, while the power dissipation on the resistor is 3.324 MW.

21.5.2.2 Case 2: Location Change of a DCPFC, Control Target Same to Case 1 The DCPFC is located at different DC lines to test the impact of location change on the power control of line B1–E1 to be −415 MW. The results are shown in Table 21.5. The line B4–B5 has the smallest value of σ which indicates the largest device power exchange or losses. It implies that locating the DCPFCs on the lightly loaded line will reduce its power rating.

Therefore, a good location of reducing the power rating of DCPFC will be on a lightly loaded DC line. However, if the voltage rating or the size of the resistor is the main factor of the cost, the DCPFC is suggested to be located at the controlled line.

The result of the VSR located on the line B3–B6 is not compliant with the physical system. The inserted resistance can only be nonnegative. This also means the VSR can only change the sign of voltage across it. The result of the VSR located on the line B3–B6 indicates that the inserted resistance is rather large (3728.4 Ω) which nearly leads to the open circuit of the line B1–B4. The control target is approaching the margin of the operating range of the VSR. Conversely, the SVS has a wider operation range. When located in the line B3–B6, it can insert a negative voltage of −1.42 kV to control P_{B1-E1}. When the SVS is located on the line B1–B4, it inserts a positive voltage to achieve the control target.

Figure 21.18 shows the operation range for the SVS and the VSR in controlling the power flow through the line B1–E1 when they are located on the line B4–B5. In Figure 21.18a, P_{B1-E1} has an approximately linear relationship with the inserted voltage. The power can be controlled from −600 MW to −100 MW by inserting a voltage between −10 kV and 30 kV. However, P_{B1-E1} has a nonlinear relation with the inserted resistance. The inserted resistance is nonnegative, which can only control P_{B1-E1} to be within −463.4 MW. When P_{B1-E1} is −330 MW, the required resistance is

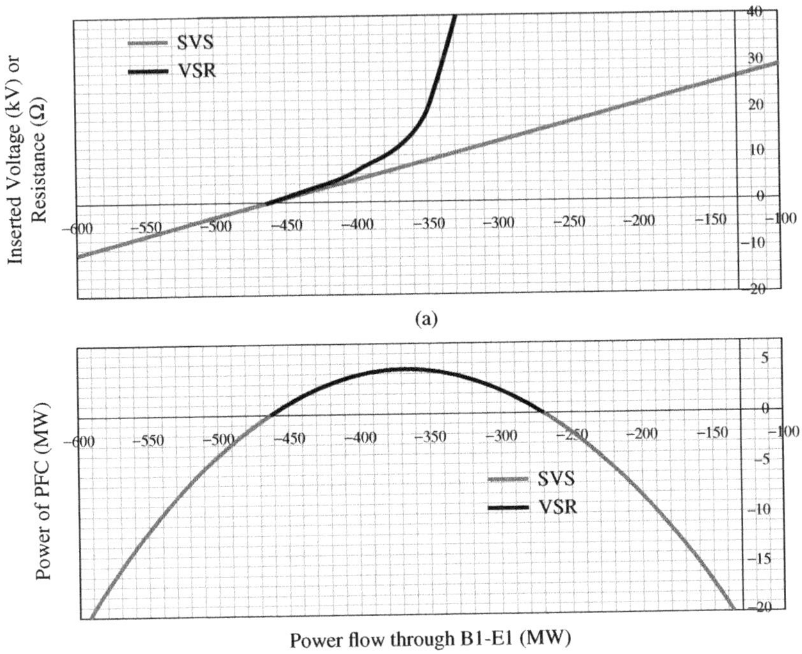

Figure 21.18 Operation range of DCPFCs.

40 Ω. Further increase of the resistance will only bring small change in P_{B1-E1}. With their control ranges the inserted voltage has relationship with inserted resistance as

$$\text{inserted resistance} = \frac{(\text{inserted voltage})^2}{\text{power exchange/losses}} \tag{21.18}$$

Figure 21.18b shows the local power of the DCPFC during the control of P_{B1-E1}. It can be seen that the VSR can only dissipate the power when controlling the P_{B1-E1} between −463.4 MW to −270 MW. Only the SVS can control P_{B1-E1} out of this range by importing the power from the AC system.

21.6 COMPARISON OF POWER CONTROL DEVICES

Comparisons of the three devices have been made in terms of the control flexibility, the power rating, and power losses. Figure 21.19 shows the control characteristics of the power flow control devices. The X axis is the current in per unit, in which 100% represents the rated current, and 0% means no current. The *Y* axis represents the control range of the devices, where 100% indicates device limits.

It can easily be seen that the series variable resistor has to change its value significantly to achieve a corresponding current adjustment. In addition, the current

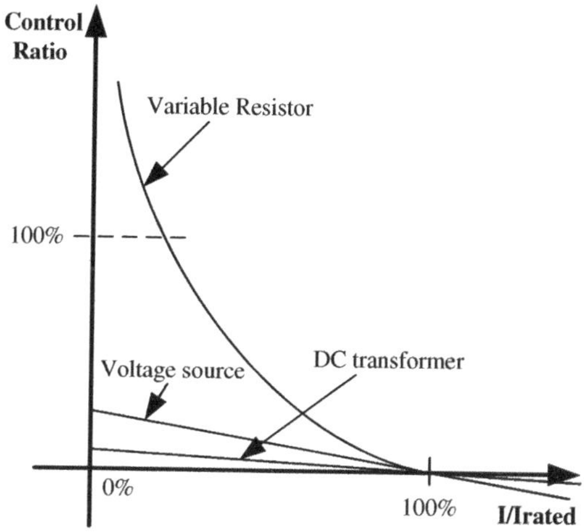

Figure 21.19 Control characteristics of three power flow control devices.

cannot decrease to 0; otherwise the series variable resistor loses its controllability over the current.

The series voltage source and the DC transformer have the linear characteristic of the current control. The rate of the series voltage source is slightly larger than that of the DC transformer. More importantly, the DC transformer and the series voltage source can adjust the current to more than the rated value or the negative value, which is impossible for the series variable resistor. In conclusion, the DC transformer and the series voltage source have larger control ranges and more effective characteristics than the series variable resistor has.

The rate of the devices is another considered aspect for applications. Even if the current of a line is reduced by 50%, the maximum resistor required is 1.8 times as much as the total line resistor. This would cause too much power losses.

In contrast, the same regulation requires only 0.025 p.u. output voltage (e.g., 5 kV in a 200-kV system) for the SVS, and for the DC transformer the required voltage is even less. Considering the topology of the SVS, the rated voltage of both converters could be 0.04 p.u., while the rated power should be 0.04 p.u. (e.g., 16 MW in a 400 MW system). However, the DC transformer is full power equipment, which has to manage all the power transferred via the line. Thus, the rated power and voltage for the DC transformer should be the same as the DC transmission system.

Finally, the losses of devices must be assessed before practical applications. The power losses of the SVS are the smallest, because there are only limited switching losses in two medium voltage converters, plus the power losses in the transformer. The VSR has fewer losses than the DC transformer. The losses of the VSR is generally 5–10% of the entire losses of the DC system, while the losses in the DC transformer including large power converters could occupy 14–15%.

In summary the VSR can only be utilized for a small range of regulation for the local current. The DC transformer has satisfied characteristics for the current regulation and the overlaid DC connection, but it has high costs and losses. In contrast, the SVS is the best solution for the power flow control because of the small losses, the large control range, and the availability of medium voltage converters.

21.7 CONCLUSIONS

Three types of DCPFC devices—DC transformers, variable series resistors (VSR), and series voltage sources (SVS)—have been proposed for the power flow control in a meshed DC grid. Their topologies, models, and control schemes were introduced and implemented in a simulation tool for the load flow and dynamic analysis.

DCPFCs can enhance the power flow controllability of the DC grid. The use of DCPFCs to control the power flow will be influenced by the control of the VSC, the system topology, line resistances, and the location of DCPFCs.

Among these three DCPFCs, SVSs are considered as the best solution in terms of the cost, power losses, and the control performance. DC transformers presented the control flexibility over the DC power flow, but their high power rating leads to the high cost. VSRs have merits of the low power rating and the simple structure, but have drawbacks of the high power losses and a limited control range. SVSs are satisfactory in all the aspects of flexibility of control, the feasible power rating, and low power losses.

REFERENCES

1. N. G. Hingorani, L. Gyugyi, and M. El-Hawary, *Understanding FACTS: Concepts and Technology of Flexible AC Transmission Systems*, vol. 1. New York: IEEE Press, 2000.
2. Q. Mu, J. Liang, Y. Li, and X. Zhou, "Power flow control devices in dc grids," in *Power and Energy Society General Meeting, 2012 IEEE*. New York: IEEE, 2012, pp. 1–7.
3. D. Jovcic, "Bidirectional, high-power dc transformer," *Power Delivery, IEEE Transactions on*, vol. 24, no. 4, pp. 2276–2283, 2009.
4. C. Meyer and R. W. De Doncker, "Design of a three-phase series resonant converter for offshore dc grids," in *Industry Applications Conference, 2007. 42nd IAS Annual Meeting. Conference Record of the 2007 IEEE*. New York: IEEE, 2007, pp. 216–223.
5. J. Robinson, D. Jovcic, and G. Joós, "Analysis and design of an offshore wind farm using a MV dc grid," *Power Delivery, IEEE Transactions on*, vol. 25, no. 4, pp. 2164–2173, 2010.
6. K. Wong, "Simple rules for the application of series voltage sources in power transmission systems," *Power Delivery, IEEE Transactions on*, vol. 13, no. 4, pp. 1488–1493, 1998.
7. M. P. Kazmierkowski and L. Malesani, "Current control techniques for three-phase voltage-source pwm converters: A survey," *Industrial Electronics, IEEE Transactions on*, vol. 45, no. 5, pp. 691–703, 1998.
8. T. K. Vrana, Y. Yang, D. Jovcic, S. Dennetière, J. Jardini, and H. Saad, "The CIGRE B4 DC grid test system," Cigré, Technical Report, 2014 [Online]. Available: http://b4.cigre.org/content/download/34038/1483266/version/1/file/CIGRE_B4_DC_Grid_Test_System_FINAL_corrected_version_with_Intro_V15.docx

CHAPTER 22

MODELING AND CONTROL OF OFFSHORE AC HUB

Xiaobo Hu,[1] Jun Liang,[2] and Jose Luis Dominguez-Garcia[3]

[1]Power System Department, China Electric Power Research Institute (CEPRI), Beijing, China

[2]Institute of Energy, School of Engineering, Cardiff University, Cardiff, Wales

[3]Research Group Wind Energy, Electrical Engineering Department, Catalonia Institute for Energy Research (IREC), Barcelona, Spain

22.1 REASONS FOR DEVELOPING AC HUB

A European SuperGrid [1–4] will allow the connection of offshore renewable energy sources in remote locations to existing onshore networks. It has been recognized that offshore wind resources will play an important role in determining the design and placement of SuperGrid infrastructure. In order to provide large power transmission capacity over long distances, it is proposed that the connection of offshore wind farms to onshore networks is performed by high-voltage direct-current (HVDC) links [5].

HVDC grid technology is one technical solution that allows the connection of multiple offshore wind farms to one or more onshore connection points. However, there are a number of technical issues and challenges to be overcome in the construction of HVDC grids [6, 7]. Knowledge of DC grid topologies and control strategies for HVDC grids as well as general operating experience of large DC systems is limited. Formal DC grid interoperability standards do not currently exist. Another crucial problem is that HVDC grids must be capable of isolating DC faults as quickly as possible to prevent over currents and DC voltage collapse. So far, high-voltage DC circuit breakers are not yet commercially available and have not been tested in the field. Some VSC topologies have been proposed that promise inherent fault blocking capability; however, these remain to be commercially proven and suffer additional complexity and potentially higher losses [8, 9]. A solution without DC circuit breakers has been proposed for multi-terminal HVDC operation [10]. However, this method needs to disconnect the entire DC backbone once a single DC fault occurs, which limits its applicability to large systems.

National Grid plc presented three design strategies for offshore electricity transmission [11], described as the "radial strategy," the "radial-plus strategy," and

HVDC Grids: For Offshore and Supergrid of the Future, First Edition. By Dirk Van Hertem, Oriol Gomis-Bellmunt, and Jun Liang.

the "integrated strategy." The radial and radial-plus strategies are effectively point-to-point configurations, whereas the "integrated strategy" interlinks neighboring offshore wind farm arrays using AC cables to form an offshore transmission network. This offers great flexibility and reliability in transmission options. The first advantage of the "integrated strategy" is that through power exchange between offshore platforms, the power injections to onshore transmission systems are controllable, which could potentially alleviate the requirement for some onshore reinforcements. Secondly, when the loss of any AC or DC cable occurs, there is still a path back to onshore grids, avoiding loss of generating capacity. Thirdly, through the coordination of AC circuit breakers equipped on the two AC sides of each point-to-point HVDC link, the requirement for fast-acting DC circuit breakers is eliminated.

An idea similar to the "integrated strategy," the AC hub, would play an important role in the initial phase (until 2020) of developing the offshore transmission grid and could also be used to interconnect different systems [12]. Based on only available technologies, AC Hub can connect long-distance HVDC lines with offshore wind farms by means of power converters and AC networks. Compared with multi-terminal HVDC solution, an AC hub arrangement can eliminate the need of DC circuit breakers by employing only point-to-point HVDC links.

This chapter is organized as follows. Section 22.2 introduces the "AC hub concept." Section 22.3 proposes and develops different economic functions to evaluate the frequency–cost dependence of AC hub components. Section 22.4 investigates the control strategy need to operate the system at variable frequency. Finally, the conclusions are summarized in Section 22.5.

22.2 WHAT IS THE AC HUB?

An AC Hub connects HVDC systems via AC to form a hybrid system, which can provide effective solutions for connecting offshore wind parks or oil and gas platforms. Figure 22.1 is a representation of an offshore AC Hub concept. In this configuration, AC circuit breakers which are located at the AC hub and onshore AC grids are used to isolate faults occurring on any individual HVDC link whilst leaving other links operational. VSCs of the HVDC links can be controlled to maintain the voltages of the AC hub.

It is worth remarking that if all wind turbine technologies within this offshore AC grid use fully rated converters, the offshore AC hub is essentially isolated from both wind turbine instantaneous dynamics and onshore grids by means of power electronic converters [13].This kind of network may run unsynchronized with onshore networks. In addition, there is no requirement for the AC hub to operate at the same nominal frequency as onshore networks. The operation at nonstandard frequencies (i.e., not 50 or 60 Hz) may confer technical and economic advantages. For instance, an operational frequency below 50 Hz can offer lower transmission losses and allow the use of longer cables, while an operational frequency above 50 Hz can reduce transformer sizes, in turn reducing offshore platform sizes.

It is an important advantage of this concept that it is largely based on technology existing today. As of today, there is much experience with DC links interconnecting

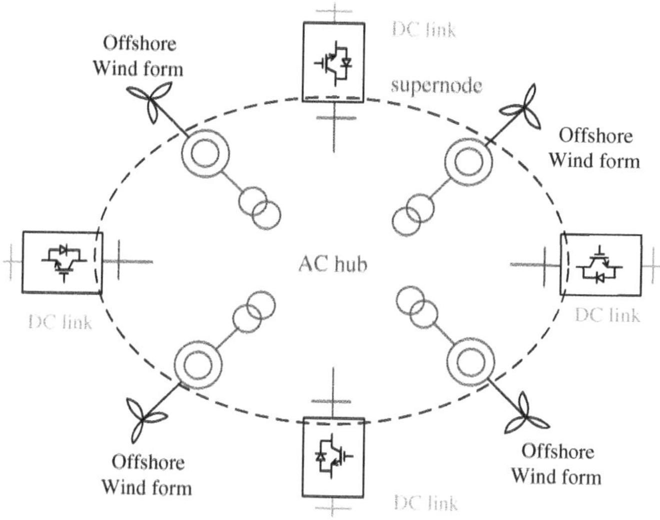

Figure 22.1 An AC hub for offshore wind farm connection.

two AC systems. There are quite a number of HVDC links operating or under construction, which are based on VSC technology. At the same time, the AC hub arrangement allows the connection of multiple wind arrays and HVDC substations, which does not require DC fault blocking capability. An example system which consists of two HVDC links is used to illustrate the difference of DC fault blocking methods between a MTDC and an AC hub system. For the four-terminal HVDC system in Figure 22.2a, fast DC circuit breakers are necessary to avoid the outage of the DC system in case of a DC fault. A permanent DC short-circuit between the DC cables of DC link1 is cleared by opening the corresponding DC circuit breakers CB1 and CB2. This results in the disconnection of DC link1 from the HVDC network. The AC hub system could realize the power sharing between two HVDC links through the offshore AC tie cable, as shown in Figure 22.2b. After the same DC fault occurs, the fault is isolated by opening AC circuit breakers CB1 and CB2 at the terminals of DC link1. So the need of DC circuit breakers could be removed in the AC hub system.

Another advantage of the AC hub arrangement is the opportunity to connect LCC–HVDC links to the AC hub where VSCs are available to establish and maintain the AC voltage. LCC–HVDCs generally offer higher power transfer capability and greater efficiency at a given cost point than VSCs, although they require significant reactive compensation and additional harmonic filtering which increases their overall space requirements.

An application of the AC hub is the "SuperNode" for the European SuperGrid, as shown in Figure 22.3 [14]. A SuperNode is an islanded AC network which interconnects a number of DC links together with wind parks. This AC network is only fed by the connected DC links and the wind parks but not connected to other AC networks.

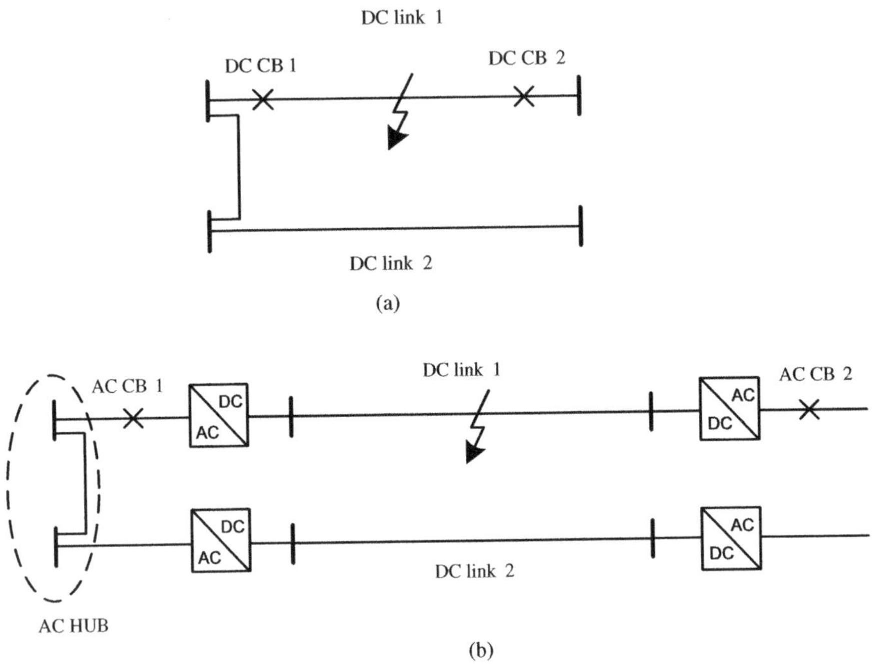

Figure 22.2 Schematic diagram of DC fault blocking. (a) Four-Terminal HVDC system. (b) AC hub system.

The AC hub will be a fully power electronics interfaced network, which leads to the lack of natural system inertia in the offshore AC grids without synchronous generators. The stability, control, and regulation of the voltage and frequency rely on the coordination of the voltage source converters of the HVDC links. The development needed to build AC hubs is mainly in the field of control and protection, which includes frequency control, power flow dispatch, and fault detection/clearing strategies.

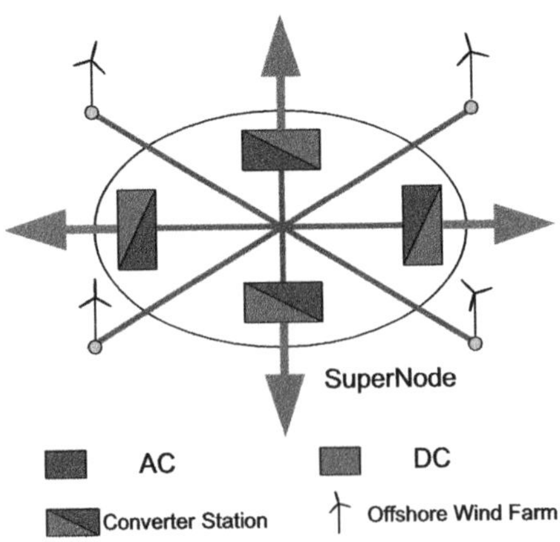

Figure 22.3 SuperNode concept.

Although the additional power converters and offshore platforms are needed and extra power losses are caused compared to a multi-terminal system, the AC hub concept is seen as a solution for the SuperGrid today [12]. It can be complemented by multi-terminal HVDC systems or HVDC grids in the future.

22.3 FREQUENCY-DEPENDENT MODELING OF AC HUB COMPONENTS

A standard frequency, 50 Hz or 60 Hz, can be used in an AC hub like an ordinary AC system. The AC components in an AC hub can use standard specifications which make it easy to design. However, an AC hub is essentially an isolated AC grid, connected to HVDC links through converters. The voltage and frequency of an AC hub are established by the converters. Therefore the operation of an AC hub is decoupled from other AC networks. A nonstandard frequency can also be used to achieve more economical design and operation, taking into account the frequency dependence of AC hub components, such as cables, transformers, reactive power compensations, and offshore platforms, which imply different costs at different frequencies.

22.3.1 Cables

Operation at a nonstandard frequency implies variation in the impedances and the ratings of the cables due to factors such as skin effect, proximity factor, and thermal capacity as discussed in IEC60287 [15]. The cable utilisation factor (ρ) is defined as [16]

$$\rho = \frac{I_{amp}^2 - I_c^2}{I_{amp}^2} \tag{22.1}$$

where I_{amp} is the ampacity of the cable and I_c is the cable charging current. The reactive current is considered to be solely the result of cable capacitance for both HV and MV cables (i.e., cable inductance is neglected [17, 18]). If cable charging current is assumed to be shared equally between both ends of the cable, I_c is maximum and equal at either end of the cable. For a fixed cable utilization factor, this assumption results in a cable twice as long when compared to a case in which the charging current is delivered only by one end of the cable (e.g., if the cable were to feed a simple resistive load [19]). A typical cable utilization factor is between 0.85 and 0.99 [20, 21].

IEC60287 provides standard formulas for calculating the maximum current capacity (ampacity) of a cable for varying frequency and for different cable geometries. In Figure 22.4, the relationship between the ampacity of several cables and the operating frequency is plotted. The dotted lines represent the frequencies above which the accuracy of IEC60287 is not assured [15]. Segmented conductors with $k_s = 0.435$ and $k_p = 0.37$ are assumed as defined in IEC60287.

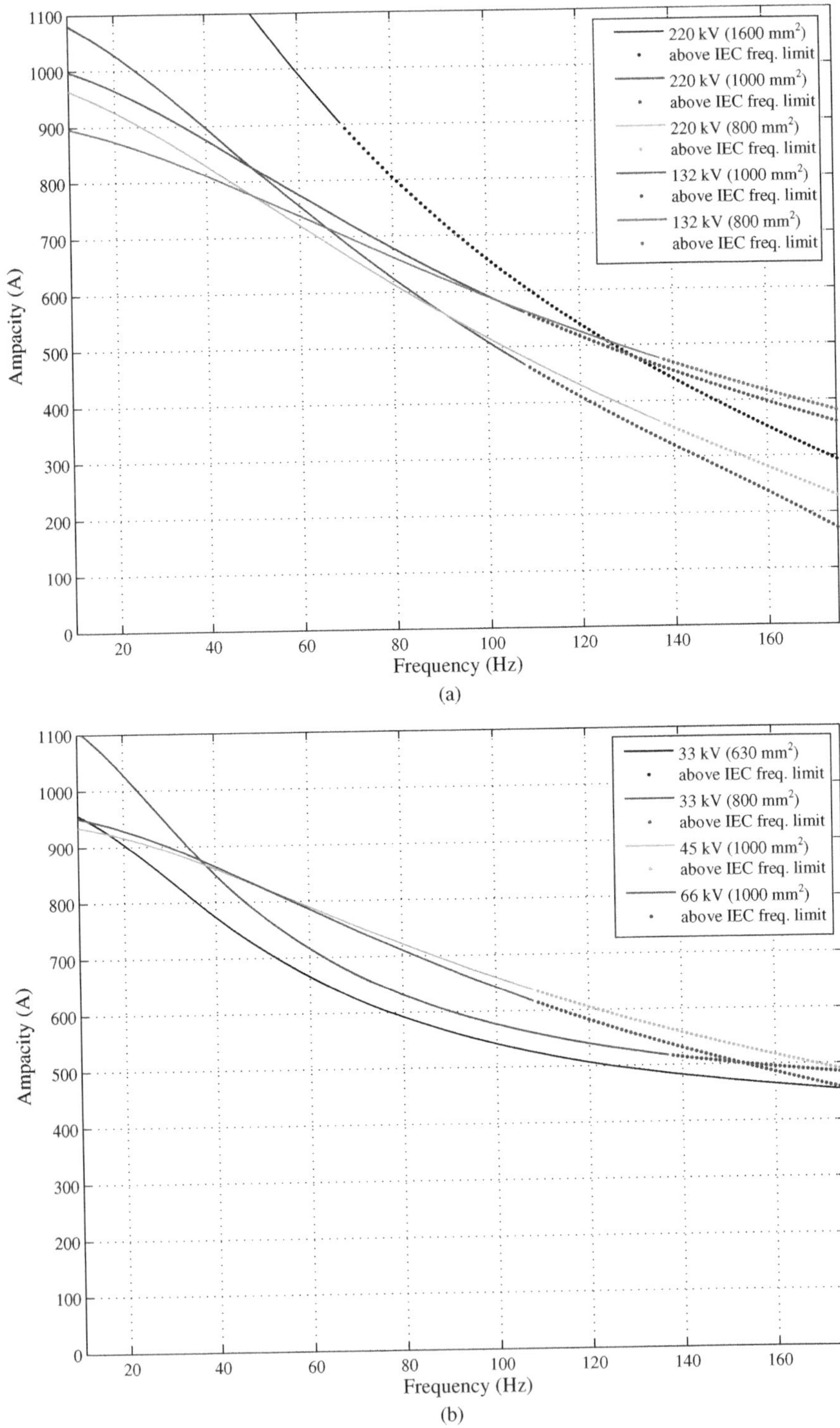

Figure 22.4 Maximum current capacity (ampacity) of HV and MV cables as function of frequency under different cable geometries. (a) HV cables ampacity. (b) MV cables ampacity.

The active power capacity of one cable (P_{cable}) depends upon the cable ampacity (I_{amp}), the desired length (l), and the frequency (f):

$$P_{cable} = \sqrt{3} \cdot U \cdot I_{amp} \cdot \sqrt{1 - \left(\frac{\pi \cdot f \cdot C_c \cdot U \cdot l}{\sqrt{3} \cdot I_{amp}} \right)^2} \tag{22.2}$$

which is clearly affected by the system frequency. The necessary number of cables (N_{cables}) to transfer an active power (P) can be calculated from

$$N_{cables} = \frac{P}{P_{cable}} \tag{22.3}$$

For offshore installations, cross-linked polyethylene (XLPE) cables are usually installed. The cost of XLPE submarine cables can be calculated with the following formula [22, 23]:

$$C_c = K_1 + K_2 \cdot e^{K_3 \cdot I_{amp}} \tag{22.4}$$

where C_c is the cost of the cable in k€/km, and the constant values (K_1, K_2, K_3) are dependent on the cable voltage rating. Therefore cable costs are affected by the nominal frequencies.

22.3.2 Transformer Design

All power transformers incorporated into the AC hub are affected by the change in system operational frequency, since the core area of a transformer (A_{core}) is related with the nominal frequency (f) by the following equation [24]:

$$A_{core} \approx \frac{E}{4.44 \cdot f \cdot N \cdot B_{sat}} \tag{22.5}$$

where E is the applied voltage, N is the number of turns, and B_{sat} is the saturation flux density of the core material. Assuming the applied voltage, the number of turns, and the saturation flux density of the core material remain constant, a relationship between core area and frequency can be determined. The resulting size–frequency scalings are shown in Table 22.1, in which core volume V_{core}, winding volume $V_{winding}$ and enclosure surface area A_{enc} scale with frequency.

TABLE 22.1 Core Volume, Winding Volume, and Enclosure Surface Area Scaling Factors of a Transformer for Different Operating Frequencies

Frequency Hz	V_{core}	$V_{winding}$	A_{enc}
25	2	2	$\sqrt[3]{2^2}$
50	1	1	1
100	$\frac{1}{2}$	$\frac{1}{2}$	$\frac{1}{\sqrt[3]{2^2}}$

Transformer core area (A_c) is inversely proportional to operating frequency. Winding DC resistance (R_w), the number of turns (N), and the winding aspect ratio α are assumed to be constant with frequency. The winding aspect ratio determines the relative sizing of core and winding such that $t_w = \alpha \cdot t_c$, where t_w is the thickness of the winding and t_c is the thickness of the core.

The winding wire thickness (t_{cu}) is related to winding DC resistance (R_w) by

$$R_w = \frac{l_w}{t_{cu}^2 \cdot \sigma} = \frac{4 \cdot (t_c + t_w) \cdot N}{t_{cu}^2 \cdot \sigma} \tag{22.6}$$

where l_w is the length of the winding, which is simply N times the average perimeter of the winding and σ is the conductivity of the winding material. The height of the winding may be found as

$$h_w = \frac{t_{cu}^2}{t_w} \cdot N = \frac{4 \cdot (t_c + t_w) \cdot N^2}{R_w \cdot \sigma \cdot t_w} = \frac{4 \cdot (1 + \frac{1}{\alpha}) \cdot N^2}{R_w \cdot \sigma} \tag{22.7}$$

which demonstrates that winding height is independent of frequency. The core volume is

$$V_c = h_w \cdot A_c \tag{22.8}$$

and the winding volume is

$$V_w = h_w \cdot \left[(t_c + 2 \cdot t_w)^2 - t_c^2 \right] = h_w \cdot 4 \cdot \alpha \cdot (\alpha + 1) \cdot t_c^2 = 4 \cdot \alpha \cdot (\alpha + 1) \cdot h_w \cdot A_c \tag{22.9}$$

Thus both core and winding volumes scale inversely proportional to frequency (as $A_c \propto 1/f$). The area of the enclosure (A_{enc}) of the transformer is also frequency-dependent. Assuming the transformer is equal in all linear dimensions, the enclosure is $A_{enc} \propto 1/f^{2/3}$. The cost of transformers designed for use at 50 Hz, depending on their size, can be calculated with the following formulas [22]:

$$LV/MV: \quad C_{Tr50} = -153 + 131 \cdot T_r^{0.447} \tag{22.10}$$

$$MV/HV: \quad C_{Tr50} = 42.7 \cdot T_r^{0.751} \tag{22.11}$$

where C_{Tr50} is the transformer cost in k€ and T_r is the rated power of the transformer in MVA. Considering the percentages of the transformer material costs [25], a variation on the total cost of different parts of the transformer can be computed depending on the frequency as follows:

$$C_{Tr} = \frac{0.325 \cdot f_r + 0.22 \cdot f_r + 0.164 \sqrt[3]{f_r^2}}{0.325 + 0.22 + 0.164} \cdot C_{Tr50} \tag{22.12}$$

where f_r is the normalized frequency, calculated as a ratio between the standard frequency and the analyzed frequency ($f_r = 50Hz/f$). It can be seen that the cost of the transformer is less at higher frequencies due to the reduction in construction material quantity.

22.3.3 Reactive Power Compensation

The reactive nature of all undersea cables is dominated by their large capacitance per unit length [26] and therefore shunt reactors may be used to provide reactive compensation. The required inductance is related to cable capacitance by

$$L = \frac{1}{C_c \cdot (2 \cdot \pi \cdot f \cdot l)^2} \tag{22.13}$$

The peak charging current of the cable $I_{c(pk)}$ is proportional to frequency [26]:

$$I_{c(pk)} = 2 \cdot \sqrt{2} \cdot \pi \cdot C_c \cdot U \cdot f \cdot l \tag{22.14}$$

The peak stored energy in the shunt reactor (E) can be calculated as

$$E = \frac{1}{2} \cdot L \cdot I_{c(pk)}^2 \tag{22.15}$$

By substituting (22.13) and (22.14) into (22.15), it can be concluded that the energy storage requirements (and hence physical size) of compensating shunt reactors per cable of a fixed length is frequency-independent. Therefore the total cost of the reactive power compensation, shunt reactors in this case, is only affected by the number of cables to be compensated, which can be evaluated as [22, 23]

$$C_{rpc} = C_{50} \cdot \frac{N_{cf}}{N_{c50}} \tag{22.16}$$

where C_{50} is the cost of shunt reactor at 50 Hz, N_{cf} is the number of required cables at a given frequency, and N_{c50} is the number of required cables at 50 Hz.

22.3.4 Offshore Substation Platform

It is assumed that the weight of transformers takes a ratio of β of the total weight of a MV/HV substation platform. Using the transformer reduction ratios obtained in Table 22.1 and assuming that transformer core and winding weight is dominant, the variation of the platform weight with frequency may be approximated as

$$W_{pf} = W_{pf50} \cdot (1 - \beta + \beta \cdot f_r) \tag{22.17}$$

where f_r is the normalized frequency, W_{pf50} is the platform weight at 50 Hz, and $1 - \beta$ represents the weight ratio of elements in the platform which are not frequency-dependent.

In order to evaluate the offshore substation platform cost C_{ss} in k€, the following equation [22] is used:

$$C_{ss} = 2530 + 88.7 \cdot P_{wt} \cdot (1 - \beta + \beta \cdot f_r) \tag{22.18}$$

where P_{wt} is the rated power of wind farms and f_r is the normalized frequency.

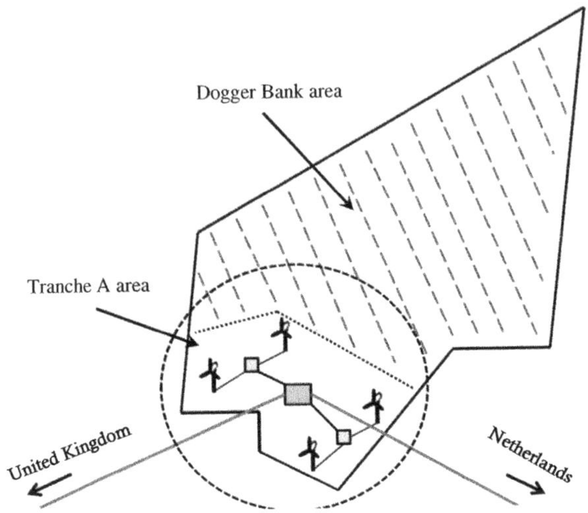

Figure 22.5 Dogger Bank geographical representation with the AC hub implemented in Tranche A.

22.3.5 Evaluation of Costs for an AC Hub: Dogger Bank Tranche A Case

An economic study based on the first stage of Dogger Bank (so-called Tranche A) is used to evaluate the use of fixed nonstandard operating frequencies. A geographical location and an AC hub design are illustrated in Figures 22.5 and 22.6. For this study, four wind farms of 1 GW each are considered.

The total cost considers not only the electrical devices (such as cables, reactive power compensation, and transformer) but also the offshore infrastructure (including only the offshore platforms required). The total capital cost evaluation for different frequencies is illustrated in Figure 22.7.

In Figure 22.7, a minimum cost for this example can be found at 85 Hz for the integer cable number curve, whereas the minimum is shifted to 93 Hz when fractional number are considered. It is worth mentioning that the use of frequencies below the standard is more expensive, whereas the consideration of frequencies above the standard leads to cheaper systems.

22.4 AC HUB CONTROL USING VARIABLE FREQUENCY

The voltage and frequency of an AC hub are established by the VSCs of HVDC links. A simple method to control an AC hub is to maintain a fixed frequency using one VSC and to regulate the power using other VSCs—that is, using a master-slave control mode. However, since it is unlikely that any commercial or domestic loads will be directly connected to the AC hub, it is not necessary to use a fixed standard frequency.

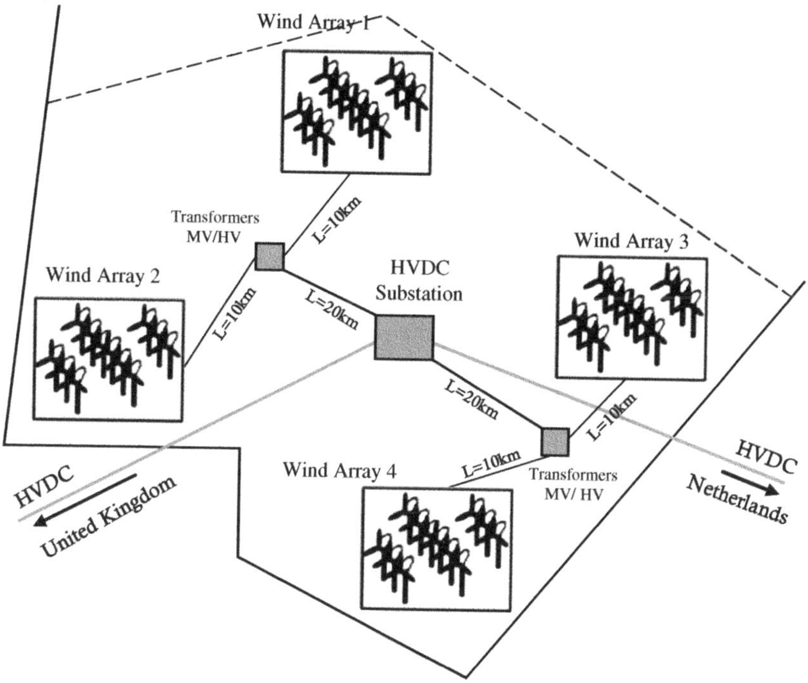

Figure 22.6 Close-up view of Tranche A showing the cable length requirements with an HVDC substation located at the center of the area.

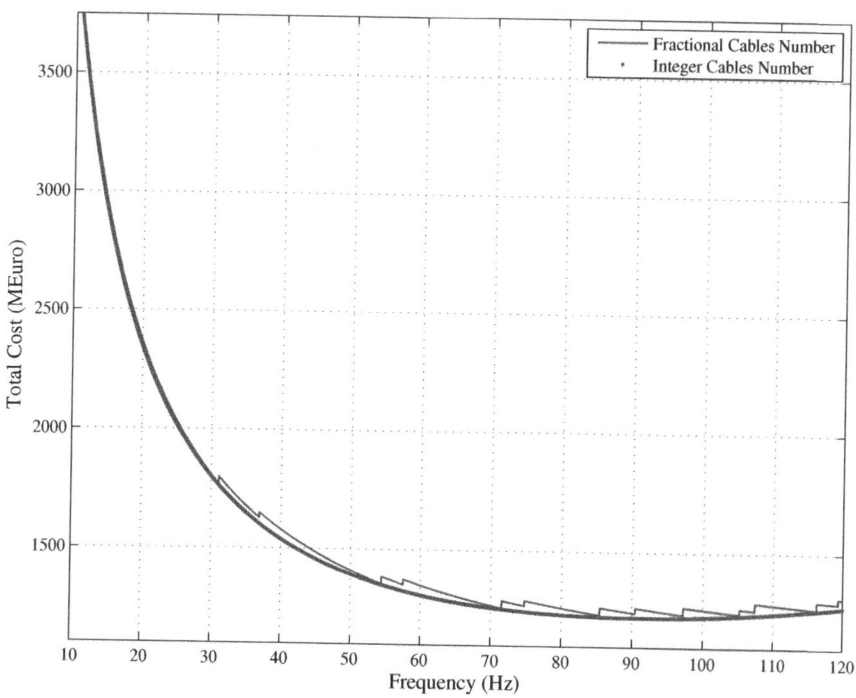

Figure 22.7 Total capital cost of the AC hub in M€ as a function of frequency.

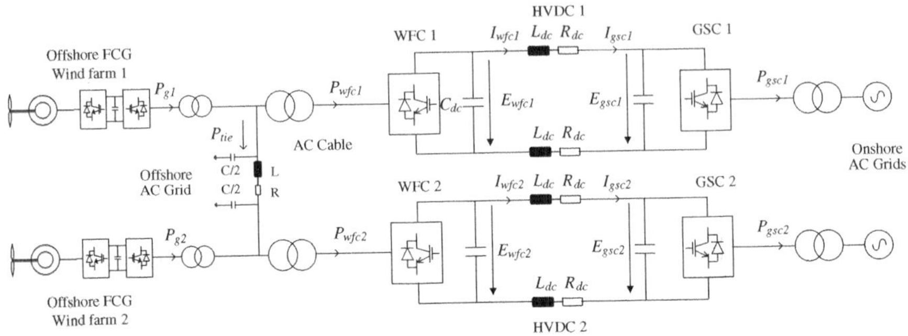

Figure 22.8 A simplified representation of a transmission system using the AC hub concept.

Therefore it becomes possible to use a variable frequency as a signal within the AC hub to achieve power scheduling and system control in order to remove the need of fast communication in the control.

A transmission system using the AC hub concept, as shown in Figure 22.8 [27], is used to demonstrate the control design using variable frequency. This system consists of two offshore wind farms and two point-to-point HVDC links. AC cables are used to connect the two wind farms to form an offshore AC grid. Each wind farm consists of fully rated converter generators (FCGs). Based on vector control methods, generator side converters of the FCGs are usually designed to track the maximum power point within their power ratings or to follow the power order from system operators. The network side converters of the FCGs are used to maintain the DC voltages of the converters and deliver wind power to the offshore AC network. The converters of the HVDC links include wind farm side voltage source converters (WFCs) and grid side voltage source converters (GSCs), which both use the three-phase, three-line, two-level configuration with IGBTs. A simple model with a resistance and an inductance connected in series is used for the DC cables, and a π-equivalent circuit model is used for the AC cables.

22.4.1 Control System Design

The AC hub frequency is used not only for fault-ride through control to reduce power outputs from each wind farm, but also for power flow control to schedule the power. The wind farm side voltage source converters of the HVDC links are operated as synchronous generators to maintain and regulate the power flows, voltages, and frequency on the network. Similar to the governor droop characteristic of synchronous generators, the wind farm side converters can coordinate their active power outputs through observation of the AC hub frequency. Primary and secondary power regulation controllers are designed for the wind farm side converters to achieve automatic power flow coordination. In addition, a "transient offshore frequency adjustment" is applied to the wind farm converters and a "power reduction strategy" is applied to the wind farms based on the AC hub frequency in order to prevent HVDC system overvoltage without a requirement for fast communication links.

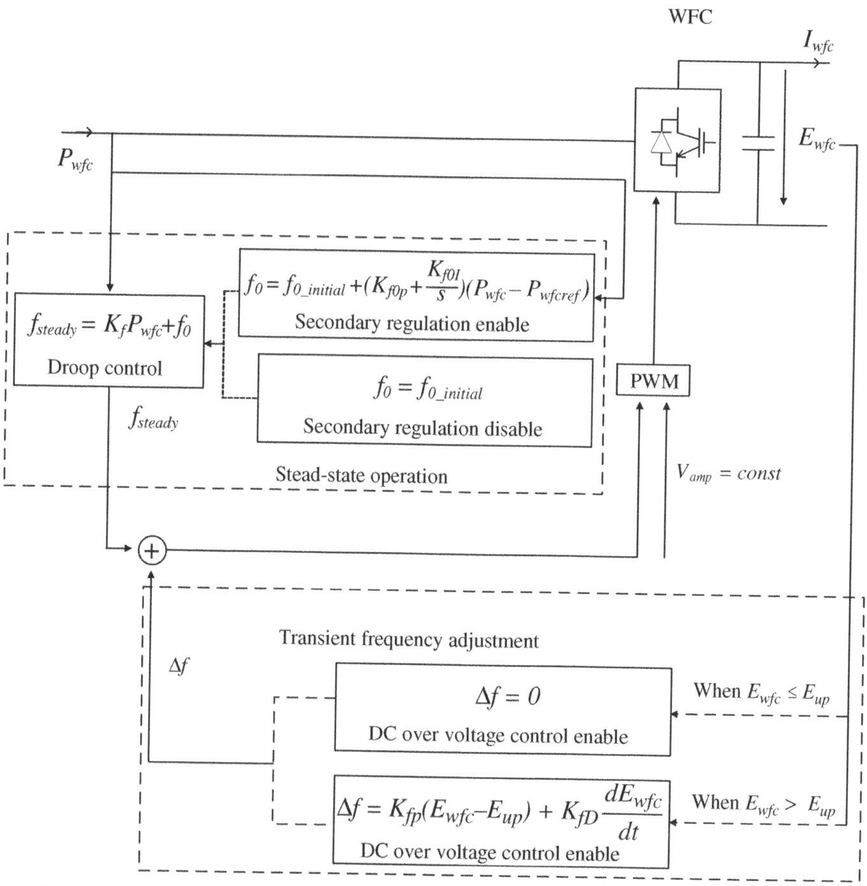

Figure 22.9 Wind farm side converter controllers of HVDC links-regulating AC hub.

A typical VSC controller, which has two hierarchical control loops, is used for GSC to regulate its DC voltage and deliver power to the onshore grid. This section focuses on the WFC control and wind turbine converter control in an AC hub.

22.4.1.1 Wind Farm Side Converter Each WFC is operated to regulate the voltage and the frequency of the AC hub. The AC voltage amplitude U_{amp} is set to be constant and the frequency f is determined by two parts: steady-state operation frequency f_{steady} and transient frequency adjustment Δf, as shown in Figure 22.9. Under steady-state operation, a primary power regulation loop, which has a power–frequency droop characteristic as given in (22.19), is designed for the WFC which resembles the primary frequency regulation of a synchronous generator.

$$f_{steady} = K_f \cdot P_{wfc} + f_0 \tag{22.19}$$

Here P_{wfc} is the active power injected into the HVDC link through this WFC. K_f and f_0 are the droop slope and frequency set-point respectively.

According to (22.19), the increase of P_{wfc} will cause WFC to increase the frequency. The phase angle differences between this wind farm side converter and other converters will increase through the integration of the frequencies. This forces some power to flow from this WFC to other nodes, which results in the decrease of P_{wfc}. This offers a negative feedback to form a stable system in which power sharing between WFCs is achieved. When the transmission system is in the steady state, the frequencies generated by WFCs should be at the same value. Ignoring the losses of the AC hub, the equations of P_{wfc1} and P_{wfc2} for the transmission system shown in Figure 22.8 can be obtained, as given by

$$P_{wfc1} - \frac{K_{f2}}{K_{f1}} \cdot P_{wfc2} + \frac{f_{01} - f_{02}}{K_{f1}} = 0 \tag{22.20}$$

$$P_{wfc1} + P_{wfc2} - P_{g1} - P_{g2} = 0 \tag{22.21}$$

Obviously, power flow dispatch is determined by the droop parameters and the steady-state power output of the two wind farms.

By resembling the secondary frequency regulation of a synchronous generator, a change in the frequency set-point f_0 shifts up/down the droop characteristic curve of the WFC and thus changes the power through the WFC, and vice versa. A proportional-integral controller is designed to determine f_0 and the regulation objective is to make P_{wfc} equal to a given reference value P_{wfcref}, as given by

$$f_0 = f_{0\_initial} + \left(K_{f0P} + \frac{K_{f0I}}{s}\right) \cdot \left(P_{wfc} - P_{wfcref}\right) \tag{22.22}$$

where $f_{0\_initial}$ is the initial value of f_0, K_{f0P} and K_{f0I} are the proportional and integral gains, respectively.

When this regulation is enabled, the power reference value will be sent to the WFC from the system operator. If P_{wfc} is smaller than P_{wfcref}, f_0 decreases which forces the WFC to receive more power and thus P_{wfc} increases. When the regulation is disabled, f_0 is kept as the initial value $f_{0\_initial}$. Under transient conditions, if the DC voltage of the WFC E_{wfc} is higher than an upper threshold E_{up}, the controller will operate in a DC overvoltage control mode. The frequency adjustment Δf is made such that

$$\Delta f = K_{fP} \cdot \left(E_{wfc} - E_{up}\right) + K_{fD} \cdot \frac{dE_{wfc}}{dt} \tag{22.23}$$

where K_{fP} and K_{fD} are the proportional and differential gains, respectively.

The frequency increase of this WFC will force power to be transferred to other WFCs so as to prevent an overvoltage of this WFC.

22.4.1.2 Power Reduction Method of Wind Turbines

If the power generated by wind farms cannot be fully transmitted to the onshore AC grids due to an onshore AC grid fault or a DC fault, power reduction is required to maintain power balance and avoid DC overvoltage of HVDC links.

The power reduction controllers of the wind turbines are shown in Figure 22.10. The active power reference of a wind turbine $P_{r-order}$ is obtained by multiplying P_{order} by K_d, where P_{order} is the power order determined by maximum power point tracking strategy (MPPT) and K_d is the corresponding power reduction coefficient. Under

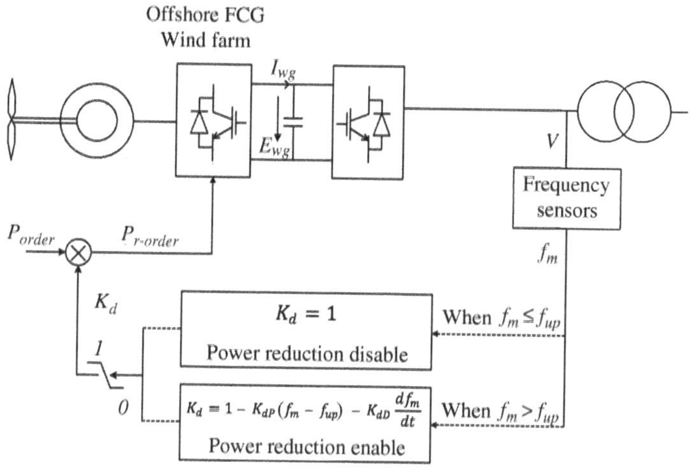

Figure 22.10 Power reduction controllers of wind farms to modify AC hub frequency.

normal operation, K_d is maintained as 1 and so maximum wind energy capture is implemented.

If an onshore grid fault close to an HVDC link occurs, the real power transmitted through the HVDC link is reduced. This results in a power imbalance and DC voltage rise of the HVDC link. Through the transient frequency adjustment of (22.23), some power is transferred to other HVDC links. However, if the initial wind power generation is high and all the remaining HVDC links are not able to transmit the wind power to the onshore AC grids, this in turn results in HVDC link voltage raise and triggers the transient frequency adjustment of the corresponding HVDC links. In this case, the frequency of the AC hub increases further. When the frequency measured at one wind turbine f_m passes an upper threshold f_{up}, the power reduction control of the wind turbine is triggered to adjust the power reduction coefficient, as given by

$$K_d = 1 - K_{dP} \cdot (f_m - f_{up}) - K_{dD} \cdot \frac{df_m}{dt} \qquad (22.24)$$
$$0 \le K_d \le 1$$

where K_{dP} is the proportional gain and K_{dD} is the differential gain. Due to the fast response of wind turbine converters, the output power of the wind turbines is reduced quickly. The resulting excess energy produced by the wind turbines can be either stored in the rotors by regulating their rotor speeds or dissipated by pitching the blades or in DC choppers attached to the DC bus [28]. Through the power reduction from the wind farms, DC overvoltage of the HVDC links is prevented.

Importantly, because the inputs of the controllers are from local frequency signals only, fast communication is not required under the proposed control strategy.

22.4.2 Case Study

The transmission system in Figure 22.8 is modeled in EMTDC/PSCAD. The rated power capacity of wind farms and transformers is 60 MW. The turn ratios are

TABLE 22.2 Parameters of the System Using the AC Hub Concept

HVDC Link		AC Tie Cable	
Capacitance C_{dc}	150 μF	Capacitance C	6 μF
Resistance R_{dc}	0.5 Ω	Resistance R	0.3 Ω
Inductance L_{dc}	5 mH	Inductance L	1.5 mH

0.69/33 kV for the generator transformers, 33/62.5 kV for the wind farm transformers, and 62.5/132 kV for the grid transformers. The rated DC voltage of HVDC links is 140 kV. The leakage reactance of each transformer is 10%. Thirty 2-MW wind turbines are lumped into one generator at each wind farm. Considering the quick regulation characteristic of the wind turbine converters, a first-order inertia response to the given power is applied to represent the generator. This simplified model imitates the active power regulation performance but neglects the internal dynamics of the generator. The time constant of the first-order inertia model is set as 0.01 s. The system parameters are listed in Table 22.2 and the control parameters are listed in Table 22.3. Here the frequency set-point f_0 corresponds to the no-load frequency and is selected to make the steady-state frequency around 50 Hz under the rated power.

Case studies are carried out to investigate power flow control capability and fault response of the system.

22.4.2.1 *Power Flow Control* This case is presented to verify the effect of the primary and secondary power regulation.

The initial active powers of wind farm 1 and 2 are both 30 MW. The secondary power regulation controller of WFC1 is enabled at 2.5 s. The power flow dispatch reference $P_{wfc1ref}$ is set as 15 MW. At 4.5 s, the power generated by wind farm 1 changes from 30 MW to 42 MW. The simulation results are shown in Figure 22.11.

The initial power of WFC1 is 30 MW, as shown in Figure 22.11a. As P_{wfc1} exceeds $P_{wfc1ref}$ at 2.5 s, the frequency set-point of WFC1 f_{01} increases quickly according to (22.22). This results in the increase of f_1, as shown in Figure 22.11b. Through the integration of frequencies, the phase angle difference between WFC1 and WFC2 increases, which causes power to flow from WFC1 to WFC2 (P_{wfc1} decreased

TABLE 22.3 Parameters of the Proposed Control System

Parameters for Droop Control and Secondary Power Regulation			
K_f	0.01	$f_{0\_initial}$	49.70
K_{f0P}	0.01	K_{f0I}	0.10
DC overvoltage Control		**Power Reduction Control**	
K_{fP}	0.250	K_{dP}	0.400
K_{fD}	0.003	K_{dD}	0.020
E_{up}	143.00	f_{up}	50.50

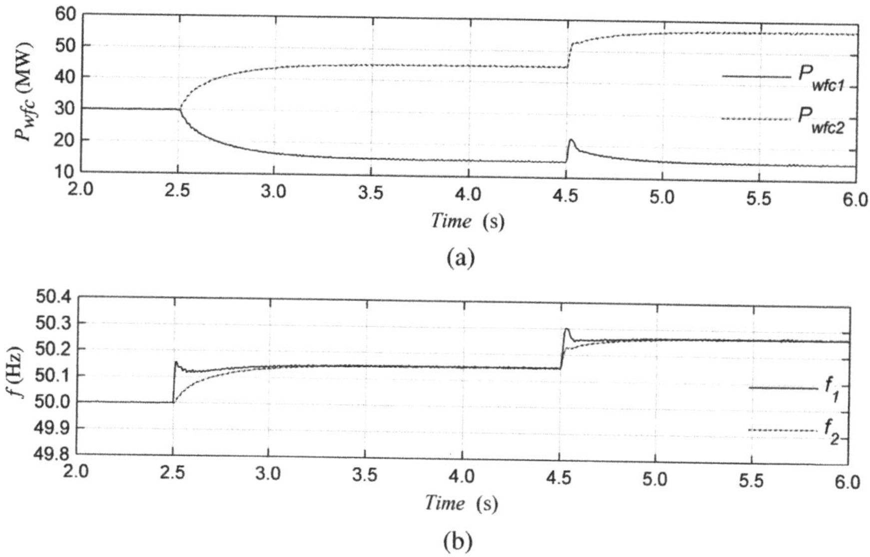

Figure 22.11 Power flow dispatch through primary and secondary power regulation. (a) Active power through WFC, P_{wfc}. (b) Frequency of AC hub, f.

and P_{wfc2} increased). According to (22.19), the variation of f_2 follows that of P_{wfc2} and so f_2 increases gradually. Through the mutual regulation of the PI controller and the power-frequency droop characteristic, the system reaches an equilibrium point. At 4.0 s, the power flows of WFC1 and WFC2 are 15 MW and 45 MW nearly. The steady-state frequency of the AC hub is 50.15 Hz.

After the step change of power generated by wind farm 1 at 4.5 s, the system response is achieved through the combination of the fast primary power regulation and the slow secondary power regulation. The desired power flow is obtained. The final power flows of WFC1 and WFC2 are 15 MW and 57 MW, respectively. The final frequency of the AC hub is 50.27 Hz.

The steady frequency deviation is quite small, which is mainly determined by the droop slope K_f. So while the proposed strategy is used to realize the desirable power flow dispatch, any changes in the AC hardware (transformers, breakers, lines, etc.) will not be required. This performance makes the proposed strategy more practical and attractive.

22.4.2.2 DC Fault Response This case is used to verify the ability of the variable frequency control when the system is subjected to a permanent DC fault. Rated wind powers are generated from the two wind farms. A permanent DC short-circuit near WFC1 occurs at 2.2 s. After 10 ms, the fault is cleared by opening the two AC circuit breakers of HVDC1. So WFC1 is disconnected from the AC hub and GSC1 is disconnected from the onshore AC grid. The simulation results are given in Figure 22.12.

During the fault, the DC voltage of WFC1 becomes zero, as shown in Figure 22.12a. This results in the quick decrease of P_{wfc1} and the quick increase of

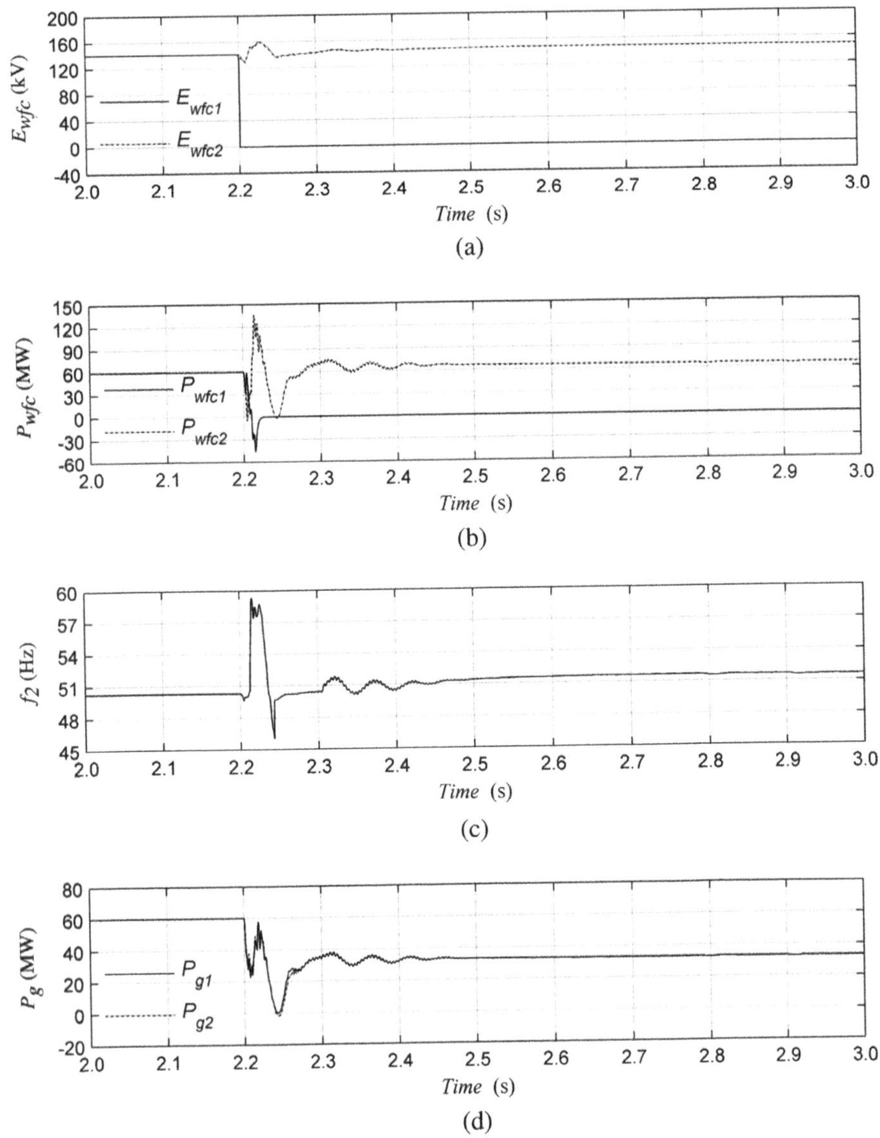

Figure 22.12 System responses subject to a permanent DC grid fault. (a) DC voltage of WFC, E_{wfc}. (b) Active power through WFC, P_{wfc}. (c) Frequency of WFC2, f_2. (d) Output wind power of wind farm, P_g.

P_{wfc2}, as shown in Figure 22.12b. After WFC1 is disconnected from the AC hub, the active power through WFC1, P_{wfc1}, becomes zero. This leads to a transfer of power to the remaining HVDC2 link and makes GSC2 reach its maximum power rating. This in turn results in the quick raise of the DC voltage of HVDC2 link. The frequency of WFC2 is regulated to a very large value, as shown in Figure 22.12c.

The high frequency triggers the power reduction control in both wind farms and the power output from the wind farms are decreased to reduce the DC overvoltage, as shown in Figure 22.12d. During the transient period, the DC voltage of WFC2 is always below 160 kV, as shown in Figure 22.12a. Finally, wind power generation of both wind farms are 33 MW, leading to a total of 66 MW wind power being transferred to the onshore AC grid through GSC2. Here the 10% overload capability of the remaining GSC2 has been fully utilized. Note that due to the disconnection of WFC1, the AC hub frequency is regulated only by WFC2. So for the AC hub system, the DC fault-blocking capability is achieved by opening the AC circuit breakers. The power transfer capability is the same as MTDC with DC circuit breakers [29].

22.5 CONCLUSIONS

The AC hub concept has been presented as an alternative to the use of MTDC systems for the aggregation of geographically close wind power resources. In comparison to the AC Hub, point-to-point configuration without an offshore AC grid lacks the redundancy and flexibility of power flow while MTDC system faces the difficulty of the unavailability of DC circuit breakers. The AC hub system using nonstandard/variable frequency can achieve the same power transfer capability and dynamic responses as MTDC systems without DC circuit breakers.

In brief, to increase the frequency of operation into the AC hub, some main advantages as the reduction on size and weight of the transformers and platforms are presented. However, some drawbacks can be found in this modification as the cable length limitations due to the reactive power increment and higher power losses owing to the requirement of extra amount of cables.

In summary, the feasibility and benefits of the AC hub using variable frequency have been proved. During steady-state operation, power-frequency droop control for the wind farm side converter of each VSC-HVDC link is used to achieve primary power regulation. Secondary power regulation may also be achieved by changing the set-point of the droop characteristic. During transient operation, the frequency of the AC hub is adjusted according to the DC voltage variation of HVDC links. An auxiliary frequency-responsive controller is defined which produces a power reduction coefficient such that when the local measured frequency exceeds a threshold, the wind turbines reduce power generation in order to avoid a DC overvoltage. The AC hub without DC circuit breakers can maintain stability by (a) rescheduling the power through variable frequency operation and (b) reduction of wind power generation through frequency-responsive power reduction control.

REFERENCES

1. S. Gordon, "Supergrid to the rescue [electricity supply security]," *Power Engineer*, vol. 20, no. 5, pp. 30–33, 2006.
2. B. Andersen, F. Groeman, D. Alvira, O. Anaya-Lara, G. Tang, J. Hanson, M. Haeusler, K. Karoui, P. Muttik, S. Nguefeu, M. Piekutowski, J. Ruiz, K. Sobczak, Bogdan Soebrink, T. Thiringer, and

D. Woodford, "Integration of large scale wind generation using hvdc and power electronics," Cigré, Technical Report 370, Feb. 2009, wG B4.39.

3. Friends of the SuperGrid, "Position paper on the ec communication for a european infrastructure package: Supergrid phase 1," Friends of the SuperGrid, Technical Report, 2010.
4. Energinet.dk, Svenksa Kraftnät and Vattenfall Europe Transmission, "An analysis of offshore grid connection at kriegers flak in the Baltic sea," Energinet.dk, Svenksa Kraftnät and Vattenfall Europe Transmission, Technical Report, 2009.
5. P. Bresesti, W. L. Kling, R. L. Hendriks, and R. Vailati, "Hvdc connection of offshore wind farms to the transmission system," *Energy Conversion, IEEE Transactions on*, vol. 22, no. 1, pp. 37–43, 2007.
6. Alstom Grid, "Multi-terminal hvdc system for large offshore wind park grid integration," Alstom Grid, Technical Report, 2010.
7. D. Van Hertem and M. Ghandhari, "Multi-terminal VSC HVDC for the european supergrid: Obstacles," *Renewable and Sustainable Energy Reviews*, vol. 14, no. 9, pp. 3156–3163, 2010.
8. J. Yang, J. E. Fletcher, and J. O'Reilly, "Multiterminal dc wind farm collection grid internal fault analysis and protection design," *Power Delivery, IEEE Transactions on*, vol. 25, no. 4, pp. 2308–2318, 2010.
9. M. Merlin, T. Green, P. Mitcheson, D. Trainer, D. Critchley, and R. Crookes, "A new hybrid multi-level voltage-source converter with dc fault blocking capability," in *AC and DC Power Transmission, 2010. ACDC. 9th IET International Conference on*. IET, 2010, pp. 1–5.
10. L. Tang and B.-T. Ooi, "Locating and isolating dc faults in multi-terminal dc systems," *Power Delivery, IEEE Transactions on*, vol. 22, no. 3, pp. 1877–1884, 2007.
11. National Grid Electricity Transmission, "Offshore development information statement 2010," National Grid, Technical Report, 2010.
12. Friends of the SuperGrid, "Roadmap to the supergrid technologies: Final report," Friends of the SuperGrid, Technical Report, 2013.
13. J. L. Domínguez-García, D. J. Rogers, C. E. Ugalde-Loo, J. Liang, and O. Gomis-Bellmunt, "Effect of non-standard operating frequencies on the economic cost of offshore ac networks," *Renewable Energy*, vol. 44, pp. 267–280, 2012.
14. F. of the SuperGrid, "Response to the uk parliament energy and climate change committee inquiry into supergrid," Friends of the SuperGrid, Technical Report, 2011.
15. IEC, "Iec60287: Electric cables—calculation of the current rating," 1993, std.
16. F. Gatta, A. Geri, S. Lauria, and M. Maccioni, "Steady-state operating conditions of very long ehvac cable lines: Two case studies," *Electric Power Systems Research*, vol. 83, no. 1, pp. 160–169, 2012.
17. G. F. Moore, *Electric Cables Handbook*. Oxford, UK: Blackwell Publishing, 1997.
18. ABB. Xple submarine cable systems: Attachment to xlpe land cable systems—users guide [Online]. Available: http://www.abb.com/cables
19. W. B1.19, "Technical and environmental issues regarding the integration of a new hv underground cable system in the network," Cigré, Technical Report 250, Feb. 2004, wG B1.19.
20. D. Lane, "Round 3 offshore wind farm connection study," The Crown Estate, Technical Report, 2010.
21. P. Nielsen, "Offshore wind energy projects, feasibility study guidelines," *SEAWIND–Altener Project—Feasibility Study Guidelines (EMD)*, 2003.
22. M. Dicorato, G. Forte, M. Pisani, and M. Trovato, "Guidelines for assessment of investment cost for offshore wind generation," *Renewable Energy*, vol. 36, no. 8, pp. 2043–2051, 2011.
23. S. Lundberg, "Performance comparison of wind park configurations," Chalmers University of Technology, Technical Report, 2003.
24. A. C. Franklin, D. P. Franklin, and S. A. Stigant, *The J & P transformer book: A practical technology of the power transformer*. Waltham, MA: Butterworths, 1983.
25. J. Olivares-Galvan, F. De León, P. Georgilakis, and R. Escarela-Perez, "Selection of copper against aluminium windings for distribution transformers," *Electric Power Applications, IET*, vol. 4, no. 6, pp. 474–485, 2010.
26. I. M. de Alegría, J. L. Martín, I. Kortabarria, J. Andreu, and P. I. Ereño, "Transmission alternatives for offshore electrical power," *Renewable and Sustainable Energy Reviews*, vol. 13, no. 5, pp. 1027–1038, 2009.

27. X. Hu, J. Liang, D. Rogers, and Y. Li, "Power flow and power reduction control using variable frequency of offshore ac grids," *Power Systems, IEEE Transactions on*, vol. 28, no. 4, pp. 3897–3905, Nov. 2013.
28. L. Harnefors, Y. Jiang-Häfner, M. Hyttinen, and T. Jonsson, "Ride-through methods for wind farms connected to the grid via a vsc-hvdc transmission," in *Proceedings of Nordic Wind Power Conference*. Risoe National Lab., DTU, Roskilde (Denmark), 2007.
29. J. Liang, O. Gomis-Bellmunt, J. Ekanayake, N. Jenkins, and W. An, "A multi-terminal hvdc transmission system for offshore wind farms with induction generators," *International Journal of Electrical Power & Energy Systems*, vol. 43, no. 1, pp. 54–62, 2012.

INDEX

HVDC Grids: For Offshore and Supergrid of the Future, First Edition. By Dirk Van Hertem, Oriol Gomis-Bellmunt, and Jun Liang.

IEEE Press Series on Power Engineering

Series Editor: M. E. El-Hawary, Dalhousie University, Halifax, Nova Scotia, Canada

The mission of IEEE Press Series on Power Engineering is to publish leadingedge books that cover the broad spectrum of current and forward-looking technologies in this fast-moving area. The series attracts highly acclaimed authors from industry/academia to provide accessible coverage of current and emerging topics in power engineering and allied fields. Our target audience includes the power engineering professional who is interested in enhancing their knowledge and perspective in their areas of interest.

1. *Principles of Electric Machines with Power Electronic Applications, Second Edition*
M. E. El-Hawary

2. *Pulse Width Modulation for Power Converters: Principles and Practice*
D. Grahame Holmes and Thomas Lipo

3. *Analysis of Electric Machinery and Drive Systems, Second Edition*
Paul C. Krause, Oleg Wasynczuk, and Scott D. Sudhoff

4. *Risk Assessment for Power Systems: Models, Methods, and Applications*
Wenyuan Li

5. *Optimization Principles: Practical Applications to the Operations of Markets of the Electric Power Industry*
Narayan S. Rau

6. *Electric Economics: Regulation and Deregulation*
Geoffrey Rothwell and Tomas Gomez

7. *Electric Power Systems: Analysis and Control*
Fabio Saccomanno

8. *Electrical Insulation for Rotating Machines: Design, Evaluation, Aging, Testing, and Repair, Second Edition*
Greg Stone, Edward A. Boulter, Ian Culbert, and Hussein Dhirani

9. *Signal Processing of Power Quality Disturbances*
Math H. J. Bollen and Irene Y. H. Gu

10. *Instantaneous Power Theory and Applications to Power Conditioning*
Hirofumi Akagi, Edson H. Watanabe, and Mauricio Aredes

11. *Maintaining Mission Critical Systems in a 24/7 Environment*
Peter M. Curtis

12. *Elements of Tidal-Electric Engineering*
Robert H. Clark

13. *Handbook of Large Turbo-Generator Operation and Maintenance, Second Edition*
Geoff Klempner and Isidor Kerszenbaum

14. *Introduction to Electrical Power Systems*
Mohamed E. El-Hawary

15. *Modeling and Control of Fuel Cells: Distributed Generation Applications*
M. Hashem Nehrir and Caisheng Wang

16. *Power Distribution System Reliability: Practical Methods and Applications*
Ali A. Chowdhury and Don O. Koval

17. *Introduction to FACTS Controllers: Theory, Modeling, and Applications*
Kalyan K. Sen and Mey Ling Sen

18. *Economic Market Design and Planning for Electric Power Systems*
James Momoh and Lamine Mili

19. *Operation and Control of Electric Energy Processing Systems*
James Momoh and Lamine Mili

20. *Restructured Electric Power Systems: Analysis of Electricity Markets with Equilibrium Models*
Xiao-Ping Zhang

21. *An Introduction to Wavelet Modulated Inverters*
S.A. Saleh and M.A. Rahman

22. *Control of Electric Machine Drive Systems*
Seung-Ki Sul

23. *Probabilistic Transmission System Planning*
Wenyuan Li

24. *Electricity Power Generation: The Changing Dimensions*
Digambar M. Tigare

25. *Electric Distribution Systems*
Abdelhay A. Sallam and Om P. Malik

26. *Practical Lighting Design with LEDs*
Ron Lenk and Carol Lenk

27. *High Voltage and Electrical Insulation Engineering*
Ravindra Arora and Wolfgang Mosch

28. *Maintaining Mission Critical Systems in a 24/7 Environment, Second Edition*
Peter Curtis

29. *Power Conversion and Control of Wind Energy Systems*
Bin Wu, Yongqiang Lang, Navid Zargari, and Samir Kouro

30. *Integration of Distributed Generation in the Power System*
Math H. Bollen and Fainan Hassan

31. *Doubly Fed Induction Machine: Modeling and Control for Wind Energy Generation Applications*
Gonzalo Abad, Jesus Lopez, Miguel Rodrigues, Luis Marroyo, and Grzegorz Iwanski

32. *High Voltage Protection for Telecommunications*
Steven W. Blume

33. *Smart Grid: Fundamentals of Design and Analysis*
James Momoh

34. *Electromechanical Motion Devices, Second Edition*
Paul C. Krause, Oleg Wasynczuk, and Steven D. Pekarek

35. *Electrical Energy Conversion and Transport: An Interactive Computer-Based Approach, Second Edition*
George G. Karady and Keith E. Holbert

36. *ARC Flash Hazard and Analysis and Mitigation*
J. C. Das

37. *Handbook of Electrical Power System Dynamics: Modeling, Stability, and Control*
Mircea Eremia and Mohammad Shahidehpour

38. *Analysis of Electric Machinery and Drive Systems, Third Edition*
Paul Krause, Oleg Wasynczuk, S. D. Sudhoff, and Steven D. Pekarek

39. *Extruded Cables for High-Voltage Direct-Current Transmission: Advances in Research and Development*
Giovanni Mazzanti and Massimo Marzinotto

40. *Power Magnetic Devices: A Multi-Objective Design Approach*
S. D. Sudhoff

41. *Risk Assessment of Power Systems: Models, Methods, and Applications, Second Edition*
Wenyuan Li

42. *Practical Power System Operation*
Ebrahim Vaahedi

43. *The Selection Process of Biomass Materials for the Production of Bio-Fuels and Co-Firing*
Najib Altawell

44. *Electrical Insulation for Rotating Machines: Design, Evaluation, Aging, Testing, and Repair, Second Edition*
Greg C. Stone, Ian Culbert, Edward A. Boulter, and Hussein Dhirani

45. *Principles of Electrical Safety*
Peter E. Sutherland

46. *Advanced Power Electronics Converters: PWM Converters Processing AC Voltages*
Euzeli Cipriano dos Santos Jr. and Edison Roberto Cabral da Silva

47. *Optimization of Power System Operation, Second Edition*
Jizhong Zhu

48. *Digital Control of High-Frequency Switched-Mode Power Converters*
Luca Corradini, Dragan Maksimovic, Paolo Mattavelli, and Regan Zane

49. *Power System Harmonics and Passive Filter Designs*
J. C. Das

50. *Industrial Power Distribution, Second Edition*
Ralph E. Fehr, III

51. *HVDC Grids: For Offshore and Supergrid of the Future*
Dirk Van Hertem, Oriol Gomis-Bellmunt, and Jun Liang